Undergraduate Texts in Mathematics

Undergraduate Texts in Mathematics

Undergraduate Texts in Mathematics are generally aimed at third- and fourth-year undergraduate mathematics students at North American universities. These texts strive to provide students and teachers with new perspectives and novel approaches. The books include motivation that guides the reader to an appreciation of interrelations among different aspects of the subject. They feature examples that illustrate key concepts as well as exercises that strengthen understanding.

More information about this series at http://www.springer.com/series/666

Omar Hijab

Introduction to Calculus and Classical Analysis

Fourth Edition

Omar Hijab
College of Science and Technology
Temple University
Philadelphia, PA, USA

Undergraduate Texts in Mathematics
ISBN 978-3-319-80345-6 ISBN 978-3-319-28400-2 (eBook)
DOI 10.1007/978-3-319-28400-2

Mathematics Subject Classification (2010): 41-XX, 40-XX, 33-XX, 05-XX
Springer Cham Heidelberg New York Dordrecht London

Softcover reprint of the hardcover 4th edition 2016

Printed on acid-free paper

Springer International Publishing AG Switzerland is part of Springer Science+Business Media (www.springer.com)

To M.A.W.

Preface

For undergraduate students, the transition from calculus to analysis is often disorienting and mysterious. What happened to the beautiful calculus formulas? Where did ϵ-δ and open sets come from? It is not until later that one integrates these seemingly distinct points of view. When teaching "advanced calculus," I always had a difficult time answering these questions.

Now, every mathematician knows that analysis arose naturally in the nineteenth century out of the calculus of the previous two centuries. Believing that it was possible to write a book reflecting, explicitly, this organic growth, I set out to do so.

I chose several of the jewels of classical eighteenth- and nineteenth-century analysis and inserted them near the end of the book, inserted the axioms for reals at the beginning, and filled in the middle with (and only with) the material necessary for clarity and logical completeness. In the process, every little piece of one-variable calculus assumed its proper place, and theory and application were interwoven throughout.

Let me describe some of the unusual features in this text, as there are other books that adopt the above point of view. First is the systematic avoidance of ϵ-δ arguments. Continuous limits are defined in terms of limits of sequences, limits of sequences are defined in terms of upper and lower limits, and upper and lower limits are defined in terms of sup and inf. Everybody thinks in terms of sequences, so why do we teach our undergraduates ϵ-δ's? (In calculus texts, especially, doing this is unconscionable.)

The second feature is the treatment of integration. We follow the standard treatment motivated by geometric measure theory, with a few twists thrown in: The area is two-dimensional Lebesgue measure, defined on all subsets of $\mathbf{R}^2$, the integral of an arbitrary[1] nonnegative function is the area under its graph, and the integral of an arbitrary integrable function is the difference of the integrals of its positive and negative parts.

[1] Not necessarily measurable.

In dealing with arbitrary subsets of $\mathbf{R}^2$ and arbitrary functions, only a few basic properties can be derived; nevertheless, surprising results are available; for example, the integral of an arbitrary integrable function over an interval is a continuous function of the endpoints.

Arbitrary functions are considered to the extent possible not because of generality for generality's sake, but because they fit naturally within the context laid out here. For example, the density theorem and maximal inequality are valid for arbitrary sets and functions, and the first fundamental theorem is valid for an integrable f if and only if f is measurable.

In Chapter 4 we restrict attention to the class of continuous functions, which is broad enough to handle the applications in Chapter 5, and derive the second fundamental theorem in the form

$$\int_a^b f(x)\,dx = F(b-) - F(a+).$$

Here a, b, $F(a+)$ or $F(b-)$ may be infinite, extending the immediate applicability, and the continuous function f need only be nonnegative or integrable.

The third feature is the treatment of the theorems involving interchange of limits and integrals. Ultimately, all these theorems depend on the monotone convergence theorem which, from our point of view, follows from the Greek mathematicians' Method of Exhaustion. Moreover, these limit theorems are stated only after a clear and nontrivial need has been elaborated. For example, differentiation under the integral sign is used to compute the Gaussian integral.

The treatment of integration presented here emphasizes geometric aspects rather than technicality. The most technical aspects, the derivation of the Method of Exhaustion in §4.5, may be skipped upon first reading, or skipped altogether, without affecting the flow.

The fourth feature is the use of real-variable techniques in Chapter 5. We do this to bring out the elementary nature of that material, which is usually presented in a complex setting using transcendental techniques. For example, included is:

- A real-variable derivation of Gauss' AGM formula motivated by the unit circle map

$$x' + iy' = \frac{\sqrt{a}x + i\sqrt{b}y}{\sqrt{a}x - i\sqrt{b}y}.$$

- A real-variable computation of the radius of convergence of the Bernoulli series, derived via the infinite product expansion of $\sinh x/x$, which is in turn derived by combinatorial real-variable methods.

- The zeta functional equation is derived via the theta functional equation, which is in turn derived via the connection to the parametrization of the AGM curve.

The fifth feature is our emphasis on computational problems. Computation, here, is often at a deeper level than expected in calculus courses and varies from the high school quadratic formula in §1.4 to $\exp(-\zeta'(0)) = \sqrt{2\pi}$ in §5.8.

Because we take the real numbers as our starting point, basic facts about the natural numbers, trigonometry, or integration are rederived in this context, either in the body of the text or as exercises. For example, while the trigonometric functions are initially defined via their Taylor series, later it is shown how they may be defined via the unit circle.

Although it is helpful for the reader to have seen calculus prior to reading this text, the development does not presume this. We feel it is important for undergraduates to see, at least once in their four years, a nonpedantic, purely logical development that really does start from scratch (rather than pretends to), is self-contained, and leads to nontrivial and striking results.

Applications include a specific transcendental number; convexity, elementary symmetric polynomial inequalities, subdifferentials, and the Legendre transform; Machin's formula; the Cantor set; the Bailey–Borwein–Plouffe series

$$\pi = \sum_{n=0}^{\infty} \frac{1}{16^n}\left(\frac{4}{8n+1} - \frac{2}{8n+4} - \frac{1}{8n+5} - \frac{1}{8n+6}\right);$$

continued fractions; Laplace and Fourier transforms; Bessel functions; Euler's constant; the AGM iteration; the gamma and beta functions; Stirling identity

$$\exp\left(\int_s^{s+1} \log \Gamma(x)\, dx\right) = s^s e^{-s}\sqrt{2\pi}, \qquad s > 0;$$

the entropy of the binomial coefficients; infinite products and Bernoulli numbers; theta functions and the AGM curve; the zeta function; the zeta series

$$\log(x!) = -\gamma x + \zeta(2)\frac{x^2}{2} - \zeta(3)\frac{x^3}{3} + \zeta(4)\frac{x^4}{4} - \dots, \qquad 1 \geq x > -1;$$

primes in arithmetic progressions; the Euler–Maclaurin formula; and the Stirling series.

After the applications, Chapter 6 develops the fundamental theorems in their general setting, based on the sunrise lemma. This material is at a more advanced level, but is included to point the reader toward twentieth-century developments.

As an aid to self-study and assimilation, there are 450 problems with all solutions at the back of the book. Every exercise can be solved using only previous material from this book. Chapters 1–4 provide the basis for a calculus or beginner undergraduate analysis course. Chapters 4 and 5 provide the basis for an undergraduate computational analysis course, while Chapters 4 and 6 provide the basis for an undergraduate real analysis course.

Philadelphia, PA, USA — Omar Hijab
Fall 2015

Contents

A Note to the Reader

This text consists of many assertions, some big, some small, some almost insignificant. These assertions are obtained from the properties of the real numbers by logical reasoning. Assertions that are especially important are called *theorems*. An assertion's importance is gauged by many factors, including its *depth*, how many other assertions it depends on, its *breadth*, how many other assertions are explained by it, and its level of *symmctry*. The later portions of the text depend on *every single assertion*, no matter how small, made in Chapter 1.

The text is self-contained, and the exercises are arranged in order: Every exercise can be done using only previous material *from this text*. No outside material is necessary.

Doing the exercises is essential for understanding the material in the text. Sections are numbered sequentially within each chapter; for example, §4.3 means the third section in Chapter 4. Equation numbers are written within parentheses and exercise numbers in bold. Theorems, equations, and exercises are numbered sequentially within each section; for example, Theorem 4.3.2 denotes the second theorem in §4.3, (4.3.1) denotes the first numbered equation in §4.3, and **4.3.3** denotes the third exercise at the end of §4.3.

Throughout, we use the abbreviation "iff" to mean "if and only if" and □ to signal the end of a derivation.

A Note to the Reader

Chapter 1
The Set of Real Numbers

1.1 Sets and Mappings

We assume the reader is familiar with the usual notions of sets and mappings, but we review them to fix the notation. The reader may wish to skip this section altogether and refer to it as needed later in the text.

In set theory, everything is a set and *set membership* is the sole primitive notion: Given sets a and A, either a *belongs to*[1] A or not. We write $a \in A$ in the first case and $a \notin A$ in the second. If a belongs to A, we say a is an *element* of A.

Let A, B be sets. If every element of A is an element of B, we say A is a *subset* of B, and we write $A \subset B$. Equivalently, we say B is a *superset* of A and we write $B \supset A$. When we write $A \subset B$ or $A \supset B$, we allow for the possibility $A = B$, i.e., $A \subset A$ and $A \supset A$.

Elements characterize sets: If A and B have the same elements, then $A = B$. More precisely, if $A \subset B$ and $B \subset A$, then $A = B$.

Given a and b, there is exactly one set, their *pair*, whose elements are a and b; this set is denoted $\{a, b\}$. In particular, given a, there is the *singleton* set $\{a, a\}$, denoted $\{a\}$. The *ordered pair* of sets a and b is the set

$$(a, b) = \{\{a\}, \{a, b\}\}.$$

Then $(a, b) = (c, d)$ iff $a = c$ and $b = d$ (Exercise **1.1.9**).

There is a set $\emptyset$ having no elements, the *empty set*. Note $\emptyset$ is a subset of every set.

The *union* of sets A and B is the set C whose elements lie in A *or* lie in B; we write $C = A \cup B$, and we say C equals A *union* B. The *intersection* of sets A and B is the set C whose elements lie in A *and* lie in B; we write $C = A \cap B$ and we say C equals A *inter* B. Similarly we write $A \cup B \cup C$, $A \cap B \cap C$ for sets A, B, C, etc.

---

[1] Alternatively, a *lies in* A or a *is in* A.

O. Hijab, *Introduction to Calculus and Classical Analysis*, Undergraduate Texts in Mathematics, DOI 10.1007/978-3-319-28400-2_1

More generally, let $\mathcal{F}$ be a set. The *union* $\bigcup \mathcal{F}$ is

$$\bigcup \mathcal{F} = \bigcup \{A : A \in \mathcal{F}\} = \{x : x \in A \text{ for some } A \in \mathcal{F}\}$$

is the set whose elements are elements of elements of $\mathcal{F}$. The *intersection* $\bigcap \mathcal{F}$ is

$$\bigcap \mathcal{F} = \bigcap \{A : A \in \mathcal{F}\} = \{x : x \in A \text{ for all } A \in \mathcal{F}\}$$

is the set whose elements lie in all the elements of $\mathcal{F}$. The cases in the previous paragraph correspond to $\mathcal{F} = \{A, B\}$ and $\mathcal{F} = \{A, B, C\}$.

The set of all elements in A, but not in B, is denoted $A \setminus B = \{x \in A : x \notin B\}$ and is called the *complement of* B *in* A. For example, when $A \subset B$, the set $A \setminus B$ is empty. Often the set A is understood from the context; in these cases, $A \setminus B$ is denoted B^c and called the *complement* of B.

Let A and B be sets. If they have no elements in common, $A \cap B = \emptyset$, we say they are *disjoint*. Note $\bigcap \mathcal{F} \subset \bigcup \mathcal{F}$ for any nonempty set $\mathcal{F}$ and $\emptyset$ is disjoint from every set.

We will have occasion to use *De Morgan's law*,

$$(A \cup B)^c = A^c \cap B^c, \qquad (A \cap B)^c = A^c \cup B^c,$$

or, more generally,

$$\left(\bigcup \{A : A \in \mathcal{F}\}\right)^c = \bigcap \{A^c : A \in \mathcal{F}\}$$
$$\left(\bigcap \{A : A \in \mathcal{F}\}\right)^c = \bigcup \{A^c : A \in \mathcal{F}\}.$$

The *power set* of a set X is the set 2^X whose elements are the subsets of X. If X, Y are sets, their *(ordered) product* is the set $X \times Y$ whose elements consist of all ordered pairs (x, y) with $x \in X$ and $y \in Y$. *Everything in this text is either an element or a subset of repeated powers or products of the set* $\mathbf{R}$ *of real numbers.*

A *relation* between two sets X and Y is a subset $f \subset X \times Y$. A *mapping* or *function* is a relation $f \subset X \times Y$, such that, for each $x \in X$, there is exactly one $y \in Y$ with $(x, y) \in f$. In this case, it is customary to write $y = f(x)$ and $f : X \to Y$.

If $f : X \to Y$ is a mapping, X is the *domain*, Y is the *codomain*, and for $A \subset X$, $f(A) = \{f(x) : x \in A\} \subset X$ is the *image* of A. In particular, the *range* is $f(X)$.

A mapping $f : X \to Y$ is *injective* if $f(x) = f(x')$ implies $x = x'$, whereas $f : X \to Y$ is *surjective* if every element y of Y equals $f(x)$ for some $x \in X$, i.e., if the range equals the codomain. A mapping that is both injective and surjective is *bijective*. Alternatively, we say f is an *injection*, a *surjection*, and a *bijection*, respectively.

If $f : X \to Y$ and $g : Y \to Z$ are mappings, their *composition* is the mapping $g \circ f : X \to Z$ given by $(g \circ f)(x) = g(f(x))$ for all $x \in X$. In general, $g \circ f \neq f \circ g$.

If $f : X \to Y$ and $g : Y \to X$ are mappings, we say they are *inverses* of each other if $g(f(x)) = x$ for all $x \in X$ and $f(g(y)) = y$ for all $y \in Y$. A mapping $f : X \to Y$ is *invertible* if it has an inverse g. It is a fact that a mapping f is invertible iff f is bijective.

Let $\mathcal{F}$ be a set such that all elements $A \in \mathcal{F}$ are nonempty. A *choice function* is a function $f : \mathcal{F} \to \bigcup \mathcal{F}$ satisfying $f(A) \in A$ for $A \in \mathcal{F}$. Given such $\mathcal{F}$, the *(unordered) product* $\prod \mathcal{F}$ is the set of all choice functions.

When X and Y are disjoint, the map $f \mapsto (f(X), f(Y))$ is a bijection between $\prod\{X, Y\}$ and $X \times Y$.

When $\mathcal{F}$ is a finite set (§1.3), $\mathcal{F} = \{A_1, \dots, A_n\}$, with $A_1, \dots, A_n$ nonempty, a choice function corresponds to a choice of an element $a_k \in A_k$ for each $1 \leq k \leq n$. In this case, the nonemptiness of $\prod \mathcal{F}$ is established in Exercise **1.3.24**. This is the *axiom of finite choice*.

We assume that $\prod \mathcal{F}$ is nonempty whenever $\mathcal{F}$ is countable (§1.7), i.e., given nonempty sets $A_1, A_2, \dots$, a choice $a_k \in A_k$ for $k \geq 1$ exists. This is the *axiom of countable choice*.

Exercises

1.1.1. Show that a mapping $f : X \to Y$ is invertible iff it is bijective.

1.1.2. Let $f : X \to Y$ be bijective. Show that the inverse $g : Y \to X$ is unique.

1.1.3. Verify De Morgan's law.

1.1.4. Show that $\bigcup\{x\} = x$ for all x and $\{\bigcup x\} = x$ iff x is a singleton set.

1.1.5. Given sets a, b, let $c = a \cup b$, $d = a \cap b$. Show that $(c \setminus a) \cup d = b$.

1.1.6. If $A \in \mathcal{F}$, then $A \subset \bigcup \mathcal{F}$ and $A \supset \bigcap \mathcal{F}$.

1.1.7. Let (a, b) be an ordered pair of sets a, b. Show that $\bigcup\bigcap(a, b) = a$, $\bigcup\bigcup(a, b) = a \cup b$, $\bigcap\bigcup(a, b) = a \cap b$, and $\bigcap\bigcap(a, b) = a$. Conclude that b may be computed from (a, b).

1.1.8. A set x is *hierarchical* if $a \in x$ implies $a \subset x$. For example, $\emptyset$ is hierarchical; let $S(x) = x \cup \{x\}$. Show that $S(x)$ is hierarchical whenever x is.

1.1.9. Given sets a, b, c, d, show that $(a, b) = (c, d)$ iff $a = c$ and $b = d$.

1.2 The Set R

We are ultimately concerned with one and only one set, the set $\mathbf{R}$ of real numbers. The properties of $\mathbf{R}$ that we use are

- The arithmetic properties,
- The ordering properties, and
- The completeness property.

Throughout, we use "real" to mean "real number," i.e., an element of $\mathbf{R}$.

The *arithmetic properties* start with the fact that reals a, b can be *added* to produce a real $a+b$, the *sum* of a and b. The rules for addition are $a+b = b+a$ and $a + (b + c) = (a + b) + c$, valid for all reals a, b, and c. There is also a real 0, called *zero*, satisfying $a + 0 = 0 + a = a$ for all reals a, and each real a has a *negative* $-a$ satisfying $a + (-a) = 0$. As usual, we write *subtraction* $a + (-b)$ as $a - b$.

Reals a, b can also be *multiplied* to produce a real $a \cdot b$, the *product* of a and b, also written ab. The rules for multiplication are $ab = ba$, $a(bc) = (ab)c$, valid for all reals a, b, and c. There is also a real 1, called *one*, satisfying $a1 = 1a = a$ for all reals a, and each real $a \neq 0$ has a *reciprocal* $1/a$ satisfying $a(1/a) = 1$. As usual, we write *division* $a(1/b)$ as a/b.

Addition and multiplication are related by the property $a(b+c) = ab+ac$ for all reals a, b, and c and the assumption $0 \neq 1$. These are the arithmetic properties of the reals.

Let us show how the arithmetic properties imply there is a unique real number 0 satisfying $0+a = a+0 = a$ for all a. If $0'$ were another real satisfying $0' + a = a + 0' = a$ for all a, then we would have $0' = 0 + 0' = 0' + 0 = 0$, hence $0 = 0'$. Also it follows that there is a unique real playing the role of one and $0a = 0$ for all a.

The *ordering properties* start with the fact that there is subset $\mathbf{R}^+$ of $\mathbf{R}$, the set of *positive* numbers, that is closed under addition and multiplication, i.e., if $a, b \in \mathbf{R}^+$, then $a+b, ab \in \mathbf{R}^+$. If a is positive, we write $a > 0$ or $0 < a$, and we say a *is greater than* 0 or 0 *is less than* a, respectively. Let $\mathbf{R}^-$ denote the set of *negative* numbers, i.e., $\mathbf{R}^- = -\mathbf{R}^+$ is the set whose elements are the negatives of the elements of $\mathbf{R}^+$. The rules for ordering are that the sets $\mathbf{R}^-$, $\{0\}$, $\mathbf{R}^+$ are pairwise disjoint and their union is all of $\mathbf{R}$. These are the ordering properties of $\mathbf{R}$.

We write $a > b$ and $b < a$ to mean $a - b > 0$. Then $0 > a$ iff a is negative and $a > b$ implies $a + c > b + c$. In particular, for any pair of reals a, b, we have $a < b$ or $a = b$ or $a > b$.

From the ordering properties, it follows, for example, that $1 > 0$, i.e., one is positive, $a < b$ and $c > 0$ imply $ac < bc$, $0 < a < b$ implies $aa < bb$, and $a < b$, $b < c$ imply $a < c$. As usual, we also write $\leq$ to mean $<$ or $=$, $\geq$ to mean $>$ or $=$, and we say a is *nonnegative* or *nonpositive* if $a \geq 0$ or $a \leq 0$.

If S is a set of reals, a number M is an *upper bound* for S if $x \leq M$ for all $x \in S$. Similarly, m is a *lower bound* for S if $m \leq x$ for all $x \in S$ (Figure 1.1).

For example, 1 and $1+1$ are upper bounds for the sets $J = \{x : 0 < x < 1\}$ and $I = \{x : 0 \le x \le 1\}$, whereas 0 and -1 are lower bounds for these sets. S is *bounded above (below)* if it has an upper (lower) bound. S is *bounded* if it is bounded above and bounded below.

Not every set of reals has an upper or a lower bound. Indeed, it is easy to see that **R** itself is neither bounded above nor bounded below. A more interesting example is the set **N** of natural numbers (next section): **N** is not bounded above.

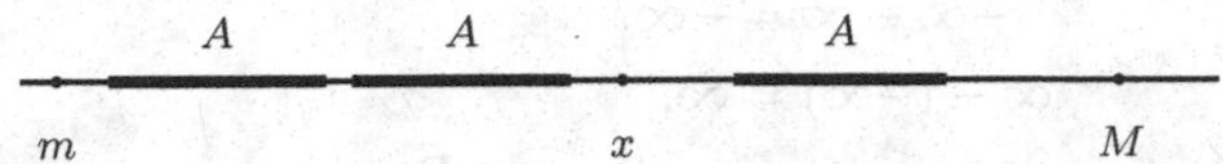

Fig. 1.1 Upper and lower bounds for A

A given set S of reals may have several upper bounds. If S has an upper bound M such that $M \le b$ for any other upper bound b of S, then we say M is a *least upper bound* or M is a *supremum* or *sup* for S, and we write $M = \sup S$.

If a is a least upper bound for S and b is an upper bound for S, then $a \le b$ since b is an upper bound and a is a least such. Similarly, if b is a least upper bound for S and a is an upper bound for S, then $b \le a$ since a is an upper bound and b is a least such. Hence, if both a and b are least upper bounds, we must have $a = b$. *Thus, the sup, whenever it exists, is uniquely determined.*

For example, consider the sets I and J defined above. If M is an upper bound for I, then $M \ge x$ for every $x \in I$, hence $M \ge 1$. Thus, 1 is the least upper bound for I, or $1 = \sup I$. The situation with the set J is only slightly more subtle: If $M < 1$, then $c = (1+M)/2$ satisfies $M < c < 1$, so $c \in J$, hence M cannot be an upper bound for J. Thus, 1 is the least upper bound for J, or $1 = \sup J$.

A real m that is a lower bound for S and satisfies $m \ge b$ for all other lower bounds b is called a *greatest lower bound* or an *infimum* or *inf* for S, and we write $m = \inf S$. Again the inf, whenever it exists, is uniquely determined. As before, it follows easily that $0 = \inf I$ and $0 = \inf J$.

The *completeness property* of **R** asserts that every nonempty set $S \subset \mathbf{R}$ that is bounded above has a sup, and every nonempty set $S \subset \mathbf{R}$ that is bounded below has an inf.

We introduce a convenient abbreviation, two symbols ∞, $-\infty$, called *infinity and minus infinity,* subject to the ordering rule $-\infty < x < \infty$ for all reals x. If a set S is not bounded above, we write $\sup S = \infty$. If S is not bounded below, we write $\inf S = -\infty$. For example, $\sup \mathbf{R} = \infty$, $\inf \mathbf{R} = -\infty$; in §1.4 we show that $\sup \mathbf{N} = \infty$. Recall that the empty set $\emptyset$ is a subset of **R**. Another convenient abbreviation is to write $\sup \emptyset = -\infty$, $\inf \emptyset = \infty$. Clearly, when S is nonempty, $\inf S \le \sup S$.

With this terminology, the completeness property asserts that *every subset of* $\mathbf{R}$*, bounded or unbounded, empty or nonempty, has a sup and has an inf; these may be reals or* $\pm\infty$.

We emphasize that ∞ and $-\infty$ are not reals but just convenient abbreviations. As mentioned above, the ordering properties of $\pm\infty$ are $-\infty < x < \infty$ for all real x; it is convenient to define the following arithmetic properties of $\pm\infty$:

$$\begin{aligned}
\infty + \infty &= \infty, \\
-\infty - \infty &= -\infty, \\
\infty - (-\infty) &= \infty, \\
\infty \pm c &= \infty, \qquad c \in \mathbf{R}, \\
-\infty \pm c &= -\infty, \qquad c \in \mathbf{R}, \\
(\pm\infty) \cdot c &= \pm\infty, \qquad c > 0, \\
\infty \cdot \infty &= \infty, \infty \cdot (-\infty) = -\infty.
\end{aligned}$$

Note that we have not defined $\infty - \infty$, $0 \cdot \infty$, ∞/∞, or $c/0$.

Let a be an upper bound for a set S. If $a \in S$, we say a is a *maximum* of S, and we write $a = \max S$. For example, with I as above, $\max I = 1$. The max of a set S need not exist; for example, according to the theorem below, $\max J$ does not exist.

Similarly, let a be a lower bound for a set S. If $a \in S$, we say a is a *minimum* of S, and we write $a = \min S$. For example, $\min I = 0$ but $\min J$ does not exist.

Theorem 1.2.1. *Let* $S \subset \mathbf{R}$ *be a set. The max of* S *and the min of* S *are uniquely determined whenever they exist. The max of* S *exists iff the sup of* S *lies in* S*, in which case the max equals the sup. The min of* S *exists iff the inf of* S *lies in* S*, in which case the min equals the inf.*

To see this, note that the first statement follows from the second since we already know that the sup and the inf are uniquely determined. To establish the second statement, suppose that $\sup S \in S$. Then since $\sup S$ is an upper bound for S, $\max S = \sup S$. Conversely, suppose that $\max S$ exists. Then $\sup S \le \max S$ since $\max S$ is an upper bound and $\sup S$ is the least such. On the other hand, $\sup S$ is an upper bound for S and $\max S \in S$. Thus, $\max S \le \sup S$. Combining $\sup S \le \max S$ and $\sup S \ge \max S$, we obtain $\max S = \sup S$. For the inf, the derivation is completely analogous. □

Because of this, when $\max S$ exists, we say *the sup is attained*. Thus, the sup for I is attained, whereas the sup for J is not. Similarly, when $\min S$ exists, we say *the inf is attained*. Thus, the inf for I is attained, whereas the inf for J is not.

Let A, B be subsets of $\mathbf{R}$, let a be real, and let $c > 0$; let

$$-A = \{-x : x \in A\}, \quad A + a = \{x + a : x \in A\}, \quad cA = \{cx : x \in A\},$$

and

$$A+B=\{x+y:x\in A,y\in B\}.$$

Here are some simple consequences of the definitions that must be checked at this stage:

- $A\subset B$ implies $\sup A\le\sup B$ and $\inf A\ge\inf B$ (monotonicity property).
- $\sup(-A)=-\inf A$, $\inf(-A)=-\sup A$ (reflection property).
- $\sup(A+a)=\sup A+a$, $\inf(A+a)=\inf A+a$ for $a\in\mathbf{R}$ (translation property).
- $\sup(cA)=c\sup A$, $\inf(cA)=c\inf A$ for $c>0$ (dilation property).
- $\sup(A+B)=\sup A+\sup B$, $\inf(A+B)=\inf A+\inf B$ (addition property), whenever the sum of the sups and the sum of the infs are defined.

These properties hold whether A and B are bounded or unbounded, empty or nonempty.

We verify the first and the last properties, leaving the others as Exercise **1.2.7**. For the monotonicity property, if A is empty, the property is immediate since $\sup A=-\infty$ and $\inf A=\infty$. If A is nonempty and $a\in A$, then $a\in B$, hence $\inf B\le a\le\sup B$. Thus, $\sup B$ and $\inf B$ are upper and lower bounds for A, respectively. Since $\sup A$ and $\inf A$ are the least and greatest such, we obtain $\inf B\le\inf A\le\sup A\le\sup B$.

Now we verify $\sup(A+B)=\sup A+\sup B$. If A is empty, then so is $A+B$; in this case, the assertion to be proved reduces to $-\infty+\sup B=-\infty$ which is true (remember we are excluding the case $\infty-\infty$). Similarly, if B is empty.

If A and B are both nonempty, then $\sup A\ge x$ for all $x\in A$, and $\sup B\ge y$ for all $y\in B$, so $\sup A+\sup B\ge x+y$ for all $x\in A$ and $y\in B$. Hence, $\sup A+\sup B\ge z$ for all $z\in A+B$, or $\sup A+\sup B$ is an upper bound for $A+B$. Since $\sup(A+B)$ is the least such, we conclude that $\sup A+\sup B\ge\sup(A+B)$. If $\sup(A+B)=\infty$, then the reverse inequality $\sup A+\sup B\le\sup(A+B)$ is immediate, yielding the result.

If, however, $\sup(A+B)<\infty$ and $x\in A$, $y\in B$, then $x+y\in A+B$, hence $x+y\le\sup(A+B)$ or, what is the same, $x\le\sup(A+B)-y$. Thus, $\sup(A+B)-y$ is an upper bound for A; since $\sup A$ is the least such, we get $\sup A\le\sup(A+B)-y$. Now this last inequality implies, first, $\sup A<\infty$ and, second, $y\le\sup(A+B)-\sup A$ for all $y\in B$. Thus, $\sup(A+B)-\sup A$ is an upper bound for B; since $\sup B$ is the least such, we conclude that $\sup B\le\sup(A+B)-\sup A$ or, what is the same, $\sup(A+B)\ge\sup A+\sup B$. Since we already know that $\sup(A+B)\le\sup A+\sup B$, we obtain $\sup(A+B)=\sup A+\sup B$.

To verify $\inf(A+B)=\inf A+\inf B$, use reflection and what we just finished to write

$$\begin{aligned}\inf(A+B)&=-\sup[-(A+B)]\\&=-\sup[(-A)+(-B)]\\&=-\sup(-A)-\sup(-B)\\&=\inf A+\inf B.\end{aligned}$$

This completes the derivation of the addition property.

Let X be a set and $f : X \to \mathbf{R}$ a function. If $A \subset X$, throughout we use the notation $\sup_A f$ and $\inf_A f$ to mean $\sup f(A)$ and $\inf f(A)$, respectively.

It is natural to ask about the existence of **R**. Where does it come from? More precisely, can a set **R** satisfying the above properties be constructed within the context of set theory as sketched in §1.1? The answer is not only that this is so but also that such a set is unique in the following sense: *If S is any set endowed with its own arithmetic, ordering, and completeness properties, as above, there is a bijection $\mathbf{R} \to S$ mapping the properties on* **R** *to the properties on S.*

Because the construction of **R** would lead us too far afield and has no impact on the content of this text, we skip it. However, perhaps the most enlightening construction of **R** is via Conway's *surreal numbers*, with the real numbers then being a specific subset of the surreal numbers.[2]

For us here, the explicit nature of the elements[3] of **R**—the real numbers—is immaterial. In summary then, *every assertion that follows in this book depends only on the arithmetic, ordering, and completeness properties of the set* **R**.

Exercises

1.2.1. Show that $a0 = 0$ for all real a.

1.2.2. Show that there is a unique real playing the role of 1. Also show that each real a has a unique negative $-a$ and each nonzero real a has a unique reciprocal.

1.2.3. Show that $-(-a) = a$ and $-a = (-1)a$.

1.2.4. Show that negative times positive is negative, negative times negative is positive, and 1 is positive.

1.2.5. Show that $a < b$ and $c \in \mathbf{R}$ imply $a + c < b + c$, $a < b$ and $c > 0$ imply $ac < bc$, $a < b$ and $b < c$ imply $a < c$, and $0 < a < b$ implies $aa < bb$.

1.2.6. Let $a, b \geq 0$. Show that $a \leq b$ iff $aa \leq bb$.

1.2.7. Verify the properties of sup and inf listed above.

---

[2] See Conway's book 7.

[3] The elements of **R** are themselves sets, since (§1.1) everything is a set.

1.3 The Subset N and the Principle of Induction

A subset $S \subset \mathbf{R}$ is *inductive* if

A. $1 \in S$ and

B. S is closed under addition by 1: $x \in S$ implies $x+1 \in S$.

For example, $\mathbf{R}^+$ is inductive. The subset $\mathbf{N} \subset \mathbf{R}$ of *natural numbers* or *naturals* is the intersection (§1.1) of all inductive subsets of $\mathbf{R}$,

$$\mathbf{N} = \bigcap \{S : S \subset \mathbf{R} \text{ inductive}\}.$$

Then $\mathbf{N}$ itself is inductive: Indeed, since $1 \in S$ for every inductive set S, we conclude that 1 is in the intersection of all the inductive sets, hence $1 \in \mathbf{N}$. Similarly, $n \in \mathbf{N}$ implies $n \in S$ for every inductive set S. Hence, $n+1 \in S$ for every inductive set S. Hence, $n+1$ is in the intersection of all the inductive sets, hence $n+1 \in \mathbf{N}$. This shows that $\mathbf{N}$ is inductive.

From the definition, we conclude that $\mathbf{N} \subset S$ for any inductive $S \subset \mathbf{R}$. For example, since $\mathbf{R}^+$ is inductive, we conclude that $\mathbf{N} \subset \mathbf{R}^+$, i.e., every natural is positive.

From the definition, we also conclude that $\mathbf{N}$ is the only inductive subset of $\mathbf{N}$. For example, $S = \{1\} \cup (\mathbf{N}+1)$ is a subset of $\mathbf{N}$, since $\mathbf{N}$ is inductive. Clearly, $1 \in S$. Moreover, $x \in S$ implies $x \in \mathbf{N}$ implies $x+1 \in \mathbf{N}+1$ implies $x+1 \in S$, so S is inductive. Hence, $S = \mathbf{N}$ or $\{1\} \cup (\mathbf{N}+1) = \mathbf{N}$, i.e., $n-1$ is a natural for every natural n other than 1. Because of this, $n \geq 1$ is often used interchangeably with $n \in \mathbf{N}$.

The conclusions above are often paraphrased by saying $\mathbf{N}$ *is the smallest inductive subset of* $\mathbf{R}$, and they are so important they deserve a name.

Theorem 1.3.1 (Principle of Induction). *If $S \subset \mathbf{R}$ is inductive, then $S \supset \mathbf{N}$. If $S \subset \mathbf{N}$ is inductive, then $S = \mathbf{N}$.* □

Let $2 = 1+1 > 1$; we show that there are no naturals n between 1 and 2, i.e., satisfying $1 < n < 2$. For this, let $S = \{1\} \cup \{n \in \mathbf{N} : n \geq 2\}$. Then $1 \in S$. If $n \in S$, there are two possibilities. Either $n = 1$ or $n \geq 2$. If $n = 1$, then $n+1 = 2 \in S$. If $n \geq 2$, then $n+1 > n \geq 2$ and $n+1 \in \mathbf{N}$, so $n+1 \in S$. Hence, S is inductive. Since $S \subset \mathbf{N}$, we conclude that $S = \mathbf{N}$. Thus, $n \geq 1$ for all $n \in \mathbf{N}$, and there are no naturals between 1 and 2. Similarly (Exercise **1.3.1**), for any $n \in \mathbf{N}$, there are no naturals between n and $n+1$.

$\mathbf{N}$ is closed under addition and multiplication by any natural. To see this, fix a natural n, and let $S = \{x : x+n \in \mathbf{N}\}$, so S is the set of all reals x whose sum with n is natural. Then $1 \in S$ since $n+1 \in \mathbf{N}$, and $x \in S$ implies $x+n \in \mathbf{N}$ implies $(x+1)+n = (x+n)+1 \in \mathbf{N}$ implies $x+1 \in S$. Thus, S is inductive. Since $\mathbf{N}$ is the smallest such set, we conclude that $\mathbf{N} \subset S$ or $m+n \in \mathbf{N}$ for all $m \in \mathbf{N}$. Thus, $\mathbf{N}$ is closed under addition. This we write simply as $\mathbf{N}+\mathbf{N} \subset \mathbf{N}$. Closure under multiplication $\mathbf{N} \cdot \mathbf{N} \subset \mathbf{N}$ is similar and left as an exercise.

In the sequel, when we apply the principle of induction, we simply say "by induction."

To show that a given set S is inductive, one needs to verify **A** and **B**. Step **B** is often referred to as *the inductive step*, even though, strictly speaking, induction is both **A** and **B**, because, usually, most of the work is in establishing **B**. Also the hypothesis in **B**, $x \in S$, is often referred to as *the inductive hypothesis*.

Let us give another example of the use of induction. A natural is *even* if it is in $2\mathbf{N} = \{2n : n \in \mathbf{N}\}$. A natural n is *odd* if $n+1$ is even. We claim that every natural is either even or odd. To see this, let S be the union of the set of even naturals and the set of odd naturals. Then $2 = 2 \cdot 1$ is even, so 1 is odd. Hence, $1 \in S$. If $n \in S$ and $n = 2k$ is even, then $n+1$ is odd since $(n+1)+1 = n+2 = 2k+2 = 2(k+1)$. Hence, $n+1 \in S$. If $n \in S$ and n is odd, then $n+1$ is even, so $n+1 \in S$. Hence, in either case, $n \in S$ implies $n+1 \in S$, i.e., S is closed under addition by 1. Thus, S is inductive. Hence, we conclude that $S = \mathbf{N}$. Thus, every natural is even or odd. Also the usual parity rules hold: Even plus even is even, etc.

Let A be a nonempty set. We say *A has n elements* and we write $\#A = n$ if there is a bijection between A and the set $\{k \in \mathbf{N} : 1 \le k \le n\}$. We often denote this last set by $\{1, 2, \dots, n\}$. If $A = \emptyset$, we say that the number of elements of A is zero. A set A is *finite* if it has n elements for some n. Otherwise, A is *infinite*. Here are some consequences of the definition that are worked out in the exercises. If A and B are disjoint and have n and m elements, respectively, then $A \cup B$ has $n+m$ elements. If A is a finite subset of $\mathbf{R}$, then $\max A$ and $\min A$ exist. In particular, we let $\max(a,b)$, $\min(a,b)$ denote the larger and the smaller of a and b.

Now we show that $\max A$ and $\min A$ may exist for certain infinite subsets of $\mathbf{R}$.

Theorem 1.3.2. *If $S \subset \mathbf{N}$ is nonempty, then* $\min S$ *exists.*

To see this, note that $c = \inf S$ is finite since S is bounded below. The goal is to establish $c \in S$. Since $c+1$ is not a lower bound, there is an $n \in S$ with $c \le n < c+1$. If $c = n$, then $c \in S$ and we are done. If $c \ne n$, then $n-1 < c < n$, and n is not a lower bound for S. Hence, there is an $m \in S$ lying between $n-1$ and n. But there are no naturals between $n-1$ and n. □

Two other subsets mentioned frequently are the *integers* $\mathbf{Z} = \mathbf{N} \cup \{0\} \cup (-\mathbf{N}) = \{0, \pm 1, \pm 2, \dots\}$ and the *rationals* $\mathbf{Q} = \{m/n : m, n \in \mathbf{Z}, n \ne 0\}$. Then $\mathbf{Z}$ is closed under subtraction (Exercise **1.3.3**), and $\mathbf{Q}$ is closed under all four arithmetic operations, except under division by zero. As for naturals, we say that the integers in $2\mathbf{Z} = \{2n : n \in \mathbf{Z}\}$ are *even*, and we say that an integer n is *odd* if $n+1$ is even.

A *sequence of reals* is a function $f : \mathbf{N} \to \mathbf{R}$. Sequences are usually written $(a_n) = (a_1, a_2, \dots)$ where $a_n = f(n)$, $n \ge 1$.

Using (an extension of) induction, one can build new sequences from old sequences as follows. Let $\mathbf{N} \times \mathbf{R}$ be the product (§1.1) and fix a function $g : \mathbf{N} \times \mathbf{R} \to \mathbf{R}$ and a real $a \in \mathbf{R}$. Then (Exercise **1.3.10**) there is a unique function $f : \mathbf{N} \to \mathbf{R}$ satisfying $f(1) = a$ and $f(n+1) = g(n, f(n))$, $n \geq 1$, or, equivalently, there is a unique sequence (a_n) satisfying $a_1 = a$ and $a_{n+1} = g(n, a_n)$, $n \geq 1$.

For example, given a sequence (a_n), there is a sequence (s_n) satisfying $s_1 = a_1$ and $s_{n+1} = a_{n+1} + s_n$, $n \geq 1$. This is usually written

$$s_n = a_1 + a_2 + \cdots + a_n = \sum_{k=1}^{n} a_k, \qquad n \geq 1,$$

and corresponds to the choices $a = a_1$ and $g(n, x) = x + a_{n+1}$. Then s_n is the *nth partial sum*, $n \geq 1$ (§1.5).

Fix a natural $N \in \mathbf{N}$ and suppose we are given a function $a : \{1, 2, \ldots, N\} \to \mathbf{R}$, i.e., we are given reals $a_1, a_2, \ldots, a_N$. Then we can extend the definition of a to all of $\mathbf{N}$ by setting $a_n = 0$ for $n > N$, and the Nth partial sum

$$a_1 + a_2 + \cdots + a_N = \sum_{n=1}^{N} a_n$$

is how one defines the sum of $a_1, a_2, \ldots, a_N$.

Similarly, given a sequence (a_n), there is a sequence (p_n) satisfying $p_1 = a_1$ and $p_{n+1} = p_n \cdot a_{n+1}$, $n \geq 1$. This is usually written

$$p_n = a_1 \cdot a_2 \cdot \cdots \cdot a_n = \prod_{k=1}^{n} a_k, \qquad n \geq 1,$$

and corresponds to the choices $a = a_1$ and $g(n, x) = x \cdot a_{n+1}$. Then p_n is the *nth partial product* (§5.6). For example, if a is a fixed real and $a_n = a$ for all $n \geq 1$, the resulting sequence is (a^n).

When $a_n = n$, $n \geq 1$, the resulting sequence of partial products satisfies $p_1 = 1$ and $p_{n+1} = p_n(n+1)$, $n \geq 1$; this is the *factorial* sequence $(n!)$. It is convenient to also define $0! = 1$.

Fix a natural $N \in \mathbf{N}$ and suppose we are given a function $a : \{1, 2, \ldots, N\} \to \mathbf{R}$, i.e., we are given reals $a_1, a_2, \ldots, a_N$. Then we can extend the definition of a to all of $\mathbf{N}$ by setting $a_n = 1$ for $n > N$, and the Nth partial product

$$a_1 \cdot a_2 \cdot \cdots \cdot a_N = \prod_{n=1}^{N} a_n$$

is how one defines the product of $a_1, a_2, \ldots, a_N$.

Now $(-1)^n$ is 1 or -1 according to whether $n \in \mathbf{N}$ is even or odd, $a > 0$ implies $a^n > 0$ for $n \in \mathbf{N}$, and $a > 1$ implies $a^n > 1$ for $n \in \mathbf{N}$. These are easily checked by induction.

If $a \neq 0$, we extend the definition of a^n to $n \in \mathbf{Z}$ by setting $a^0 = 1$ and $a^{-n} = 1/a^n$ for $n \in \mathbf{N}$. Then (Exercise **1.3.11**), $a^{n+m} = a^n a^m$ and $(a^n)^m = a^{nm}$ for all integers n, m.

Let $a > 1$. Then $a^n = a^m$ with $n, m \in \mathbf{Z}$ only when $n = m$. Indeed, $n - m \in \mathbf{Z}$, and $a^{n-m} = a^n a^{-m} = a^n/a^m = 1$. But $a^k > 1$ for $k \in \mathbf{N}$, and $a^k = 1/a^{-k} < 1$ for $k \in -\mathbf{N}$. Hence, $n - m = 0$ or $n = m$. This shows that powers are unique.

As another application of induction, we establish, simultaneously, the validity of the inequalities $1 < 2^n$ and $n < 2^n$ for all naturals n. This time, we do this without mentioning the set S explicitly, as follows: The inequalities in question are true for $n = 1$ since $1 < 2^1 = 2$. Moreover, if the inequalities $1 < 2^n$ and $n < 2^n$ are true for a particular n (the inductive hypothesis), then $1 < 2^n < 2^n + 2^n = 2^n 2 = 2^{n+1}$, so the first inequality is true for $n+1$. Adding the inequalities valid for n yields $n + 1 < 2^n + 2^n = 2^n 2 = 2^{n+1}$, so the second inequality is true for $n + 1$. This establishes the inductive step. Hence, by induction, the two inequalities are true for all $n \in \mathbf{N}$. Explicitly, the set S here is $S = \{n \in \mathbf{N} : 1 < 2^n, n < 2^n\}$.

Using these inequalities, we show that every nonzero $n \in \mathbf{Z}$ is of the form $2^k p$ for a uniquely determined $k \in \mathbf{N} \cup \{0\}$ and an odd $p \in \mathbf{Z}$. We call k *the number of factors of* 2 *in* n.

If $2^k p = 2^j q$ with $k > j$ and odd integers p, q, then $q = 2^{k-j} p = 2 \cdot 2^{k-j-1} p$ is even, a contradiction. On the other hand, if $j > k$, then p is even. Hence, we must have $k = j$. This establishes the uniqueness of k.

To show the existence of k, by multiplying by a minus, if necessary, we may assume $n \in \mathbf{N}$. If n is odd, we may take $k = 0$ and $p = n$. If n is even, then $n_1 = n/2$ is a natural $< 2^{n-1}$. If n_1 is odd, we take $k = 1$ and $p = n_1$. If n_1 is even, then $n_2 = n_1/2$ is a natural $< 2^{n-2}$. If n_2 is odd, we take $k = 2$ and $p = n_2$. If n_2 is even, we continue this procedure by dividing n_2 by 2. Continuing in this manner, we obtain $n_1, n_2, \dots$ naturals with $n_j < 2^{n-j}$. Since this procedure ends in fewer than n steps, there is some k natural or 0 for which $p = n/2^k$ is odd.

The final issue we take up here concerns square roots. Given a real a, a *square root* of a, denoted $\sqrt{a}$, is any real x whose square is a, $x^2 = a$. For example, 1 has the square roots ± 1, and 0 has the square root 0. On the other hand, not every real has a square root. For example, -1 does not have a square root, i.e., there is no real x satisfying $x^2 = -1$, since $x^2 + 1 > 0$. In fact a similar argument shows that negative numbers never have square roots.

At this point, we do not know whether 2 has a square root. First, we show that $\sqrt{2}$ cannot be rational. See also Exercises **1.4.12** and **1.4.13**.

Theorem 1.3.3. *There is no rational a satisfying $a^2 = 2$.*

We argue by contradiction. Suppose that $a = m/n$ is a rational whose square is 2. Then $(m/n)^2 = 2$ or $m^2 = 2n^2$, i.e., there is a natural N, such that $N = m^2$ and $N = 2n^2$. Then $m = 2^k p$ with odd p and $k \in \mathbf{N} \cup \{0\}$, so $N = m^2 = 2^{2k} p^2$. Since p^2 is odd, we conclude that $2k$ is the number of factors of 2 in N. Similarly $n = 2^j q$ with odd q and $j \in \mathbf{N} \cup \{0\}$, so $N = 2n^2 = 22^{2j} q^2 = 2^{2j+1} q^2$. Since q^2 is odd, we conclude that $2j+1$ is the number of factors of 2 in N. Since $2k \neq 2j+1$, we arrive at a contradiction. □

Note that **Q** satisfies the arithmetic and ordering properties. The completeness property is all that distinguishes **Q** and **R**.

As usual, in the following, a *digit* means either 0, 1, 2, or $3 = 2+1$, $4 = 3+1$, $5 = 4+1$, $6 = 5+1$, $7 = 6+1$, $8 = 7+1$, or $9 = 8+1$. Also the letters n, m, i, j will usually denote integers, so $n \geq 1$ will be used interchangeably with $n \in \mathbf{N}$, with similar remarks for m, i, j.

We say that a nonzero $n \in \mathbf{Z}$ *divides* $m \in \mathbf{Z}$ if $m/n \in \mathbf{Z}$. Alternatively, we say that m is *divisible* by n, and we write $n \mid m$. A natural n is *composite* if $n = jk$ for some $j, k \in \mathbf{N}$ with $j > 1$ and $k > 1$. A natural is *prime* if it is not composite and is not 1. Thus, a natural is prime if it is not divisible by any smaller natural other than 1.

Let $n \in \mathbf{Z}$. A *factor* of n is an integer p dividing n. We say integers n and m *have no common factor* if the only natural dividing n and m simultaneously is 1.

For $a \geq 1$, let $\lfloor a \rfloor = \max\{n \in \mathbf{N} : n \leq a\}$ denote the *greatest integer* $\leq a$ (Exercises **1.3.7** and **1.3.9**). Then $\lfloor a \rfloor \leq a < \lfloor a \rfloor + 1$; the *fractional part* of a is $\{a\} = a - \lfloor a \rfloor$. Note that the fractional part is a real in $[0,1)$. More generally, $\lfloor a \rfloor \in \mathbf{Z}$ and $0 \leq \{a\} < 1$ are defined[4] for all $a \in \mathbf{R}$.

Exercises

1.3.1. Let n be a natural. Show that there are no naturals between n and $n+1$.

1.3.2. Show that the product of naturals is natural, $\mathbf{N} \cdot \mathbf{N} \subset \mathbf{N}$.

1.3.3. If $m > n$ are naturals, then $m - n \in \mathbf{N}$. Conclude that **Z** is closed under subtraction.

1.3.4. Show that no integer is both even and odd. Also show that even times even is even, even times odd is even, and odd times odd is odd.

1.3.5. If n, m are naturals and there is a bijection between $\{1, 2, \ldots, n\}$ and $\{1, 2, \ldots, m\}$, then $n = m$ (use induction on n). Conclude that the number of elements $\#A$ of a nonempty set A is well defined. Also show that $\#A = n$, $\#B = m$, and $A \cap B = \emptyset$ imply $\#(A \cup B) = n + m$.

---

[4] $\{n \in \mathbf{Z} : n \leq a\}$ is nonempty since $\inf \mathbf{Z} = -\infty$ (§1.4).

1.3.6. If $A \subset \mathbf{R}$ is finite and nonempty, then show that $\max A$ and $\min A$ exist (use induction).

1.3.7. If $S \subset \mathbf{Z}$ is nonempty and bounded above, then show that S has a max.

1.3.8. For $x > 0$ real, let $S = \{n \in \mathbf{N} : nx \in \mathbf{N}\}$. Show that S is nonempty iff $x \in \mathbf{Q}$. Show that $x = n/d$ with $n, d \in \mathbf{N}$ having no common factor iff $d = \min S$ and $n = dx$. If we set $D(x) = d$, show that $D(x) = D(x + k)$ for $k \in \mathbf{N}$.

1.3.9. If $x \geq y > 0$ are reals, then show that $x = yq + r$ with $q \in \mathbf{N}$, $r \in \mathbf{R}^+ \cup \{0\}$, and $r < y$. (Look at the sup of $\{q \in \mathbf{N} : yq \leq x\}$.)

1.3.10. Let X be a set and let $g : \mathbf{N} \times X \to X$ be a mapping and fix $a \in X$. A set $f \subset \mathbf{N} \times X$ is *inductive* if

A. $(1, a) \in f$,
B. $(n, x) \in f$ implies $(n + 1, g(n, x)) \in f$.

For example, $\mathbf{N} \times X$ is inductive[5] for any g and any a. Now let f be the intersection of all inductive subsets of $\mathbf{N} \times X$

$$f = \bigcap \{h \subset \mathbf{N} \times X : h \text{ inductive}\}$$

and let $A = \{n \in \mathbf{N} : (n, x) \in f \text{ for some } x \in X\}$. Show

- $A = \mathbf{N}$,
- f is a mapping (§1.1) with domain $\mathbf{N}$ and codomain X,
- $f(1) = a$ and $f(n + 1) = g(n, f(n))$ for all $n \geq 1$.

Show also that there is a unique such function. (Given $n \in \mathbf{N}$, let $B_n = \{x : (n, x) \in f\}$. Then f is a mapping iff $\#B_n = 1$ for all n. Let $B = \{n : \#B_n > 1\}$ and use Theorem 1.3.2 to show B is empty.)

1.3.11. Let a be a nonzero real. By induction, show that $a^n a^m = a^{n+m}$ and $(a^n)^m = a^{nm}$ for all integers n, m.

1.3.12. Using induction, show that

$$1 + 2 + \cdots + n = \frac{n(n + 1)}{2}, \qquad n \geq 1.$$

1.3.13. Let $p > 1$ be a natural. Show that for each nonzero $n \in \mathbf{Z}$, there is a unique $k \in \mathbf{N} \cup \{0\}$ and an integer m not divisible by p (i.e., m/p is not in $\mathbf{Z}$), such that $n = p^k m$.

---

[5] Note here that *inductive* depends on g and on a.

1.3.14. Let $S \subset \mathbf{R}$ satisfy

- $1 \in S$
- $n \in S$ whenever $k \in S$ for all naturals $k < n$.

Show that $S \supset \mathbf{N}$. This is an alternate form of induction.

1.3.15. Fix $a > 0$ real, and let $S_a = \{n \in \mathbf{N} : na \in \mathbf{N}\}$. If S_a is nonempty, $m \in S_a$, and $p = \min S_a$, show that p divides m (Exercise **1.3.9**).

1.3.16. Let n, m be naturals and suppose that a prime p divides the product nm. Show that p divides n or m. (Consider $a = n/p$, and show that $\min S_a = 1$ or $\min S_a = p$.)

1.3.17. (Fundamental Theorem of Arithmetic) By induction, show that every natural n either is 1 or is a product of primes, $n = p_1 \dots p_r$, with the p_j's unique except, possibly, for the ordering. (Given n, either n is prime or $n = pm$ for some natural $1 < m < n$; use induction as in Exercise **1.3.14**.)

1.3.18. Given $0 < x < 1$, let $r_0 = x$. Define naturals a_n and remainders r_n by setting

$$\frac{1}{r_n} = a_{n+1} + r_{n+1}, \qquad n \geq 0.$$

Thus, $a_{n+1} = \lfloor 1/r_n \rfloor$ is the integer part of $1/r_n$ and $r_{n+1} = \{1/r_n\}$ is the fractional part of $1/r_n$, and

$$x = \cfrac{1}{a_1 + \cfrac{1}{\ddots \, a_{n-1} + \cfrac{1}{a_n + r_n}}}$$

is a *continued fraction*.[6] This algorithm stops the first time $r_n = 0$. Then the continued fraction is *finite* and we write $x = [a_1, a_2, a_3, \dots, a_n]$. If this never happens, this algorithm does not end, and the continued fraction is *infinite* and we write $x = [a_1, a_2, a_3, \dots]$. Show that the algorithm stops iff $x \in \mathbf{Q}$. Computer code generating $[a_1, a_2, a_3, \dots]$ is in Exercise **5.2.11**.

1.3.19. Let $f : \mathbf{N} \to \mathbf{N}$ be injective, and for $n \geq 1$, let $A_n = \{m \in \mathbf{N} : f(m) \leq n\}$. Show by induction that A_n is bounded above for all $n \in \mathbf{N}$.

1.3.20. Show that $2^{n-1} \leq n!$, $n \geq 1$.

1.3.21. Show that $n! \leq n^n \leq (n!)^2$ for all naturals n. For the second inequality, rearrange the factors in $(n!)^2$ into pairs.

1.3.22. Show that $(1 + a)^n \leq 1 + (2^n - 1)a$ for $n \geq 1$ and $0 \leq a \leq 1$. Also show that $(1 + a)^n \geq 1 + na$ for $n \geq 1$ and $a \geq -1$.

---

[6] Because the numerators are all equal to 1, this is a *simple* continued fraction.

1.3.23. Let $0 < x < y < z < 1$. Express x, y, z as continued fractions as in Exercise **1.3.18**: $x = [a_1, a_2, \dots]$, $y = [b_1, b_2, \dots]$, $z = [c_1, c_2, \dots]$. If $a_j = c_j$, $j = 1, \dots, n$, then $a_j = b_j = c_j$, $j = 1, \dots, n$. Use induction on n.

1.3.24. Let X be a finite set with $\emptyset \notin X$. Show there is a function $f : X \to \bigcup X$ such that $f(x) \in x$ for $x \in X$. This is the *axiom of finite choice* (use induction on $\#X$). Equivalently, given $A_1, \dots, A_n$ nonempty, there are $a_1, \dots, a_n$ with $a_k \in A_k$, $1 \le k \le n$.

1.4 The Completeness Property

We begin by showing that $\mathbf{N}$ has no upper bound. Indeed, if $\mathbf{N}$ has an upper bound, then $\mathbf{N}$ has a (finite) sup, and call it c. Then c is an upper bound for $\mathbf{N}$, whereas $c - 1$ is not an upper bound for $\mathbf{N}$, since c is the least such. Thus, there is an $n \ge 1$, satisfying $n > c - 1$, which gives $n + 1 > c$ and $n + 1 \in \mathbf{N}$. But this contradicts the fact that c is an upper bound. Hence, $\mathbf{N}$ is not bounded above. In the notation of §1.2, $\sup \mathbf{N} = \infty$.

Let $S = \{1/n : n \in \mathbf{N}\}$ be the reciprocals of all naturals. Then S is bounded below by 0, hence S has an inf. We show that $\inf S = 0$. First, since 0 is a lower bound, by definition of inf, $\inf S \ge 0$. Second, let $c > 0$. Since $\sup \mathbf{N} = \infty$, there is some natural, call it k, satisfying $k > 1/c$. Multiplying this inequality by the positive c/k, we obtain $c > 1/k$. Since $1/k$ is an element of S, this shows that c is not a lower bound for S. Thus, any lower bound for S must be less or equal to 0. Hence, $\inf S = 0$.

The two results just derived are so important we state them again.

Theorem 1.4.1. $\sup \mathbf{N} = \infty$, *and* $\inf\{1/n : n \in \mathbf{N}\} = 0$. □

As a consequence, since $\mathbf{Z} \supset \mathbf{N}$, it follows that $\sup \mathbf{Z} = \infty$. Since $\mathbf{Z} \supset (-\mathbf{N})$ and $\inf(A) = -\sup(-A)$, it follows that $\inf \mathbf{Z} \le \inf(-\mathbf{N}) = -\sup \mathbf{N} = -\infty$, hence $\inf \mathbf{Z} = -\infty$.

An *interval* is a subset of $\mathbf{R}$ of the following form:

$$\begin{aligned}(a, b) &= \{x : a < x < b\},\\ [a, b] &= \{x : a \le x \le b\},\\ [a, b) &= \{x : a \le x < b\},\\ (a, b] &= \{x : a < x \le b\}.\end{aligned}$$

Intervals of the form (a, b), (a, ∞), $(-\infty, b)$, $(-\infty, \infty)$ are *open*, whereas those of the form $[a, b]$, $[a, \infty)$, $(-\infty, b]$ are *closed*. When $-\infty < a < b < \infty$, the interval $[a, b]$ is *compact*. Thus, $(a, \infty) = \{x : x > a\}$, $(-\infty, b] = \{x : x \le b\}$, and $(-\infty, \infty) = \mathbf{R}$.

For $x \in \mathbf{R}$, we define $|x|$, the *absolute value* of x, by $|x| = \max(x, -x)$. Then $x \le |x|$ for all x, and, for $a > 0$, $\{x : -a < x < a\} = \{x : |x| < a\} = \{x : x < a\} \cap \{x : x > -a\}$, $\{x : x < -a\} \cup \{x : x > a\} = \{x : |x| > a\}$.

The absolute value satisfies the following properties:

A. $|x| > 0$ for all nonzero x, and $|0| = 0$,
B. $|x|\,|y| = |xy|$ for all x, y,
C. $|x + y| \leq |x| + |y|$ for all x, y.

We leave the first two as exercises. The third, the *triangle inequality*, is derived using $|x|^2 = x^2$ as follows:

$$\begin{aligned}|x + y|^2 &= (x + y)^2 = x^2 + 2xy + y^2 \\ &\leq |x|^2 + 2|xy| + |y|^2 = |x|^2 + 2|x|\,|y| + |y|^2 = (|x| + |y|)^2.\end{aligned}$$

Since $a \leq b$ iff $a^2 \leq b^2$ for a, b nonnegative (Exercise **1.2.6**), the triangle inequality is established.

Frequently, the triangle inequality is used in alternate forms, one of which is

$$|x - y| \geq |x| - |y|.$$

This follows by writing $|x| = |(x - y) + y| \leq |x - y| + |y|$ and transposing $|y|$ to the other side. Another form is

$$|a_1 + a_2 + \cdots + a_n| \leq |a_1| + |a_2| + \cdots + |a_n|, \quad n \geq 1.$$

We show how the completeness property can be used to derive the existence of $\sqrt{2}$ within $\mathbf{R}$.

Theorem 1.4.2. *There is a real a satisfying $a^2 = 2$.*

To see this, let $S = \{x : x \geq 1 \text{ and } x^2 < 2\}$. Since $1 \in S$, S is nonempty. Also $x \in S$ implies $x = x1 \leq xx = x^2 < 2$, hence S is bounded above by 2, hence S has a sup, call it a. We claim that $a^2 = 2$. We establish this claim by ruling out the cases $a^2 < 2$ and $a^2 > 2$, leaving us with the desired conclusion (remember every real is positive or negative or zero).

So suppose that $a^2 < 2$. If we find a natural n with $(a + 1/n)^2 < 2$, then $a + 1/n \in S$, hence the real a could not have been an upper bound for S, much less the least such. To see how to find such an n, note that

$$\left(a + \frac{1}{n}\right)^2 = a^2 + \frac{2a}{n} + \frac{1}{n^2} \leq a^2 + \frac{2a}{n} + \frac{1}{n} = a^2 + \frac{2a+1}{n}.$$

But this last quantity is < 2 if $(2a+1)/n < 2-a^2$, i.e., if $n > (2a+1)/(2-a^2)$. Since $a^2 < 2$, $b = (2a + 1)/(2 - a^2) > 0$; since $\sup \mathbf{N} = \infty$, such a natural $n > b$ can always be found. This rules out $a^2 < 2$.

Before we rule out $a^2 > 2$, we note that S is bounded above by any positive b satisfying $b^2 > 2$ since, for b and x positive, $b^2 > x^2$ iff $b > x$.

Now suppose that $a^2 > 2$. Then $b = (a^2 - 2)/2a$ is positive, hence there is a natural n satisfying $1/n < b$ which implies $a^2 - 2a/n > 2$. Hence,

$$\left(a - \frac{1}{n}\right)^2 = a^2 - \frac{2a}{n} + \frac{1}{n^2} > 2,$$

so $a - 1/n$ is an upper bound for S. This shows that a is not the *least* upper bound, contradicting the definition of a. Thus, we are forced to conclude that $a^2 = 2$. □

A real a satisfying $a^2 = 2$ is called a *square root* of 2. Since $(-x)^2 = x^2$, there are two square roots of 2, one positive and one negative. From now on, the positive square root is denoted $\sqrt{2}$. Similarly, every positive a has a positive square root, which we denote $\sqrt{a}$. In the next chapter, after we have developed more material, a simpler proof of this fact will be derived.

More generally, for every $b > 0$ and $n \geq 1$, there is a unique $a > 0$ satisfying $a^n = b$, the *nth root* $a = b^{1/n}$ of b. Now for $n \geq 1$, $k \geq 1$, and $m \in \mathbf{Z}$,

$$\left[(b^m)^{1/n}\right]^{nk} = \left\{\left[(b^m)^{1/n}\right]^n\right\}^k = (b^m)^k = b^{mk};$$

hence, by uniqueness of roots, $(b^m)^{1/n} = (b^{mk})^{1/nk}$. Thus, for $r = m/n$ rational, we may set $b^r = (b^m)^{1/n}$, defining rational powers of positive reals.

Since $\sqrt{2} \notin \mathbf{Q}$, $\mathbf{R} \setminus \mathbf{Q}$ is not empty. The reals in $\mathbf{R} \setminus \mathbf{Q}$ are the *irrationals*. In fact, both the rationals and the irrationals have an interlacing or *density* property.

Theorem 1.4.3. *If $a < b$ are any two reals, there is a rational s between them, $a < s < b$, and there is an irrational t between them, $a < t < b$.*

To see this, first, choose a natural n satisfying $1/n < b - a$. Second let $S = \{m \in \mathbf{N} : na < m\}$, and let $k = \inf S = \min S$. Since $k \in S$, $na < k$. Since $k - 1 \notin S$, $k - 1 \leq na$. Hence, $s = k/n$ satisfies

$$a < s \leq a + \frac{1}{n} < b.$$

For the second assertion, choose a natural n satisfying $1/n\sqrt{2} < b - a$, let $T = \{m \in \mathbf{N} : \sqrt{2}na < m\}$, and let $k = \min T$. Since $k \in T$, $k > \sqrt{2}na$. Since $k - 1 \notin T$, $k - 1 \leq \sqrt{2}na$. Hence, $t = k/(n\sqrt{2})$ satisfies

$$a < t \leq a + \frac{1}{n\sqrt{2}} < b.$$

Moreover, t is necessarily irrational. □

Approximation of reals by rationals is discussed further in the exercises.

Exercises

1.4.1. Show that $x \leq |x|$ for all x and, for $a > 0$, $\{x : -a < x < a\} = \{x : |x| < a\} = \{x : x < a\} \cap \{x : x > -a\}$, $\{x : x < -a\} \cup \{x : x > a\} = \{x : |x| > a\}$.

1.4.2. For all $x \in \mathbf{R}$, $|x| \geq 0$, $|x| > 0$ if $x \neq 0$, and $|x|\,|y| = |xy|$ for all $x, y \in \mathbf{R}$.

1.4.3. By induction, show that $|a_1 + a_2 + \cdots + a_n| \leq |a_1| + |a_2| + \cdots + |a_n|$ for $n \geq 1$.

1.4.4. Show that every $a > 0$ real has a unique positive square root.

1.4.5. Show that $ax^2 + bx + c = 0$, $a \neq 0$, has two, one, or no solutions in $\mathbf{R}$ according to whether $b^2 - 4ac$ is positive, zero, or negative. When there are solutions, they are given by $x = (-b \pm \sqrt{b^2 - 4ac})/2a$.

1.4.6. For $a, b \geq 0$, show that $a^n \geq b^n$ iff $a \geq b$. Also show that every $b > 0$ has a unique positive nth root for all $n \geq 1$ (use Exercise **1.3.22** and modify the derivation for $\sqrt{2}$).

1.4.7. Show that the real t constructed in the derivation of Theorem 1.4.3 is irrational.

1.4.8. Let a be any real. Show that, for each $\epsilon > 0$, no matter how small, there are integers $n \neq 0$, m satisfying

$$\left| a - \frac{m}{n} \right| < \frac{\epsilon}{n}.$$

(Let $\{a\}$ denote the fractional part of a, consider $S = \{a\}, \{2a\}, \{3a\}, \dots$, and divide $[0, 1]$ into finitely many subintervals of length less than ϵ. Since there are infinitely many terms in S, at least two of them must lie in the same subinterval.)

1.4.9. Show that $a = \sqrt{2}$ satisfies

$$\left| a - \frac{m}{n} \right| \geq \frac{1}{(2\sqrt{2} + 1)n^2}, \qquad n, m \geq 1.$$

(Consider the two cases $|a - m/n| \geq 1$ and $|a - m/n| \leq 1$, separately, and look at the minimum of $n^2|f(m/n)|$ with $f(x) = x^2 - 2$.)

1.4.10. Let $a = \sqrt{1 + \sqrt{2}}$. Then a is irrational, and there is a positive real c satisfying

$$\left| a - \frac{m}{n} \right| \geq \frac{c}{n^4}, \qquad n, m \geq 1.$$

(Factor $f(a) = a^4 - 2a^2 - 1 = 0$, and proceed as in the previous exercise.)

1.4.11. For $n \in \mathbf{Z} \setminus \{0\}$, define $|n|_2 = 1/2^k$ where k is the number of factors of 2 in n. Also define $|0|_2 = 0$. For $n/m \in \mathbf{Q}$ define

$$|n/m|_2 = |n|_2/|m|_2.$$

Show that $|\cdot|_2 : \mathbf{Q} \to \mathbf{R}$ is well defined and satisfies the absolute value properties **A**, **B**, and **C**.

1.4.12. Show that $\sqrt{2}$ is not rational by showing $x = \sqrt{2} - 1$ satisfies $x = 1/(2+x)$, hence

$$\sqrt{2} = 1 + \cfrac{1}{2 + \cfrac{1}{2 + \dots}}$$

(Exercise **1.3.18**).

1.4.13. For $n \in \mathbf{N}$ let $x = \sqrt{n}$. Then either $x \in \mathbf{N}$ or $x \notin \mathbf{Q}$. This is another proof of the irrationality of $\sqrt{2}$. (Consider $d' = d(x - \lfloor x \rfloor)$ where $d = D(x)$ is as in Exercise **1.3.8**.)

1.5 Sequences and Limits

A *sequence* of real numbers is a function $f : \mathbf{N} \to \mathbf{R}$. A *finite sequence* is a function $f : \{1, \dots, N\} \to \mathbf{R}$ for some $N \geq 1$. Usually, we write a sequence as (a_n) where $a_n = f(n)$ is the nth term. For example, the formulas $a_n = n$, $b_n = 2n$, $c_n = 2^n$, and $d_n = 2^{-n} + 5n$ yield sequences (a_n), (b_n), (c_n), and (d_n). Later we will consider sequences of sets (Q_n) and sequences of functions (f_n), but now we discuss only sequences of reals.

It is important to distinguish between the *sequence* (a_n) (the function f) and the *set* $\{a_n\}$ (the range $f(\mathbf{N})$ of f). In fact, a sequence is an *ordered* set $(a_1, a_2, a_3, \dots)$ and not just a set $\{a_1, a_2, a_3, \dots\}$. Sometimes it is more convenient to start sequences from the index $n = 0$, i.e., to consider a sequence as a function on $\mathbf{N} \cup \{0\}$. For example, the sequence $(1, 2, 4, 8, \dots)$ can be written $a_n = 2^n, n \geq 0$. Specific examples of sequences are usually constructed by induction as in Exercise **1.3.10**. However, we will not repeat the construction carried out there for each sequence we encounter.

In this section, we are interested in the behavior of sequences as the index n increases without bound. Often this is referred to as the "limiting behavior" of sequences. For example, consider the sequences

$$\begin{aligned}
(a_n) &= (1/2, 2/3, 3/4, 4/5, \dots), \\
(b_n) &= (1, -1, 1, -1, \dots), \\
(c_n) &= \left(2, \sqrt{2}, \sqrt{\sqrt{2}}, \sqrt{\sqrt{\sqrt{2}}}, \dots\right), \\
(d_n) &= (2, 3/2, 17/12, 577/408, \dots),
\end{aligned}$$

where, in the last[7] sequence, $d_1 = 2$, $d_2 = (d_1 + 2/d_1)/2$, $d_3 = (d_2 + 2/d_2)/2$, $d_4 = (d_3 + 2/d_3)/2$, and so on. What are the limiting behaviors of these sequences?

As n increases, the terms in (a_n) are arranged in increasing order, and $a_n \leq 1$ for all $n \geq 1$. However, if we increase n sufficiently, the terms $a_n = (n-1)/n = 1 - 1/n$ become arbitrarily close to 1, since $\sup\{1 - 1/n : n \geq 1\} = 1$ (§1.4). Thus, it seems reasonable to say that (a_n) approaches one or the limit of the sequence (a_n) equals one.

On the other hand, the sequence (b_n) does not seem to approach any single real, as it flips back and forth between 1 and -1. Indeed, one is tempted to say that (b_n) has two limits, 1 and -1.

The third sequence is more subtle. Since we have $\sqrt{x} < x$ for $x > 1$, the terms are arranged in decreasing order. Because of this, it seems reasonable that (c_n) approaches its "bottom," i.e., (c_n) approaches $L = \inf\{c_n : n \geq 1\}$. Although, in fact, this turns out to be so, it is not immediately clear just what L equals.

The limiting behavior of the fourth sequence is not at all clear. If one computes the first nine terms, it is clear that this sequence approaches something quickly. However, since such a computation is approximate, at the outset, we cannot be sure there is a single real number that qualifies as "the limit" of (d_n). The sequence (d_n) is discussed in Exercise **1.5.12** and in Exercise **1.6.4**.

It is important to realize that the questions

A. What does "limit" mean?
B. Does the limit exist?
C. How do we compute the limit?

are very different. When the situation is sufficiently simple, say, as in (a_n) or (b_n) above, we may feel that the notion of "limit" is self-evident and needs no elaboration. Then we may choose to deal with more complicated situations on a case-by-case basis and not worry about a "general" definition of limit. Historically, however, mathematicians have run into trouble using this ad hoc approach. Because of this, a more systematic approach was adopted in which a single definition of "limit" is applied. This approach was so successful that it is universally followed today.

Below, we define the concept of limit in two stages: first, for monotone sequences and then for general sequences. To deal with situations where sequences approach more than one real, the auxiliary concept of a "limit point" is introduced in Exercise **1.5.9**. Now we turn to the formal development.

Let (a_n) be any sequence. We say (a_n) is *decreasing* if $a_n \geq a_{n+1}$ for all natural n. If $L = \inf\{a_n : n \geq 1\}$, in this case, we say (a_n) *approaches* L *as* $n \nearrow \infty$, and we write $a_n \searrow L$ as $n \nearrow \infty$. Usually, we drop the phrase "as $n \nearrow \infty$" and simply write $a_n \searrow L$. We say a sequence (a_n) is *increasing* if

[7] Decimal notation, e.g., $17 = (9 + 1) + 7$, is reviewed in the next section.

$a_n \le a_{n+1}$ for all $n \ge 1$. If $L = \sup\{a_n : n \ge 1\}$, in this case, we say (a_n) *approaches* L *as* $n \nearrow \infty$, and we write $a_n \nearrow L$ as $n \nearrow \infty$. Usually, we drop the phrase "as $n \nearrow \infty$" and simply write $a_n \nearrow L$. Alternatively, in either case, we say *the limit of* (a_n) *is* L, and we write

$$\lim_{n \nearrow \infty} a_n = L.$$

Note that since sups and infs are uniquely determined, we say "the limit" instead of "a limit." Thus,

$$\lim_{n \nearrow \infty} \left(1 - \frac{1}{n}\right) = 1$$

since $\sup\{1 - 1/n : n \ge 1\} = 1$,

$$\lim_{n \nearrow \infty} \frac{1}{n} = 0$$

since $\inf\{1/n : n \ge 1\} = 0$, and

$$\lim_{n \nearrow \infty} n^2 = \infty$$

since $\sup\{n^2 : n \ge 1\} = \infty$.

We say a sequence is *monotone* if the sequence is either increasing or decreasing. Thus, the concept of *limit* is now defined for every monotone sequence. We say a sequence is *constant* if it is both decreasing and increasing, i.e., it is of the form $(a, a, \dots)$ where a is a fixed real.

If (a_n) *is a monotone sequence approaching a nonzero limit* a*, then there is a natural* N *beyond which* $a_n \neq 0$ *for* $n \ge N$. To see this, suppose that (a_n) is increasing and $a > 0$. Then by definition $a = \sup\{a_n : n \ge 1\}$, hence $a/2$ is not an upper bound for (a_n). Thus, there is a natural N with $a_N > a/2 > 0$. Since the sequence is increasing, we conclude that $a_n \ge a_N > 0$ for $n \ge N$. If (a_n) is increasing and $a < 0$, then $a_n \le a < 0$ for all $n \ge 1$. If the sequence is decreasing, the reasoning is similar.

Before we define limits for arbitrary sequences, we show that every sequence (a_n) lies between a decreasing sequence (a_n^*) and an increasing sequence (a_{n*}) in a simple and systematic fashion.

Let (a_n) be any sequence. Let $a_1^* = \sup\{a_k : k \ge 1\}$, $a_2^* = \sup\{a_k : k \ge 2\}$, and for each natural n, let $a_n^* = \sup\{a_k : k \ge n\}$. Thus, a_n^* is the sup of all the terms starting from the nth term. Since $\{a_k : k \ge n+1\} \subset \{a_k : k \ge n\}$ and the sup is monotone (§1.2), $a_{n+1}^* \le a_n^*$. Moreover, it is clear from the definition that

$$a_n^* = \max(a_n, a_{n+1}^*) \ge a_{n+1}^*, \qquad n \ge 1,$$

holds for every $n \geq 1$. Thus, (a_n^*) is decreasing and $a_n \leq a_n^*$ since $a_n \in \{a_k : k \geq n\}$. Similarly, we set $a_{n*} = \inf\{a_k : k \geq n\}$ for each $n \geq 1$. Then (a_{n*}) is increasing and $a_n \geq a_{n*}$. (a_n^*) is the *upper sequence*, and (a_{n*}) is the *lower sequence* of the sequence (a_n) (Figure 1.2).

Let us look at the sequence (a_n^*) more closely and consider the following question: When might the sup be attained in the definition of a_n^*? To be specific, suppose the sup is attained in a_9^*, i.e., suppose

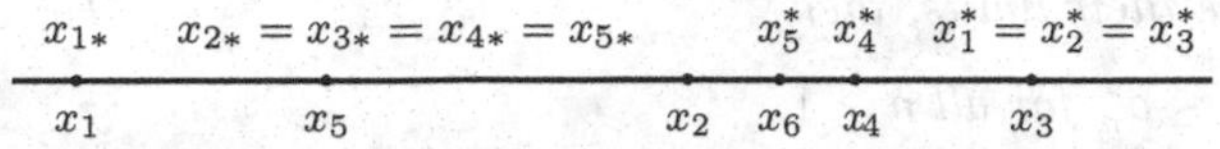

Fig. 1.2 Upper and lower sequences with $x_n = x_6, n \geq 6$

$$a_9^* = \sup\{a_n : n \geq 9\} = \max\{a_n : n \geq 9\}.$$

This means the set $\{a_n : n \geq 9\}$ has a greatest element. Then since $a_8^* = \max(a_8, a_9^*)$, it follows that the set $\{a_n : n \geq 8\}$ has a greatest element or that the sup is attained in a_8^*. Continuing in this way, it follows that all the suprema in a_n^*, for $1 \leq n \leq 9$, are attained. We conclude that if the sup is attained in a_n^* for some particular n, then the sups are attained in a_m^* for all $m < n$. Equivalently, if the sup is *not* attained in a_n^* for a particular n, then the suprema are not attained for all subsequent terms a_m^*, $m > n$.

Now suppose $a_n^* > a_{n+1}^*$ for a particular n, say $a_8^* > a_9^*$. Since $a_8^* = \max(a_8, a_9^*)$, this implies $a_8^* = a_8$, which implies the sup is attained in a_8^*. Equivalently, if the sup is *not* attained in a_8^*, then neither is it attained in a_9^*, a_{10}^*, $\dots$, and moreover we have $a_8^* = a_9^* = a_{10}^* = \dots$.

Summarizing, we conclude: For any sequence (a_n), there is an $1 \leq N \leq \infty$ such that the terms a_n^*, $1 \leq n < N$ are maxima, $a_n^* = \max\{a_k : k \geq n\}$, rather than suprema, and the sequence $(a_N^*, a_{N+1}^*, \dots)$ is constant. When $N = 1$, the whole sequence (a_n^*) is constant, and when $N = \infty$, all terms in the sequence (a_n^*) are maxima.

Let us now return to the main development.

If the sequence (a_n) is any sequence, then the sequences (a_n^*), (a_{n*}) are monotone; hence, they have limits,

$$a_n^* \searrow a^*, \qquad a_{n*} \nearrow a_*.$$

In fact, $a_* \leq a^*$. To see this, fix a natural $N \geq 1$. Then

$$a_{N*} \leq a_{n*} \leq a_n \leq a_n^* \leq a_N^*, \qquad n \geq N.$$

But since (a_{n*}) is increasing, a_{1*}, a_{2*}, $\dots$, a_{N*} are all $\leq a_{N*}$, hence

$$a_{n*} \leq a_N^*, \qquad n \geq 1.$$

Hence, a_N^* is an upper bound for the set $\{a_{n*} : n \geq 1\}$. Since a_* is the sup of this set, we must have $a_* \leq a_N^*$. But this is true for every natural N. Since a^* is the inf of the set $\{a_N^* : N \geq 1\}$, we conclude that $a_* \leq a^*$.

Theorem 1.5.1. *Let (a_n) be a sequence, and let*

$$a_n^* = \sup\{a_k : k \geq n\}, \qquad a_{n*} = \inf\{a_k : k \geq n\}, \qquad n \geq 1.$$

Then (a_n^*) and (a_{n*}) are decreasing and increasing, respectively. Moreover, if a^* and a_* are their limits, then

A. $a_{n*} \leq a_n \leq a_n^*$ *for all* $n \geq 1$,
B. $a_n^* \searrow a^*$,
C. $a_{n*} \nearrow a_*$, *and*
D. $-\infty \leq a_* \leq a^* \leq \infty$.

A sequence (a_n) is *bounded* if $\{a_k : k \geq 1\}$ is a bounded subset of $\mathbf{R}$. Otherwise, (a_n) is *unbounded.* We caution the reader that some of the terms a_n^*, a_{n*}, as well as the limits a^*, a_* may equal $\pm\infty$, when (a_n) is unbounded. Keeping this possibility in mind, the theorem is correct as it stands. □

If the sequence (a_n) happens to be increasing, then $a_n^* = a^*$ and $a_{n*} = a_n$ for all $n \geq 1$. If (a_n) happens to be decreasing, then $a_n^* = a_n$ and $a_{n*} = a_*$ for all $n \geq 1$.

If N is a fixed natural and (a_n) is a sequence, let (a_{N+n}) be the sequence $(a_{N+1}, a_{N+2}, \dots)$. Then $a_n \nearrow a_*$ iff $a_{N+n} \nearrow a_*$, and $a_n \searrow a^*$ iff $a_{N+n} \searrow a^*$. Also by the sup reflection property (§1.2), $b_n = -a_n$ for all $n \geq 1$ implies $b_n^* = -a_{n*}$, $b_{n*} = -a_n^*$ for all $n \geq 1$. Hence, $b^* = -a_*$, $b_* = -a^*$.

Now we define the limit of an arbitrary sequence. Let (a_n) be any sequence, and let (a_n^*), (a_{n*}), a^*, a_* be the upper and lower sequences together with their limits. We call a^* the *upper limit* of the sequence (a_n) and a_* the *lower limit* of the sequence (a_n). If they are equal, $a^* = a_*$, we say that $L = a^* = a_*$ is *the limit* of (a_n), and we write

$$\lim_{n \nearrow \infty} a_n = L.$$

Alternatively, we say a_n *approaches* L or a_n *converges* to L, and we write $a_n \to L$ as $n \nearrow \infty$ or just $a_n \to L$. If they are not equal, $a^* \neq a_*$, we say that (a_n) *does not have a limit.*

Since the upper limit is the limit of suprema, and the lower limit is the limit of infima, they are also called the *limsup* and *liminf* of the sequence, and we write

$$a^* = \limsup_{n\to\infty} a_n, \quad a_* = \liminf_{n\to\infty} a_n.$$

If (a_n) is monotone, let L be its limit as a monotone sequence. Then its upper and lower sequences are equal to itself and the constant sequence $(L, L, \dots)$. Thus, its upper limit is L, and its lower limit is L. Hence L is its limit according to the second definition. In other words, *the two definitions are consistent.*

Clearly a constant sequence $(a, a, a, \ldots)$ approaches a in any of the above senses, as $a_n^* = a$ and $a_{n*} = a$ for all $n \geq 1$.

Let us look at an example. Take $a_n = (-1)^n/n$, $n \geq 1$, or

$$(a_n) = \left(-1, \frac{1}{2}, -\frac{1}{3}, \frac{1}{4}, \ldots\right).$$

Then

$$(a_n^*) = \left(\frac{1}{2}, \frac{1}{2}, \frac{1}{4}, \frac{1}{4}, \ldots\right),$$

$$(a_{n*}) = \left(-1, -\frac{1}{3}, -\frac{1}{3}, -\frac{1}{5}, -\frac{1}{5}, \ldots\right).$$

Hence $a^* = a_* = 0$; thus, $a_n \to 0$.

Not every sequence has a limit. For example, $(1, 0, 1, 0, 1, 0, \ldots)$ does not have a limit. Indeed, here, $a_n^* = 1$ and $a_{n*} = 0$ for all $n \geq 1$, hence $a_* = 0 < 1 = a^*$.

Limits of sequences satisfy simple properties. For example, $a_n \to a$ implies $-a_n \to -a$, and $a_n \to L$ iff $a_{N+n} \to L$. Thus, in a very real sense, the limiting behavior of a sequence does not depend on the first N terms of the sequence, for any $N \geq 1$. Here is the ordering property for sequences.

Theorem 1.5.2. *Suppose that (a_n), (b_n), and (c_n) are sequences with $a_n \leq b_n \leq c_n$ for all $n \geq 1$. If $b_n \to K$ and $c_n \to L$, then $K \leq L$. If $a_n \to L$ and $c_n \to L$, then $b_n \to L$.*

Note that, in the second assertion, the existence of the limit of (b_n) is not assumed, but is rather part of the conclusion. Why is this theorem true? Well, c_1^* is an upper bound for the set $\{c_k : k \geq 1\}$. Since $b_k \leq c_k$ for all k, c_1^* is an upper bound for $\{b_k : k \geq 1\}$. Since b_1^* is the least such, $b_1^* \leq c_1^*$. Repeating this argument with k starting at n, instead of at 1, yields $b_n^* \leq c_n^*$ for all $n \geq 1$. Repeating the same reasoning again yields $b^* \leq c^*$. If $b_n \to K$ and $c_n \to L$, then $b^* = K$ and $c^* = L$, so $K \leq L$, establishing the first assertion. To establish the second, we know that $b^* \leq c^*$. Now set $C_n = -a_n$ and $B_n = -b_n$ for all $n \geq 1$. Then $B_n \leq C_n$ for all $n \geq 1$, so by what we just learned, $B^* \leq C^*$. But $B^* = -b_*$ and $C^* = -a_*$, so $a_* \leq b_*$. We conclude that $a_* \leq b_* \leq b^* \leq c^*$. If $a_n \to L$ and $c_n \to L$, then $a_* = L$ and $c^* = L$, hence $b_* = b^* = L$. □

As an application, $2^{-n} \to 0$ as $n \nearrow \infty$ since $0 < 2^{-n} < 1/n$ for all $n \geq 1$. Similarly, $\lim_{n \nearrow \infty} \left(\frac{1}{n} - \frac{1}{n^2}\right) = 0$ since

$$-\frac{1}{n} \leq -\frac{1}{n^2} \leq \frac{1}{n} - \frac{1}{n^2} \leq \frac{1}{n}$$

for all $n \geq 1$ and $\pm 1/n \to 0$ as $n \nearrow \infty$.

Let (a_n) be a sequence with nonnegative terms. Often the ordering property is used to show that $a_n \to 0$ by finding a sequence (e_n) satisfying $0 \le a_n \le e_n$ for all $n \ge 1$ and $e_n \to 0$.

Below and throughout the text, we will use the following easily checked fact: *If a and b are reals and $a < b + \epsilon$ for all real $\epsilon > 0$, then $a \le b$.* Indeed, either $a \le b$ or $a > b$. If the latter case occurs, we may choose $\epsilon = (a-b)/2 > 0$, yielding the contradiction $a = b + (a-b) > b + \epsilon$. Thus, the former case must occur, or $a \le b$. Moreover, *if a and b are reals and $b \le a < b + \epsilon$ for all $\epsilon > 0$, then $a = b$.*

Throughout the text, ϵ will denote a positive real number.

Theorem 1.5.3. *If $a_n \to a$ and $b_n \to b$ with a, b real, then* $\max(a_n, b_n) \to \max(a, b)$ *and* $\min(a_n, b_n) \to \min(a, b)$. *Moreover, for any sequence (a_n) and L real, $a_n \to L$ iff $a_n - L \to 0$ iff $|a_n - L| \to 0$.*

Let $c_n = \max(a_n, b_n)$, $n \ge 1$, $c = \max(a, b)$, and let us assume, first, that the sequences (a_n), (b_n) are decreasing. Then their limits are their infs, and $c_n = \max(a_n, b_n) \ge \max(a, b) = c$. Hence setting $c_* = \inf\{c_n : n \ge 1\}$, we conclude that $c_* \ge c$. On the other hand, given $\epsilon > 0$, there are n and m satisfying $a_n < a + \epsilon$ and $b_m < b + \epsilon$, so $c_{n+m} = \max(a_{n+m}, b_{n+m}) \le \max(a_n, b_m) < \max(a+\epsilon, b+\epsilon) = c + \epsilon$. Thus, $c_* < c + \epsilon$. Since $\epsilon > 0$ is arbitrary and we already know $c_* \ge c$, we conclude that $c_* = c$. Since (c_n) is decreasing, we have shown that $c_n \to c$.

Now assume (a_n), (b_n) are increasing. Then their limits are their sups, and $c_n = \max(a_n, b_n) \le \max(a, b) = c$. Hence setting $c^* = \sup\{c_n : n \ge 1\}$, we conclude that $c^* \le c$. On the other hand, given $\epsilon > 0$, there are n and m satisfying $a_n > a - \epsilon$ and $b_m > b - \epsilon$, so $c_{n+m} = \max(a_{n+m}, b_{n+m}) \ge \max(a_n, b_m) > \max(a-\epsilon, b-\epsilon) = c - \epsilon$. Thus, $c^* > c - \epsilon$. Since $\epsilon > 0$ is arbitrary and we already know $c^* \le c$, we conclude that $c^* = c$. Since (c_n) is increasing, we have shown that $c_n \to c$.

Now for a general sequence (a_n), we have (a_n^*) decreasing, (a_{n*}) increasing, and

$$\max(a_{n*}, b_{n*}) \le c_n \le \max(a_n^*, b_n^*), \qquad n \ge 1.$$

Thus, (c_n) lies between two sequences converging to $c = \max(a, b)$. By the ordering property, we conclude that $c_n \to \max(a, b)$.

Since $\min(a, b) = -\max(-a, -b)$, the second assertion follows from the first.

For the third assertion, assume, first, $a_n \to L$, and set $b_n = a_n - L$. Since $\sup(A - a) = \sup A - a$ and $\inf(A - a) = \inf A - a$, $b_n^* = a_n^* - L$, and $b_{n*} = a_{n*} - L$. Hence $b^* = a^* - L = 0$, and $b_* = a_* - L = 0$. Thus, $a_n - L \to 0$. If $a_n - L \to 0$, then $L - a_n \to 0$. Hence $|a_n - L| = \max(a_n - L, L - a_n) \to 0$ by the first assertion. Conversely, since

$$-|a_n - L| \le a_n - L \le |a_n - L|, \qquad n \ge 1,$$

$|a_n - L| \to 0$ implies $a_n - L \to 0$, by the ordering property. Since $a_n = (a_n - L) + L$, this implies $a_n \to L$. □

Often this theorem will be used to show that $a_n \to L$ by finding a sequence (e_n) satisfying $|a_n - L| \le e_n$ and $e_n \to 0$. For example, let $A \subset \mathbf{R}$ be bounded above. Then $\sup A - 1/n$ is not an upper bound for A; hence, for each $n \ge 1$, there is a real $x_n \in A$ satisfying $\sup A - 1/n < x_n \le \sup A$, hence $|x_n - \sup A| < 1/n$. By the above, we conclude that $x_n \to \sup A$. When A is not bounded above, for each $n \ge 1$, there is a real $x_n \in A$ satisfying $x_n > n$. Then $x_n \to \infty = \sup A$. In either case, we conclude, if $A \subset \mathbf{R}$, there is a sequence $(x_n) \subset A$ with $x_n \to \sup A$. Similarly, if $A \subset \mathbf{R}$, there is a sequence $(x_n) \subset A$ with $x_n \to \inf A$. We have shown

Theorem 1.5.4. *If A is a subset of $\mathbf{R}$, bounded or unbounded, there is a sequence (x_n) in A converging to $\sup A$, $x_n \to \sup A$, and there is a sequence (x_n) in A converging to $\inf A$, $x_n \to \inf A$.* □

Now we derive the arithmetic properties of limits.

Theorem 1.5.5. *If $a_n \to a$ and c is real, then $ca_n \to ca$. Let a, b be real. If $a_n \to a$, $b_n \to b$, then $a_n + b_n \to a + b$ and $a_n b_n \to ab$. Moreover, if $b \ne 0$, then $b_n \ne 0$ for n sufficiently large and $a_n/b_n \to a/b$.*

If $c = 0$, there is nothing to show. If $c > 0$, set $b_n = ca_n$. Since $\sup(cA) = c \sup A$ and $\inf(cA) = c \inf A$, $b_n^* = ca_n^*$, $b_{n*} = ca_{n*}$, $b^* = ca^*$, $b_* = ca_*$. Hence $a_n \to a$ implies $ca_n \to ca$. Since $(-c)a_n = -(ca_n)$, the case with c negative follows.

To derive the additive property, assume, first, that $a = b = 0$. We have to show that $a_n + b_n \to 0$. Then

$$2\min(a_n, b_n) \le a_n + b_n \le 2\max(a_n, b_n), \quad n \ge 1.$$

Thus, $a_n + b_n$ lies between two sequences approaching 0, so $a_n + b_n \to 0$. For general a, b, apply the previous to $a_n' = a_n - a$, $b_n' = b_n - b$.

To derive the multiplicative property, first, note that $a_{1*} \le a_n \le a_1^*$, so $|a_n| \le k$ for some k, i.e., (a_n) is bounded. Use the triangle inequality to get

$$\begin{aligned} |a_n b_n - ab| = |(a_n - a)b + (b_n - b)a_n| &\le |b|\,|a_n - a| + |a_n|\,|b_n - b| \\ &\le |b|\,|a_n - a| + k|b_n - b|, \quad n \ge 1. \end{aligned}$$

Now the result follows from the additive and ordering properties.

To obtain the division property, assume $b > 0$. From the above, $a_n b - ab_n \to 0$. Since $b_n \to b$, $b_{n*} \nearrow b$, so there exists $N \ge 1$ beyond which $b_n \ge b_{N*} > 0$ for $n \ge N$. Thus,

$$0 \le \left| \frac{a_n}{b_n} - \frac{a}{b} \right| = \frac{|a_n b - ab_n|}{|b_n|\,|b|} \le \frac{|a_n b - ab_n|}{b_{N*}\, b}, \qquad n \ge N.$$

Thus, $|a_n/b_n - a/b|$ lies between zero and a sequence approaching zero. The case $b < 0$ is entirely similar. □

In fact, although we do not derive this, this theorem remains true when a or b is infinite, as long as we do not allow undefined expressions, such as $\infty - \infty$ (the allowable expressions are defined in §1.2).

As an application of this theorem,

$$\lim_{n\nearrow\infty} \frac{2n^2+1}{n^2-2n+1} = \lim_{n\nearrow\infty} \frac{2+\frac{1}{n^2}}{1-\frac{2}{n}+\frac{1}{n^2}} = \frac{2+0}{1-2\cdot 0+0} = 2.$$

If (a_n) is a sequence with positive terms and $b_n = 1/a_n$, then $a_n \to 0$ iff $b_n \to \infty$ (Exercise **1.5.3**). Now let $a > 1$ and set $b = a - 1$. Then $a^n = (1+b)^n \geq 1+nb$ for all $n \geq 1$ (Exercise **1.3.22**). Hence $a^n \nearrow \infty$. If $0 < a < 1$, then $a = 1/b$ with $b > 1$, so $a^n = 1/b^n \searrow 0$. Summarizing,

A. If $a > 1$, then $a^n \nearrow \infty$,
B. If $a = 1$, then $a^n = 1$ for all $n \geq 1$, and
C. If $0 \leq a < 1$, then $a^n \to 0$.

Sometimes we say that a sequence (a_n) *converges to* L if $a_n \to L$. If the specific limit is not relevant, we say that the sequence *converges* or is *convergent*. If a sequence has no limit, we say it *diverges*. More precisely, if the sequence (a_n) does not approach L, we say that it *diverges from* L, and we write $a_n \not\to L$. *From the definition of* $a_n \to L$*, we see that* $a_n \not\to L$ *means either* $a^* \neq L$ *or* $a_* \neq L$. This is so whether L is real or $\pm\infty$.

Typically, divergence is *oscillatory behavior*, e.g., $a_n = (-1)^n$. Here the sequence goes back and forth never settling on anything, not even ∞ or $-\infty$. Nevertheless, this sequence is bounded. Of course a sequence may be oscillatory and unbounded, e.g., $a_n = (-1)^n n$.

Let (a_n) be a sequence, and suppose that $1 \leq k_1 < k_2 < k_3 < \dots$ is an increasing sequence of distinct *naturals*. Set $b_n = a_{k_n}, n \geq 1$. Then $(b_n) = (a_{k_n})$ is a *subsequence* of (a_n). If $a_n \to L$, then $a_{k_n} \to L$. Conversely, if (a_n) is monotone, $a_{k_n} \to L$ implies $a_n \to L$ (Exercise **1.5.4**).

Generally, if a sequence (x_n) has a subsequence (x_{k_n}) converging to L, we say that (x_n) *subconverges to* L.

Let (a_n) converge to a (finite) real limit L, and let $\epsilon > 0$ be given. Since (a_{n*}) is increasing to L, there must exist a natural N_*, such that $a_{n*} > L - \epsilon$ for $n \geq N_*$. Similarly, there must exist N^* beyond which we have $a_n^* < L+\epsilon$. Since $a_{n*} \leq a_n \leq a_n^*$ for all $n \geq 1$, we obtain $L - \epsilon < a_n < L + \epsilon$ for $n \geq N = \max(N^*, N_*)$. Thus, all but finitely many terms of the sequence lie in $(L-\epsilon, L+\epsilon)$ (Figure 1.3).

Note that choosing a smaller $\epsilon > 0$ is a more stringent condition on the terms. As such, it leads to (in general) a larger N, i.e., the number of terms that fall outside the interval $(L-\epsilon, L+\epsilon)$ depends on the choice of $\epsilon > 0$.

Conversely, suppose that $L - \epsilon < a_n < L + \epsilon$ for all but finitely many terms, for every $\epsilon > 0$. Then for a given $\epsilon > 0$, by the ordering property,

$L - \epsilon \leq a_{n*} \leq a_n^* \leq L + \epsilon$ for all but finitely terms. Hence $L - \epsilon \leq a_* \leq a^* \leq L + \epsilon$. Since this holds for every $\epsilon > 0$, we conclude that $a_* = a^* = L$, i.e., $a_n \to L$. We have derived the following:

Theorem 1.5.6. *Let (a_n) be a sequence and let L be real. If $a_n \to L$, then all but finitely many terms of the sequence lie within the interval $(L - \epsilon, L + \epsilon)$, for all $\epsilon > 0$. Conversely, if all but finitely many terms lie in the interval $(L - \epsilon, L + \epsilon)$, for all $\epsilon > 0$, then $a_n \to L$.* □

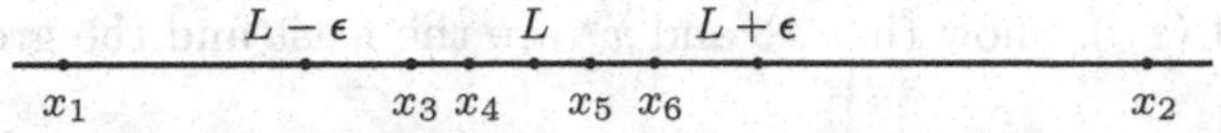

Fig. 1.3 Convergence to L

From this, we conclude that *if $a_n \to L$ and $L \neq 0$, then $a_n \neq 0$ for all but finitely many n.*

Here is a useful application of the above.

Theorem 1.5.7. *Suppose that $f : \mathbf{N} \to \mathbf{N}$ is injective, i.e., suppose that $(f(n)) = (a_n)$ is a sequence consisting of* distinct *naturals. Then $f(n) \to \infty$.*

To see this, let $k_n = \max\{k : f(k) < n\}$ (Exercise **1.3.19**). Then $k > k_n$ implies $f(k) \geq n$ which implies $f_*(k_n + 1) = \inf\{f(k) : k > k_n\} \geq n$. Hence $f_*(k_n + 1) \nearrow \infty$, as $n \nearrow \infty$. Since $(f_*(n))$ is monotone and $(f_*(k_n + 1))$ is a subsequence of $(f_*(n))$, it follows that $f_*(n) \nearrow \infty$, as $n \nearrow \infty$ (Exercise **1.5.4**). Since $f(n) \geq f_*(n)$, we conclude that $f(n) \to \infty$. □

Exercises

1.5.1. Fix $N \geq 1$ and (a_n). Let (a_{N+n}) be the sequence $(a_{N+1}, a_{N+2}, \dots)$. Then $a_n \nearrow L$ iff $a_{N+n} \nearrow L$, and $a_n \searrow L$ iff $a_{N+n} \searrow L$. Conclude that $a_n \to L$ iff $a_{n+N} \to L$.

1.5.2. If $a_n \to L$, then $-a_n \to -L$.

1.5.3. If $A \subset \mathbf{R}^+$ is nonempty and $1/A = \{1/x : x \in A\}$, then $\inf(1/A) = 1/\sup A$, where $1/\infty$ is interpreted here as 0. If (a_n) is a sequence with positive terms and $b_n = 1/a_n$, then $a_n \to 0$ iff $b_n \to \infty$.

1.5.4. If $a_n \to L$ and (a_{k_n}) is a subsequence, then $a_{k_n} \to L$. If (a_n) is monotone and $a_{k_n} \to L$, then $a_n \to L$.

1.5.5. If $a_n \to L$ and $L \neq 0$, then $a_n \neq 0$ for all but finitely many n.

1.5.6. Let $a_n = \sqrt{n+1} - \sqrt{n}$, $n \ge 1$. Compute (a_n^*), (a_{n*}), a^*, and a_*. Does (a_n) converge?

1.5.7. Let (a_n) be any sequence with upper and lower limits a^* and a_*. Show that (a_n) subconverges to a^* and subconverges to a_*, i.e., there are subsequences (a_{k_n}) and (a_{j_n}) satisfying $a_{k_n} \to a^*$ and $a_{j_n} \to a_*$.

1.5.8. Suppose that (a_n) diverges from $L \in \mathbf{R}$. Show that there is an $\epsilon > 0$ and a subsequence (a_{k_n}) satisfying $|a_{k_n} - L| \ge \epsilon$ for all $n \ge 1$.

1.5.9. Let (x_n) be a sequence. If (x_n) subconverges to L, we say that L is a *limit point* of (x_n). Show that x_* and x^* are the least and the greatest limit points.

1.5.10. Show that a sequence (x_n) converges iff (x_n) has exactly one limit point.

1.5.11. Given $f : (a, b) \to \mathbf{R}$, let $M = \sup\{f(x) : a < x < b\}$. Show that there is a sequence (x_n) with $f(x_n) \to M$. (Consider the cases $M < \infty$ and $M = \infty$.)

1.5.12. Define (d_n) by $d_1 = 2$ and

$$d_{n+1} = \frac{1}{2}\left(d_n + \frac{2}{d_n}\right), \qquad n \ge 1,$$

and set $e_n = d_n - \sqrt{2}$, $n \ge 1$. By induction, show that $e_n \ge 0$ for $n \ge 1$. Also show that

$$e_{n+1} \le \frac{e_n^2}{2\sqrt{2}}, \qquad \text{for all } n \ge 1.$$

(First, check that, for any real $x > 0$, one has $(x + 2/x)/2 \ge \sqrt{2}$.)

1.5.13. Let $0 < x < 1$ be irrational, and let (a_n) be as in Exercise **1.3.18**. Let

$$x_n = \cfrac{1}{a_1 + \cfrac{1}{\ddots \cfrac{1}{a_{n-1} + \cfrac{1}{a_n}}}}.$$

Let x' and x'_n be the unique reals satisfying $x = 1/(a_1 + x')$ and $x_n = 1/(a_1 + x'_n)$, respectively. Then $0 < x_n, x', x'_n < 1$ and x' and x'_n are obtained by "peeling off the top layer." Similarly, let $x^{(2)} = x'' = (x')'$, $x^{(3)} = (x^{(2)})'$, ..., $x_n^{(2)} = x''_n = (x'_n)'$, $x_n^{(3)} = (x_n^{(2)})'$, ... Then $x_n^{(n-1)} = 1/a_n$, $n \ge 1$.

A. Show that $|x - x_n| = x x_n |x' - x'_n|$, $n \ge 2$.
B. Iterate **A** to show that $|x - x_n| \le 1/a_n$, $n \ge 1$.

C. Show that $x \le (a_2+1)/(a_2+2)$, $x' \le (a_3+1)/(a_3+2)$, etc.
D. If N of the a_k's are bounded by c, iterate **A** and use **C** to obtain

$$|x - x_n| \le \left(\frac{c+1}{c+2}\right)^n,$$

for all but finitely many n.

Conclude that $|x - x_n| \to 0$ as $n \nearrow \infty$ (either $a_n \to \infty$ or $a_n \not\to \infty$).

1.6 Nonnegative Series and Decimal Expansions

Let (a_n) be a sequence of reals. The *series* formed from the sequence (a_n) is the sequence (s_n) with terms $s_1 = a_1$, $s_2 = a_1 + a_2$, and for any $n \ge 1$, $s_n = a_1 + a_2 + \cdots + a_n$ (this sum was defined in §1.3). The sequence (s_n) is the sequence of *partial sums*. The terms a_n are called *summands*, and the series is *nonnegative* if $a_n \ge 0$ for all $n \ge 1$. We often use sigma notation, and write $s_n = \sum_{k=1}^n a_k$. Series are often written

$$a_1 + a_2 + \dots.$$

In sigma notation, $\sum a_n$ or $\sum_{n=1}^\infty a_n$. If the sequence of partial sums (s_n) has a limit L, then we say the series *sums* or *converges* to L, and we write

$$L = a_1 + a_2 + \cdots = \sum_{n=1}^\infty a_n.$$

Then L is the *sum* of the series. *By convention, we do not allow* $\pm\infty$ *as limits for series, only reals.* Nevertheless, for nonnegative series, we write $\sum a_n = \infty$ to mean $\sum a_n$ diverges and $\sum a_n < \infty$ to mean $\sum a_n$ converges. As with sequences, sometimes it is more convenient to start a series from $n = 0$. In this case, we write $\sum_{n=0}^\infty a_n$.

Let $L = \sum_{n=1}^\infty a_n$ be a convergent series and let s_n denote its nth partial sum. The *nth tail* of the series is $L - s_n = \sum_{k=n+1}^\infty a_k$. Since the nth tail is the difference between the nth partial sum and the sum, we see that *the nth tail of a convergent series goes to zero*:

$$\lim_{n\nearrow\infty} \sum_{k=n+1}^\infty a_k = 0. \tag{1.6.1}$$

Let a be real. Our first series is the *geometric* series

$$1 + a + a^2 + \cdots = \sum_{n=0}^\infty a^n.$$

Here the nth partial sum $s_n = 1 + a + \cdots + a^n$ is computed as follows:

$$as_n = a(1 + a + \cdots + a^n) = a + a^2 + \cdots + a^{n+1} = s_n + a^{n+1} - 1.$$

Hence

$$s_n = \frac{1 - a^{n+1}}{1 - a}, \qquad a \neq 1.$$

If $a = 1$, then $s_n = n$, so $s_n \nearrow \infty$. If $|a| < 1$, $a^n \to 0$, so $s_n \to 1/(1-a)$. If $a > 1$, then $a^n \nearrow \infty$, so the series equals ∞ and hence diverges. If $a < -1$, then (a^n) diverges, so the series diverges. If $a = -1$, s_n equals 0 or 1 (depending on n), hence diverges; hence, the series diverges. We have shown

$$\sum_{n=0}^{\infty} a^n = \frac{1}{1-a}, \qquad \text{if } |a| < 1,$$

and $\sum_{n=0}^{\infty} a^n$ diverges if $|a| \geq 1$.

To study more general series, we need their arithmetic and ordering properties.

Theorem 1.6.1. *If $\sum a_n = L$ and $\sum b_n = M$, then $\sum(a_n + b_n) = L + M$. If $\sum a_n = L$, $c \in \mathbf{R}$, and $b_n = ca_n$, then $\sum b_n = cL = c(\sum a_n)$. If $a_n \leq b_n \leq c_n$ and $\sum a_n = L = \sum c_n$, then $\sum b_n = L$.*

To see the first property, if s_n, t_n, and r_n denote the partial sums of $\sum a_n$, $\sum b_n$, and $\sum c_n$, then $s_n + t_n$ equals the partial sum of $\sum(a_n + b_n)$. Hence the result follows from the corresponding arithmetic property of sequences. For the second property, note that $t_n = cs_n$. Hence the result follows from the corresponding arithmetic property of sequences. The third property follows from the ordering property of sequences, since $s_n \leq t_n \leq r_n$, $s_n \to L$, and $r_n \to L$. □

Now we describe the comparison test which we use below to obtain the decimal expansions of reals.

Theorem 1.6.2 (Comparison Test). *Let $\sum a_n$, $\sum b_n$ be nonnegative series with $a_n \leq b_n$ for all $n \geq 1$. If $\sum b_n < \infty$, then $\sum a_n < \infty$. If $\sum a_n = \infty$, then $\sum b_n = \infty$.*

Stated this way, the theorem follows from the ordering property for sequences and looks too simple to be of any serious use. □ In fact, we use it to express every real as a sequence of naturals.

Theorem 1.6.3. *Let $\mathbf{b} = 9 + 1$. If $d_1, d_2, \ldots$ is a sequence of digits (§1.3), then*

$$\sum_{n=1}^{\infty} d_n \mathbf{b}^{-n}$$

sums to a real x, $0 \leq x \leq 1$. Conversely, if $0 \leq x \leq 1$, there is a sequence of digits $d_1, d_2, \ldots$, such that the series sums to x.

The first statement follows by comparison, since

$$\sum_{n=1}^{\infty} d_n \mathbf{b}^{-n} \leq \sum_{n=1}^{\infty} 9\mathbf{b}^{-n}$$
$$= \sum_{n=0}^{\infty} \frac{9}{\mathbf{b}} \mathbf{b}^{-n} = \frac{9}{\mathbf{b}} \sum_{n=0}^{\infty} \mathbf{b}^{-n} = \frac{9}{\mathbf{b}} \cdot \frac{1}{1-(1/\mathbf{b})} = 1.$$

To establish the second statement, if $x = 1$, we simply take $d_n = 9$ for all $n \geq 1$. If $0 \leq x < 1$, let d_1 be the largest integer $\leq x\mathbf{b}$. Then $d_1 \geq 0$. This way we obtain a digit d_1 (since $x < 1$) satisfying $d_1 \leq x\mathbf{b} < d_1 + 1$. Now set $x_1 = x\mathbf{b} - d_1$. Then $0 \leq x_1 < 1$. Repeating the above process, we obtain a digit d_2 satisfying $d_2 \leq x_1\mathbf{b} < d_2 + 1$. Substituting yields $d_2 + \mathbf{b}d_1 \leq \mathbf{b}^2 x < d_2 + \mathbf{b}d_1 + 1$ or $d_2\mathbf{b}^{-2} + d_1\mathbf{b}^{-1} \leq x < d_2\mathbf{b}^{-2} + d_1\mathbf{b}^{-1} + \mathbf{b}^{-2}$. Continuing in this manner yields a sequence of digits (d_n) satisfying

$$\left(\sum_{k=1}^{n} d_k \mathbf{b}^{-k}\right) \leq x < \left(\sum_{k=1}^{n} d_k \mathbf{b}^{-k}\right) + \mathbf{b}^{-n}, \quad n \geq 1.$$

Thus, x lies between two sequences converging to the same limit. □

The sequence (d_n) is the *decimal expansion* of the real $0 \leq x \leq 1$. As usual, we write

$$x = .d_1 d_2 d_3 \dots .$$

To extend the decimal notation to any nonnegative real, for each $x \geq 1$, there is a smallest natural N, such that $\mathbf{b}^{-N} x < 1$. As usual, if $\mathbf{b}^{-N} x = .d_1 d_2 \dots$, we write

$$x = d_1 d_2 \dots d_N . d_{N+1} d_{N+2} \dots ,$$

the *decimal point* (.) moved N places. For example, $1 = 1.00\dots$ and $\mathbf{b} = 10.00\dots$. In fact, x is a natural iff $x = d_1 d_2 \dots d_N .00\dots$. Thus, for naturals, we drop the decimal point and the trailing zeros, e.g., $1 = 1$, $\mathbf{b} = 10$.

As an illustration, $x = 10/8 = 1.25$ since $x/10 = 1/8 < 1$, and $y = 1/8 = .125$ since $1 \leq 10y < 2$ so $d_1 = 1$, $z = 10y - 1 = 10/8 - 1 = 2/8$ satisfies $2 \leq 10z < 3$ so $d_2 = 2$, $t = 10z - d_2 = 10 * 2/8 - 2 = 4/8$ satisfies $5 = 10t < 6$ so $d_3 = 5$ and $d_4 = 0$.

Note that we have two decimal representations of 1, $1 = .99\cdots = 1.00\dots$. This is not an accident. In fact, two distinct sequences of digits yield the same real in $[0, 1]$ under only very special circumstances (Exercise **1.6.2**).

The natural $\mathbf{b}$, the *base* of the expansion, can be replaced by any natural > 1. Then the digits are $(0, 1, \dots, \mathbf{b} - 1)$, and we would obtain $\mathbf{b}$-*ary expansions*. In §4.1, we use $\mathbf{b} = 2$ with digits $(0, 1)$ leading to *binary expansions* and $\mathbf{b} = 3$ with digits $(0, 1, 2)$ leading to *ternary expansions*. In §5.2, we mention $\mathbf{b} = 16$ with digits $(0, 1, \dots, 9, a, b, c, d, e, f)$ leading to *hexadecimal expansions*. This completes our discussion of decimal expansions.

How can one tell if a given series converges by inspecting the individual terms? Here is a necessary condition.

Theorem 1.6.4 (nth-Term Test). *If $\sum a_n = L \in \mathbf{R}$, then $a_n \to 0$.*

In particular, it follows that the terms of a series must be bounded: There is a real $C > 0$ with $|a_n| \le C$ for all $n \ge 1$.

To see this, we know that $s_n \to L$, and so $s_{n-1} \to L$. By the triangle inequality,

$$\begin{aligned}|a_n| &= |s_n - s_{n-1}| \\ &= |(s_n - L) + (L - s_{n-1})| \\ &\le |s_n - L| + |s_{n-1} - L| \to 0.\square\end{aligned}$$

However, a series whose nth term approaches zero need not converge. For example, the *harmonic* series

$$\sum_{n=1}^{\infty} \frac{1}{n} = 1 + \frac{1}{2} + \frac{1}{3} + \cdots = \infty.$$

To see this, use comparison as follows,

$$\begin{aligned}\sum_{n=1}^{\infty} \frac{1}{n} &= 1 + \frac{1}{2} + \frac{1}{3} + \frac{1}{4} + \frac{1}{5} + \frac{1}{6} + \frac{1}{7} + \frac{1}{8} + \dots \\ &\ge 1 + \frac{1}{2} + \frac{1}{4} + \frac{1}{4} + \frac{1}{8} + \frac{1}{8} + \frac{1}{8} + \frac{1}{8} + \dots \\ &= 1 + \frac{1}{2} + \frac{1}{2} + \frac{1}{2} + \cdots = \infty.\end{aligned}$$

On the other hand, the nonnegative series

$$\frac{1}{0!} + \frac{1}{1!} + \frac{1}{2!} + \frac{1}{3!} + \dots$$

converges. To see this, check that $2^{n-1} \le n!$ by induction. Thus,

$$1 + \sum_{n=1}^{\infty} \frac{1}{n!} \le 1 + \sum_{n=1}^{\infty} 2^{-n+1} = 3$$

and hence is convergent. Since the third partial sum is $s_3 = 2.5$, we see that the sum lies in the interval $(2.5, 3]$.

A series is *telescoping* if it is a sum of differences, i.e., of the form

$$\sum_{n=1}^{\infty} (a_n - a_{n+1}) = (a_1 - a_2) + (a_2 - a_3) + (a_3 - a_4) + \dots.$$

In this case, the following is true.

Theorem 1.6.5. *If (a_n) is any sequence converging to zero, then the corresponding telescoping series converges, and its sum is a_1.*

This follows since the partial sums are

$$s_n = (a_1 - a_2) + (a_2 - a_3) + \cdots + (a_n - a_{n+1}) = a_1 - a_{n+1}$$

and $a_{n+1} \to 0$. □

As an application, note that

$$\frac{1}{1\cdot 2} + \frac{1}{2\cdot 3} + \frac{1}{3\cdot 4} + \cdots = 1$$

since

$$\sum_{n=1}^{\infty} \frac{1}{n(n+1)} = \sum_{n=1}^{\infty}\left(\frac{1}{n} - \frac{1}{n+1}\right) = 1.$$

Another application is to establish the convergence of

$$\sum_{n=1}^{\infty} \frac{1}{n^2} = 1 + \sum_{n=2}^{\infty} \frac{1}{n^2} < 1 + \sum_{n=2}^{\infty} \frac{1}{n(n-1)} = 1 + \sum_{n=1}^{\infty} \frac{1}{n(n+1)} = 2.$$

Thus,

$$\frac{1}{1^2} + \frac{1}{2^2} + \frac{1}{3^2} + \cdots < 2.$$

Expressing this sum in terms of familiar quantities is a question of a totally different magnitude. Later (§5.6), we will see how this is done.

More generally,

$$\sum_{n=1}^{\infty} \frac{1}{n^{1+1/N}} < \infty$$

follows in a similar manner, for any $N \geq 1$.

Exercises

1.6.1. Let $0 < x < 1$ be real. Then $x = .d_1 d_2 \ldots$ is in $\mathbf{Q}$ iff there are $n, m \geq 1$, such that $d_{j+n} = d_j$ for $j > m$, i.e., the sequence of digits repeats every n digits, from the $(m+1)$st digit on.

1.6.2. Suppose that $.d_1 d_2 \cdots = .e_1 e_2 \ldots$ are distinct decimal expansions for the same real, and let N be the first natural with $d_N \neq e_N$. Then either $d_N = e_N + 1$, $d_k = 0$, and $e_k = 9$ for $k > N$, or $e_N = d_N + 1$, $e_k = 0$, and $d_k = 9$ for $k > N$. Conclude that $x > 0$ has more than one decimal expansion iff $10^N x \in \mathbf{N}$ for some $N \in \mathbf{N} \cup \{0\}$.

1.6.3. Fix $N \geq 1$. Show that

A. $\left(\frac{n}{n+1}\right)^{1/N} \leq 1 - \frac{1}{N(n+1)}$, $n \geq 1$,

B. $(n+1)^{1/N} - n^{1/N} \geq \frac{1}{N(n+1)^{(N-1)/N}}$, $n \geq 1$, and

C. $\sum_{n=2}^{\infty} \frac{1}{n^{1+1/N}} \leq N \sum_{n=1}^{\infty}(a_n - a_{n+1})$ where $a_n = 1/(n^{1/N})$,$n \geq 1$.

Conclude that $\sum_{n=1}^{\infty} \frac{1}{n^{1+1/N}} < \infty$. (Use Exercises **1.3.22** and **1.4.6** for **A**.)

1.6.4. Let (d_n) and (e_n) be as in Exercise **1.5.12**. By induction, show that

$$e_{n+2} \leq 10^{-2^n}, \qquad n \geq 1.$$

This shows that the decimal expansion of d_{n+2} agrees[8] with that of $\sqrt{2}$ to 2^n places. For example, d_9 yields $\sqrt{2}$ to at least 128 decimal places. (First, show that $e_3 \leq 1/100$. Since the point here is to compute the decimal expansion of $\sqrt{2}$, do not use it in your derivation. Use only $1 < \sqrt{2} < 2$ and $(\sqrt{2})^2 = 2$.)

1.6.5. Let $C \subset [0,1]$ be the set of reals $x = .d_1d_2d_3\ldots$ whose decimal digits $d_n, n \geq 1$, are zero or odd. Show that (§1.2) $C + C = [0,2]$.

1.7 Signed Series and Cauchy Sequences

A series is *signed* if its first term is positive and at least one of its terms is negative. A series is *alternating* if it is of the form

$$\sum_{n=1}^{\infty}(-1)^{n-1}a_n = a_1 - a_2 + a_3 \cdots + (-1)^{n-1}a_n + \ldots$$

with a_n positive for all $n \geq 1$. Alternating series are particularly tractable, but, first, we need a new concept.

A sequence (not a series!) (a_n) is *Cauchy* if its terms approach each other, i.e., if $|a_m - a_k|$ is small when m and k are large. We make this precise by defining (a_n) to be Cauchy if there is a positive sequence (e_n) converging to zero, such that

$$|a_m - a_k| \leq e_n, \qquad \text{for all } m, k \geq n, \text{ for all } n \geq 1.$$

If a sequence is Cauchy, there are many choices for (e_n). Any such sequence (e_n) is an *error sequence* for the Cauchy sequence (a_n).

[8] This algorithm was known to the Babylonians.

It follows from the definition that every Cauchy sequence is bounded, $|a_m| \le |a_m - a_1| + |a_1| \le e_1 + |a_1|$ for all $m \ge 1$.

It is easy to see that a convergent sequence is Cauchy. Indeed, if (a_n) converges to L, then $b_n = |a_n - L| \to 0$, so (S1.5), $b_n^* \to 0$. Hence by the triangle inequality

$$|a_m - a_k| \le |a_m - L| + |a_k - L| \le b_n^* + b_n^*, \quad m, k \ge n \ge 1.$$

Since $2b_n^* \to 0$, $(2b_n^*)$ is an error sequence for (a_n), so (a_n) is Cauchy.

The following theorem shows that if the terms of a sequence "approach each other," then they "approach something." To see that this is not a self-evident assertion, consider the following example. Let a_n be the rational given by the first n places in the decimal expansion of $\sqrt{2}$. Then $|a_n - \sqrt{2}| \le 10^{-n}$, hence $a_n \to \sqrt{2}$, hence (a_n) is Cauchy. But, as far as $\mathbf{Q}$ is concerned, there is no limit, since $\sqrt{2} \notin \mathbf{Q}$. In other words, to actually establish the existence of the limit, one needs an additional property not enjoyed by $\mathbf{Q}$, the completeness property of $\mathbf{R}$.

Theorem 1.7.1. *A Cauchy sequence (a_n) is convergent.*

With the notation of §1.5, we need to show that $a_* = a^*$. But this follows since the sequence is Cauchy. Indeed, let (e_n) be any error sequence. Then for all $n \ge 1$, $m \ge n$, $j \ge n$, we have $a_m - a_j \le e_n$. For n and j fixed, this inequality is true for all $m \ge n$. Taking the sup over all $m \ge n$ yields

$$a_n^* - a_j \le e_n$$

for all $j \ge n \ge 1$. Now for n fixed, this inequality is true for all $j \ge n$. Taking the sup over all $j \ge n$ and using $\sup(-A) = -\inf A$ yields

$$0 \le a_n^* - a_{n*} \le e_n, \qquad n \ge 1.$$

Letting $n \nearrow \infty$ yields $0 \le a^* - a_* \le 0$, hence $a^* = a_*$. □

A series $\sum a_n$ is said to be *absolutely convergent* if $\sum |a_n|$ converges. For example, below, we will see that $\sum(-1)^{n-1}/n$ converges. Since $\sum 1/n$ diverges, however, $\sum(-1)^{n-1}/n$ does not converge absolutely. A convergent series that is not absolutely convergent is *conditionally convergent*. On the other hand, every nonnegative convergent series is absolutely convergent.

Let $\sum a_n$ be an absolutely convergent series and let $e_n = \sum_{k=n}^{\infty} |a_k|$. Since e_n is the tail of a convergent series, we have $e_n \to 0$. *Then (e_n) is an error sequence for the sequence (s_n) of partial sums of $\sum a_n$,* because

$$|s_m - s_k| = |a_{k+1} + \cdots + a_m| \le |a_{k+1}| + \cdots + |a_m| \le e_n, \qquad m > k \ge n.$$

Thus, (s_n) is Cauchy; hence, (s_n) is convergent. We have shown

Theorem 1.7.2. *If $\sum |a_n|$ converges, then $\sum a_n$ converges, and*

$$\left|\sum a_n\right| \le \sum |a_n|.$$

The inequality follows from the triangle inequality. □

A typical application of this result is as follows.

If (a_n) is a sequence of positive reals decreasing to zero and (b_n) is bounded, then

$$\sum_{n=1}^{\infty}(a_n - a_{n+1})b_n \tag{1.7.1}$$

converges absolutely. Indeed, if $|b_n| \le C$, $n \ge 1$, is a bound for (b_n), then

$$\sum_{n=1}^{\infty}|(a_n - a_{n+1})b_n| \le \sum_{n=1}^{\infty}(a_n - a_{n+1})C = Ca_1 < \infty$$

since the last series is telescoping.

To extend the scope of this last result, we will need the following elementary formula:

$$a_1b_1 + \sum_{n=2}^{N} a_n(b_n - b_{n-1}) = \sum_{n=1}^{N-1}(a_n - a_{n+1})b_n + a_Nb_N. \tag{1.7.2}$$

This important identity, easily verified by decomposing the sums, is called *summation by parts*.

Theorem 1.7.3 (Dirichlet Test). *If (a_n) is a positive sequence decreasing to zero and (c_n) is such that the sequence $b_n = c_1 + c_2 + \cdots + c_n$, $n \ge 1$, is bounded, then $\sum_{n=1}^{\infty} a_nc_n$ converges and*

$$\sum_{n=1}^{\infty} a_nc_n = \sum_{n=1}^{\infty}(a_n - a_{n+1})b_n. \tag{1.7.3}$$

This is an immediate consequence of letting $N \nearrow \infty$ in (1.7.2) since $b_n - b_{n-1} = c_n$ for $n \ge 2$. □ An important aspect of the Dirichlet test is that the right side of (1.7.3) is, from above, absolutely convergent, whereas the left side is often only conditionally convergent. For example, taking $a_n = 1/n$ and $c_n = (-1)^{n-1}$, $n \ge 1$, yields $(b_n) = (1, 0, 1, 0, \dots)$. Hence we conclude not only that

$$1 - \frac{1}{2} + \frac{1}{3} - \frac{1}{4} + \dots$$

converges but also that its sum equals the sum of the absolutely convergent series obtained by grouping the terms in pairs.

Now we can state the situation with alternating series.

Theorem 1.7.4 (Leibnitz Test). *If (a_n) is a positive decreasing sequence with $a_n \searrow 0$, then*

$$a_1 - a_2 + a_3 - a_4 + \dots \tag{1.7.4}$$

converges to a limit L satisfying $0 \le L \le a_1$. If, in addition, (a_n) is strictly decreasing, then $0 < L < a_1$. Moreover, if s_n denotes the nth partial sum, $n \ge 1$, then the error $|L - s_n|$ at the nth stage is no greater than the $(n+1)$st term a_{n+1}, with $L \ge s_n$ or $L \le s_n$ according to whether the $(n+1)$st term is added or subtracted.

For example,

$$L = 1 - \frac{1}{3} + \frac{1}{5} - \frac{1}{7} + \frac{1}{9} - \cdots \tag{1.7.5}$$

converges, and $0 < L < 1$. In fact, since $s_2 = 2/3$ and $s_3 = 13/15$, $2/3 < L < 13/15$.

In the previous section, estimating the sum of

$$1 + \frac{1}{1!} + \frac{1}{2!} + \frac{1}{3!} + \dots$$

to one decimal place involved estimating the entire series. Here the situation is markedly different: The absolute error between the sum and the nth partial sum is no larger than the next term a_{n+1}.

To derive the Leibnitz test, clearly, the convergence of (1.7.4) follows by taking $c_n = (-1)^{n-1}$ and applying the Dirichlet test, as above. Now the differences $a_n - a_{n+1}$, $n = 1, 3, 5, \dots$, are nonnegative. Grouping the terms in (1.7.4) in pairs, we obtain $L \ge 0$. Similarly, the differences $-a_n + a_{n+1}$, $n = 2, 4, 6, \dots$, are nonpositive. Grouping the terms in (1.7.4) in pairs, we obtain $L \le a_1$. Thus, $0 \le L \le a_1$. But

$$(-1)^n(L - s_n) = a_{n+1} - a_{n+2} + a_{n+3} - a_{n+4} + \dots, \qquad n \ge 1.$$

Repeating the above reasoning, we obtain $0 \le (-1)^n(L - s_n) \le a_{n+1}$, which implies the rest of the statement. If, in addition, (a_n) is strictly decreasing, this reasoning yields $0 < L < a_1$. □

If $a_1 + a_2 + a_3 + \dots$ is absolutely convergent, its *alternating version* is $a_1 - a_2 + a_3 - \dots$. For example, the alternating version of

$$\frac{1}{1-x} = 1 + x + x^2 + \dots$$

equals

$$\frac{1}{1+x} = 1 - x + x^2 - \dots.$$

Clearly, the alternating version is also absolutely convergent and the alternating version of the alternating version of a series is itself. Note that the alternating version of a series $\sum a_n$ need not be an alternating series. This happens iff $\sum a_n$ is nonnegative.

Our next topic is the rearrangement of series. A series $\sum A_n$ is a *rearrangement* of a series $\sum a_n$ if there is a bijection (§1.1) $f : \mathbf{N} \to \mathbf{N}$, such that $A_n = a_{f(n)}$ for all $n \geq 1$.

Theorem 1.7.5. *If $\sum a_n$ is nonnegative, any rearrangement $\sum A_n$ converges to the same limit. If $\sum a_n$ is absolutely convergent, any rearrangement $\sum A_n$ converges absolutely to the same limit.*

Nonnegative convergence and absolute convergence are stated separately because the case $\sum a_n = \infty$ is included in nonnegative convergence.

To see this, first assume $\sum |a_n| < \infty$, and let $e_n = \sum_{j=n}^{\infty} |a_j|$, $n \geq 1$. Then (e_n) is an error sequence for the Cauchy sequence of partial sums of $\sum |a_n|$. Since (§1.5) $(f(n))$ is a sequence of distinct naturals, $f(n) \to \infty$. In fact (§1.5), if we let $f_*(n) = \inf\{f(k) : k \geq n\}$, then $f_*(n) \nearrow \infty$. To show that $\sum |A_n|$ is convergent, it is enough to show that $\sum |A_n|$ is Cauchy.

To this end, for $m > k \geq n$,

$$\begin{aligned} |A_k| + |A_{k+1}| + \ldots + |A_m| &= |a_{f(k)}| + |a_{f(k+1)}| + \cdots + |a_{f(m)}| \\ &\leq \left|a_{f_*(k)}\right| + \left|a_{f_*(k+1)}\right| + \left|a_{f_*(k+2)}\right| + \cdots \leq e_{f_*(n)} \end{aligned}$$

which approaches zero, as $n \nearrow \infty$. Thus, $\sum |A_n|$ is Cauchy, hence convergent. Hence $\sum A_n$ is absolutely convergent.

Now let s_n, S_n denote the partial sums of $\sum a_n$ and $\sum A_n$, respectively. Let $E_n = \sum_{k=n}^{\infty} |A_k|$, $n \geq 1$. Then (E_n) is an error sequence for the Cauchy sequence of partial sums of $\sum |A_n|$. Now in the difference $S_n - s_n$, there will be cancellation, the only terms remaining being of one of two forms, either $A_k = a_{f(k)}$ with $f(k) > n$ or a_k with $k = f(j)$ with $j > n$ (this is where surjectivity of f is used). Hence in either case, the absolute values of the remaining terms in $S_n - s_n$ are summands in the series $e_n + E_n$, so

$$|S_n - s_n| \leq e_n + E_n \to 0, \qquad \text{as } n \nearrow \infty.$$

This completes the derivation of the theorem when $\sum |a_n| < \infty$. When $\sum a_n$ is nonnegative and infinite, then any rearrangement must also be so, since (by what we just showed!) $\sum A_n < \infty$ implies $\sum a_n < \infty$. This completes the derivation of the nonnegative case. □

The situation with conditionally convergent series is strikingly different.

Theorem 1.7.6. *If $\sum a_n$ is conditionally convergent and c is any real number, then there is a rearrangement $\sum A_n$ converging to c.*

Let (a_n^+), (a_n^-) denote the nonnegative and the negative terms in the series $\sum a_n$. Then we must have $\sum a_n^+ = \infty$ and $\sum a_n^- = -\infty$. Otherwise, $\sum a_n$ would converge absolutely. Moreover, $a_n^+ \to 0$ and $a_n^- \to 0$ since $a_n \to 0$. We construct a rearrangement as follows: Take the minimum number of terms a_n^+ whose sum s_1^+ is greater than c, then take the minimum number of terms a_n^- whose sum s_1^- with s_1^+ is less than c, then take the minimum number of additional terms a_n^+ whose sum s_2^+ with s_1^- is greater than c, then take the minimum number of additional terms a_n^- whose sum s_2^- with s_2^+ is less than c, etc. Because $a_n^+ \to 0$, $a_n^- \to 0$, $\sum a_n^+ = \infty$, and $\sum a_n^- = -\infty$, this rearrangement of the terms produces a series converging to c. Of course, if $c < 0$, one starts with the negative terms. □

We can use the fact that the sum of a nonnegative or absolutely convergent series is unchanged under rearrangements to study series over other sets. For example, let $\mathbf{N}^2 = \mathbf{N} \times \mathbf{N}$ be (§1.1) the set of ordered pairs of naturals (m, n), and set

$$\sum_{(m,n)\in\mathbf{N}^2} \frac{1}{m^3 + n^3}. \tag{1.7.6}$$

What do we mean by such a series? To answer this, we begin with a definition.

A set A is *countable* if there is a bijection $f : \mathbf{N} \to A$, i.e., the elements of A form a sequence. If there is no such f, we say that A is *uncountable*. Let us show that $\mathbf{N}^2$ is countable:

$$(1,1), (1,2), (2,1), (1,3), (2,2), (3,1), (1,4), (2,3), (3,2), (4,1), \dots.$$

Here we are listing the pairs (m, n) according to the sum $m+n$ of their entries. It turns out that $\mathbf{Q}$ is countable (Exercise **1.7.2**), but $\mathbf{R}$ is uncountable (Exercise **1.7.4**).

Every subset of $\mathbf{N}$ is countable or finite (Exercise **1.7.1**). Thus, if $f : A \to B$ is an injection and B is countable, then A is finite or countable. Indeed, choosing a bijection $g : B \to \mathbf{N}$ yields a bijection $g \circ f$ of A with the subset $(g \circ f)(A) \subset \mathbf{N}$.

Similarly, if $f : A \to B$ is a surjection and A is countable, then B is countable or finite. To see this, choose a bijection $g : \mathbf{N} \to A$. Then $f \circ g : \mathbf{N} \to B$ is a surjection, so we may define $h : B \to \mathbf{N}$ by setting $h(b)$ equal to the least n satisfying $f[g(n)] = b$, $h(b) = \min\{n \in \mathbf{N} : f[g(n)] = b\}$. Then h is an injection, and thus, B is finite or countable.

Let A be a countable set. Given a nonnegative function $f : A \to \mathbf{R}$, we define the *sum of the series over* A

$$\sum_{a\in A} f(a) \tag{1.7.7}$$

as the sum of $\sum_{n=1}^{\infty} f(a_n)$ obtained by taking any bijection of A with $\mathbf{N}$. Since the sum of a nonnegative series is unchanged by rearrangement, this is well defined. As an exercise, we leave it to be shown that (1.7.6) converges.

Series $\sum_{(m,n)} a_{mn}$ over $\mathbf{N}^2$ are called *double series* over $\mathbf{N}$. A useful arrangement of a double series follows the order of $\mathbf{N}^2$ displayed above,

$$\sum_{(m,n)\in\mathbf{N}^2} a_{mn} = \sum_{n=1}^{\infty} \left(\sum_{i+j=n+1} a_{ij} \right). \tag{1.7.8}$$

This is the *Cauchy order*.

Let $\sum a_{mn}$ be a nonnegative double series. Then we have two corresponding *iterated series*

$$\sum_{n=1}^{\infty}\left(\sum_{m=1}^{\infty} a_{mn}\right) \text{ and } \sum_{m=1}^{\infty}\left(\sum_{n=1}^{\infty} a_{mn}\right). \tag{1.7.9}$$

We clarify the meaning of the series on the left, since the series on the right is obtained by switching the indices m and n.

For each $n \geq 1$, the series $S_n = \sum_m a_{mn}$ is a nonnegative real or ∞. If $S_n < \infty$ for all $n \geq 1$, then $S = \sum_n S_n$ is well defined. If $S < \infty$, it is reasonable to set $\sum_n (\sum_m a_{mn}) = S$. Otherwise, either S or at least one of the terms S_n is infinite; in this case, it makes sense to set $\sum_n (\sum_m a_{mn}) = \infty$. With these clarifications, we have the following result.

Theorem 1.7.7. *If a double series $\sum a_{mn}$ is nonnegative, then it equals either iterated series:*

$$\sum_{(m,n)\in\mathbf{N}^2} a_{mn} = \sum_{k=1}^{\infty}\left(\sum_{i+j=k+1} a_{ij}\right) = \sum_{n=1}^{\infty}\left(\sum_{m=1}^{\infty} a_{mn}\right) = \sum_{m=1}^{\infty}\left(\sum_{n=1}^{\infty} a_{mn}\right). \tag{1.7.10}$$

Note this result is valid whether $\sum a_{mn} < \infty$ or $\sum a_{mn} = \infty$. To see this, recall that the first equality is due to the fact that a nonnegative double series may be summed in any order. Since the third and fourth sums are similar, it is enough to derive the second equality.

For any natural K, the set $A_K \subset \mathbf{N}^2$ of pairs (i,j) with $i + j \leq K + 1$ is contained in the set $B_{MN} \subset \mathbf{N}^2$ of pairs (m,n) with $m \leq M$, $n \leq N$, for M, N large enough (Figure 1.4). Hence

$$\sum_{k=1}^{K}\left(\sum_{i+j=k+1} a_{ij}\right) \leq \sum_{n=1}^{N}\left(\sum_{m=1}^{M} a_{mn}\right) \leq \sum_{n=1}^{\infty}\left(\sum_{m=1}^{\infty} a_{mn}\right).$$

Letting $K \nearrow \infty$, we obtain

$$\sum_{k=1}^{\infty}\left(\sum_{i+j=k+1} a_{ij}\right) \leq \sum_{n=1}^{\infty}\left(\sum_{m=1}^{\infty} a_{mn}\right).$$

Conversely, for any M, N, $B_{MN} \subset A_K$ for K large enough, hence

$$\sum_{n=1}^{N}\left(\sum_{m=1}^{M} a_{mn}\right) \leq \sum_{k=1}^{K}\left(\sum_{i+j=k+1} a_{ij}\right) \leq \sum_{k=1}^{\infty}\left(\sum_{i+j=k+1} a_{ij}\right).$$

Letting $M \nearrow \infty$,

$$\sum_{k=1}^{\infty}\left(\sum_{i+j=k+1} a_{ij}\right) \geq \sum_{n=1}^{N}\left(\sum_{m=1}^{\infty} a_{mn}\right).$$

Letting $N \nearrow \infty$,

$$\sum_{k=1}^{\infty}\left(\sum_{i+j=k+1} a_{ij}\right) \geq \sum_{n=1}^{\infty}\left(\sum_{m=1}^{\infty} a_{mn}\right).$$

This yields (1.7.10). □

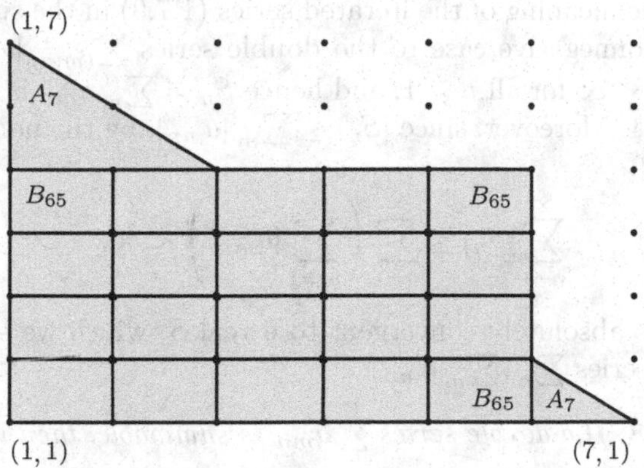

Fig. 1.4 The sets B_{MN} and A_K

To give an application of this, note that since $\sum 1/n^{1+1/N}$ converges, by comparison, so does

$$Z(s) = \sum_{n=2}^{\infty} \frac{1}{n^s}, \qquad s > 1.$$

(In the next chapter, we will know what n^s means for s real. Now think of s as rational.) Then (1.7.10) can be used to show that

$$\sum_{n=2}^{\infty} \frac{1}{n^s - 1} = Z(s) + Z(2s) + Z(3s) + \dots . \tag{1.7.11}$$

Let A be a countable set. Given a signed function $f : A \to \mathbf{R}$, we say (1.7.7) is *summable*[9] if

$$\sum_{a \in A} |f(a)| < \infty.$$

[9] The nonnegative/summable separation is analogous to the nonnegative/integrable separation appearing in Chapters 4 and 5.

In this case, we define (1.7.7) as the sum of $\sum_{n=1}^{\infty} f(a_n)$ obtained by taking any bijection of A with $\mathbf{N}$. Since the sum of an absolutely convergent series is unchanged by rearrangement, this is well defined.

Turning to signed series over $\mathbf{N}^2$, note that when a double series is summable, the series arranged in Cauchy order (1.7.8) is absolutely convergent, since

$$\sum_{n=1}^{\infty}\left|\sum_{i+j=n+1} a_{ij}\right| \le \sum_{n=1}^{\infty}\left(\sum_{i+j=n+1} |a_{ij}|\right) < \infty$$

by (1.7.10).

We clarify the meaning of the iterated series (1.7.9) in the summable case. Applying the nonnegative case to the double series $\sum_{(m,n)} |a_{mn}| < \infty$, we have $\sum_m |a_{mn}| < \infty$ for all $n \ge 1$, and hence $S_n = \sum_m a_{mn}$ is a well-defined real for all $n \ge 1$. Moreover, since $|S_n| \le \sum_m |a_{mn}|$, by the nonnegative case again,

$$\sum_{n=1}^{\infty} |S_n| \le \sum_{n=1}^{\infty}\left(\sum_{m=1}^{\infty} |a_{mn}|\right) < \infty;$$

hence, $\sum_n S_n$ is absolutely convergent to a real S, which we set as the sum of the iterated series $\sum_n \left(\sum_m a_{mn}\right)$.

Theorem 1.7.8. *If a double series* $\sum a_{mn}$ *is summable, then it equals either iterated series:*

$$\sum_{(m,n)\in\mathbf{N}^2} a_{mn} = \sum_{k=1}^{\infty}\left(\sum_{i+j=k+1} a_{ij}\right) = \sum_{n=1}^{\infty}\left(\sum_{m=1}^{\infty} a_{mn}\right) = \sum_{m=1}^{\infty}\left(\sum_{n=1}^{\infty} a_{mn}\right). \tag{1.7.12}$$

To see this, recall that the first equality is due to the fact that a summable double series may be summed in any order. Since the third and fourth sums are similar, it is enough to derive the second equality.

First, since the iterated series converges absolutely (1.7.10), the tail

$$A_L = \sum_{n=L}^{\infty}\left(\sum_{m=1}^{\infty} |a_{mn}|\right) \to 0 \text{ as } L \to \infty.$$

Similarly, and since we know the iterated series are equal in the nonnegative case, the tail

$$B_L = \sum_{n=1}^{\infty}\left(\sum_{m=L}^{\infty} |a_{mn}|\right) = \sum_{m=L}^{\infty}\left(\sum_{n=1}^{\infty} |a_{mn}|\right) \to 0 \text{ as } L \to \infty.$$

Moreover, $m + n \ge 2L$ implies $m \ge L$ or $n \ge L$; hence for $P \ge 2L$,

$$\sum_{k=2L}^{P}\left(\sum_{i+j=k+1}|a_{ij}|\right) \le \sum_{n=L}^{P}\left(\sum_{m=1}^{P}|a_{mn}|\right)+\sum_{n=1}^{P}\left(\sum_{m=L}^{P}|a_{mn}|\right)$$
$$\le \sum_{n=L}^{P}\left(\sum_{m=1}^{\infty}|a_{mn}|\right)+\sum_{n=1}^{P}\left(\sum_{m=L}^{\infty}|a_{mn}|\right)$$
$$\le \sum_{n=L}^{\infty}\left(\sum_{m=1}^{\infty}|a_{mn}|\right)+\sum_{n=1}^{\infty}\left(\sum_{m=L}^{\infty}|a_{mn}|\right) = A_L + B_L;$$

letting $P \to \infty$, we see the tail

$$C_L = \sum_{k=2L}^{\infty}\left(\sum_{i+j=k+1}|a_{ij}|\right) \le A_L + B_L \to 0 \text{ as } L \to \infty.$$

But the absolute value of the difference between the series $\sum_n (\sum_m a_{mn})$ and the finite sum

$$\sum_{n=1}^{L}\left(\sum_{m=1}^{L} a_{mn}\right)$$

is no larger than $A_L + B_L$, hence vanishes as $L \to \infty$. Also, the absolute value of the difference between the double series and the finite sum

$$\sum_{k=1}^{2L}\left(\sum_{i+j=k+1} a_{ij}\right)$$

is no larger than C_L, hence vanishes as $L \to \infty$. Finally, the absolute value of the difference of the finite sums themselves is no larger than $A_L + B_L + C_L$, hence vanishes as $L \to \infty$. □

Above we described double series over $\mathbf{N}$. We could have just as easily started the index from zero, describing double series over $\mathbf{N} \cup \{0\}$. This is important for Taylor series (§3.5), where the index starts from $n = 0$.

As an application, we describe the product of two series $\sum a_n$ and $\sum b_n$. We do this for series with the index starting from zero. The *product* or *Cauchy product* of series $\sum_{n=0}^{\infty} a_n$ and $\sum_{n=0}^{\infty} b_n$ is the series

$$\sum_{n=0}^{\infty} c_n = \sum_{n=0}^{\infty}\left(\sum_{i+j=n} a_i b_j\right) = a_0b_0 + (a_0b_1 + a_1b_0) + (a_0b_2 + a_1b_1 + a_2b_0) + \dots .$$

Theorem 1.7.9. *If $a = \sum_{n=0}^{\infty} a_n$ and $b = \sum_{n=0}^{\infty} b_n$ are both nonnegative or both absolutely convergent and $c_n = \sum_{i+j=n} a_i b_j$, $n \ge 1$, then $c = \sum_{n=0}^{\infty} c_n$ is nonnegative or absolutely convergent and $ab = c$.* □

Exercises

1.7.1. If $A \subset B$ and B is countable, then A is countable or finite. (If $B = \mathbf{N}$, look at the smallest element in A, then the next smallest, and so on.)

1.7.2. Show that $\mathbf{Q}$ is countable.

1.7.3. If A and B are countable, so is $A \times B$. Conclude that $\mathbf{Q} \times \mathbf{Q}$ is countable.

1.7.4. Show that $[0,1]$ and $\mathbf{R}$ are uncountable. (Assume $[0,1]$ is countable. List the elements as $a_1, a_2, \dots$. Using the decimal expansions of a_1, a_2, $\dots$, construct a decimal expansion not in the list.)

1.7.5. Show that (1.7.6) converges.

1.7.6. Derive (1.7.11).

1.7.7. Let $\sum a_n$ and $\sum b_n$ be absolutely convergent. Then the product of the alternating versions of $\sum a_n$ and $\sum b_n$ is the alternating version of the product of $\sum a_n$ and $\sum b_n$.

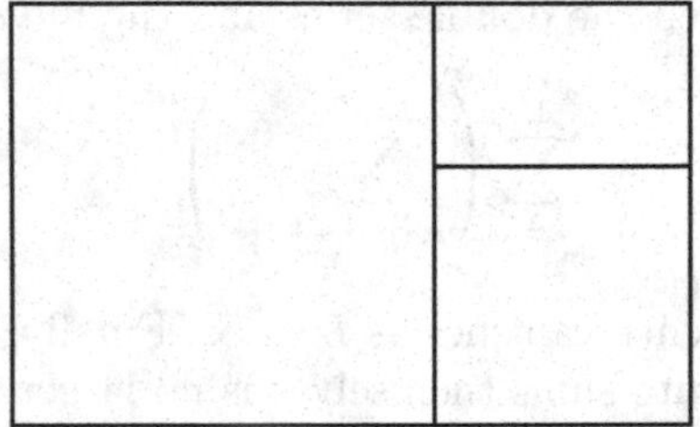

Fig. 1.5 The golden mean x. The ratios of the sides of the rectangles are all $x : 1$

1.7.8. Given a sequence (a_n) of naturals, let x_n be as in Exercise **1.5.13**. Show that (x_n) is Cauchy, hence convergent to an irrational x. Thus, continued fractions yield a bijection between sequences of naturals and irrationals in $(0,1)$. From this point of view, the continued fraction (Figure 1.5)

$$x = \frac{1+\sqrt{5}}{2} = 1 + \cfrac{1}{1 + \cfrac{1}{1 + \cfrac{1}{1 + \cfrac{1}{1 + \dots}}}}$$

is special. This real x is the *golden mean*; x satisfies

$$\frac{x}{1} = \frac{1}{x-1}$$

which is reflected in the infinite decreasing sequence of rectangles in Figure 1.5. Show in fact this continued fraction *does* converge to $(1+\sqrt{5})/2$.

Chapter 2
Continuity

2.1 Compactness

An *open interval* is a set of reals of the form $(a, b) = \{x : a < x < b\}$. As in §1.4, we are allowing $a = -\infty$ or $b = \infty$ or both. A *compact interval* is a set of reals of the form $[a, b] = \{x : a \le x \le b\}$, where a, b are real. The *length* of $[a, b]$ is $b - a$. Recall (§1.5) that a sequence *subconverges* to L if it has a subsequence converging to L.

Recall a subset $K \subset \mathbf{R}$ is *bounded* if $\sup K$ and $\inf K$ are finite. We say K is *closed* if $(x_n) \subset K$ and $x_n \to c$ implies $c \in K$. For example, from the comparison property of sequences, *a compact interval is closed and bounded.*

Theorem 2.1.1. *Let $K \subset \mathbf{R}$ be closed and bounded and let (x_n) be any sequence in K. Then (x_n) subconverges to some c in K.*

To derive this result, since K is bounded, we may choose $[a, b]$ with $K \subset [a, b]$. Divide the interval $I = [a, b]$ into 10 subintervals (of the same length), and order them from left to right (Figure 2.1), $I_0, I_1, \dots, I_9$. Pick one of them, say I_{d_1}, containing infinitely many terms of (x_n), i.e., $\{n : x_n \in I_{d_1}\}$ is infinite, and select one of the terms of the sequence in I_{d_1} and call it x'_1. Then the length of $J_1 \equiv I_{d_1}$ is $(b - a)/10$. Now divide J_1 into 10 subintervals again ordered left to right and called $I_{d_1 0}, \dots, I_{d_1 9}$. Select[1] one of them, say $I_{d_1 d_2}$, containing infinitely many terms of the sequence, and pick one of the terms (beyond x'_1) in the sequence in $I_{d_1 d_2}$ and call it x'_2. The length of $J_2 \equiv I_{d_1 d_2}$ is $(b - a)/100$. Continuing by induction, this yields

$$I \supset J_1 \supset J_2 \supset J_3 \supset \dots$$

[1] The choice can be avoided by selecting the leftmost interval at each stage.

O. Hijab, *Introduction to Calculus and Classical Analysis*, Undergraduate Texts in Mathematics, DOI 10.1007/978-3-319-28400-2_2

and a subsequence (x'_n), where the length of J_n is $(b-a)10^{-n}$ and $x'_n \in J_n$ for all $n \geq 1$. But, by construction, the real

$$c = a + (b-a) \cdot .d_1 d_2 d_3 \ldots$$

lies in all the intervals J_n, $n \geq 1$ (it may help to momentarily replace $[a,b]$ by $[0,1]$). Hence,

$$|x'_n - c| \leq (b-a)10^{-n} \to 0.$$

Thus (x_n) subconverges to c. Since K is closed, $c \in K$. □

$I_{d_1 d_2}$ I_{d_1}

Fig. 2.1 The intervals $I_{d_1 d_2 \ldots d_n}$

Thus this theorem is equivalent to, more or less, the existence of decimal expansions.

If K is replaced by an open interval (a,b), the theorem is false as it stands; hence, the theorem needs to be modified. A useful modification is the following.

Theorem 2.1.2. *If (x_n) is a sequence of reals in (a,b), then (x_n) subconverges to some $a < c < b$ or to a or to b.*

To see this, since $a = \inf(a,b)$, there is (Theorem 1.5.4) a sequence (c_n) in (a,b) satisfying $c_n \to a$. Similarly, since $b = \sup(a,b)$, there is a sequence (d_n) in (a,b) satisfying $d_n \to b$. Now there is either an $m \geq 1$ with (x_n) in $[c_m, d_m]$ or not. If so, the result follows from Theorem 2.1.1. If not, for every $m \geq 1$, there is an x_{n_m} not in $[c_m, d_m]$. Let (y_m) be the subsequence of (x_{n_m}) obtained by restricting attention to terms satisfying $x_{n_m} > d_m$, and let (z_m) be the subsequence of (x_{n_m}) obtained by restricting attention to terms satisfying $x_{n_m} < c_m$. Then at least one of the sequences (y_m) or (z_m) is infinite, so either $y_m \to b$ or $z_m \to a$ (or both) as $m \to \infty$. Thus (x_n) subconverges to a or to b. □

Note this result holds even when $a = -\infty$ or $b = \infty$. *The remainder of this section is used only in §6.6 and may be skipped until then.*

We say a set $K \subset \mathbf{R}$ is *sequentially compact* if every sequence $(x_n) \subset K$ subconverges to some $c \in K$. Thus we conclude *every closed and bounded set is sequentially compact.*

A set $U \subset \mathbf{R}$ is *open in* $\mathbf{R}$ if for every $c \in U$, there is an open interval I containing c and contained in U. Clearly, an open interval is an open set.

A *collection of open sets in* $\mathbf{R}$ is a set $\mathcal{U}$ whose elements are open sets in $\mathbf{R}$. Then by Exercise **2.1.3**,

$$\bigcup \mathcal{U} = \{x : x \in U \text{ for some } U \in \mathcal{U}\}$$

is open in $\mathbf{R}$.

Let X be a set. A *sequence of sets* in X is a function $f : \mathbf{N} \to 2^X$. This is written $(A_n) = (A_1, A_2, \dots)$, where $f(n) = A_n$, $n \ge 1$. If (A_n) is a sequence of sets, its union and intersection are denoted

$$\bigcup_{n=1}^{\infty} A_n = \bigcup\{A_n : n \ge 1\}, \qquad \bigcap_{n=1}^{\infty} A_n = \bigcap\{A_n : n \ge 1\}.$$

Given a sequence of sets (A_n), using Exercise **1.3.10**, one can construct by induction

$$A_1 \cup \ldots \cup A_n = \bigcup_{k=1}^{n} A_k, \qquad A_1 \cap \ldots \cap A_n = \bigcap_{k=1}^{n} A_k$$

for $n \ge 1$, by choosing $g(n, A) = A \cup A_{n+1}$ and $g(n, A) = A \cap A_{n+1}$.

Let $K \subset \mathbf{R}$ be any set. An *open cover* of K is a collection $\mathcal{U}$ of open sets in $\mathbf{R}$ whose union contains K, $K \subset \bigcup \mathcal{U}$. If $\mathcal{U}$ and $\mathcal{U}'$ are open covers and $\mathcal{U} \subset \mathcal{U}'$, we say $\mathcal{U} \subset \mathcal{U}'$ is a *subcover*. If $\mathcal{U}$ is countable, we say $\mathcal{U}$ is a *countable open cover*. If $\mathcal{U}$ is finite, we say $\mathcal{U}$ is a *finite open cover*.

Theorem 2.1.3. *If $K \subset \mathbf{R}$ is sequentially compact, then every countable open cover has a finite subcover.*

To see this, argue by contradiction. Suppose this was not so, and let

$$\mathcal{U} = (U_1, U_2, \dots).$$

Then for each $n \ge 1$, $U_1 \cup \ldots \cup U_n$ does not contain K. Since $K \setminus U_1 \cup \ldots \cup U_n$ is closed and bounded (Exercise **2.1.3**),

$$x_n \equiv \inf(K \setminus U_1 \cup \ldots \cup U_n) \in K \setminus U_1 \cup \ldots \cup U_n, \qquad n \ge 1.$$

Then $(x_n) \subset K$ so (x_n) subconverges to some $c \in K$. Now select U_N with $c \in U_N$. Then $x_n \in U_N$ for infinitely many n, contradicting the construction of x_n, $n \ge 1$. □

We say K is *countably compact* if every countable open cover has a finite subcover. Thus we conclude *every sequentially compact set is countably compact.*

Theorem 2.1.4. *If $K \subset \mathbf{R}$ is countably compact, then every open cover has a finite subcover.*

To see this, let $\mathcal{U}$ be an open cover, and let $\mathcal{I}$ be the collection of open sets I such that

- I is an open interval with rational endpoints, and
- $I \subset U$ for some $U \in \mathcal{U}$.

Then $\mathcal{I}$ is countable and $\bigcup\mathcal{I} = \bigcup\mathcal{U}$. Thus $\mathcal{I}$ is a countable open cover; hence, there is a finite subcover $\{I_1, \dots, I_N\} \subset \mathcal{I}$. For each $k = 1, \dots, N$, select[2] an open set $U = U_k \in \mathcal{U}$ containing I_k. Then $\{U_1, \dots, U_N\} \subset \mathcal{U}$ is a finite subcover. □

We say $K \subset \mathbf{R}$ is *compact* if every open cover has a finite subcover. Thus we conclude *every countably compact set is compact.*

We summarize the results of this section.

Theorem 2.1.5. *For $K \subset \mathbf{R}$, the following are equivalent:*

- *K is closed and bounded,*
- *K is sequentially compact,*
- *K is countably compact,*
- *K is compact.*

To complete the proof of this, it remains to show compactness implies closed and bounded. So suppose K is compact. Then $\mathcal{U} = \{(-n, n) : n \geq 1\}$ is an open cover and hence has a finite subcover. Thus K is bounded. If $(x_n) \subset \mathbf{R}$ converges to $c \notin K$, let $U_n = \{x : |x - c| > 1/n\}$, $n \geq 1$, and let $\mathcal{U} = \{U_1, U_2, \dots\}$. Then $\mathcal{U}$ is an open cover and hence has a finite subcover. This implies (x_n) is not wholly contained in K, which implies K is closed. □

In particular, a compact interval $[a, b]$ is a compact set. This is used in §6.6.

Exercises

2.1.1. Let (a_n, b_n), $n \geq 1$, be a sequence in $\mathbf{R}^2$. We say (a_n, b_n), $n \geq 1$, *subconverges* to $(a, b) \in \mathbf{R}^2$ if there is a sequence of naturals (n_k) such that (a_{n_k}) converges to a and (b_{n_k}) converges to b. Show that if (a_n) and (b_n) are bounded, then (a_n, b_n) subconverges to some (a, b).

2.1.2. In the derivation of the first theorem, suppose that the intervals are chosen, at each stage, to be the leftmost interval containing infinitely many terms. In other words, suppose that I_{d_1} is the leftmost of the intervals I_j containing infinitely many terms, $I_{d_1 d_2}$ is the leftmost of the intervals $I_{d_1 j}$ containing infinitely many terms, etc. In this case, show that the limiting point obtained is x_*.

2.1.3. If $\mathcal{U}$ is a collection of open sets in $\mathbf{R}$, then $\bigcup\mathcal{U}$ is open in $\mathbf{R}$. Also if K is closed and U is open, then $K \setminus U$ is closed.

---

[2] This uses the axiom of finite choice (Exercise **1.3.24**).

2.2 Continuous Limits

Let (a, b) be an open interval, and let $a < c < b$. The interval (a, b), *punctured* at c, is the set $(a, b) \setminus \{c\} = \{x : a < x < b, x \neq c\}$.

Let f be a function defined on an interval (a, b) punctured at c, $a < c < b$. We say L is the *limit of f at c*,, and we write

$$\lim_{x \to c} f(x) = L$$

or $f(x) \to L$ as $x \to c$, if, for *every* sequence $(x_n) \subset (a, b)$ satisfying $x_n \neq c$ for all $n \geq 1$ and converging to c, $f(x_n) \to L$.

For example, let $f(x) = x^2$, and let $(a, b) = \mathbf{R}$. If $x_n \to c$, then (§1.5), $x_n^2 \to c^2$. This holds true no matter what sequence (x_n) is chosen, as long as $x_n \to c$. Hence, in this case, $\lim_{x \to c} f(x) = c^2$.

Going back to the general definition, suppose that f is also defined at c. Then *the value $f(c)$ has no bearing on* $\lim_{x \to c} f(x)$ (Figure 2.2). For example, if $f(x) = 0$ for $x \neq 0$ and $f(0)$ is defined arbitrarily, then $\lim_{x \to 0} f(x) = 0$. For a more dramatic example of this phenomenon, see Exercise **2.2.1**.

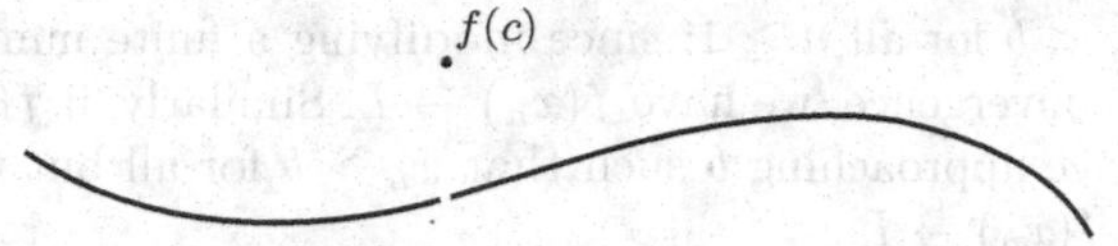

Fig. 2.2 The value $f(c)$ has no bearing on the limit at c

Of course, not every function has limits. For example, set $f(x) = 1$ if $x \in \mathbf{Q}$ and $f(x) = 0$ if $x \in \mathbf{R} \setminus \mathbf{Q}$. Choose any c in $(a, b) = \mathbf{R}$. Since (§1.4) there is a rational and an irrational between any two reals, for each $n \geq 1$ we can find $r_n \in \mathbf{Q}$ and $i_n \in \mathbf{R} \setminus \mathbf{Q}$ with $c < r_n < c + 1/n$ and $c < i_n < c + 1/n$. Thus $r_n \to c$ and $i_n \to c$, but $f(r_n) = 1$ and $f(i_n) = 0$ for all $n \geq 1$. Hence, f has no limit anywhere on $\mathbf{R}$.

Let f be a function defined on an interval (a, b) punctured at c, $a < c < b$. Let $(x_n) \subset (a, b)$ be a sequence satisfying $x_n \neq c$ for all $n \geq 1$ and converging to c. If $x_n \to c$, then $(f(x_n))$ may have several limit points (Exercise **1.5.9**). We say L is a *limit point of f at c* if for some sequence $x_n \to c$, L is a limit point of $(f(x_n))$. Then the limit of f at c exists iff all limit points of f at c are equal.

By analogy with sequences, the *upper limit of f at c* and *lower limit of f at c* are[3]

$$L^* = \inf_{\delta > 0} \sup_{0 < |x - c| < \delta} f(x), \qquad L_* = \sup_{\delta > 0} \inf_{0 < |x - c| < \delta} f(x).$$

[3] $\sup_A f$ and $\inf_A f$ are alternative notations for $\sup f(A)$ and $\inf f(A)$.

Then (Exercise **2.2.8**) L^* and L_* are the greatest and least limit points of f at c.

Let (x_n) be a sequence approaching b. If $x_n < b$ for all $n \geq 1$, we write $x_n \to b-$. Let f be defined on (a, b). We say L is the *limit* of f at b *from the left*, and we write

$$\lim_{x \to b-} f(x) = L,$$

if $x_n \to b-$ implies $f(x_n) \to L$. In this case, we also write $f(b-) = L$. If $b = \infty$, we write, instead, $\lim_{x\to\infty} f(x) = L$, $f(\infty) = L$, i.e., we drop the minus.

Let (x_n) be a sequence approaching a. If $x_n > a$ for all $n \geq 1$, we write $x_n \to a+$. Let f be defined on (a, b). We say L is the *limit* of f at a *from the right*, and we write

$$\lim_{x \to a+} f(x) = L,$$

if $x_n \to a+$ implies $f(x_n) \to L$. In this case, we also write $f(a+) = L$. If $a = -\infty$, we write, instead, $\lim_{x\to-\infty} f(x) = L$, $f(-\infty) = L$, i.e., we drop the plus.

Suppose $f(b-) = L$ and (x_n) is a sequence approaching b such that $x_n < b$ for all but finitely many $n \geq 1$. Then we may modify finitely many terms in (x_n) so that $x_n < b$ for all $n \geq 1$; since modifying a finite number of terms does not affect convergence, we have $f(x_n) \to L$. Similarly, if $f(b+) = L$ and (x_n) is a sequence approaching b such that $x_n > b$ for all but finitely many $n \geq 1$, we have $f(x_n) \to L$.

Of course, L above is either a real or $\pm\infty$.

Theorem 2.2.1. *Let f be defined on an interval (a, b) punctured at c, $a < c < b$. Then $\lim_{x\to c} f(x)$ exists and equals L iff $f(c+)$ and $f(c-)$ both exist and equal L.*

If $\lim_{x\to c} f(x) = L$, then $f(x_n) \to L$ for *any* sequence $x_n \to c$, whether the sequence is to the right, the left, or neither. Hence, $f(c-) = L$ and $f(c+) = L$.

Conversely, suppose that $f(c-) = f(c+) = L$, and let $x_n \to c$ with $x_n \neq c$ for all $n \geq 1$. We have to show that $f(x_n) \to L$.

Let (y_n) denote the terms in (x_n) that are greater than c, and let (z_n) denote the terms in (x_n) that are less than c, arranged in their given order. If (y_n) is finite, then all but finitely many terms of (x_n) are less than c; thus, $f(x_n) \to L$. If (z_n) is finite, then all but finitely many terms of (x_n) are greater than c; thus, $f(x_n) \to L$. Hence, we may assume both (y_n) and (z_n) are infinite sequences with $y_n \to c+$ and $z_n \to c-$. Since $f(c+) = L$, it follows that $f(y_n) \to L$; since $f(c-) = L$, it follows that $f(z_n) \to L$.

Let f^* and f_* denote the upper and lower limits of the sequence $(f(x_n))$, and set $f_n^* = \sup\{f(x_k) : k \geq n\}$. Then $f_n^* \searrow f^*$. Hence, for any subsequence $(f_{k_n}^*)$, we have $f_{k_n}^* \searrow f^*$. The goal is to show that $f^* = L = f_*$.

Since $f(y_n) \to L$, its upper sequence converges to L, $\sup_{i\geq n} f(y_i) \searrow L$; since $f(z_n) \to L$, its upper sequence converges to L, $\sup_{i\geq n} f(z_i) \searrow L$.

For each $m \geq 1$, let x_{k_m} denote the term in (x_n) corresponding to y_m, if the term y_m appears after the term z_m in (x_n). Otherwise, if z_m appears after y_m, let x_{k_m} denote the term in (x_n) corresponding to z_m. In other words, if $y_n = x_{i_n}$ and $z_n = x_{j_n}$, $x_{k_n} = x_{\max(i_n, j_n)}$. Thus for each $n \geq 1$, if $k \geq k_n$, we must have x_k equal to y_i or z_i with $i \geq n$, so

$$\{x_k : k \geq k_n\} \subset \{y_i : i \geq n\} \cup \{z_i : i \geq n\}.$$

Hence,

$$f^*_{k_n} = \sup_{k \geq k_n} f(x_k) \leq \max \left[\sup_{i \geq n} f(y_i), \sup_{i \geq n} f(z_i) \right], \qquad n \geq 1.$$

Now both sequences on the right are decreasing in $n \geq 1$ to L, and the sequence on the left decreases to f^* as $n \nearrow \infty$. Thus $f^* \leq L$. Now let $g = -f$. Since $g(c+) = g(c-) = -L$, by what we have just learned, we conclude that the upper limit of $(g(x_n))$ is $\leq -L$. But the upper limit of $(g(x_n))$ equals minus the lower limit f_* of $(f(x_n))$. Hence, $f_* \geq L$, so $f^* = f_* = L$. □

A limit point of f at c is a *left limit point of f at c* if it is a limit point of $(f(x_n))$ for some sequence $x_n \to c-$. Similarly, if $x_n \to c+$, we have *right limit points.* Every limit point at c is a left limit point at c or a right limit point at c. Then $f(c+)$ exists iff all right limit points of f at c are equal, and $f(c-)$ exists iff all left limit points of f at c are equal. From the above result, *the limit of f at c exists iff all left and right limit points of f at c are equal.*

Define L^*_+ to be the greatest of the right limit points of f at c, L_{*+} the least of the right limit points of f at c, L^*_- the greatest of the left limit points of f at c, and L_{*-} the least of the right limit points of f at c. These are the *upper and lower left and right limits of f at c.* We conclude *the limit of f at c exists iff the four quantities $L^*_+, L_{*+}, L^*_-, L_{*-}$ are equal.*

Since continuous limits are defined in terms of limits of sequences, they enjoy the same arithmetic and ordering properties. For example,

$$\lim_{x \to a} [f(x) + g(x)] = \lim_{x \to a} f(x) + \lim_{x \to a} g(x),$$
$$\lim_{x \to a} [f(x) \cdot g(x)] = \lim_{x \to a} f(x) \cdot \lim_{x \to a} g(x).$$

These properties will be used without comment.

A function f is *increasing (decreasing)* if $x \leq x'$ implies $f(x) \leq f(x')$ ($f(x) \geq f(x')$, respectively), for all x, x' in the domain of f. The function f is *strictly increasing (strictly decreasing)* if $x < x'$ implies $f(x) < f(x')$ ($f(x) > f(x')$, respectively), for all x, x' in the domain of f. If f is increasing or decreasing, we say f is *monotone*. If f is strictly increasing or strictly decreasing, we say f is *strictly monotone*.

Exercises

2.2.1. Define $f : \mathbf{R} \to \mathbf{R}$ by setting $f(m/n) = 1/n$, for $m/n \in \mathbf{Q}$ with no common factor in m and $n > 0$, and $f(x) = 0$, $x \notin \mathbf{Q}$. Show that $\lim_{x\to c} f(x) = 0$ for all $c \in \mathbf{R}$.

2.2.2. Let f be increasing on (a, b). Then $f(a+)$ (exists and) equals $\inf\{f(x) : a < x < b\}$, and $f(b-)$ equals $\sup\{f(x) : a < x < b\}$.

2.2.3. If f is monotone on (a, b), then $f(c+)$ and $f(c-)$ exist, and $f(c)$ is between $f(c-)$ and $f(c+)$, for all $c \in (a, b)$. Show also that, for each $\delta > 0$, there are, at most, countably many points $c \in (a, b)$ where $|f(c+) - f(c-)| \geq \delta$. Conclude that there are, at most, countably many points c in (a, b) at which $f(c+) \neq f(c-)$.

2.2.4. Let f be defined on $[a, b]$, and let $I_k = (c_k, d_k)$, $1 \leq k \leq N$, be disjoint open intervals in (a, b). The *variation* of f over these intervals is

$$\sum_{k=1}^{N} |f(d_k) - f(c_k)| \tag{2.2.1}$$

and the *total variation* $v_f(a, b)$ is the supremum of variations of f in (a, b) over all such disjoint unions of open intervals in (a, b). We say that f is *bounded variation* on $[a, b]$ if $v_f(a, b)$ is finite. Show bounded variation on $[a, b]$ implies bounded on $[a, b]$.

2.2.5. If f is increasing on an interval $[a, b]$, then f is bounded variation on $[a, b]$ and $v_f(a, b) = f(b) - f(a)$. If $f = g - h$ with g, h increasing on $[a, b]$, then f is bounded variation on $[a, b]$.

2.2.6. Let f be bounded variation on $[a, b]$, and, for $a \leq x \leq b$, let $v(x) = v_f(a, x)$. Show

$$v(x) + |f(y) - f(x)| \leq v(y), \qquad a \leq x < y \leq b,$$

hence, v and $v - f$ are increasing on $[a, b]$. Conclude that f is of bounded variation on $[a, b]$ iff f is the difference of two increasing functions on $[a, b]$. If moreover f is continuous, so are v and $v - f$.

2.2.7. Show that the f in Exercise **2.2.1** is not bounded variation on $[0, 2]$ (remember that $\sum 1/n = \infty$).

2.2.8. Show that the upper limit and lower limit of f at c are the greatest and least limit points of f at c, respectively.

2.3 Continuous Functions

Let f be defined on (a, b), and choose $a < c < b$. We say that f *is continuous at* c if

$$\lim_{x \to c} f(x) = f(c).$$

If f is continuous at every real c in (a, b), then we say that f *is continuous on* (a, b) or, if (a, b) is understood from the context, f *is continuous.*

Recalling the definition of $\lim_{x \to c}$, we see that f is continuous at c iff, for all sequences (x_n) satisfying $x_n \to c$ and $x_n \neq c$, $n \geq 1$, $f(x_n) \to f(c)$. In fact, f is continuous at c iff $x_n \to c$ implies $f(x_n) \to f(c)$, i.e., the condition $x_n \neq c$, $n \geq 1$, is superfluous. To see this, suppose that f is continuous at c, and suppose that $x_n \to c$, but $f(x_n) \not\to f(c)$. Since $f(x_n) \not\to f(c)$, by Exercise **1.5.8**, there is an $\epsilon > 0$ and a subsequence (x'_n), such that $|f(x'_n) - f(c)| \geq \epsilon$ and $x'_n \to c$, for $n \geq 1$. But, then $f(x'_n) \neq f(c)$ for all $n \geq 1$; hence, $x'_n \neq c$ for all $n \geq 1$. Since $x'_n \to c$, by the continuity at c, we obtain $f(x'_n) \to f(c)$, contradicting $|f(x'_n) - f(c)| \geq \epsilon$. Thus f *is continuous at* c *iff* $x_n \to c$ *implies* $f(x_n) \to f(c)$.

In the previous section, we saw that $f(x) = x^2$ is continuous at c. Since this works for any c, f is continuous. Repeating this argument, one can show that $f(x) = x^4$ is continuous, since $x^4 = x^2 x^2$. A simpler example is to choose a real k and to set $f(x) = k$ for all x. Here $f(x_n) = k$, and $f(c) = k$ for all sequences (x_n) and all c, so f is continuous. Another example is $f : (0, \infty) \to \mathbf{R}$ given by $f(x) = 1/x$. By the division property of sequences, $x_n \to c$ implies $1/x_n \to 1/c$ for $c > 0$, so f is continuous.

Functions can be continuous at various points and not continuous at other points. For example, the function f in Exercise **2.2.1** is continuous at every irrational c and not continuous at every rational c. On the other hand, the function $f : \mathbf{R} \to \mathbf{R}$, given by (§2.2)

$$f(x) = \begin{cases} 1, & x \in \mathbf{Q} \\ 0, & x \notin \mathbf{Q}, \end{cases}$$

is continuous at no point.

Continuous functions have very simple arithmetic and ordering properties. If f and g are defined on (a, b) and k is real, we have functions $f + g$, kf, fg, $\max(f, g)$, $\min(f, g)$ defined on (a, b) by setting, for $a < x < b$,

$$\begin{aligned} (f+g)(x) &= f(x) + g(x), \\ (kf)(x) &= kf(x), \\ (fg)(x) &= f(x)g(x), \\ \max(f,g)(x) &= \max[f(x), g(x)], \\ \min(f,g)(x) &= \min[f(x), g(x)]. \end{aligned}$$

If g is *nonzero* on (a, b), i.e., $g(x) \neq 0$ for all $a < x < b$, define f/g by setting

$$(f/g)(x) = \frac{f(x)}{g(x)}, \qquad a < x < b.$$

Theorem 2.3.1. *If f and g are continuous, then so are $f + g$, kf, fg, $\max(f, g)$, and $\min(f, g)$. Moreover, if g is nonzero, then f/g is continuous.*

This is an immediate consequence of the arithmetic and ordering properties of sequences: If $a < c < b$ and $x_n \to c$, then $f(x_n) \to f(c)$ and $g(x_n) \to g(c)$. Hence, $f(x_n) + g(x_n) \to f(c) + g(c)$, $kf(x_n) \to kf(c)$, $f(x_n)g(x_n) \to f(c)g(c)$, $\max[f(x_n), g(x_n)] \to \max[f(c), g(c)]$, and $\min[f(x_n), g(x_n)] \to \min[f(c), g(c)]$. If $g(c) \neq 0$, then $f(x_n)/g(x_n) \to f(c)/g(c)$. □

For example, we see immediately that $f(x) = |x|$ is continuous on $\mathbf{R}$ since $|x| = \max(x, -x)$.

Let us prove, by induction, that, for all $k \geq 1$, the *monomials* $f_k(x) = x^k$ are continuous (on $\mathbf{R}$). For $k = 1$, this is so since $x_n \to c$ implies $f_1(x_n) = x_n \to c = f_1(c)$. Assuming that this is true for k, $f_{k+1} = f_k f_1$ since $x^{k+1} = x^k x$. Hence, the result follows from the arithmetic properties of continuous functions.

A *polynomial* $f : \mathbf{R} \to \mathbf{R}$ is a linear combination of monomials, i.e., a polynomial has the form

$$f(x) = a_0 x^d + a_1 x^{d-1} + a_2 x^{d-2} + \cdots + a_{d-1} x + a_d.$$

If $a_0 \neq 0$, we call d the *degree* of f. The reals $a_0, a_1, \ldots, a_d$ are the *coefficients* of the polynomial.

Let f be a polynomial of degree $d > 0$, and let $a \in \mathbf{R}$. Then there is a polynomial g of degree $d - 1$ satisfying[4]

$$\frac{f(x) - f(a)}{x - a} = g(x), \qquad x \neq a. \tag{2.3.1}$$

To see this, since every polynomial is a linear combination of monomials, it is enough to check (2.3.1) on monomials. But, for $f(x) = x^n$,

$$\frac{x^n - a^n}{x - a} = x^{n-1} + x^{n-2}a + \cdots + xa^{n-2} + a^{n-1}, \qquad x \neq a, \tag{2.3.2}$$

which can be checked[5] by cross multiplying. This establishes (2.3.1).

Since a monomial is continuous and a polynomial is a linear combination of monomials, by induction on the degree, we obtain the following.

Theorem 2.3.2. *Every polynomial f is continuous on $\mathbf{R}$. Moreover, if d is its degree, there are, at most, d real numbers x satisfying $f(x) = 0$.*

---

[4] g also depends on a.

[5] (2.3.2) with $x = 1$ was used to sum the geometric series in §1.6.

A real x satisfying $f(x) = 0$ is called a *zero* or a *root* of f. Thus every polynomial f has, at most, d roots. To see this, proceed by induction on the degree of f. If $d = 1$, $f(x) = a_0x + a_1$, so f has one root $x = -a_1/a_0$. Now suppose that every dth-degree polynomial has, at most, d roots, and let f be a polynomial of degree $d+1$. We have to show that the number of roots of f is at most $d+1$. If f has no roots, we are done. Otherwise, let a be a root, $f(a) = 0$. Then by (2.3.1) there is a polynomial g of degree d such that $f(x) = (x-a)g(x)$. Thus any root $b \neq a$ of f must satisfy $g(b) = 0$. Since by the inductive hypothesis g has, at most, d roots, we see that f has, at most, $d+1$ roots. □

A polynomial may have no roots, e.g., $f(x) = x^2 + 1$. However, every polynomial of odd degree has at least one root (Exercise **2.3.1**).

A *rational function* is a quotient $f = p/q$ of two polynomials. The natural domain of f is $\mathbf{R} \setminus Z(q)$, where $Z(q)$ denotes the set of roots of q. Since $Z(q)$ is a finite set, the natural domain of f is a finite union of open intervals. We conclude that *every rational function is continuous where it is defined.*

Let $f : (a,b) \to \mathbf{R}$. If f is not continuous at $c \in (a,b)$, we say that f is *discontinuous* at c. There are "mild" discontinuities, and there are "wild" discontinuities. The mildest situation (Figure 2.3) is when the limits $f(c+)$ and $f(c-)$ exist and are equal, but not equal to $f(c)$. This can be easily remedied by modifying the value of $f(c)$ to equal $f(c+) = f(c-)$. With this modification, the resulting function then is continuous at c. Because of this, such a point c is called a *removable discontinuity*. For example, the function f in Exercise **2.2.1** has removable discontinuities at every rational.

The next level of complexity is when $f(c+)$ and $f(c-)$ exist but may or may not be equal. In this case, we say that f has a *jump discontinuity* (Figure 2.3) or a *mild discontinuity* at c. For example, every monotone function has (at worst) jump discontinuities. In fact, every function of bounded variation has (at worst) jump discontinuities (Exercise **2.3.18**). The (amount of) *jump* at c, a real number, is $f(c+) - f(c-)$. In particular, a jump discontinuity of jump zero is nothing more than a removable discontinuity.

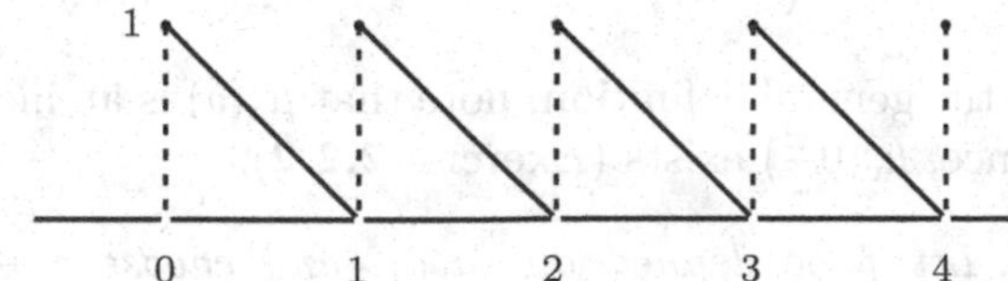

Fig. 2.3 A jump of 1 at each integer

Any discontinuity that is not a jump is called a *wild discontinuity* (Figure 2.4). If f has a wild discontinuity at c, then from above f cannot be of bounded variation on any open interval surrounding c. The converse of this statement is false. It is possible for f to have mild discontinuities but not be of bounded variation (Exercise **2.2.7**).

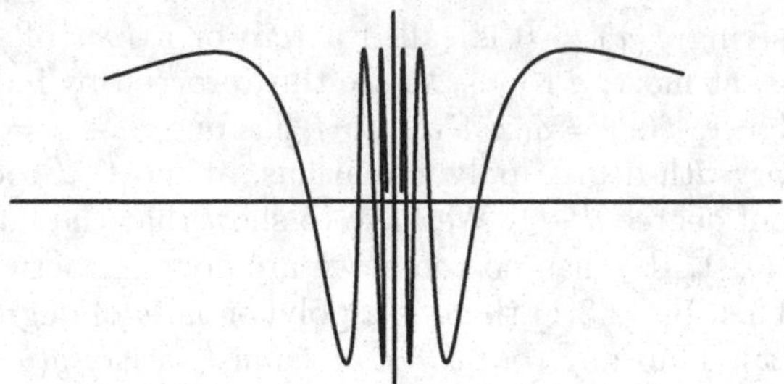

Fig. 2.4 A wild discontinuity

An alternate and useful description of continuity is in terms of a modulus of continuity. Let $f:(a,b)\to\mathbf{R}$, and fix $a<c<b$. For $\delta>0$, let

$$\mu_c(\delta)=\sup\{|f(x)-f(c)|:|x-c|<\delta,a<x<b\}.$$

Since the sup, here, is possibly that of an unbounded set, we may have $\mu_c(\delta)=\infty$. The function $\mu_c:(0,\infty)\to[0,\infty)\cup\{\infty\}$ is the *modulus of continuity of f at c* (Figure 2.5).

For example, let $f:(1,10)\to\mathbf{R}$ be given by $f(x)=x^2$ and pick $c=9$. Since x^2 is monotone over any interval not containing zero, the maximum value of $|x^2-81|$ over any interval not containing zero is obtained by plugging in the endpoints. Hence, $\mu_9(\delta)$ is obtained by plugging in $x=9\pm\delta$, leading to $\mu_9(\delta)=\delta(\delta+18)$. In fact, this is correct only if $0<\delta\le 1$. If $1\le\delta\le 8$, the interval under consideration is $(9-\delta,9+\delta)\cap(1,10)=(9-\delta,10)$. Here plugging in the endpoints leads to $\mu_9(\delta)=\max(19,18\delta-\delta^2)$. If $\delta\ge 8$, then $(9-\delta,9+\delta)$ contains $(1,10)$, and hence, $\mu_9(\delta)=80$. Summarizing, for $f(x)=x^2$, $c=9$, and $(a,b)=(1,10)$,

$$\mu_c(\delta)=\begin{cases}\delta(\delta+18), & 0<\delta\le 1,\\ \max(19,18\delta-\delta^2), & 1\le\delta\le 8,\\ 80, & \delta\ge 8.\end{cases}$$

Going back to the general definition, note that $\mu_c(\delta)$ is an increasing function of δ, and hence, $\mu_c(0+)$ exists (Exercise **2.2.2**).

Theorem 2.3.3. *Let f be defined on (a,b), and choose $c\in(a,b)$. The following are equivalent.*

A. *f is continuous at c.*
B. $\mu_c(0+)=0$.
C. *For all $\epsilon>0$, there exists $\delta>0$, such that*

$$|x-c|<\delta \textit{ implies } |f(x)-f(c)|<\epsilon.$$

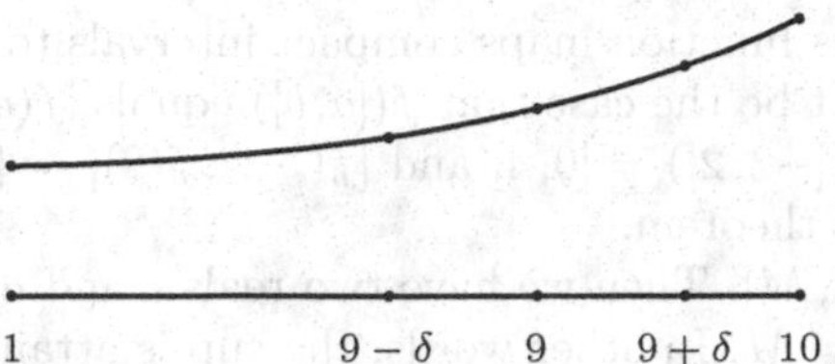

Fig. 2.5 Computing the modulus of continuity

That **A** implies **B** is left as Exercise **2.3.2**. Now assume **B**, and suppose that $\epsilon > 0$ is given. Since $\mu_c(0+) = 0$, there exists a $\delta > 0$ with $\mu_c(\delta) < \epsilon$. Then by definition of μ_c, $|x - c| < \delta$ implies $|f(x) - f(c)| \le \mu_c(\delta) < \epsilon$, which establishes **C**. Now assume the ϵ-δ criterion **C**, and let $x_n \to c$. Then for all but a finite number of terms, $|x_n - c| < \delta$. Hence, for all but a finite number of terms, $f(c) - \epsilon < f(x_n) < f(c) + \epsilon$. Let $y_n = f(x_n)$, $n \ge 1$. By the ordering properties of sup and inf, $f(c) - \epsilon \le y_{n*} \le y_n^* \le f(c) + \epsilon$. By the ordering properties of sequences, $f(c) - \epsilon \le y_* \le y^* \le f(c) + \epsilon$. Since $\epsilon > 0$ is arbitrary, $y^* = y_* = f(c)$. Thus $y_n = f(x_n) \to f(c)$. Since (x_n) was any sequence converging to c, $\lim_{x \to c} f(x) = f(c)$, i.e., **A**. □

Thus in practice, one needs to compute $\mu_c(\delta)$ only for δ small enough, since it is the behavior of μ_c near zero that counts. For example, to check continuity of $f(x) = x^2$ at $c = 9$, it is enough to note that $\mu_9(\delta) = \delta(\delta + 18)$ for small enough δ, which clearly approaches zero as $\delta \to 0+$.

To check the continuity of $f(x) = x^2$ at $c = 9$ using the ϵ-δ criterion **C**, given $\epsilon > 0$, it is enough to exhibit a $\delta > 0$ with $\mu_9(\delta) < \epsilon$. Such a δ is the lesser of $\epsilon/20$ and 1, $\delta = \min(\epsilon/20, 1)$. To see this, first, note that $\delta(\delta + 18) \le 19$ for this δ. Then $\epsilon \le 19$ implies $\delta(\delta + 18) \le (\epsilon/20)(1 + 18) = (19/20)\epsilon < \epsilon$, whereas $\epsilon > 19$ implies $\delta(\delta + 18) < \epsilon$. Hence, in either case, $\mu_9(\delta) < \epsilon$, establishing **C**.

Now we turn to the mapping properties of a continuous function. First, we define one-sided continuity. Let f be defined on $(a, b]$. We say that f is *continuous at* b *from the left* if $f(b-) = f(b)$. In addition, if f is continuous on (a, b), we say that f is *continuous on* $(a, b]$. Let f be defined on $[a, b)$. We say that f is *continuous at* a *from the right* if $f(a+) = f(a)$. In addition, if f is continuous on (a, b), we say that f is *continuous on* $[a, b)$.

Note by Theorem 2.2.1 that a function f is continuous at a particular point c iff f is continuous at c from the right and continuous at c from the left.

Let f be defined on $[a, b]$. We say that f is *continuous on* $[a, b]$ if f is continuous on $[a, b)$ and $(a, b]$. Checking the definitions, we see f is continuous on A if, for every $c \in A$ and every sequence $(x_n) \subset A$ converging to c, $f(x_n) \to f(c)$, whether A is (a, b), $(a, b]$, $[a, b)$, or $[a, b]$.

Theorem 2.3.4. *Let* f *be continuous on a compact interval* $[a, b]$*. Then* $f([a, b])$ *is a compact interval* $[m, M]$*.*

Thus a continuous function maps compact intervals to compact intervals. Of course, it may not be the case that $f([a,b])$ equals $[f(a),f(b)]$. For example, if $f(x)=x^2$, $f([-2,2])=[0,4]$ and $[f(-2),f(2)]=\{4\}$. We derive two consequences of this theorem.

Let $f([a,b])=[m,M]$. Then we have two reals c and d in $[a,b]$, such that $f(c)=m$ and $f(d)=M$. In other words, the sup is attained in

$$M=\sup\{f(x):a\le x\le b\}=\max\{f(x):a\le x\le b\}$$

and the inf is attained in

$$m=\inf\{f(x):a\le x\le b\}=\min\{f(x):a\le x\le b\}.$$

More succinctly, M is a max and m is a min for the set $f([a,b])$.

Theorem 2.3.5. *Let f be continuous on $[a,b]$. Then f achieves its max and its min over $[a,b]$.*

Of course, this is not generally true on noncompact intervals since $f(x)=1/x$ has no max on $(0,1]$.

A second consequence is: Suppose that L is an intermediate value between $f(a)$ and $f(b)$. Then there must be a c, $a<c<b$, satisfying $f(c)=L$. This follows since $f(a)$ and $f(b)$ are two reals in $f([a,b])$ and $f([a,b])$ is an interval. This is the *intermediate value property.*

Theorem 2.3.6 (Intermediate Value Property). *Let f be continuous on $[a,b]$ and suppose $f(a)<L<f(b)$. Then there is $c\in(a,b)$ with $f(c)=L$.*

On the other hand, the two consequences, the existence of the max and the min and the intermediate value property, combine to yield Theorem 2.3.4. To see this, let $m=f(c)$ and $M=f(d)$ denote the max and the min, with $c,d\in[a,b]$. If $m=M$, f is constant; hence, $f([a,b])=[m,M]$. If $m<M$ and $m<L<M$, apply the intermediate value property to conclude that there is an x between c and d with $f(x)=L$. Hence, $f([a,b])=[m,M]$. Thus to derive the theorem, it is enough to derive the two consequences.

For the first, let $M=\sup f([a,b])$. By Theorem 1.5.4, there is a sequence (x_n) in $[a,b]$ such that $f(x_n)\to M$. But Theorem 2.1.1, (x_n) subconverges to some $c\in[a,b]$. By continuity, $(f(x_n))$ subconverges to $f(c)$. Since $(f(x_n))$ also converges to M, $M=f(c)$, so f has a max. Proceed similarly for the min. This establishes Theorem 2.3.5.

For the second, suppose that $f(a)<f(b)$, and let L be an intermediate value, $f(a)<L<f(b)$. We proceed as in the construction of $\sqrt{2}$ in §1.4. Let $S=\{x\in[a,b]:f(x)<L\}$, and let $c=\sup S$. S is nonempty since $a\in S$, and S is clearly bounded. By Theorem 1.5.4, select a sequence (x_n) in S converging to c, $x_n\to c$. By continuity, it follows that $f(x_n)\to f(c)$. Since $f(x_n)<L$ for all $n\ge 1$, we obtain $f(c)\le L$. On the other hand, $c+1/n$ is not in S; hence, $f(c+1/n)\ge L$. Since $c+1/n\to c$, we obtain $f(c)\ge L$. Thus

$f(c) = L$. The case $f(a) > f(b)$ is similar or is established by applying the previous to $-f$. This establishes Theorem 2.3.6 and hence Theorem 2.3.4. □

From this theorem, it follows that a continuous function maps open intervals to intervals. However, they need not be open. For example, with $f(x) = x^2$, $f((-2,2)) = [0,4)$. However, a function that is continuous and strictly monotone maps open intervals to open intervals (Exercise **2.3.3**).

The above theorem is the result of compactness mixed with continuity. This mixture yields other dividends. Let $f : (a,b) \to \mathbf{R}$ be given, and fix a subset $A \subset (a,b)$. For $\delta > 0$, set

$$\mu_A(\delta) = \sup\{\mu_c(\delta) : c \in A\}.$$

This is the *uniform modulus of continuity of f on A*. Since $\mu_c(\delta)$ is an increasing function of δ for each $c \in A$, it follows that $\mu_A(\delta)$ is an increasing function of δ, and hence $\mu_A(0+)$ exists. We say f is *uniformly continuous on A* if $\mu_A(0+) = 0$. When $A = (a,b)$ equals the whole domain of the function, we delete the subscript A and write $\mu(\delta)$ for the uniform modulus of continuity of f on its domain.

Whereas continuity is a property pertaining to the behavior of a function at (or near) a given point c, uniform continuity is a property pertaining to the behavior of f near a given set A. Moreover, since $\mu_c(\delta) \le \mu_A(\delta)$, uniform continuity on A implies continuity at every point $c \in A$.

Inserting the definition of $\mu_c(\delta)$ in $\mu_A(\delta)$ yields

$$\mu_A(\delta) = \sup\{|f(x) - f(c)| : |x - c| < \delta, a < x < b, c \in A\},$$

where, now, the sup is over both x and c.

For example, for $f(x) = x^2$, the uniform modulus $\mu_A(\delta)$ over $A = (1,10)$ equals the sup of $|x^2 - y^2|$ over all $1 < x < y < 10$ with $y - x < \delta$. But this is largest when $y = x + \delta$; hence, $\mu_A(\delta)$ is the sup of $\delta^2 + 2x\delta$ over $1 < x < 10 - \delta$ which yields $\mu_A(\delta) = 20\delta - \delta^2$. In fact, this is correct only if $0 < \delta \le 9$. For $\delta = 9$, the sup is already over all of $(1,10)$ and hence cannot get any bigger. Hence, $\mu_A(\delta) = 99$ for $\delta \ge 9$. Summarizing, for $f(x) = x^2$ and $A = (1,10)$,

$$\mu_A(\delta) = \begin{cases} 20\delta - \delta^2, & 0 < \delta \le 9, \\ 99, & \delta \ge 9. \end{cases}$$

Since f is uniformly continuous on A if $\mu_A(0+) = 0$, in practice one needs to compute $\mu_A(\delta)$ only for δ small enough. For example, to check uniform continuity of $f(x) = x^2$ over $A = (1,10)$, it is enough to note that $\mu_A(\delta) = 20\delta - \delta^2$ for small enough δ, which clearly approaches zero as $\delta \to 0+$.

Now let $f : (a,b) \to \mathbf{R}$ be continuous, and fix $A \subset (a,b)$. What additional conditions on f are needed to guarantee uniform continuity on A? When A is a finite set $\{c_1, \dots, c_N\}$,

$$\mu_A(\delta) = \max\left[\mu_{c_1}(\delta), \dots, \mu_{c_N}(\delta)\right],$$

and hence f is necessarily uniformly continuous on A.

When A is an infinite set, this need not be so. For example, with $f(x) = x^2$ and $B = (0, \infty)$, $\mu_B(\delta)$ equals the sup of $\mu_c(\delta) = 2c\delta + \delta^2$ over $0 < c < \infty$, or $\mu_B(\delta) = \infty$, for each $\delta > 0$. Hence, f is not uniformly continuous on B.

It turns out that continuity on a compact interval is sufficient for uniform continuity.

Theorem 2.3.7. *If f is continuous on $[a, b]$, then f is uniformly continuous on (a, b). Conversely, if f is uniformly continuous on (a, b), then f extends to a continuous function on $[a, b]$.*

To see this, suppose that $\mu(0+) = \mu_{(a,b)}(0+) > 0$, and set $\epsilon = \mu(0+)/2$. Since μ is increasing, $\mu(1/n) \ge 2\epsilon$, $n \ge 1$. Hence, for each $n \ge 1$, by the definition of the sup in the definition of $\mu(1/n)$, there is a $c_n \in (a, b)$ with $\mu_{c_n}(1/n) > \epsilon$. Now by the definition of the sup in $\mu_{c_n}(1/n)$, for each $n \ge 1$, there is an $x_n \in (a, b)$ with $|x_n - c_n| < 1/n$ and $|f(x_n) - f(c_n)| > \epsilon$. By compactness, (x_n) subconverges to some $x \in [a, b]$. Since $|x_n - c_n| < 1/n$ for all $n \ge 1$, (c_n) subconverges to the same x. Hence, by continuity, $(|f(x_n) - f(c_n)|)$ subconverges to $|f(x) - f(x)| = 0$, which contradicts the fact that this last sequence is bounded below by $\epsilon > 0$.

Conversely, let $f : (a, b) \to \mathbf{R}$ be uniformly continuous with modulus of continuity μ, and suppose $x_n \to a+$. Then (x_n) is Cauchy, so let (e_n) be an error sequence for (x_n). Since

$$\sup_{m,k \ge n} |f(x_k) - f(x_m)| \le \mu(|x_k - x_m|) \le \mu(e_n) \to 0, \qquad n \to \infty,$$

it follows $(f(x_n))$ is Cauchy and hence converges. If $x'_n \to a+$,

$$|f(x_n) - f(x'_n)| \le \mu(|x_n - x'_n|) \to 0, \quad n \to \infty,$$

hence, $f(a+)$ exists. Similarly, $f(b-)$ exists. □

The conclusion may be false if f is continuous on (a, b) but not on $[a, b]$ (see Exercise **2.3.23**). One way to understand the difference between continuity and uniform continuity is as follows.

Let f be a continuous function defined on an interval (a, b), and pick $c \in (a, b)$. Then by definition of μ_c, $|f(x) - f(c)| \le \mu_c(\delta)$ whenever x lies in the interval $(c - \delta, c + \delta)$. Setting $g(x) = f(c)$ for $x \in (c - \delta, c + \delta)$, we see that, for any error tolerance ϵ, by choosing δ satisfying $\mu_c(\delta) < \epsilon$, we obtain a constant function g approximating f to within ϵ, at least in the interval $(c - \delta, c + \delta)$. Of course, in general, we do not expect to approximate f closely by one and the same constant function over the whole interval (a, b). Instead, we use piecewise constant functions.

If (a, b) is an open interval, a *partition of* (a, b) is a choice of points $a = x_0 < x_1 < \cdots < x_{n-1} < x_n = b$ in (a, b), where we denote the endpoints a

and b by x_0 and x_n, respectively (even when they are infinite). We use the same notation for compact intervals, i.e., a partition of $[a,b]$ is a partition of (a,b) (Figure 2.6).

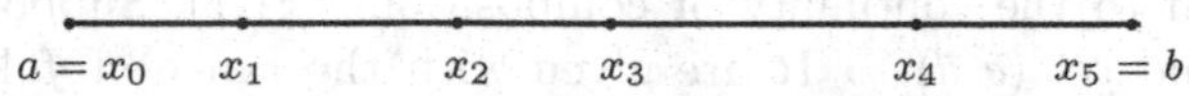

$a = x_0 \quad x_1 \quad x_2 \quad x_3 \quad x_4 \quad x_5 = b$

Fig. 2.6 A partition of (a,b)

We say $g:(a,b) \to \mathbf{R}$ is *piecewise constant* if there is a partition $a = x_0 < x_1 < \cdots < x_n = b$, such that g restricted to (x_{i-1}, x_i) is constant for $i = 1, \ldots, n$ (in this definition, the values of g at the points x_i are not restricted in any way). The *mesh* δ of the partition $a = x_0 < x_1 < \cdots < x_n = b$, by definition, is the largest length of the subintervals, $\delta = \max_{1 \le i \le n} |x_i - x_{i-1}|$. Note that an interval has partitions of arbitrarily small mesh iff the interval is bounded.

Let $f:[a,b] \to \mathbf{R}$ be continuous. Then from above, f is uniformly continuous on (a,b). Given a partition $a = x_0 < x_1 < \cdots < x_n = b$ with mesh δ, choose $x_i^\#$ in (x_{i-1}, x_i) arbitrarily, $i = 1, \ldots, n$. Then by definition of μ, $|f(x) - f(x_i^\#)| \le \mu(\delta)$ for $x \in (x_{i-1}, x_i)$. If we set $g(x) = f(x_i^\#)$ for $x \in (x_{i-1}, x_i)$, $i = 1, \ldots, n$, and $g(x_i) = f(x_i)$, $i = 0, 1, \ldots, n$, we obtain a piecewise constant function $g:[a,b] \to \mathbf{R}$ satisfying $|f(x) - g(x)| \le \mu(\delta)$ for every $x \in [a,b]$. Since f is uniformly continuous, $\mu(0+) = 0$. Hence, for any error tolerance $\epsilon > 0$, we can find a mesh δ, such that $\mu(\delta) < \epsilon$. We have derived the following (Figure 2.7).

Theorem 2.3.8. *If f is continuous on $[a,b]$, then for each $\epsilon > 0$, there is a piecewise constant function f_ϵ on $[a,b]$ such that*

$$|f(x) - f_\epsilon(x)| \le \epsilon, \qquad a \le x \le b. \square$$

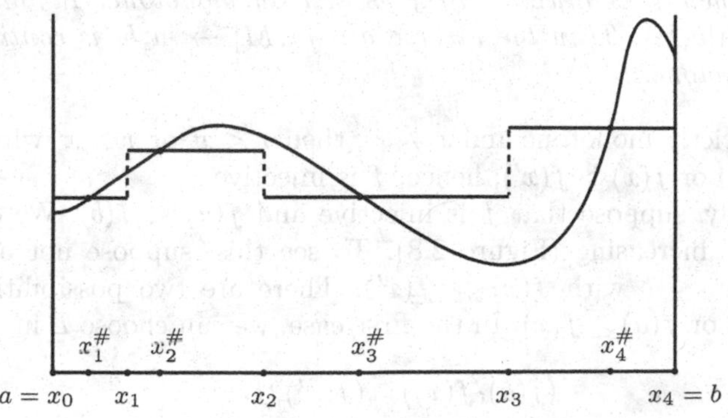

Fig. 2.7 Piecewise constant approximation

If f is continuous on an open interval, this result may be false. For example, $f(x) = 1/x$, $0 < x < 1$, cannot be approximated as above by a piecewise constant function (unless infinitely many subintervals are used), precisely because f "shoots up to ∞" near 0.

Let us turn to the continuity of compositions (§1.1). Suppose that $f : (a,b) \to \mathbf{R}$ and $g : (c,d) \to \mathbf{R}$ are given with the range of f lying in the domain of g, $f[(a,b)] \subset (c,d)$. Then the *composition* $g \circ f : (a,b) \to \mathbf{R}$ is given by $(g \circ f)(x) = g[f(x)]$, $a < x < b$.

Theorem 2.3.9. *If f and g are continuous, so is $g \circ f$.*

Since f is continuous, $x_n \to c$ implies $f(x_n) \to f(c)$. Since g is continuous, $(g \circ f)(x_n) = g[f(x_n)] \to g[f(c)] = (g \circ f)(c)$. □

This result can be written as

$$\lim_{x\to c} g[f(x)] = g\left[\lim_{x\to c} f(x)\right].$$

Since $g(x) = |x|$ is continuous, this implies

$$\lim_{x\to c} |f(x)| = \left|\lim_{x\to c} f(x)\right|.$$

The final issue is the invertibility of continuous functions. Let $f : [a,b] \to [m,M]$ be a continuous function. When is there an inverse (§1.1) $g : [m,M] \to [a,b]$? If it exists, is the inverse g necessarily continuous? It turns out that the answers to these questions are related to the monotonicity properties (§2.2) of the continuous function. For example, if f is continuous and increasing on $[a,b]$ and $A \subset [a,b]$, $\sup f(A) = f(\sup A)$, and $\inf f(A) = f(\inf A)$ (Exercise **2.3.4**). It follows that the upper and lower limits of $(f(x_n))$ are $f(x^*)$ and $f(x_*)$, respectively, where x^*, x_* are the upper and lower limits of (x_n) (Exercise **2.3.5**).

Theorem 2.3.10 (Inverse Function Theorem). *Let f be continuous on $[a,b]$. Then f is injective iff f is strictly monotone. In this case, let $[m,M] = f([a,b])$. Then the inverse $g : [m,M] \to [a,b]$ is continuous and strictly monotone.*

If f is strictly monotone and $x \neq x'$, then $x < x'$ or $x > x'$ which implies $f(x) < f(x')$ or $f(x) > f(x')$; hence, f is injective.

Conversely, suppose that f is injective and $f(a) < f(b)$. We claim that f is strictly increasing (Figure 2.8). To see this, suppose not and choose $a \le x < x' \le b$ with $f(x) > f(x')$. There are two possibilities: Either $f(a) < f(x)$ or $f(a) \ge f(x)$. In the first case, we can choose L in

$$(f(a), f(x)) \cap (f(x'), f(x)).$$

By the intermediate value property, there are c, d with $a < c < x < d < x'$ with $f(c) = L = f(d)$. Since f is injective, this cannot happen, ruling out the first case. In the second case, we must have $f(x') < f(b)$; hence, $x' < b$, so we choose L in

$$(f(x'), f(x)) \cap (f(x'), f(b)).$$

By the intermediate value property, there are c, d with $x < c < x' < d < b$ with $f(c) = L = f(d)$. Since f is injective, this cannot happen, ruling out the second case. Thus f is strictly increasing. If $f(a) > f(b)$, applying what we just learned to $-f$ yields $-f$ strictly increasing or f strictly decreasing. Thus in either case, f is strictly monotone.

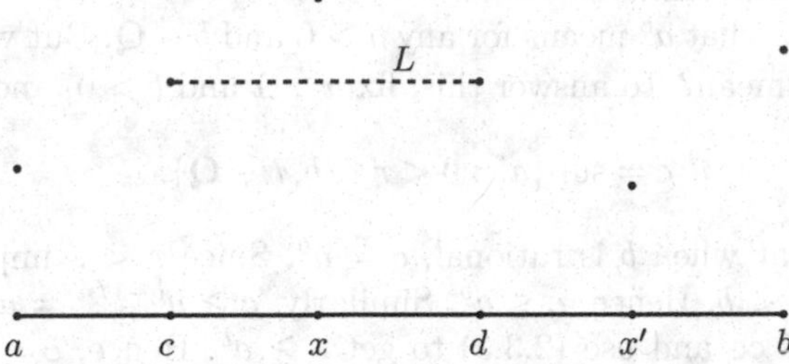

Fig. 2.8 Derivation of the IFT when $f(a) < f(b)$

Clearly strict monotonicity of f implies that of g. Now assume that f is strictly increasing, the case with f strictly decreasing being entirely similar. We have to show that g is continuous. Suppose that $(y_n) \subset [m, M]$ with $y_n \to y$. Let $x = g(y)$, let $x_n = g(y_n)$, $n \geq 1$, and let x^* and x_* denote the upper and lower limits of (x_n). We have to show $g(y_n) = x_n \to x = g(y)$. Since f is continuous and increasing, $f(x^*)$ and $f(x_*)$ are the upper and lower limits of $y_n = f(x_n)$ (Exercise **2.3.5**). Hence, $f(x^*) = y = f(x_*)$. Hence, by injectivity, $x^* = x = x_*$. □

As an application, note that $f(x) = x^2$ is strictly increasing on $[0, n]$ and hence has an inverse $g_n(x) = \sqrt{x}$ on $[0, n^2]$, for each $n \geq 1$. By uniqueness of inverses (Exercise **1.1.4**), the functions g_n, $n \geq 1$, agree wherever their domains overlap, hence yielding a single, continuous, strictly monotone $g : [0, \infty) \to [0, \infty)$ satisfying $g(x) = \sqrt{x}$, $x \geq 0$. Similarly, for each $n \geq 1$, $f(x) = x^n$ is strictly increasing on $[0, \infty)$. Thus every positive real x has a unique positive nth root $x^{1/n}$, and, moreover, the function $g(x) = x^{1/n}$ is continuous on $[0, \infty)$. By composition, it follows that $f(x) = x^{m/n} = (x^m)^{1/n}$ is continuous and strictly monotone on $[0, \infty)$ for all naturals m, n. Since $x^{-a} = 1/x^a$ for $a \in \mathbf{Q}$, we see that the *power functions* $f(x) = x^r$ are defined, strictly increasing, and continuous on $(0, \infty)$ for all rationals r. Moreover, $x^{r+s} = x^r x^s$, $(x^r)^s = x^{rs}$ for r, s rational, and, for $r > 0$ rational, $x^r \to 0$ as $x \to 0$ and $x^r \to \infty$ as $x \to \infty$. The following limit is important: For $x > 0$,

$$\lim_{n \nearrow \infty} x^{1/n} = 1. \tag{2.3.3}$$

To derive this, assume $x \geq 1$. Then $x \leq xx^{1/n} = x^{(n+1)/n}$, so $x^{1/(n+1)} \leq x^{1/n}$, so the sequence $(x^{1/n})$ is decreasing and bounded below by 1; hence, its limit $L \geq 1$ exists. Since $L \leq x^{1/2n}$, $L^2 \leq x^{2/2n} = x^{1/n}$; hence, $L^2 \leq L$ or $L \leq 1$. We conclude that $L = 1$. If $0 < x < 1$, then $1/x > 1$, so $x^{1/n} = 1/(1/x)^{1/n} \to 1$ as $n \nearrow \infty$.

Any function that can be obtained from polynomials or rational functions by arithmetic operations and/or the taking of roots is called a *(constructible) algebraic function*. For example,

$$f(x) = \frac{1}{\sqrt{x(1-x)}}, \qquad 0 < x < 1,$$

is an algebraic function.

We now know what a^b means for any $a > 0$ and $b \in \mathbf{Q}$. But what if $b \notin \mathbf{Q}$? What does $2^{\sqrt{2}}$ mean? To answer this, fix $a > 1$ and $b > 0$, and let

$$c = \sup\{a^r : 0 < r < b, r \in \mathbf{Q}\}.$$

Let us check that when b is rational, $c = a^b$. Since $r < s$ implies $a^r < a^s$, $a^r \leq a^b$ when $r < b$. Hence, $c \leq a^b$. Similarly, $c \geq a^{b-1/n} = a^b/a^{1/n}$ for all $n \geq 1$. Let $n \nearrow \infty$ and use (2.3.3) to get $c \geq a^b$. Hence, $c = a^b$ when b is rational. Thus it is consistent to *define*, for any $a > 1$ and *real* $b > 0$,

$$a^b = \sup\{a^r : 0 < r < b, r \in \mathbf{Q}\},$$

$a^0 = 1$, and $a^{-b} = 1/a^b$. For all b real, we define $1^b = 1$, whereas for $0 < a < 1$, we define $a^b = 1/(1/a)^b$. This defines $a^b > 0$ for all positive real a and all real b. Moreover (Exercise **2.3.7**),

$$a^b = \inf\{a^s : s > b, s \in \mathbf{Q}\}.$$

Theorem 2.3.11. a^b *satisfies the usual rules:*

A. *For* $a > 1$ *and* $0 < b < c$ *real,* $1 < a^b < a^c$.
B. *For* $0 < a < 1$ *and* $0 < b < c$ *real,* $a^b > a^c$.
C. *For* $0 < a < b$ *and* $c > 0$ *real,* $a^c b^c = (ab)^c$, $(b/a)^c = b^c/a^c$, *and* $a^c < b^c$.
D. *For* $a > 0$ *and* b, c *real,* $a^{b+c} = a^b a^c$.
E. *For* $a > 0$, b, c *real,* $a^{bc} = \left(a^b\right)^c$.

Since $A \subset B$ implies $\sup A \leq \sup B$, $a^b \leq a^c$ when $a > 1$ and $b < c$. Since, for any $b < c$, there is an $r \in \mathbf{Q} \cap (b, c)$, $a^b < a^c$, thus the first assertion. Since, for $0 < a < 1$, $a^b = 1/(1/a)^b$, applying the first assertion to $1/a$ yields $(1/a)^b < (1/a)^c$ or $a^b > a^c$, yielding the second assertion. For the third, assume $a > 1$. If $0 < r < c$ is in $\mathbf{Q}$, then $a^r < a^c$ and $b^r < b^c$ yields

$$(ab)^r = a^r b^r < a^c b^c.$$

Taking the sup over $r < c$ yields $(ab)^c \le a^c b^c$. If $r < c$ and $s < c$ are positive rationals, let t denote their max. Then

$$a^r b^s \le a^t b^t = (ab)^t < (ab)^c.$$

Taking the sup of this last inequality over all $0 < r < c$, first, then over all $0 < s < c$ yields $a^c b^c \le (ab)^c$. Hence, $(ab)^c = a^c b^c$ for $b > a > 1$. Using this, we obtain $(b/a)^c a^c = b^c$ or $(b/a)^c = b^c/a^c$. Since $b/a > 1$ implies $(b/a)^c > 1$, we also obtain $a^c < b^c$. The cases $a < b < 1$ and $a < 1 < b$ follow from the case $b > a > 1$. This establishes the third. For the fourth, the case $0 < a < 1$ follows from the case $a > 1$, so assume $a > 1$, $b > 0$, and $c > 0$. If $r < b$ and $s < c$ are positive rationals, then

$$a^{b+c} \ge a^{r+s} = a^r a^s.$$

Taking the sups over r and s yields $a^{b+c} \ge a^b a^c$. If $r < b + c$ is rational, let $d = (b+c-r)/3 > 0$. Pick rationals t and s with $b > t > b-d$, $c > s > c-d$. Then $t + s > b + c - 2d > r$, so

$$a^r < a^{t+s} = a^t a^s \le a^b a^c.$$

Taking the sup over all such r, we obtain $a^{b+c} \le a^b a^c$. This establishes the fourth when b and c are positive. The cases $b \le 0$ or $c \le 0$ follow from the positive case. The fifth involves approximating b and c by rationals, and we leave it to the reader. □

As an application, we define the power function with an irrational exponent. This is a nonalgebraic or *transcendental* function. Some of the transcendental functions in this book are the power function x^a (when a is irrational), the exponential function a^x, the logarithm $\log_a x$, the trigonometric functions and their inverses, and the gamma function. The trigonometric functions are discussed in §3.5, the gamma function in §5.1, whereas the power, exponential, and logarithm functions are discussed below.

Theorem 2.3.12. *Let a be real, and let $f(x) = x^a$ on $(0, \infty)$. For $a > 0$, f is strictly increasing and continuous with $f(0+) = 0$ and $f(\infty) = \infty$. For $a < 0$, f is strictly decreasing and continuous with $f(0+) = \infty$ and $f(\infty) = 0$.*

Since $x^{-a} = 1/x^a$, the second part follows from the first, so assume $a > 0$. Let r, s be positive rationals with $r < a < s$, and let $x_n \to c$. We have to show that $x_n^a \to c^a$. But the sequence (x_n^a) lies between (x_n^r) and (x_n^s). Since we already know that the rational power function is continuous, we conclude that the upper and lower limits L^*, L_* of (x_n^a) satisfy $c^r \le L_* \le L^* \le c^s$. Taking the sup over all r rational and the inf over all s rational, with $r < a < s$, gives $L^* = L_* = c^a$. Thus f is continuous. Also since $x^r \to \infty$ as $x \to \infty$ and $x^r \le x^a$ for $r < a$, $f(\infty) = \infty$. Since $x^a \le x^s$ for $s > a$ and $x^s \to 0$ as $x \to 0+$, $f(0+) = 0$. □

Now we vary b and fix a in a^b.

Theorem 2.3.13. *Fix $a > 1$. Then the function $f(x) = a^x$, $x \in \mathbf{R}$ is strictly increasing and continuous. Moreover,*

$$f(x + x') = f(x)f(x'), \tag{2.3.4}$$

$f(-\infty) = 0$, $f(0) = 1$, and $f(\infty) = \infty$.

From Theorem 2.3.11, we know that f is strictly increasing. Since $a^n \nearrow \infty$ as $n \nearrow \infty$, $f(\infty) = \infty$. Since $f(-x) = 1/f(x)$, $f(-\infty) = 0$. Continuity remains to be shown. If $x_n \searrow c$, then (a^{x_n}) is decreasing and $a^{x_n} \ge a^c$, so its limit L is $\ge a^c$. On the other hand, for $d > 0$, the sequence is eventually below $a^{c+d} = a^c a^d$; hence, $L \le a^c a^d$. Choosing $d = 1/n$, we obtain $a^c \le L \le a^c a^{1/n}$. Let $n \nearrow \infty$ to get $L = a^c$. Thus, $a^{x_n} \searrow a^c$. If $x_n \to c+$ is not necessarily decreasing, then $x_n^* \searrow c$; hence, $a^{x_n^*} \to a^c$. But $x_n^* \ge x_n$ for all $n \ge 1$; hence, $a^{x_n^*} \ge a^{x_n} \ge a^c$, so $a^{x_n} \to a^c$. Proceed similarly from the left. □

The function $f(x) = a^x$ is the *exponential function with base* $a > 1$. In fact, the exponential is the unique continuous function f on $\mathbf{R}$ satisfying the functional equation (2.3.4) and $f(1) = a$.

By the inverse function theorem, f has an inverse g on any compact interval and hence on $\mathbf{R}$. We call g the *logarithm with base* $a > 1$ and write $g(x) = \log_a x$. By definition of inverse, $a^{\log_a x} = x$, for $x > 0$, and $\log_a(a^x) = x$, for $x \in \mathbf{R}$. The following is an immediate consequence of the above.

Theorem 2.3.14. *The inverse of the exponential $f(x) = a^x$ with base $a > 1$ is the logarithm with base $a > 1$, $g(x) = \log_a x$. The logarithm is continuous and strictly increasing on $(0, \infty)$. The domain of $\log_a$ is $(0, \infty)$, the range is $\mathbf{R}$, $\log_a(0+) = -\infty$, $\log_a 1 = 0$, $\log_a \infty = \infty$, and*

$$\log_a(bc) = \log_a b + \log_a c, \qquad \log_a(b^c) = c \log_a b,$$

for $b > 0$, $c > 0$. □

Exercises

2.3.1. If f is a polynomial of odd degree, then $f(\pm\infty) = \pm\infty$ or $f(\pm\infty) = \mp\infty$, and there is at least one real c with $f(c) = 0$.

2.3.2. If f is continuous at c, then[6] $\mu_c(0+) = 0$.

2.3.3. If $f : (a, b) \to \mathbf{R}$ is continuous, then $f((a, b))$ is an interval. In addition, if f is strictly monotone, $f((a, b))$ is an open interval.

---

[6] This uses the axiom of countable choice.

2.3.4. If f is continuous and increasing on $[a,b]$ and $A \subset [a,b]$, then $\sup f(A) = f(\sup A)$ and $\inf f(A) = f(\inf A)$.

2.3.5. With f as in Exercise **2.3.4**, let x^* and x_* be the upper and lower limits of a sequence (x_n). Then $f(x^*)$ and $f(x_*)$ are the upper and lower limits of $(f(x_n))$.

2.3.6. With $r, s \in \mathbf{Q}$ and $x > 0$, show that $(x^r)^s = x^{rs}$ and $x^{r+s} = x^r x^s$.

2.3.7. Show that $a^b = \inf\{a^s : s > b, s \in \mathbf{Q}\}$.

2.3.8. With b and c real and $a > 0$, show that $(a^b)^c = a^{bc}$.

2.3.9. Fix $a > 0$. If $f : \mathbf{R} \to \mathbf{R}$ is continuous, $f(1) = a$, and $f(x + x') = f(x)f(x')$ for $x, x' \in \mathbf{R}$, then $f(x) = a^x$.

2.3.10. Use the ϵ-δ criterion to show that $f(x) = 1/x$ is continuous at $x = 1$.

2.3.11. A real x is *algebraic* if x is a root of a polynomial of degree $d \ge 1$,

$$a_0 x^d + a_1 x^{d-1} + \cdots + a_{d-1} x + a_d = 0,$$

with rational coefficients $a_0, a_1, \ldots, a_d$. A real is *transcendental* if it is not algebraic. For example, every rational is algebraic. Show that the set of algebraic numbers is countable (§1.7). Conclude that the set of transcendental numbers is uncountable.

2.3.12. Let a be an algebraic number. If $f(a) = 0$ for some polynomial f with rational coefficients, but $g(a) \ne 0$ for any polynomial g with rational coefficients of lesser degree, then f is a *minimal polynomial* for a, and the degree of f is the *algebraic order* of a. Now suppose that a is algebraic of order $d \ge 2$. Show that all the roots of a minimal polynomial f are irrational.

2.3.13. Suppose that the algebraic order of a is $d \ge 2$. Then there is a $c > 0$, such that

$$\left| a - \frac{m}{n} \right| \ge \frac{c}{n^d}, \qquad n, m \ge 1.$$

(See Exercise **1.4.9**. Here you will need the modulus of continuity μ_a at a of $g(x) = f(x)/(x - a)$, where f is a minimal polynomial of a.)

2.3.14. Use the previous exercise to show that

$$.1100010\ldots010\cdots = \frac{1}{10} + \frac{1}{10^2} + \frac{1}{10^6} + \cdots = \sum_{n=1}^{\infty} \frac{1}{10^{n!}}$$

is transcendental.

2.3.15. For $s > 1$ real, $\sum_{n=1}^{\infty} n^{-s}$ converges.

2.3.16. If $a > 1$, $b > 0$, and $c > 0$, then $b^{\log_a c} = c^{\log_a b}$, and $\sum_{n=1}^{\infty} 5^{-\log_3 n}$ converges.

2.3.17. Give an example of an $f : [0,1] \to [0,1]$ that is invertible but not monotone.

2.3.18. Let f be of bounded variation (Exercise **2.2.4**) on (a,b). Then the set of points at which f is not continuous is at most countable. Moreover, every discontinuity, at worst, is a jump.

2.3.19. Let $f : (a,b) \to \mathbf{R}$ be continuous and let $M = \sup\{f(x) : a < x < b\}$. Assume $f(a+)$ exists with $f(a+) < M$ and $f(b-)$ exists with $f(b-) < M$. Then $\sup\{f(x) : a < x < b\}$ is attained. Use Theorem 2.1.2.

2.3.20. If $f : \mathbf{R} \to \mathbf{R}$ satisfies

$$\lim_{x \to \pm\infty} \frac{f(x)}{|x|} = +\infty,$$

we say that f is *superlinear*. If f is superlinear and continuous, then the sup is attained in

$$g(y) = \sup_{-\infty < x < \infty} [xy - f(x)] = \max_{-\infty < x < \infty} [xy - f(x)],$$

and g is superlinear. Use Exercise **2.3.19**.

2.3.21. If $f : \mathbf{R} \to \mathbf{R}$ is superlinear and continuous and g is as above, then g is also continuous. (Modify the logic of the previous solution.)

2.3.22. Let $f(x) = 1 + \lfloor x \rfloor - x$, $x \in \mathbf{R}$, where $\lfloor x \rfloor$ denotes the greatest integer $\leq x$ (Figure 2.3). Compute

$$\lim_{n \nearrow \infty} \left(\lim_{m \nearrow \infty} [f(n!x)]^m \right)$$

for $x \in \mathbf{Q}$ and for $x \notin \mathbf{Q}$.

2.3.23. Let $f(x) = 1/x$, $0 < x < 1$. Compute $\mu_c(\delta)$ explicitly for $0 < c < 1$ and $\delta > 0$. With $I = (0,1)$, show that $\mu_I(\delta) = \infty$ for all $\delta > 0$. Conclude that f is not uniformly continuous on $(0,1)$. (There are two cases, $c \leq \delta$ and $c > \delta$.)

2.3.24. Let $f : \mathbf{R} \to \mathbf{R}$ be continuous, and suppose that $f(\infty)$ and $f(-\infty)$ exist and are finite. Show that f is uniformly continuous on $\mathbf{R}$.

2.3.25. Use $\sqrt{2}^{\sqrt{2}}$ to show that there are irrationals a, b, such that a^b is rational. (Consider the two cases $\sqrt{2}^{\sqrt{2}} \in \mathbf{Q}$ and $\sqrt{2}^{\sqrt{2}} \notin \mathbf{Q}$.)

Chapter 3
Differentiation

3.1 Derivatives

Let f be defined on (a,b), and choose $c \in (a,b)$. We say that f is *differentiable at* c if

$$\lim_{x\to c} \frac{f(x)-f(c)}{x-c}$$

exists as a real, i.e., exists and is not $\pm\infty$. If it exists, we denote this limit $f'(c)$ or $\dfrac{df}{dx}(c)$, and we say that $f'(c)$ is the *derivative* of f at c. If f is differentiable at c for all $a < c < b$, we say that f is *differentiable* on (a,b) or, if it is clear from the context, *differentiable*. In this case, the derivative $f' : (a,b) \to \mathbf{R}$ is a function defined on all of (a,b).

For example, the function $f(x) = mx + b$ is differentiable on $\mathbf{R}$ with derivative $f'(c) = m$ for all c since

$$\lim_{x\to c} \frac{(mx+b)-(mc+b)}{x-c} = \lim_{x\to c} m = m.$$

Since its graph is a line, the derivative of $f(x) = mx + b$ (at any real) is the slope of its graph. In particular, the derivative of a constant function $f(x) = b$ for all x is zero.

If $f(x) = x^2$, then f is differentiable with derivative

$$f'(c) = \lim_{x\to c} \frac{x^2-c^2}{x-c} = \lim_{x\to c} \frac{(x-c)(x+c)}{x-c} = \lim_{x\to c}(x+c) = 2c.$$

O. Hijab, *Introduction to Calculus and Classical Analysis*, Undergraduate Texts in Mathematics, DOI 10.1007/978-3-319-28400-2_3

If f is differentiable at c, then

$$\begin{aligned}\lim_{x\to c} f(x) &= \lim_{x\to c}\left[\left(\frac{f(x)-f(c)}{x-c}\right)(x-c)+f(c)\right]\\ &= \lim_{x\to c}\left(\frac{f(x)-f(c)}{x-c}\right)\lim_{x\to c}(x-c)+f(c) = f'(c)\cdot 0+f(c) = f(c).\end{aligned}$$

So f is continuous at c. Hence, *a differentiable function is continuous.*

However, $f(x)=|x|$ is continuous at 0 but not differentiable there since

$$\lim_{x\to 0+}\frac{|x|-|0|}{x-0}=1,$$

whereas

$$\lim_{x\to 0-}\frac{|x|-|0|}{x-0}=-1.$$

However,

$$(|x|)'=\frac{x}{|x|}, \qquad x\neq 0,$$

since $|x|=x$; hence, $(|x|)'=1$ on $(0,\infty)$, and $|x|=-x$; hence, $(|x|)'=-1$ on $(-\infty,0)$.

Derivatives are computed using their arithmetic properties.

Theorem 3.1.1. *If f and g are differentiable on (a,b), and k is real, so are $f+g$, kf, fg, and, for $a<x<b$,*

$$\begin{aligned}(f+g)'(x) &= f'(x)+g'(x),\\ (kf)'(x) &= kf'(x),\\ (fg)'(x) &= f'(x)g(x)+f(x)g'(x).\end{aligned}$$

Moreover, if g is nonzero on (a,b), then f/g is differentiable and

$$\left(\frac{f}{g}\right)'(x)=\frac{f'(x)g(x)-f(x)g'(x)}{g(x)^2}, \qquad a<x<b.$$

The first and second identities are *linearity* of the derivative, the third is the *product rule*, and the last is the *quotient rule*. To derive these rules, let $a<c<b$. For sums,

$$\begin{aligned}(f+g)'(c) &= \lim_{x\to c}\frac{(f(x)+g(x))-(f(c)+g(c))}{x-c}\\ &= \lim_{x\to c}\frac{f(x)-f(c)}{x-c}+\lim_{x\to c}\frac{g(x)-g(c)}{x-c}\\ &= f'(c)+g'(c).\end{aligned}$$

For scalar multiplication,

$$\begin{aligned}(kf)'(c) &= \lim_{x\to c} \frac{kf(x)-kf(c)}{x-c} \\ &= k\lim_{x\to c} \frac{f(x)-f(c)}{x-c} = kf'(c).\end{aligned}$$

For products,

$$\begin{aligned}(fg)'(c) &= \lim_{x\to c} \frac{f(x)g(x)-f(c)g(c)}{x-c} \\ &= \lim_{x\to c} \frac{f(x)g(x)-f(c)g(x)+f(c)g(x)-f(c)g(c)}{x-c} \\ &= \lim_{x\to c} \frac{f(x)-f(c)}{x-c} \lim_{x\to c} g(x) + f(c)\lim_{x\to c} \frac{g(x)-g(c)}{x-c} \\ &= f'(c)g(c)+f(c)g'(c).\end{aligned}$$

For quotients,

$$\begin{aligned}\left(\frac{f}{g}\right)'(c) &= \lim_{x\to c} \frac{f(x)/g(x)-f(c)/g(c)}{x-c} \\ &= \lim_{x\to c} \frac{f(x)g(c)-f(c)g(x)}{(x-c)g(x)g(c)} \\ &= \lim_{x\to c} \frac{\left(\dfrac{f(x)-f(c)}{x-c}\right)\cdot g(c) - f(c)\cdot \left(\dfrac{g(x)-g(c)}{x-c}\right)}{g(x)g(c)} \\ &= \frac{f'(c)g(c)-f(c)g'(c)}{g(c)^2}. \square\end{aligned}$$

Above, we saw that the derivative of $f(x)=x$ is $f'(x)=1$. By induction, we show that the derivative of the monomial $f(x)=x^n$ is nx^{n-1}. Since this is true for $n=1$, assume it is true for $n\ge 1$. Then by the product rule if $f(x)=x^{n+1}$,

$$f'(x) = \left(x^{n+1}\right)' = (x^n x)' = (x^n)'\,x + x^n (x)' = nx^{n-1}x + x^n(1) = (n+1)x^n.$$

This establishes that $(x^n)' = nx^{n-1}$ for all $n\ge 1$. Explicitly, this means $(x^n-c^n)/(x-c)$ converges to nc^{n-1} as $x\to c$, for all c real. A more vivid description of this convergence is given in (3.4.4).

Since polynomials are linear combinations of monomials, they are differentiable everywhere. For example,

$$(x^3+5x+1)' = (x^3)' + (5x)' + (1)' = 3x^2+5.$$

Moreover,

$$(x^n)' = nx^{n-1}, \qquad n \in \mathbf{Z}, x \neq 0. \tag{3.1.1}$$

This is clear for $n = 0$, whereas, for $n \geq 1$, using the quotient rule, we find that

$$\begin{aligned}(x^{-n})' = \left(\frac{1}{x^n}\right)' &= \frac{(1)'x^n - 1(x^n)'}{(x^n)^2} \\ &= \frac{0 \cdot x^n - nx^{n-1}}{x^{2n}} = -\frac{n}{x^{n+1}} = -nx^{-n-1}.\end{aligned}$$

This establishes (3.1.1). Another consequence of the quotient rule is that a rational function is differentiable wherever it is defined. For example,

$$\left(\frac{x^2-1}{x^2+1}\right)' = \frac{(2x)(x^2+1) - (x^2-1)(2x)}{(x^2+1)^2} = \frac{4x}{(x^2+1)^2}.$$

We say that a function g is *tangent to* f *at* c if the difference $f(x) - g(x)$ vanishes faster than first order in $x - c$, i.e., if

$$\lim_{x \to c} \frac{f(x) - g(x)}{x - c} = 0.$$

Suppose that $g(x) = mx + b$ is tangent to f at c. Since the graph of g is a line, it is reasonable to call it the *line tangent to* f *at* $(c, f(c))$ or, more simply, the *tangent line at* c. Note two lines are tangent to each other iff they coincide. Thus, a function f can have, at most, one tangent line at a given real c.

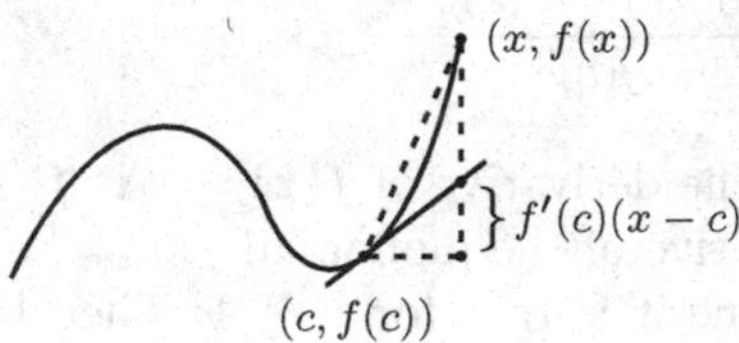

Fig. 3.1 The derivative is the slope of the tangent line

If f is differentiable at c, then $g(x) = f'(c)(x - c) + f(c)$ is tangent to f at c, since

$$\begin{aligned}\lim_{x \to c} \frac{f(x) - g(x)}{x - c} &= \lim_{x \to c} \frac{f(x) - f(c) - f'(c)(x - c)}{x - c} \\ &= \lim_{x \to c} \frac{f(x) - f(c)}{x - c} - f'(c) = 0.\end{aligned}$$

Hence, *the derivative* $f'(c)$ *of* f *at* c *is the slope of the tangent line at* c *(Figure 3.1).*

If f is differentiable at c, there is a positive k and some interval $(c-d, c+d)$ about c on which

$$|f(x) - f(c)| \leq k|x - c|, \qquad c - d < x < c + d. \tag{3.1.2}$$

Indeed, if this were not so for each $n \geq 1$, we would find a real $x_n \in (c - 1/n, c + 1/n)$ contradicting this claim, i.e., satisfying

$$\left| \frac{f(x_n) - f(c)}{x_n - c} \right| > n.$$

But then $x_n \to c$, and hence, this inequality would contradict differentiability at c.

The following describes the behavior of derivatives under composition.

Theorem 3.1.2 (Chain Rule). *Let f, g be differentiable on (a,b), (c,d), respectively. If $f((a,b)) \subset (c,d)$, then $g \circ f$ is differentiable on (a,b) with*

$$(g \circ f)'(x) = g'(f(x))f'(x), \qquad a < x < b.$$

To see this, let $a < c < b$, and assume, first, $f'(c) \neq 0$. Then $x_n \to c$ and $x_n \neq c$ for all $n \geq 1$ imply $f(x_n) \to f(c)$ and $(f(x_n) - f(c))/(x_n - c) \to f'(c) \neq 0$. Hence, there is an $N \geq 1$, such that $f(x_n) - f(c) \neq 0$ for $n \geq N$. Thus,

$$\begin{aligned}
\lim_{n\nearrow\infty} \frac{g(f(x_n)) - g(f(c))}{x_n - c} &= \lim_{n\nearrow\infty} \frac{g(f(x_n)) - g(f(c))}{f(x_n) - f(c)} \cdot \frac{f(x_n) - f(c)}{x_n - c} \\
&= \lim_{n\nearrow\infty} \frac{g(f(x_n)) - g(f(c))}{f(x_n) - f(c)} \lim_{n\nearrow\infty} \frac{f(x_n) - f(c)}{x_n - c} \\
&= g'(f(c))f'(c).
\end{aligned}$$

Since $x_n \to c$ and $x_n \neq c$ for all $n \geq 1$, by definition of $\lim_{x\to c}$ (§2.2),

$$(g \circ f)'(c) = \lim_{x\to c} \frac{g(f(x)) - g(f(c))}{x - c} = g'(f(c))f'(c).$$

This establishes the result when $f'(c) \neq 0$. If $f'(c) = 0$, by (3.1.2) there is a k with

$$|g(y) - g(f(c))| \leq k|y - f(c)|$$

for y near $f(c)$. Since $x \to c$ implies $f(x) \to f(c)$, in this case, we obtain

$$\begin{aligned}
|(g \circ f)'(c)| &= \lim_{x\to c} \left| \frac{g(f(x)) - g(f(c))}{x - c} \right| \\
&\leq \lim_{x\to c} \frac{k|f(x) - f(c)|}{|x - c|} \\
&= k|f'(c)| = 0.
\end{aligned}$$

Hence, $(g \circ f)'(c) = 0 = g'(f(c))f'(c)$. □

For example,

$$\left(\left(1 - \frac{x}{n}\right)^n\right)' = n\left(1 - \frac{x}{n}\right)^{n-1} \cdot \left(-\frac{1}{n}\right) = -\left(1 - \frac{x}{n}\right)^{n-1}$$

follows by choosing $g(x) = x^n$ and $f(x) = 1 - x/n$, $0 < x < n$.

If we set $u = f(x)$ and $y = g(u) = g(f(x))$, then the chain rule takes the easily remembered form

$$\frac{dy}{dx} = \frac{dy}{du} \cdot \frac{du}{dx}.$$

We say that $f : (a, b) \to \mathbf{R}$ has a *local* maximum at $c \in (a, b)$ if, for some $\delta > 0$, $f(x) \leq f(c)$ on $(c - \delta, c + \delta)$. Similarly, we say that f has a local minimum at $c \in (a, b)$ if, for some $\delta > 0$, $f(x) \geq f(c)$ on $(c - \delta, c + \delta)$. Alternatively, we say that c is a local max or a local min of f. If, instead, these inequalities hold for all x in (a, b), then we say that c is a (*global*) maximum or a (global) minimum of f on (a, b). It is possible for a function to have a local maximum at every rational (see Exercise **3.1.9**).

A *critical point* of a differentiable f is a real c with $f'(c) = 0$. A *critical value* of f is a real d, such that $d = f(c)$ for some critical point c.

Let f be defined on (a, b). Suppose that f has a local minimum at c and is differentiable there. Then for $x > c$ near c, $f(x) \geq f(c)$, so

$$f'(c) = \lim_{x \to c+} \frac{f(x) - f(c)}{x - c} \geq 0.$$

For $x < c$ near c, $f(x) \geq f(c)$, so

$$f'(c) = \lim_{x \to c-} \frac{f(x) - f(c)}{x - c} \leq 0.$$

Hence, $f'(c) = 0$. Applying this result to $g = -f$, we see that if f has a local maximum at c, then $f'(c) = 0$. We conclude that *a local maximum or a local minimum is a critical point.* The converse is not, generally, true since $c = 0$ is a critical point of $f(x) = x^3$ but is neither a local maximum nor a local minimum.

Using critical points, one can maximize and minimize functions over their domains: To compute

$$M = \sup_{a<x<b} f(x),$$

either the sup is attained at some $a < x < b$ or $M = f(a+)$ or $M = f(b-)$, assuming these exist (Exercise **2.3.19**). When f is differentiable, it is enough therefore to compute the critical values of f and compare them with $f(a+)$ and $f(b-)$. If the largest of these values is $f(c)$ for some critical point $c \in (a, b)$, then f is maximized at c. If the largest of these values is $f(b-)$ or $f(a+)$, then f has a sup but no maximum over (a, b). For example,

$$\max_{-\infty<x<\infty}(6x-x^2)=9$$

since the only critical point of $f(x) = 6x - x^2$ is at $x = 3$ and $f(\infty) = f(-\infty) = -\infty$. Proceed similarly for computing minima.

Theorem 3.1.3 (Mean Value Theorem). *If f is continuous on $[a,b]$ and differentiable on (a,b), then there is a c in (a,b) with*

$$f'(c)=\frac{f(b)-f(a)}{b-a}.$$

To see this (Figure 3.2), first we subtract a line from f by setting

$$g(x)=f(x)-\left\{\left[\frac{f(b)-f(a)}{b-a}\right](x-a)+f(a)\right\},\qquad a\le x\le b.$$

Then g is continuous on $[a,b]$, differentiable on (a,b), and $g(a)=g(b)=0$. If $g(x)=0$ everywhere on $[a,b]$, let $a<c<b$ be any real. If $g(x)>0$ somewhere in (a,b), let c be a real at which g is maximized. If $g(x)<0$ somewhere in (a,b), let c be a real at which g is minimized. In all three cases, we obtain $g'(c)=0$. Since

$$g'(c)=f'(c)-\frac{f(b)-f(a)}{b-a},$$

we are done. □

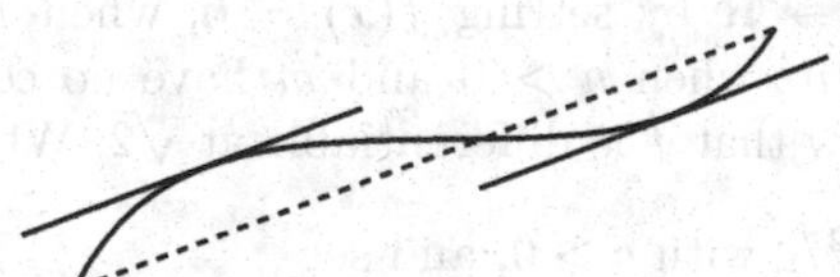

Fig. 3.2 The mean value theorem

For example, choose $f(x)=(1-x/n)^n$, $a=0$, $b>0$. Then $f'(x) = -(1-x/n)^{n-1}$ is between -1 and 0 when $0<x<n$. By the mean value theorem, we conclude that

$$0\le\frac{1-(1-b/n)^n}{b}\le 1,\qquad 0<b<n, n\ge 1,$$

since the ratio equals the negative of $(f(b)-f(0))/(b-0)$. The point of this inequality is that, when $b>0$ is small, the numerator is small enough to compensate for the smallness of the denominator, yielding a quotient bounded between 0 and 1.

As a consequence of the mean value theorem, *if f and g are differentiable on (a,b) and $f'(x)=g'(x)$ for all x, then f and g differ by a constant;* $f(x) =$

$g(x) + C$. To see this, note that $h(x) = f(x) - g(x)$ satisfies $h'(x) = 0$, so by the mean value theorem $(h(c) - h(d))/(c - d)$ equals h' at some intermediate real. Hence, $h(c) = h(d)$; hence, h is a constant function.

Let $(-b, b)$ be an interval symmetric about 0. Given a function $f : (-b, b) \to \mathbf{R}$, its *even part* f^e is the function

$$f^e(x) = \frac{f(x) + f(-x)}{2},$$

and its *odd part* f^o is

$$f^o(x) = \frac{f(x) - f(-x)}{2}.$$

Clearly, $f = f^e + f^o$.

A function f is *even* over $(-b, b)$ if $f = f^e$ on $(-b, b)$ and *odd* over $(-b, b)$ if $f = f^o$ on $(-b, b)$. Thus, an even function satisfies $f(-x) = f(x)$ on $(-b, b)$, whereas an odd function satisfies $f(-x) = -f(x)$ on $(-b, b)$. For example, x^n is even or odd on $\mathbf{R}$ according to whether n is even or odd.

Exercises

3.1.1. Let $a > 0$ and define $f(x) = |x|^a$. Show that f is differentiable at 0 iff $a > 1$.

3.1.2. Define $f : \mathbf{R} \to \mathbf{R}$ by setting $f(x) = 0$, when x is irrational, and setting $f(m/n) = 1/n^3$ when $n > 0$ and m have no common factor. Use Exercise **1.4.9** to show that f is differentiable at $\sqrt{2}$. What is $f'(\sqrt{2})$?

3.1.3. Let $f(x) = ax^2/2$ with $a > 0$, and set

$$g(y) = \sup_{-\infty < x < \infty} (xy - f(x)), \qquad y \in \mathbf{R}. \tag{3.1.3}$$

By direct computation, show that $g(y) = y^2/2a$ and f' and g' are inverses.

3.1.4. If $g : \mathbf{R} \to \mathbf{R}$ is superlinear (Exercise **2.3.20**) and differentiable, then $g'(\mathbf{R})$ is unbounded above and below; $\sup g'(\mathbf{R}) = \infty$ and $\inf g'(\mathbf{R}) = -\infty$. (Argue by contradiction, and use the mean value theorem.)

3.1.5. Suppose that f is continuous on (a, b), differentiable on (a, b) punctured at c, $a < c < b$, and $\lim_{x \to c} f'(x) = L$ exists. Show that $f'(c)$ exists and equals L.

3.1.6. Suppose that $f : (a, b) \to \mathbf{R}$ is differentiable, $a < c < b$, and $f'(c+)$ and $f'(c-)$ exist. Show that $f'(c+) = f'(c) = f'(c-)$. (As opposed to the previous exercise, here, we assume that $f'(c)$ exists.)

3.1.7. Suppose that f is differentiable on a bounded interval (a, b) with $|f'| \leq I$. Show that f is of bounded variation (Exercise **2.2.4**) over (a, b) with total variation $\leq I(b-a)$.

3.1.8. Show that the function $f : \mathbf{R} \to \mathbf{R}$ in Exercise **2.2.1** has a local maximum at every $c \in \mathbf{Q}$.

3.1.9. Suppose that $f : (-b, b) \to \mathbf{R}$ is differentiable. Then f' is even or odd if f is odd or even, respectively.

3.1.10. Suppose that $f : \mathbf{R} \to \mathbf{R}$ is continuous on $\mathbf{R}$ and f is differentiable at $r \in \mathbf{R}$. We say r is a *root* of f if $f(r) = 0$. Show that r is a root of f iff $f(x) = (x - r)g(x)$ for some continuous function $g : \mathbf{R} \to \mathbf{R}$.

3.1.11. Suppose $f : \mathbf{R} \to \mathbf{R}$ is continuous, and suppose f is differentiable at d distinct reals $r_1, \ldots, r_d$. Show that $r_1, r_2, \ldots, r_d$ are roots of f iff

$$f(x) = (x - r_1)(x - r_2)\ldots(x - r_d)g(x)$$

for some continuous function $g : \mathbf{R} \to \mathbf{R}$.

3.1.12. Let $f : \mathbf{R} \to \mathbf{R}$ be differentiable. Show that if f has d distinct roots $r_1, \ldots, r_d$, then f' has $d - 1$ distinct roots $s_1, \ldots, s_{d-1}$, where the s_j's are distinct from the r_j's.

3.2 Mapping Properties

To differentiate roots, we need to know how derivatives of inverses behave. But continuous functions are invertible iff they are strictly monotone (§2.3), so we begin by using the derivative to identify monotonicity.

Theorem 3.2.1. *Let f be differentiable on (a, b). If $f'(x) \neq 0$ for $a < x < b$, then f is strictly monotone on (a, b) and $f'(x) > 0$ on (a, b) or $f'(x) < 0$ on (a, b). Moreover, $f'(x) \geq 0$ on (a, b) iff f is increasing, and $f'(x) \leq 0$ on (a, b) iff f is decreasing.*

By the mean value theorem, given $a < x < y < b$, there is a c in (x, y) satisfying

$$f(y) - f(x) = f'(c)(y - x).$$

If f' is never zero, this shows that f is injective, hence, strictly monotone by the inverse function theorem (§2.3). This also shows that $f' \geq 0$ on (a, b) implies f is increasing and $f' \leq 0$ on (a, b) implies f is decreasing. Conversely, increasing f implies $f(x) \geq f(c)$ for $x > c$, so

$$f'(c) = \lim_{x \to c+} \frac{f(x) - f(c)}{x - c} \geq 0,$$

for all $a < c < b$. Similarly, if f is decreasing. In particular, we conclude that if f' is never zero and f is monotone, we must have $f' > 0$ on (a, b) or $f' < 0$ on (a, b). □

It is not, generally, true that strict monotonicity implies the nonvanishing of f'. For example, $f(x) = x^3$ is strictly increasing on $\mathbf{R}$, but $f'(0) = 0$.

Since its derivative was computed in the previous section, the function

$$f(x) = \frac{x^2 - 1}{x^2 + 1}$$

is strictly increasing on $(0, \infty)$ and strictly decreasing on $(-\infty, 0)$. Thus, the critical point $x = 0$ is a minimum of f on $\mathbf{R}$.

A useful consequence of this theorem is the following: *If f and g are differentiable on (a, b), continuous on $[a, b]$, $f(a) = g(a)$, and $f'(x) \geq g'(x)$ on (a, b), then $f(x) \geq g(x)$ on $[a, b]$.* This follows by applying the theorem to $h = f - g$.

Another consequence is that derivative functions, although themselves not necessarily continuous, satisfy the intermediate value property (Exercise **3.2.8**).

Now we can state the inverse function theorem for differentiable functions.

Theorem 3.2.2 (Inverse Function Theorem). *Let f be continuous on $[a, b]$ and differentiable on (a, b), and suppose that $f'(x) \neq 0$ on (a, b). Let $[m, M] = f([a, b])$. Then $f : [a, b] \to [m, M]$ is invertible, and its inverse g is continuous on $[m, M]$, differentiable on (m, M), and $g'(y) \neq 0$ on (m, M). Moreover,*

$$g'(y) = \frac{1}{f'(g(y))}, \qquad m < y < M.$$

Note, first, that $f' > 0$ on (a, b) or $f' < 0$ on (a, b) by the previous theorem. Suppose that $f' > 0$ on (a, b), the case $f' < 0$ being entirely similar. Then f is strictly increasing; hence, the range $[m, M]$ must equal $[f(a), f(b)]$, f is invertible, and its inverse g is strictly increasing and continuous. If $a < c < b$ and $y_n \to f(c)$, $y_n \neq f(c)$ for all $n \geq 1$, then $x_n = g(y_n) \to g(f(c)) = c$ and $x_n \neq c$ for all $n \geq 1$, so $y_n = f(x_n)$, $n \geq 1$, and

$$\lim_{n \nearrow \infty} \frac{g(y_n) - g(f(c))}{y_n - f(c)} = \lim_{n \nearrow \infty} \frac{x_n - c}{f(x_n) - f(c)} = \frac{1}{f'(c)}.$$

Since (y_n) is any sequence converging to $f(c)$, this implies

$$g'(f(c)) = \lim_{y \to f(c)} \frac{g(y) - g(f(c))}{y - f(c)} = \frac{1}{f'(c)}.$$

Since $y = f(c)$ iff $c = g(y)$, the result follows. □

This result is false if the hypothesis is weakened to the nonvanishing of the derivative at a single point: It is possible for a differentiable function to satisfy $f'(0) \neq 0$ but not be injective on any interval containing zero (Exercise **3.6.7**).

As an application, let $b > 0$. Since for $n > 0$, the function $f(x) = x^n$ is continuous on $[0,b]$ and $f'(x) = nx^{n-1} \neq 0$ on $(0,b)$, its inverse $g(y) = y^{1/n}$ is continuous on $[0,b^n]$ and differentiable on $(0,b^n)$ with

$$g'(y) = \frac{1}{f'(g(y))} = \frac{1}{n(g(y))^{n-1}} = \frac{1}{ny^{(n-1)/n}} = \frac{1}{n}y^{(1/n)-1}.$$

Since $b > 0$ is arbitrary, this is valid on $(0,\infty)$. Similarly, this holds on $(0,\infty)$ for $n < 0$.

By applying the chain rule, for all rationals $r = m/n$, the power functions $f(x) = x^r = x^{m/n} = (x^m)^{1/n}$ are differentiable on $(0,\infty)$ with derivative $f'(x) = rx^{r-1}$, since

$$\begin{aligned} f'(x) &= \left((x^m)^{1/n}\right)' = \frac{1}{n}(x^m)^{1/n-1}(x^m)' \\ &= \frac{1}{n}x^{(m/n)-m}mx^{m-1} = \frac{m}{n}x^{(m/n)-1} = rx^{r-1}. \end{aligned}$$

Thus, the derivative of $f(x) = x^r$ is $f'(x) = rx^{r-1}$ for $x > 0$, for all $r \in \mathbf{Q}$.

Using the chain rule, we now know how to differentiate any algebraic function. For example, the derivative of

$$f(x) = \sqrt{\frac{1-x^2}{1+x^2}}, \qquad 0 < x < 1,$$

is

$$f'(x) = \frac{1}{2}\left(\frac{1-x^2}{1+x^2}\right)^{-1/2} \cdot \left(\frac{-4x}{(1+x^2)^2}\right) = \frac{-2x}{(1+x^2)\sqrt{1-x^4}}, \qquad 0 < x < 1.$$

To compute the derivative of $f(x) = x^a$ when $a > 0$ is not in $\mathbf{Q}$, let $r < s$ be rationals with $r < a < s$, and consider the limit

$$\lim_{x\to 1+} \frac{x^a - 1}{x - 1}. \tag{3.2.1}$$

Since for any $x_n \to 1+$, the sequence $B_n = (x_n^a - 1)/(x_n - 1)$ lies between the sequences $A_n = (x_n^r - 1)/(x_n - 1)$ and $C_n = (x_n^s - 1)/(x_n - 1)$, the upper and lower limits of (B_n) lie between $\lim_{n\nearrow\infty} A_n = r1^{r-1} = r$ and $\lim_{n\nearrow\infty} C_n = s1^{s-1} = s$. Since $r < a < s$ are arbitrary, the upper and lower limits both equal a; hence,

$$B_n = (x_n^a - 1)/(x_n - 1) \to a;$$

thus, the limit (3.2.1) equals a. Since $f(x) = 1/x$ is continuous at $x = 1$, $x_n \to 1-$ implies $y_n = 1/x_n \to 1+$, so

$$\lim_{n\nearrow\infty}\frac{x_n^a-1}{x_n-1}=\lim_{n\nearrow\infty}\frac{1/y_n^a-1}{1/y_n-1}=\lim_{n\nearrow\infty}y_n^{1-a}\cdot\frac{y_n^a-1}{y_n-1}=1\cdot a=a.$$

Thus,

$$\lim_{x\to1-}\frac{x^a-1}{x-1}=a.$$

Hence, $f(x)=x^a$ is differentiable at $x=1$ with $f'(1)=a$. Since

$$\lim_{x\to c}\frac{x^a-c^a}{x-c}=c^{a-1}\lim_{x/c\to1}\frac{(x/c)^a-1}{(x/c)-1}=ac^{a-1},$$

f is differentiable on $(0,\infty)$ with $f'(c)=ac^{a-1}$. Thus, for all real $a>0$, the derivative of $f(x)=x^a$ at $x>0$ is $f'(x)=ax^{a-1}$. Using the quotient rule, the same result holds for real $a\le0$.

As an application, let v be any real greater than 1. Then by the chain rule, the derivative of $f(x)=(1+x)^v-1-vx$ is $f'(x)=v(1+x)^{v-1}-v$; hence, the only critical point is $x=0$. Since $f(-1)=-1+v>0=f(0)$ and $f(\infty)=\infty$, the minimum of f over $(-1,\infty)$ is $f(0)=0$. Hence,

$$(1+b)^v\ge1+vb,\qquad b\ge-1. \tag{3.2.2}$$

We already knew this for v a natural (Exercise **1.3.22**), but now we know this for any real $v\ge1$.

Now we compute the derivative of the exponential function $f(x)=a^x$ with base $a>1$. We begin with finding $f'(0)$.

If $0<x\le y$ and $a>1$, then insert $v=y/x\ge1$ and $b=a^x-1>0$ in (3.2.2), and rearrange to get

$$\frac{a^x-1}{x}\le\frac{a^y-1}{y},\qquad 0<x\le y.$$

Thus,

$$m_+=\lim_{x\to0+}\frac{a^x-1}{x}$$

exists since it equals

$$\inf\{(a^x-1)/x:x>0\}$$

(Exercise **2.2.2**). Moreover, $m_+\ge0$ since $a^x>1$ for $x>0$. Also

$$m_-=\lim_{x\to0-}\frac{a^x-1}{x}=\lim_{x\to0+}\frac{a^{-x}-1}{-x}=\lim_{x\to0+}a^{-x}\cdot\frac{a^x-1}{x}=1\cdot m_+=m_+.$$

Hence, the exponential with base $a>1$ is differentiable at $x=0$, and we denote its derivative there by $m(a)$. Since $a^x=a^ca^{x-c}$,

$$\lim_{x\to c}\frac{a^x-a^c}{x-c}=a^c\lim_{x\to c}\frac{a^{x-c}-1}{x-c}=a^cm(a).$$

Hence, $f(x) = a^x$ is differentiable on $\mathbf{R}$, and $f'(x) = a^x m(a)$ with $m(a) \geq 0$. If $m(a) = 0$, then $f'(x) = 0$ for all x; hence, a^x is constant, a contradiction. Hence, $m(a) > 0$. Also for $b > 1$ and $a > 1$,

$$\begin{aligned} m(b) &= \lim_{x \to 0} \frac{b^x - 1}{x} = \lim_{x \to 0} \frac{(a^{\log_a b})^x - 1}{x} \\ &= \lim_{x \to 0} \frac{a^{x \log_a b} - 1}{x} = m(a) \log_a b, \end{aligned}$$

by the chain rule. By fixing a and varying b, we see that m is a continuous, strictly increasing function with $m(\infty) = \infty$ and $m(1+) = 0$. By the intermediate value property §2.2, we conclude that $m((1, \infty)) = (0, \infty)$.

Thus, and this is very important, there is a unique real $e > 1$ with $m(e) = 1$. The exponential and logarithm functions with base e are called *natural*. Throughout the book, e denotes this particular number. We summarize the results.

Theorem 3.2.3. *For all $a > 0$, the exponential $f(x) = a^x$ is differentiable on $\mathbf{R}$. There is a unique real $e > 1$, such that $f(x) = e^x$ implies $f'(x) = e^x$. More generally, $f(x) = a^x$ implies $f'(x) = a^x \log_e a$.*

For $a > 1$, this was derived above. To derive the theorem for $0 < a < 1$, use $a^x = (1/a)^{-x}$ and the chain rule. □

In the sequel, $\log x$ will denote $\log_e x$, i.e., we drop the e when writing the natural logarithm. Then

$$e^{\log x} = x, \qquad \log e^x = x.$$

We end with the derivative of $f(x) = \log_a x$. Since this is the inverse of the exponential,

$$f'(x) = \frac{1}{a^{f(x)} \log a} = \frac{1}{x \log a}.$$

Thus, $f(x) = \log_a x$ implies $f'(x) = 1/x \log a$, $x > 0$. In particular $\log e = 1$, so $f(x) = \log x$ implies $f'(x) = 1/x$, $x > 0$.

For example, combining the above with the chain rule,

$$(\log |x|)' = \frac{1}{x}, \qquad x \neq 0.$$

Another example is ($x \neq \pm 1$)

$$\left[\log\left(\frac{x-1}{x+1}\right)\right]' = \frac{1}{\left(\dfrac{x-1}{x+1}\right)} \cdot \left(\frac{x-1}{x+1}\right)' = \left(\frac{x+1}{x-1}\right) \cdot \frac{2}{(x+1)^2} = \frac{2}{x^2 - 1}.$$

We will need the following in §3.5.

Theorem 3.2.4 (Generalized Mean Value Theorem). *If f and g are continuous on $[a,b]$, differentiable on (a,b), and $g'(x) \neq 0$ on (a,b), there exists a c in (a,b), such that*[1]

$$\frac{f(b)-f(a)}{g(b)-g(a)} = \frac{f'(c)}{g'(c)}.$$

Either $g' > 0$ on (a,b) or $g' < 0$ on (a,b). Assume $g' > 0$ on (a,b). To see the theorem, let h denote the inverse function of g, so $h(g(x)) = x$, and set $F(x) = f(h(x))$. Then $F(g(x)) = f(x)$; F is continuous on $[g(a), g(b)]$ and differentiable on $(g(a), g(b))$. So applying the mean value theorem, the chain rule, and the inverse function theorem, there is a d in $(g(a), g(b))$, such that

$$\begin{aligned}\frac{f(b)-f(a)}{g(b)-g(a)} &= \frac{F(g(b))-F(g(a))}{g(b)-g(a)}\\ &= F'(d) = f'(h(d))h'(d)\\ &= \frac{f'(h(d))}{g'(h(d))}.\end{aligned}$$

Now let $c = h(d)$. Then c is in (a,b). The case $g' < 0$ on (a,b) is similar. □

We end the section with l'Hôpital's rule.

Theorem 3.2.5 (L'Hôpital's Rule). *Let f and g be differentiable on an open interval (a,b) punctured at c, $a < c < b$. Suppose that $\lim_{x\to c} f(x) = 0$ and $\lim_{x\to c} g(x) = 0$. Then $g'(x) \neq 0$ for $x \neq c$ and*

$$\lim_{x\to c} \frac{f'(x)}{g'(x)} = L$$

imply[2]

$$\lim_{x\to c} \frac{f(x)}{g(x)} = L. \tag{3.2.3}$$

To obtain this, define f and g at c by setting $f(c) = g(c) = 0$. Then f and g are continuous on (a,b). Now let $x_n \to c+$. Apply the generalized mean value theorem on (c, x_n) for each $n \geq 1$. Then

$$\frac{f(x_n)}{g(x_n)} = \frac{f(x_n)-f(c)}{g(x_n)-g(c)} = \frac{f'(d_n)}{g'(d_n)} \to L,$$

since $c < d_n < x_n$. Similarly, this also holds when $x_n \to c-$, and thus, this holds for $x_n \to c$, which establishes (3.2.3). □

The above deals with the "indeterminate form" $f(x)/g(x) \to 0/0$. The case $f(x)/g(x) \to \infty/\infty$ can be handled by turning the fraction $f(x)/g(x)$ upside down and applying the above. We do not state this case as we do not use it.

---

[1] $g(b) - g(a)$ is not zero because it equals $g'(d)(b-a)$ for some $a < d < b$.

[2] $g(x) \neq 0$ for $x \neq c$ since $g(x) = g(x) - g(c) = g'(d)(x-c)$.

Exercises

3.2.1. Show that $1+x \le e^x$ for $x \ge 0$.

3.2.2. Use the generalized mean value theorem to show $(1+x)^\alpha \le 1+x^\alpha$ for $x \ge 0$ and $0<\alpha<1$. By induction, conclude

$$(a_1+\cdots+a_n)^\alpha \le a_1^\alpha+\cdots+a_n^\alpha$$

for $a_j \ge 0$, $j=1,\dots,n$.

3.2.3. Show that $\lim_{x\to 0}\log(1+x)/x = 1$ and

$$\lim_{n\nearrow\infty}(1+a/n)^n = e^a, \qquad a\in\mathbf{R}$$

(take the log of both sides). If $a_n \to a$, show also that $\lim_{n\nearrow\infty}(1+a_n/n)^n = e^a$.

3.2.4. Let $x\in\mathbf{R}$. Show that the sequence $(1+x/n)^n$, $n \ge |x|$ increases to e^x as $n\nearrow\infty$ (use (3.2.2)).

3.2.5. Use the mean value theorem to show that

$$1-\frac{1}{\sqrt{1+x^2}} \le x^2, \qquad x>0.$$

3.2.6. Use the mean value theorem to show that

$$\frac{1}{(2j-1)^x}-\frac{1}{(2j)^x} \le \frac{x}{(2j-1)^{x+1}}, \qquad j\ge 1, x>0.$$

3.2.7. Let $f:\mathbf{R}\to\mathbf{R}$ be differentiable with $f(0)=0$ and $|f'(x)| \le |f(x)|$ for all x. Show that f is identically zero.

3.2.8. If $f:(a,b)\to\mathbf{R}$ is differentiable, then $f':(a,b)\to\mathbf{R}$ satisfies the *intermediate value property*: If $a<c<d<b$ and $f'(c)<L<f'(d)$, then $L=f'(x)$ for some $c<x<d$. (Start with $L=0$; then consider $g(x)=f(x)-Lx$. Here the point is that f' need not be continuous.)

3.2.9. If $f:\mathbf{R}\to\mathbf{R}$ is superlinear and differentiable, then $f'(\mathbf{R})=\mathbf{R}$, i.e., f' is surjective (Exercise **3.1.4**).

3.2.10. For $d\ge 2$, let

$$f_d(t)=\frac{d-1}{d}\cdot t^{(d-1)/d}+\frac{1}{d}\cdot t^{-1/d}, \qquad t\ge 1.$$

Show that

$$f_d'(t) \le (d-1)^2/d^2, \qquad t \ge 1.$$

Conclude that

$$f_d(t) - 1 \le \left(\frac{d-1}{d}\right)^2 (t-1), \qquad t \ge 1.$$

3.3 Graphing Techniques

Let f be differentiable on (a,b). If $f' = df/dx$ is differentiable on (a,b), we denote its derivative by $f'' = (f')'$; f'' is the *second derivative* of f. If f'' is differentiable on (a,b), $f''' = (f'')'$ is the *third derivative* of f. In general, we let $f^{(n)}$ denote the *nth derivative* or the derivative *of order* n, where, by convention, we take $f^{(0)} = f$. If f has all derivatives, $f', f'', f''', \dots$, we say f is *smooth* on (a,b).

An alternate and useful notation for higher derivatives is obtained by thinking of $f' = df/dx$ as the result of applying d/dx to f, i.e., $df/dx = (d/dx)f$. From this point of view, d/dx signifies the operation of differentiation. Thus, applying d/dx twice, we obtain

$$f'' = \left(\frac{d}{dx}\right)\left(\frac{d}{dx}\right) f = \left(\frac{d^2}{dx^2}\right) f = \frac{d^2 f}{dx^2}.$$

Similarly, third derivatives may be denoted

$$f''' = \left(\frac{d}{dx}\right)\left(\frac{d^2 f}{dx^2}\right) = \frac{d^3 f}{dx^3}.$$

For example, $f(x) = x^2$ has $f'(x) = 2x$, $f''(x) = 2$, and $f^{(n)}(x) = 0$ for $n \ge 3$. More generally, by induction, $f(x) = x^n$; $n \ge 0$ has derivatives

$$f^{(k)}(x) = \frac{d^k f}{dx^k} = \begin{cases} \dfrac{n!}{(n-k)!} x^{n-k}, & 0 \le k \le n, \\ 0, & k > n, \end{cases} \tag{3.3.1}$$

so $f(x) = x^n$ is smooth. By the arithmetic properties of derivatives, it follows that rational functions are smooth wherever they are defined.

Not all functions are smooth. The function $f(x) = |x|$ is not differentiable at zero. Using this, one can show that $f(x) = x^n|x|$ is n times differentiable on $\mathbf{R}$, but $f^{(n)}$ is not differentiable at zero. More generally, for f, g differentiable, we do not expect $\max(f,g)$ to be differentiable. However, since $f(x) = x^{1/n}$ is smooth on $(0,\infty)$, algebraic functions are smooth on any open interval of definition. Also the functions x^a, a^x, and $\log_a x$ are smooth on $(0,\infty)$.

We know the sign of f' determines the monotonicity of f, in the sense that $f' \geq 0$ iff f is increasing and $f' \leq 0$ iff f is decreasing. How is the sign of f'' reflected in the graph of f? Since $f'' = (f')'$, we see that $f'' \geq 0$ iff f' is increasing and $f'' \leq 0$ iff f' is decreasing.

More precisely, we say f is *convex* on (a, b) if, for all $a < x < y < b$,

$$f((1-t)x + ty) \leq (1-t)f(x) + tf(y), \qquad 0 \leq t \leq 1.$$

Take any two points on the graph of f and join them by a chord or line segment. Then f is convex if the chord lies on or above the graph (Figure 3.3). We say f is *concave* on (a, b) if, for all $a < x < y < b$,

$$f((1-t)x + ty) \geq (1-t)f(x) + tf(y), \qquad 0 \leq t \leq 1.$$

Take any two points on the graph of f and join them by a chord. Then f is concave if the chord lies on or below the graph.

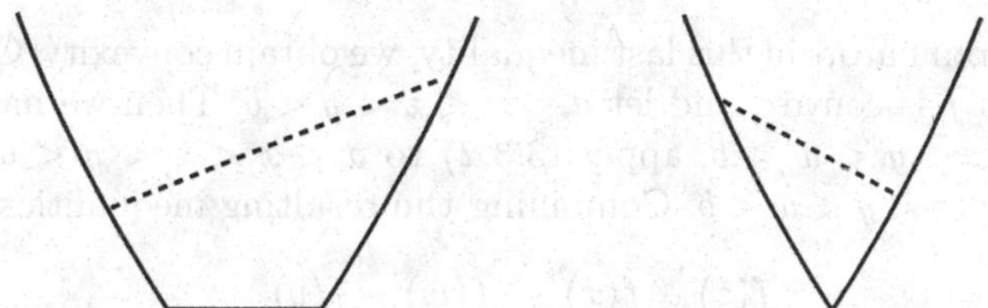

Fig. 3.3 Examples of convex and strictly convex functions

For example, $f(x) = x^2$ is convex and $f(x) = -x^2$ is concave. A function $f : (a, b) \to \mathbf{R}$ that is both convex and concave is called *affine*. It is easy to see that $f : (a, b) \to \mathbf{R}$ is affine iff $f' = m$ is constant on (a, b).

Similarly, we say that f is *strictly convex* on (a, b) if, for all $a < x < y < b$,

$$f((1-t)x + ty) < (1-t)f(x) + tf(y), \qquad 0 < t < 1.$$

Take any two points on the graph of f and join them by a chord. Then f is strictly convex if the chord lies strictly above the graph. Similarly, we define *strictly concave*.

Note that a strictly convex $f : (a, b) \to \mathbf{R}$ cannot attain its infimum m at more than one point in (a, b). Indeed, if f had two minima at x and x' and $x'' = (x + x')/2$, then $f(x'') < [f(x) + f(x')]/2 = (m + m)/2 = m$, contradicting the fact that m is a minimum.

The negative of a (strictly) convex function is (strictly) concave.

Theorem 3.3.1. *Suppose that f is differentiable on (a, b). Then f is convex iff f' is increasing, and f is concave iff f' is decreasing. Moreover, f is strictly convex iff f' is strictly increasing, and f is strictly concave iff f' is strictly decreasing. If f is twice differentiable on (a, b), then f is convex iff $f'' \geq 0$,* *and f is concave iff $f'' \le 0$. Moreover, f is strictly convex if $f'' > 0$, and f is strictly concave if $f'' < 0$.*

Since $-f$ is convex iff f is concave, we derive only the convex part. First, suppose that f' is increasing. If $a < x < y < b$ and $0 < t < 1$, let $z = (1-t)x + ty$. Then

$$\frac{f(z) - f(x)}{z - x} = f'(c)$$

for some $x < c < z$. Also

$$\frac{f(y) - f(z)}{y - z} = f'(d)$$

for some $z < d < y$. Since $f'(c) \le f'(d)$,

$$\frac{f(z) - f(x)}{z - x} \le \frac{f(y) - f(z)}{y - z}. \tag{3.3.2}$$

Clearing denominators in this last inequality, we obtain convexity. Conversely, suppose that f is convex, and let $a < x < z < y < b$. Then we have (3.3.2). If $a < x < z < y < w < b$, apply (3.3.2) to $a < x < z < y < b$ and then (3.3.2) to $a < z < y < w < b$. Combining the resulting inequalities yields

$$\frac{f(z) - f(x)}{z - x} \le \frac{f(w) - f(y)}{w - y}.$$

Fixing x, y, and w and letting $z \to x$ yields

$$f'(x) \le \frac{f(w) - f(y)}{w - y}.$$

Let $w \to y$ to obtain $f'(x) \le f'(y)$; hence, f' is increasing. If f' is strictly increasing, then the inequality (3.3.2) is strict; hence, f is strictly convex. Conversely, if f' is increasing but $f'(c) = f'(d)$ for some $c < d$, then f' is constant on $[c, d]$. Hence, f is affine on $[c, d]$ contradicting strict convexity. This shows that f is strictly convex iff f' is strictly increasing.

When f is twice differentiable, $f'' \ge 0$ iff f' is increasing, hence, the third statement. Since $f'' > 0$ implies f' strictly increasing, we also have the fourth statement. □

A key feature of convexity (Figure 3.4) is that the graph of a convex function lies above any of its tangent lines (Exercise **3.3.7**).

A real c is an *inflection point* of f if f is convex on one side of c and concave on the other. For example, $c = 0$ is an inflection point of $f(x) = x^3$ since f is convex on $x > 0$ and concave on $x < 0$. From the theorem, we see that $f''(c) = 0$ at any inflection point c where f is twice differentiable.

If c is a critical point and $f''(c) > 0$, then f' is strictly increasing near c; hence, $f'(x) < 0$ for $x < c$ near c and $f'(x) > 0$ for $x > c$ near c. Thus,

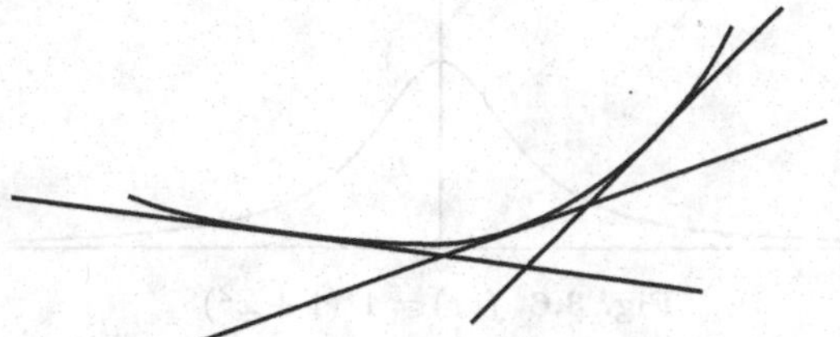

Fig. 3.4 A convex function lies above any of its tangents

$f'(c) = 0$ and $f''(c) > 0$ implies c is a local minimum. Similarly, $f'(c) = 0$ and $f''(c) < 0$ implies c is a local maximum. The converses are not, generally, true since $c = 0$ is a minimum of $f(x) = x^4$, but $f'(0) = f''(0) = 0$.

For example, $f(x) = a^x$, $a > 1$ satisfies $f'(x) = a^x \log a$ and $f''(x) = a^x(\log a)^2$. Since $\log a > 0$, a^x is increasing and strictly convex everywhere. Also $f(x) = \log_a x$ has $f'(x) = 1/x \log a$ and $f''(x) = -1/x^2 \log a$, so $\log_a x$ is increasing and strictly concave everywhere. The graphs are shown in Figure 3.5.

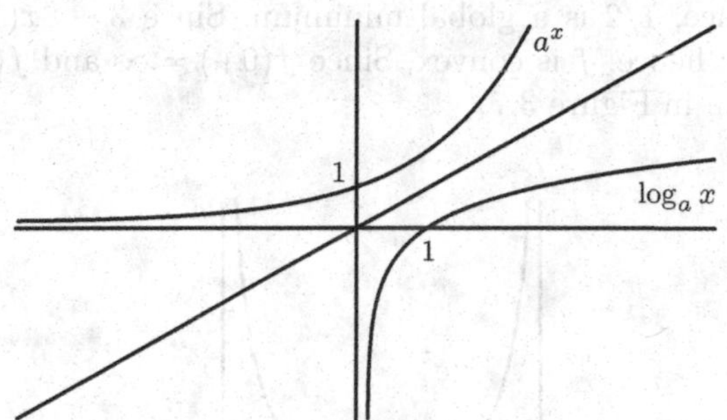

Fig. 3.5 The exponential and logarithm functions

In the following, we sketch the graphs of some twice differentiable functions on an interval (a, b), using knowledge of the critical points, the inflection points, the signs of f' and f'', and $f(a+)$, $f(b-)$.

If $f(x) = 1/(x^2+1)$, $-\infty < x < \infty$, then $f'(x) = -2x/(x^2+1)^2$. Hence, $f'(x) < 0$ for $x > 0$ and $f'(x) > 0$ for $x < 0$, so f is increasing for $x < 0$ and decreasing for $x > 0$. Hence, 0 is a global maximum. Moreover,

$$f''(x) = \left(\frac{-2x}{(x^2+1)^2}\right)' = \frac{6x^2-2}{(x^2+1)^3},$$

so $f''(0) < 0$ which is consistent with 0 being a maximum. Now $f''(x) < 0$ on $|x| < 1/\sqrt{3}$ and $f''(x) > 0$ on $|x| > 1/\sqrt{3}$. Hence, $x = \pm 1/\sqrt{3}$ are inflection points. Since $f(0) = 1$ and $f(\infty) = f(-\infty) = 0$, we obtain the graph in Figure 3.6.

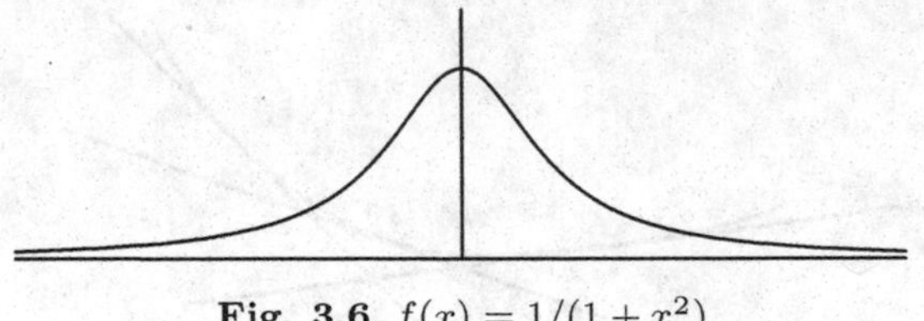

Fig. 3.6 $f(x) = 1/(1+x^2)$

Let $f(x) = 1/\sqrt{x(1-x)}$, $0 < x < 1$. Then

$$f'(x) = \frac{2x-1}{2(x(1-x))^{3/2}}$$

and

$$f''(x) = \frac{3-8x(1-x)}{4(x(1-x))^{5/2}}.$$

Thus, $x = 1/2$ is a critical point. Since $f''(1/2) > 0$, $x = 1/2$ is a local minimum. In fact, f is decreasing to the left of $1/2$ and increasing to the right of $1/2$; hence, $1/2$ is a global minimum. Since $3 - 8x(1-x) > 0$ on $(0,1)$, $f''(x) > 0$; hence, f is convex. Since $f(0+) = \infty$ and $f(1-) = \infty$, the graph is as shown in Figure 3.7.

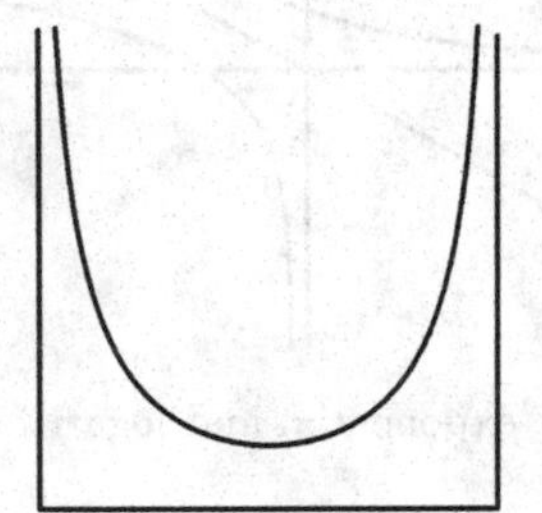

Fig. 3.7 $f(x) = 1/\sqrt{x(1-x)}$

Let $f(x) = \dfrac{3x+1}{x(1-x)}$. This rational function is defined away from $x = 0$ and $x = 1$. Thus, we graph f on the intervals $(-\infty, 0)$, $(0,1)$, $(1,\infty)$. Computing,

$$f'(x) = \frac{3x^2+2x-1}{x^2(1-x)^2}.$$

Solving $3x^2 + 2x - 1 = 0$, $x = -1, 1/3$ are the critical points. Moreover, $f(\infty) = 0$, $f(1+) = -\infty$, $f(1-) = \infty$, $f(0+) = \infty$, $f(0-) = -\infty$, and $f(-\infty) = 0$. Since there are no critical points in $(1,\infty)$, f is increasing on $(1,\infty)$. Moreover,

$$f''(x) = \frac{6x^3 - 6x(1-x) + 2}{x^3(1-x)^3},$$

so f is concave on $(1, \infty)$. Moreover, the numerator in $f''(x)$ is $\geq 1/2$ on $(0, 1)$. Hence, $f''(x) > 0$ on $(0, 1)$. Hence, f is convex in $(0, 1)$ and $x = 1/3$ is a minimum of f. Since $f''(-1) = -1 < 0$, $x = -1$ is a maximum. Since $f''(x) \to 0+$ as $x \to -\infty$, there is an inflection point in $(-\infty, -1)$. Thus, the graph is as shown in Figure 3.8.

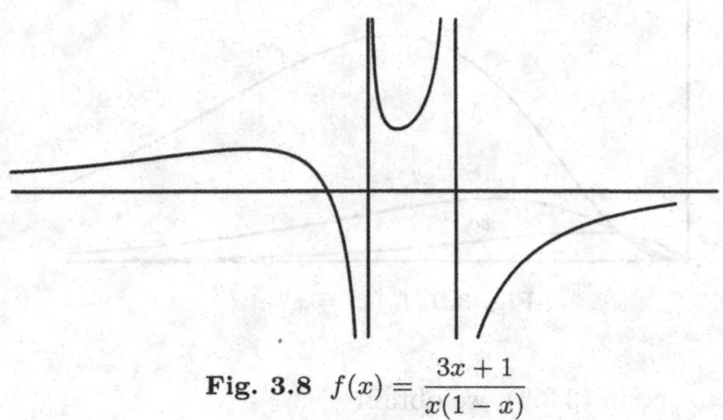

Fig. 3.8 $f(x) = \dfrac{3x+1}{x(1-x)}$

To graph $x^n e^{-x}$, $x \geq 0$, we will need to show that

$$\lim_{x \to \infty} x^n e^{-x} = 0. \tag{3.3.3}$$

To establish the limit, we first show that

$$e^x \geq 1 + x + \frac{x^2}{2!} + \cdots + \frac{x^n}{n!}, \qquad x \geq 0, \tag{3.3.4}$$

for all $n \geq 1$. We do this by induction. For $n = 1$, let $f(x) = e^x$, $g(x) = 1 + x$. Then $f'(x) = e^x \geq 1 = g'(x)$ on $x > 0$ and $f(0) = g(0)$; hence, $f(x) \geq g(x)$ establishing (3.3.4) for $n = 1$. Now let $g_n(x)$ denote the right side of (3.3.4), and suppose that (3.3.4) is true for some $n \geq 1$. Since $f'(x) = f(x) \geq g_n(x) = g'_{n+1}(x)$ and $f(0) = g_{n+1}(0)$, we conclude that $f(x) \geq g_{n+1}(x)$, establishing (3.3.4) for $n + 1$. By induction, (3.3.4) is true for all $n \geq 1$. Now (3.3.4) with $n + 1$ replacing n implies $e^x \geq x^{n+1}/(n+1)!$ which implies

$$x^n e^{-x} \leq \frac{(n+1)!}{x}, \qquad x > 0,$$

which implies (3.3.3).

Setting $f_n(x) = x^n e^{-x}$, $n \geq 1$, $f_n(0) = 0$, and $f_n(\infty) = 0$. Moreover,

$$f'_n(x) = x^{n-1}(n - x)e^{-x}$$

and

$$f_n''(x) = x^{n-2}\left[x^2 - 2nx + n(n-1)\right]e^{-x}.$$

Thus, the critical point is $x = n$, and f_n is increasing on $(0, n)$ and decreasing on (n, ∞). Hence, $x = n$ is a max. The reals $x = n \pm \sqrt{n}$ are inflection points. Between them, f_n is concave and elsewhere convex. The graph is as shown in Figure 3.9.

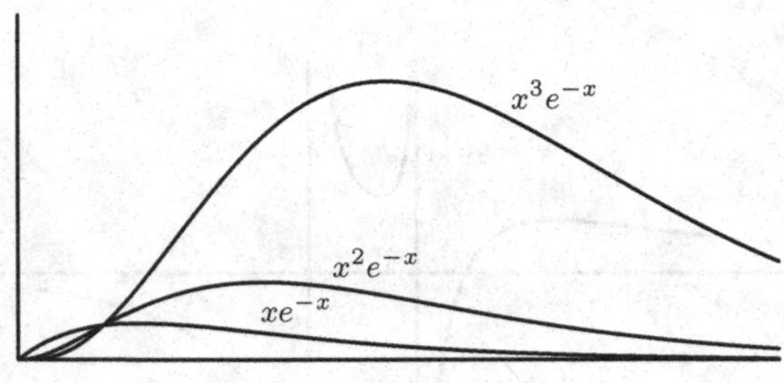

Fig. 3.9 $f_n(x) = x^n e^{-x}$

If we let $n \nearrow \infty$ in (3.3.4), we obtain

$$e^x \geq \sum_{n=0}^{\infty} \frac{x^n}{n!}, \qquad x \geq 0. \tag{3.3.5}$$

As a consequence (nth term test §1.5),

$$\lim_{n\nearrow\infty} \frac{x^n}{n!} = 0 \tag{3.3.6}$$

for all $x \geq 0$, hence, for all x. Using (3.3.3), we can derive other limits.

Theorem 3.3.2. *For $a > 0$, $b > 0$, and $c > 0$,*

A. $\lim_{x\to\infty} x^a e^{-bx} = 0$;
B. $\lim_{t\to 0+} t^b(-\log t)^a = 0$; *in particular,* $t \log t \to 0$ *as* $t \to 0+$;
C. $\lim_{n\nearrow\infty} (\log n)^a / n^b = 0$; *in particular,* $\log n / n \to 0$ *as* $n \nearrow \infty$;
D. *If* $c < 1$, $\lim_{n\nearrow\infty} n^a c^n = 0$. *If* $c > 1$, $\lim_{n\nearrow\infty} n^{-a} c^n = \infty$;
E. $\lim_{n\nearrow\infty} n^{1/n} = 1$.

To obtain the first limit, choose $n > a$, and let $y = bx$. Then $x \to \infty$ implies $y \to \infty$; hence, $x^a e^{-bx} = y^a e^{-y}/b^a < y^n e^{-y}/b^a \to 0$ by (3.3.3). Substituting $t = e^{-x}$ in the first yields the second since $e^{-x} \to 0+$, as $x \to \infty$. Substituting $t = 1/n$ in the second yields the third. For the fourth, in the first, replace x by n and e^{-b} by c, if $c < 1$. If $c > 1$, $n^{-a}c^n = 1/n^a(1/c)^n \to \infty$ by what we just derived. For the fifth, take the exponential of both sides of the third with $a = b = 1$. Since e^x is continuous, we obtain the fifth. □

The moral of the theorem is $\log n << n << e^n$ as $n \nearrow \infty$, where $A << B$ means that any positive power of A is much smaller than B.

Let us use (3.3.1) to derive the binomial theorem. If $n \geq 1$, then $(c+x)^n$ is a polynomial of degree n; hence, there are numbers $a_0, \dots, a_n$ with

$$f(x) = (c+x)^n = a_n x^n + a_{n-1}x^{n-1} + \cdots + a_1 x + a_0. \tag{3.3.7}$$

Let us compute the derivatives of f by using either of the expressions in (3.3.7). The left expression and (3.3.1) and the chain rule yield $f^{(k)}(0) = n!/(n-k)!c^{n-k}$. The right expression yields $f^{(k)}(0) = k!a_k$. Hence, $a_k = n!c^{n-k}/(n-k)!k!$. Now define the *binomial coefficient* (read "n-choose-k")

$$\binom{n}{k} = \frac{n!}{(n-k)!k!} = \frac{n(n-1)\cdots(n-k+1)}{k!}, \qquad 0 \leq k \leq n.$$

Then we obtain the following.

Theorem 3.3.3 (Binomial Theorem). *If $n \geq 1$ and $a, b \in \mathbf{R}$, then*

$$(a+b)^n = a^n + \binom{n}{1}a^{n-1}b + \cdots + \binom{n}{n-1}ab^{n-1} + b^n$$
$$= \sum_{j=0}^{n} \binom{n}{j} a^{n-j} b^j . \square$$

As an application, note

$$\binom{n}{j} = \frac{n(n-1)\cdots(n-j+1)}{j!} \leq \frac{n^j}{j!},$$

which implies

$$\left(1+\frac{x}{n}\right)^n = \sum_{j=0}^{n} \binom{n}{j}\left(\frac{x}{n}\right)^j \leq \sum_{j=0}^{n} \frac{x^j}{j!}. \tag{3.3.8}$$

Let $n \to \infty$ and use Exercise **3.2.3** to obtain

$$e^x \leq \sum_{n=0}^{\infty} \frac{x^n}{n!}, \qquad x \geq 0.$$

Coupling this with (3.3.5), we arrive at

$$e^x = \sum_{n=0}^{\infty} \frac{x^n}{n!}, \qquad x \geq 0. \tag{3.3.9}$$

We say $f : \mathbf{R} \to \mathbf{R}$ is *superlinear* if

$$\lim_{x \to \pm\infty} \frac{f(x)}{|x|} = +\infty.$$

Given $f : \mathbf{R} \to \mathbf{R}$, its *Legendre transform* is the function

$$g(y) = \sup_{-\infty<x<\infty} (xy - f(x)), \qquad y \in \mathbf{R}. \tag{3.3.10}$$

Exercise **2.3.20** shows that the sup is attained; hence, g is well defined when f is superlinear and continuous. Note $g(y)$ can be thought of as a sup of lines $y \mapsto xy - f(x)$, one for each x (compare with Exercise **3.3.11**).

Below is a set of exercises that show that the Legendre transform of a convex superlinear function is well defined, and derive the result that the Legendre transform of the Legendre transform of a convex superlinear function f is f: If g is the Legendre transform of f, then f is the Legendre transform of the Legendre transform of g,

$$f(x) = \sup_{-\infty<y<\infty} (xy - g(y)), \qquad x \in \mathbf{R}. \tag{3.3.11}$$

Examples of Legendre transforms are given in Exercises **3.1.3** and **3.3.14**. The perfect symmetry between f and its Legendre transform g is exhibited in Exercises **3.3.16**, **3.3.20**, and **3.3.23**.

Exercises

3.3.1. Graph $f(x) = (x + 2/x)/2$ for $x > 0$.

3.3.2. Show that $f : (a, b) \to \mathbf{R}$ is affine iff $f'(x) = m$ is a constant on (a, b).

3.3.3. Suppose $f : (a, b) \to (c, d)$ is convex and $g : (c, d) \to \mathbf{R}$ is increasing and convex. Show that the composition $g \circ f : (a, b) \to \mathbf{R}$ is convex.

3.3.4. Let $f : \mathbf{R} \to \mathbf{R}$ be convex, and for $b \neq a$, let

$$s[a, b] = \frac{f(b) - f(a)}{b - a}.$$

Show that $a < b < c$ implies $s[a, b] \le s[a, c] \le s[b, c]$.

3.3.5. Suppose that $f : (a, b) \to \mathbf{R}$ is convex. Then for all $c \in (a, b)$,

$$f'_\pm(c) = \lim_{x\to c\pm} \frac{f(x) - f(c)}{x - c}$$

both exist, and

$$f'_+(c) \le \frac{f(x) - f(c)}{x - c} \le \frac{f(d) - f(x)}{d - x} \le f'_-(d), \qquad a < c < x < d < b. \tag{3.3.12}$$

Moreover $f'_- \le f'_+$ and both f'_+ and f'_- are increasing on (a, b).

3.3.6. If $f : (a, b) \to \mathbf{R}$ is convex, then f is continuous on (a, b).

3.3.7. Suppose that $f : (a, b) \to \mathbf{R}$ is convex and let $a < c < b$. Then

$$f(x) \geq f(c) + f'_{\pm}(c)(x - c), \qquad a < x < b.$$

In particular, if f is differentiable at c, then the graph of f lies above its tangent line at c,

$$f(x) \geq f(c) + f'(c)(x - c), \qquad a < x < b.$$

3.3.8. Let $f : (a, b) \to \mathbf{R}$ and let $a < c < b$. A *subdifferential of* f *at* c is a real p satisfying

$$f(x) \geq f(c) + p(x - c), \qquad a < x < b.$$

Show that when $f'_{\pm}(c)$ both exist, we have $f'_{-}(c) \leq p \leq f'_{+}(c)$. Show also that when f is convex, the set of subdifferentials of f at c exactly equals the interval $[f'_{-}(c), f'_{+}(c)]$.

3.3.9. (Maximum Principle) Suppose that $f : (a, b) \to \mathbf{R}$ is convex and has a maximum at some c in (a, b). Then f is constant (use subdifferentials).

3.3.10. Suppose that $f : (a, b) \to \mathbf{R}$ is convex, $g : (a, b) \to \mathbf{R}$ is differentiable, and $f - g$ attains its supremum at some $a < c < b$. Show that $f'(c)$ exists and equals $g'(c)$ (use subdifferentials).

3.3.11. If $f_1, \dots, f_n$ are convex on (a, b), then so is

$$f = \max(f_1, \dots, f_n).$$

In particular, if $f_1, \dots, f_n$ are lines, then f is convex. Exercise **3.3.12** shows that this is also sometimes true for infinitely many lines.

3.3.12. If $f : \mathbf{R} \to \mathbf{R}$ is superlinear and continuous, then the sup is attained in (3.3.10), the sup is attained in (3.3.11) (Exercise **2.3.20** and Exercise **2.3.21**), and the Legendre transform g is convex. Moreover, for each y, if x attains the sup in the definition (3.3.10) of $g(y)$, then x is a subdifferential of g at y.

3.3.13. If $f : \mathbf{R} \to \mathbf{R}$ is superlinear and even, then its Legendre transform is even, and $g(y)$, $y \geq 0$ can be computed by restricting over $x \geq 0$:

$$g(y) = \sup_{x \geq 0}(xy - f(x)), \qquad y \geq 0.$$

3.3.14. Let $p > 1$ and $q > 1$ satisfy $(1/p) + (1/q) = 1$, and let $f = |x|^p/p$. Show, by direct computation, that the Legendre transform is $g(y) = |y|^q/q$. Also show f' and g' are inverses. (Use Exercise **3.3.13**.)

3.3.15. Let $f(x) = e^{|x|} - 1$. Show, by direct computation, that the Legendre transform is $g(y) = |y|\log|y| - |y| + 1$ for $|y| \geq 1$ and $g(y) = 0$ for $|y| \leq 1$ (Exercise **3.3.13**).

3.3.16. Suppose that $f : \mathbf{R} \to \mathbf{R}$ is superlinear and convex. Show that the Legendre transform g of f is superlinear and convex and f is the Legendre transform of g, i.e., (3.3.11) holds. Show also that this result is false if f is not assumed convex.

3.3.17. If f is superlinear and convex, then for each y, the sup in the definition (3.3.10) of $g(y)$ is attained at x iff x is a subdifferential of g at y. Show also that this result is false if f is not assumed convex.

3.3.18. If $f : (a, b) \to \mathbf{R}$ is convex and differentiable, then $f' : (a, b) \to \mathbf{R}$ is continuous (recall f' is then increasing).

3.3.19. If $f : \mathbf{R} \to \mathbf{R}$ is superlinear and strictly convex, then its Legendre transform g is differentiable and g' is continuous (start by showing that the sup in the definition (3.3.10) of $g(y)$ is attained at a unique x).

3.3.20. If $f : \mathbf{R} \to \mathbf{R}$ is superlinear, differentiable, and strictly convex, then its Legendre transform g is superlinear, differentiable, and strictly convex, and f' is the inverse of g'.

3.3.21. Graph f, g, f', and g' where f and g are as in Exercise **3.3.15**.

3.3.22. Show that $f(x) = e^x$ is convex on $\mathbf{R}$. Deduce the inequality $a^t b^{1-t} \leq ta + (1-t)b$ valid for $a > 0$, $b > 0$, and $0 \leq t \leq 1$.

3.3.23. If f is superlinear, smooth, and strictly convex, then the Legendre transform g of f is superlinear, smooth, and strictly convex iff $f''(x) > 0$ for all $x \in \mathbf{R}$. In this case, we have

$$g''(y) = \frac{1}{f''(x)}$$

whenever $y = f'(x)$ or equivalently $x = g'(y)$. Also give an example of a superlinear, smooth, and strictly convex f with a non-smooth Legendre transform g.

3.3.24. Suppose $f : \mathbf{R} \to \mathbf{R}$ is smooth. We say r *is a root of* f *of order* n if $f(r) = f'(r) = \cdots = f^{(n-1)}(r) = 0$. We say f *has* n *roots* if there are distinct reals $r_1, \ldots, r_k$ and naturals $n_1, \ldots, n_k$ such that $n_1 + \cdots + n_k = n$ and r_j is a root of f of order n_j, $j = 1, \ldots, k$. Show that if f has n roots in an interval (a, b), then f' has $n - 1$ roots in the same interval (a, b) (Use Exercise **3.1.12**).

3.3.25. Show that a degree n polynomial f has n roots iff

$$f(x) = C(x-r_1)^{n_1}(x-r_2)^{n_2}\dots(x-r_k)^{n_k}$$

for some distinct reals $r_1,\dots,r_k$ and naturals $n_1,\dots,n_k$ satisfying $n_1+\dots+n_k=n$ (use induction on n for the only if part).

3.3.26. If f is a degree n polynomial with n negative roots, then $g(x) = x^n f(1/x)$ is a degree n polynomial with n negative roots.

3.3.27. Given positive reals $a_1,\dots,a_n$, not necessarily distinct, let the reals $p_1, \dots, p_n$ be the coefficients[3] given by

$$(x+a_1)(x+a_2)\dots(x+a_n) = x^n + \binom{n}{1}p_1x^{n-1} + \dots + \binom{n}{n-1}p_{n-1}x + p_n.$$

Let $f(x)$ denote the polynomial on the right. Show the following.

A. f has n negative roots.

B. Differentiating f $n-k-1$ times $(1 \le k \le n-1)$ yields a $(k+1)$-degree polynomial g with $k+1$ negative roots.

C. Show that $h(x) = x^{k+1}g(1/x)$ is a degree $k+1$ polynomial with $k+1$ negative roots.

D. Differentiating h $k-1$ times, there are two roots for the quadratic polynomial

$$p(x) = \frac{1}{2}n!\left(p_{k-1} + 2p_kx + p_{k+1}x^2\right).$$

Conclude that $p_k^2 \ge p_{k-1}p_{k+1}$ (Exercise **1.4.5**). This result is due to Newton.

3.3.28. With $p_1,\dots,p_n$ as in the previous exercise and with $a_1,\dots,a_n$ positive, show that $p_1 \ge p_2^{1/2} \ge \dots \ge p_n^{1/n}$, with equality throughout only if all the a_i's are equal. This result is due to Maclaurin.

3.3.29. Let a be an integer, and let $p(x)$ be a polynomial with integers coefficients satisfying $p(a)=0$. Set $f(t,x) = \exp(tp(x))$. Show that

$$f^{(k)}(t,a) = \frac{\partial^k f}{\partial x^k}(t,a)$$

is a polynomial in t with integer coefficients for all $k \ge 0$. Here derivatives are taken with respect to x.

3.3.30. Let a and b be naturals and set $p = a/b \in \mathbf{Q}$. For $n \ge 0$, let

$$g_n(x) = \frac{(bx)^n(a-bx)^n}{n!}, \qquad 0 \le x \le p.$$

Show that $g_n^{(k)}(0)$ and $g_n^{(k)}(p)$ are integers for all $k \ge 0$ and $n \ge 0$. (Apply the previous exercise to $p(x) = x(a-x)$ and let $g(t,x) = f(t,bx) = \exp(tp(bx))$.)

---

[3] $p_1,\dots,p_n$ are the *normalized elementary symmetric polynomials* in $a_1,\dots,a_n$.

3.4 Power Series

A *power series* is a series of the form

$$\sum_{n=0}^{\infty} a_n x^n = a_0 + a_1 x + a_2 x^2 + \dots .$$

Since this is a series involving a variable x, it may converge at some x's and diverge at other x's. For example (§1.6),

$$1 + x + x^2 + \dots \begin{cases} = \dfrac{1}{1-x}, & |x| < 1, \\ \text{diverges}, & |x| \geq 1. \end{cases}$$

Note that the set of x's for which this series converges is an interval centered at zero. This is not an accident (Figure 3.10).

Theorem 3.4.1. *Let $\sum a_n x^n$ be a power series. Either the series converges absolutely for all x, the series converges only at $x = 0$, or there is an $R > 0$, such that the series converges absolutely if $|x| < R$ and diverges if $|x| > R$.*

To derive the theorem, let $R = \sup\{|x| : \sum a_n x^n \text{ converges}\}$. If $R = 0$, the series converges only for $x = 0$. If $R > 0$ and $|x| < R$, choose c with $|x| < |c| < R$ and $\sum a_n c^n$ convergent. Then $\{a_n c^n\}$ is a bounded sequence by the nth term test, say $|a_n c^n| \leq C$, $n \geq 0$, and it follows that

$$\sum |a_n x^n| = \sum |a_n c^n| \cdot |x/c|^n \leq C \sum |x/c|^n < \infty,$$

since $|x/c| < 1$ and the last series is geometric. This shows that $\sum a_n x^n$ converges absolutely for all x in $(-R, R)$. On the other hand, if $R < \infty$, the definition of R shows that $\sum a_n x^n$ diverges for $|x| > R$. Finally, if $R = \infty$, this shows that the series converges absolutely for all x. □

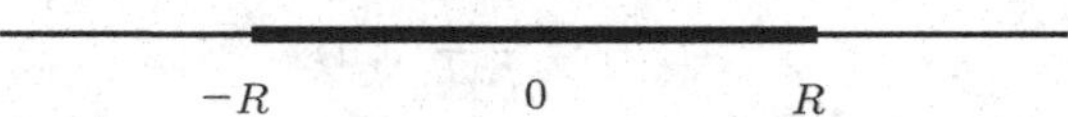

Fig. 3.10 Region of convergence of a power series

Note that the theorem says nothing about $x = R$ and $x = -R$. At these two points, anything can happen. For example, the power series

$$f(x) = 1 - \frac{x}{1} + \frac{x^2}{2} - \frac{x^3}{3} + \dots$$

has radius $R = 1$ and converges for $x = 1$ but diverges for $x = -1$. On the other hand, the series $f(-x)$ has $R = 1$ and converges for $x = -1$ but diverges for $x = 1$. The geometric series has $R = 1$ but diverges at $x = 1$ and $x = -1$, whereas

$$1 + \frac{x}{1^2} + \frac{x^2}{2^2} + \frac{x^3}{3^2} + \dots$$

has $R = 1$ and converges at $x = 1$ and $x = -1$. Because the interval of convergence is $(-R, R)$, the number R is called the *radius of convergence*. If the series converges only at $x = 0$, we say $R = 0$, whereas if the series converges absolutely for all x, we say $R = \infty$.

Here are two useful formulas for the radius R.

Theorem 3.4.2 (Root Test). *Let $\sum a_n x^n$ be a power series and let L denote the upper limit of the sequence $(|a_n|^{1/n})$,*

$$L = \limsup_{n\to\infty} |a_n|^{1/n}.$$

Then $R = 1/L$ where we take $1/0 = \infty$ and $1/\infty = 0$.

To derive this, it is enough to show that $LR = 1$. If $|x| < R$, then $\sum a_n x^n$ converges absolutely; hence, $|a_n| \cdot |x|^n = |a_n x^n| \le C$ for some C (possibly depending on x); thus, $|a_n|^{1/n}|x| \le C^{1/n}$. Taking the upper limit of both sides, we conclude that $L|x| \le 1$. Since $|x|$ may be as close as desired to R, we obtain $LR \le 1$. On the other hand, if $x > R$, then the sequence $(a_n x^n)$ is unbounded (otherwise, the series converges on $(-x, x)$ contradicting the definition of R). Hence, some subsequence of $(|a_n||x|^n)$ is bounded below by 1, which implies that some subsequence of $(|a_n|^{1/n}|x|)$ is bounded below by 1. We conclude that $L|x| \ge 1$. Since $|x|$ may be as close as desired to R, we obtain $LR \ge 1$. Hence, $LR = 1$. □

Theorem 3.4.3 (Ratio Test). *Let (a_n) be a nonzero sequence, and suppose that*

$$\rho = \lim_{n\nearrow\infty} |a_n|/|a_{n+1}|$$

exists. Then ρ equals the radius of convergence R of $\sum a_n x^n$.

To show that $\rho = R$, we show that $\sum a_n x^n$ converges when $|x| < \rho$ and diverges when $|x| > \rho$. If $|x| < \rho$, choose c with $|x| < |c| < \rho$. Then $|c| \le |a_n|/|a_{n+1}|$ for $n \ge N$. Hence,

$$\left|a_{n+1}x^{n+1}\right| = \left|\frac{a_{n+1}}{a_n}\right| \cdot |c| \cdot |a_n x^n| \cdot \left|\frac{x}{c}\right| \le |a_n x^n| \cdot \left|\frac{x}{c}\right|, \quad n \ge N.$$

Iterating this, we obtain $|a_{N+2}x^{N+2}| \le |x/c|^2 |a_N x^N|$. Continuing in this manner, we obtain $|a_n x^n| \le C|x/c|^{n-N}$, $n \ge N$, for some constant C. Since $\sum |x/c|^n$ converges, this shows that $\sum |a_n x^n|$ converges. On the other hand, if $|x| > \rho$, then the same argument shows that $\left|a_{n+1}x^{n+1}\right| \ge |a_n x^n|$ for $n \ge N$; hence, $a_n x^n \not\to 0$. By the nth term test, $\sum a_n x^n$ diverges. Thus, $\rho = R$. □

In the previous section, we saw (3.3.9) that the series

$$\exp(x) = \sum_{n=0}^{\infty} \frac{x^n}{n!}$$

converges to e^x for $x \geq 0$. Since the region of convergence of a power series is an interval centered at zero, it follows that $\exp(x)$ converges absolutely for all real x. Now for x and y real, the Cauchy product $\exp(x)\exp(y)$ (§1.7) of $\exp(x)$ and $\exp(y)$ is

$$\sum_{n=0}^{\infty} \sum_{i+j=n} \frac{x^i y^j}{i!j!} = \sum_{n=0}^{\infty} \frac{1}{n!} \sum_{i+j=n} \binom{n}{i} x^i y^j = \sum_{n=0}^{\infty} \frac{(x+y)^n}{n!} = \exp(x+y).$$

By selecting $y = -x$ for x positive, $e^x \exp(-x) = \exp(x)\exp(-x) = \exp(0) = 1$; hence, $\exp(x) = e^x$ for x negative as well.

As an application, using the formula for the radius of convergence together with the fact that exp converges everywhere yields

$$\lim_{n \nearrow \infty} (n!)^{1/n} = \infty. \tag{3.4.1}$$

This can also be derived directly.

We have shown that

$$e^x = 1 + \frac{x}{1!} + \frac{x^2}{2!} + \frac{x^3}{3!} + \dots, \qquad x \in \mathbf{R}. \tag{3.4.2}$$

In particular, we have arrived at a series for e,

$$e = 1 + \frac{1}{1!} + \frac{1}{2!} + \frac{1}{3!} + \dots.$$

In §1.6, we obtained $2.5 < e \leq 3$. In fact, by the addition of sufficiently many terms, e can be computed to arbitrarily many places.

What other functions can be expressed as power series? Two examples are the even and the odd parts of exp.

The even and odd parts (§3.1) of exp are the *hyperbolic cosine* cosh and the *hyperbolic sine* sinh (these are pronounced to rhyme with "gosh" and "cinch"). Thus,

$$\sinh x = \frac{e^x - e^{-x}}{2} = x + \frac{x^3}{3!} + \frac{x^5}{5!} + \dots,$$

and

$$\cosh x = \frac{e^x + e^{-x}}{2} = 1 + \frac{x^2}{2!} + \frac{x^4}{4!} + \dots.$$

Since

$$\sum_{n=0}^{\infty} |x^n/n!| = \sum_{n=0}^{\infty} |x|^n/n! = \exp|x|,$$

the series exp is absolutely convergent on all of **R**. By comparison with exp, the series for sinh and cosh are also absolutely convergent on all of **R**. From their definitions, $\cosh' = \sinh$ and $\sinh' = \cosh$.

Note that the series for $\sinh x$ involves only odd powers of x and the series for $\cosh x$ involves only even powers of x. This holds for all odd and even functions (Exercise **3.4.7**).

To obtain other examples, we write alternating versions (§1.7) of the last two series obtaining

$$\sin x = x - \frac{x^3}{3!} + \frac{x^5}{5!} - \dots,$$

and

$$\cos x = 1 - \frac{x^2}{2!} + \frac{x^4}{4!} - \dots.$$

These functions are studied in §3.6. Again, by comparison with exp, sin and cos are absolutely convergent series on all of **R**. However, unlike exp, cosh, and sinh, we do not as yet know $\sin'$ and $\cos'$.

It turns out that functions constructed from power series are smooth in their interval of convergence. They have derivatives of all orders.

Theorem 3.4.4. *Let $f(x) = \sum a_n x^n$ be a power series with radius of convergence $R > 0$. Then*

$$\sum_{n=1}^{\infty} n a_n x^{n-1} = a_1 + 2a_2 x + 3a_3 x^2 + \dots \qquad (3.4.3)$$

has radius of convergence R, f is differentiable on $(-R, R)$, and $f'(x)$ equals (3.4.3) for all x in $(-R, R)$.

In other words, to obtain the derivative of a power series, one needs only to differentiate the series term by term. To see this, we first show that the radius of the power series $\sum(n+1)a_{n+1}x^n$ is R. Here the nth coefficient is $b_n = (n+1)a_{n+1}$, so

$$|b_n|^{1/n} = (n+1)^{1/n}|a_{n+1}|^{1/n} = (n+1)^{1/n}\left[|a_{n+1}|^{1/(n+1)}\right]^{(n+1)/n},$$

so the upper limit of $(|b_n|^{1/n})$ equals the upper limit of $(|a_n|^{1/n})$ since $(n+1)^{1/n} \to 1$ (§3.3) and $(n+1)/n \to 1$.

Now we show that $f'(c)$ exists and equals $\sum n a_n c^{n-1}$, where $-R < c < R$ is fixed. To do this, let us consider only a single term in the series, i.e., let us consider x^n with n fixed, and pick c real. Then by the binomial theorem (§3.3),

$$x^n = [c + (x - c)]^n = \sum_{j=0}^{n} \binom{n}{j} c^{n-j}(x - c)^j$$
$$= c^n + nc^{n-1}(x - c) + \sum_{j=2}^{n} \binom{n}{j} c^{n-j}(x - c)^j.$$

Thus,

$$\frac{x^n - c^n}{x - c} - nc^{n-1} = \sum_{j=2}^{n} \binom{n}{j} c^{n-j}(x - c)^{j-1}, \qquad x \neq c.$$

Now choose any $d > 0$; then for x satisfying $0 < |x - c| < d$,

$$\begin{aligned}\left|\frac{x^n - c^n}{x - c} - nc^{n-1}\right| &= \left|\sum_{j=2}^{n} \binom{n}{j} c^{n-j}(x - c)^{j-1}\right| \\ &\leq |x - c| \sum_{j=2}^{n} \binom{n}{j} |c|^{n-j}|x - c|^{j-2} \\ &\leq |x - c| \sum_{j=2}^{n} \binom{n}{j} |c|^{n-j} d^{j-2} \\ &= \frac{|x - c|}{d^2} \sum_{j=2}^{n} \binom{n}{j} |c|^{n-j} d^j \leq \frac{|x - c|}{d^2}(|c| + d)^n,\end{aligned}$$

where we have used the binomial theorem again. To summarize,

$$\left|\frac{x^n - c^n}{x - c} - nc^{n-1}\right| \leq \frac{|x - c|}{d^2}(|c| + d)^n, \qquad 0 < |x - c| < d. \tag{3.4.4}$$

Now assume $|c| < R$ and choose $d < R - |c|$. Then by the triangle inequality, $|x - c| < d$ implies $|x| < R$ since $|x| \leq |x - c| + |c| < d + |c| \leq R$. Assume also temporarily that the coefficients a_n are nonnegative, $a_n \geq 0$, $n \geq 0$. Multiplying (3.4.4) by a_n and summing over $n \geq 0$ yields

$$\left|\frac{f(x) - f(c)}{x - c} - \sum_{n=1}^{\infty} na_n c^{n-1}\right| \leq \frac{|x - c|}{d^2} f(|c| + d), \qquad 0 < |x - c| < d.$$

Letting $x \to c$ in the last inequality establishes the result when $a_n \geq 0$, $n \geq 0$. If this is not so, we obtain instead

$$\left|\frac{f(x) - f(c)}{x - c} - \sum_{n=1}^{\infty} na_n c^{n-1}\right| \leq \frac{|x - c|}{d^2} g(|c| + d), \qquad 0 < |x - c| < d.$$

where $g(x) = \sum_{n=0}^{\infty} |a_n| x^n$, so the same argument works. □

Since this theorem can be applied repeatedly to $f, f', f'', \ldots$, *every power series with radius of convergence R determines a smooth function f on* $(-R, R)$. For example, $f(x) = \sum a_n x^n$ implies

$$f''(x) = \sum_{n=2}^{\infty} n(n-1)a_n x^{n-2}$$

on the interval of convergence. More generally, on $(-R, R)$,

$$f^{(j)}(x) = \sum_{n=j}^{\infty} n(n-1)\ldots(n-j+1)a_n x^{n-j}, \qquad j \geq 0.$$

In particular, sin and cos are smooth functions on $\mathbf{R}$ (we already knew this for exp, cosh, and sinh). Moreover, differentiating the series for sin and cos, term by term, yields $\sin' = \cos$ and $\cos' = -\sin$.

Exercises

3.4.1. Use the exponential series to compute e to four decimal places, justifying your reasoning.

3.4.2. Show directly that $\lim_{n\nearrow\infty}(n!)^{1/n} = \infty$. (First, show that the lower limit is ≥ 100.)

3.4.3. Suppose that $\sum a_n x^n$ and $\sum b_n x^n$ both converge to $f(x)$ on $(-R, R)$. Show that $a_n = b_n$ for $n \geq 0$. Thus, the coefficients of a power series are uniquely determined.

3.4.4. Show that the inverse $\operatorname{arcsinh} : \mathbf{R} \to \mathbf{R}$ of $\sinh : \mathbf{R} \to \mathbf{R}$ exists and is smooth, and compute $\operatorname{arcsinh}'$. Show that $f(x) = \cosh x$ is superlinear, smooth, and strictly convex. Compute the Legendre transform $g(y)$ (Exercise **3.3.16**), and check that g is smooth.

3.4.5. Compute the radius of convergence of

$$\sum_{n=0}^{\infty} (-1)^n x^n / 4^n (n!)^2.$$

3.4.6. What is the radius of convergence of

$$\sum_{n=0}^{\infty} x^{n!} = 1 + x + x^2 + x^6 + x^{24} + x^{120} + \ldots?$$

3.4.7. Let $f(x) = a_0 + a_1x + a_2x^2 + \dots$ converge for $-R < x < R$. Then f is even iff the coefficients a_{2n-1}, $n \geq 1$ of the odd powers vanish, and f is odd iff the coefficients a_{2n}, $n \geq 1$ of the even powers vanish.

3.4.8. Let $k \geq 0$. Show that there are integers a_j, $0 \leq j \leq k$, such that

$$\left(x\frac{d}{dx}\right)^k \left(\frac{1}{1-x}\right) = \sum_{j=0}^k \frac{a_j}{(1-x)^{j+1}}.$$

Use this to show the sum

$$\sum_{n=1}^\infty \frac{n^k}{2^n}$$

is a natural.

3.5 Taylor Series

A function f is *analytic at* 0 with radius of convergence R if it can be expressed as a convergent power series

$$f(x) = \sum_{n=0}^\infty a_n x^n, \qquad |x| < R, \tag{3.5.1}$$

on $(-R, R)$. A function is *entire at* 0 if it is analytic at 0 with infinite radius of convergence. For example, exp, sinh, cosh, sin, and cos are analytic at 0 and are in fact entire.

If f and g are analytic at 0 with radius of convergence R, then so is $f + g$. Since the Cauchy product (§1.7) of absolutely convergent series is absolutely convergent, if f and g are analytic at 0 with radius of convergence R, then so is fg. In general, however, it is not true that the quotient f/g is analytic at 0 with the same radius of convergence R. For example, the quotient of the entire functions $f(x) = 1$ and $g(x) = 1 - x$ is not entire.

Let f be analytic at 0 with radius of convergence R. In §3.4, we saw that this implies the smoothness of f on $(-R, R)$; the derivatives of f are also analytic at 0, with the same radius of convergence, and all of the form

$$f^{(j)}(x) = \sum_{n=j}^\infty n(n-1)\dots(n-j+1)a_n x^{n-j}, \qquad j \geq 0. \tag{3.5.2}$$

Inserting $x = 0$ in these formulas, we obtain $j!a_j = f^{(j)}(0)$ for $j \geq 0$; hence,

$$f(x) = \sum_{n=0}^\infty \frac{f^{(n)}(0)}{n!} x^n = f(0) + f'(0)x + \frac{f''(0)}{2!}x^2 + \dots \tag{3.5.3}$$

on $(-R, R)$. The power series in (3.5.3) is the *Taylor series of f centered at zero.*[4]

Let c be a real. A function f is *analytic at c* with radius of convergence R if $f(x+c)$ is analytic at 0 with radius of convergence R. Equivalently, f is *analytic at c* with radius of convergence R if

$$f(x) = \sum_{n=0}^{\infty} a_n (x-c)^n, \qquad |x-c| < R.$$

A function is *entire at c* if it is analytic at c with infinite radius of convergence.

Now suppose f is analytic at 0 with radius R and fix $0 \le c \le x < R$. Suppose also that the coefficients a_n in (3.5.1) are nonnegative for all $n \ge 0$. Using (3.5.2), the binomial theorem, (1.7.10), and the substitution $n = k+j$,

$$\begin{aligned}
f(x) = \sum_{m=0}^{\infty} a_m x^m &= \sum_{m=0}^{\infty} a_m \left(\sum_{j=0}^{m} \binom{m}{j} c^{m-j} (x-c)^j \right) \\
&= \sum_{m=0}^{\infty} \left(\sum_{k+j=m} a_{k+j} \binom{k+j}{j} c^k (x-c)^j \right) \\
&= \sum_{(j,k)} a_{k+j} \binom{k+j}{j} c^k (x-c)^j \\
&= \sum_{j=0}^{\infty} \left(\sum_{k=0}^{\infty} \binom{k+j}{j} a_{k+j} c^k (x-c)^j \right) \\
&= \sum_{j=0}^{\infty} \left(\sum_{n=j}^{\infty} \binom{n}{j} a_n c^{n-j} (x-c)^j \right) \\
&= \sum_{j=0}^{\infty} \left(\sum_{n=j}^{\infty} n(n-1)\cdots(n-j+1) a_n c^{n-j} \right) \frac{(x-c)^j}{j!} \\
&= \sum_{j=0}^{\infty} \frac{f^{(j)}(c)}{j!} (x-c)^j
\end{aligned} \tag{3.5.4}$$

This last series is the *Taylor series of f centered at c.* We arrive at the following.

Theorem 3.5.1 (Taylor series). *Let f be analytic at* 0 *with radius of convergence R. Then, for all* $|c| < R$, *f is analytic at c with radius of convergence* $R - |c|$, *and*

[4] Also called the *Maclaurin series of f*

$$f(x) = \sum_{n=0}^{\infty} \frac{f^{(n)}(c)}{n!}(x-c)^n = f(c) + f'(c)(x-c) + \frac{f''(c)}{2!}(x-c)^2 + \dots \quad (3.5.5)$$

for $|x - c| < R - |c|$.

This result follows as soon as we establish the validity of the result (3.5.4). We assumed $a_n \geq 0$, $n \geq 0$, and $0 \leq c \leq x < R$ so that all terms in the computation leading to (3.5.4) are nonnegative, allowing us to apply the result (1.7.10) that nonnegative double series equal either iterated series.

More generally, when $|c| < R$ and $|x - c| < R - |c|$ and (a_n) is signed, to establish the result (3.5.4), we apply the result (1.7.12) that summable double series equal either iterated series. To this end, we must first establish the summability of the double series

$$\sum_{(j,k)} \binom{k+j}{j} a_{k+j} c^k (x-c)^j. \quad (3.5.6)$$

Let $g(x) = \sum |a_n| x^n$. If $|c| < R$ and $|x - c| < R - |c|$, then $0 \leq |c| \leq |c| + |x - c| < R$ and $(|a_n|)$ is nonnegative, so the computation leading to (3.5.4) is valid if we replace c by $|c|$, x by $|c| + |x - c|$, a_n by $|a_n|$, and f by g. Thus, we obtain

$$g(|c| + |x - c|) = \sum_{(j,k)} \binom{k+j}{j} |a_{k+j}| |c|^k |x - c|^j. \quad (3.5.7)$$

Since $g(|c| + |x - c|)$ is finite, this establishes the summability of (3.5.6); hence, we may go back and apply (1.7.12), establishing the validity of the result (3.5.4) when $|c| < R$ and $|x - c| < R - |c|$. □

In particular, f entire at 0 implies f entire at c for all real c. Hence, for entire functions, we drop the qualifiers "at 0" and "at c" and simply call such functions *entire*.

We now turn to the converse situation. Starting with a smooth function f on $(-R, R)$, what can we say about its Taylor series (3.5.5)? It turns out the Taylor series of a smooth function may or may not converge for a given x. When it does converge, its sum need not equal the function. For example, (Exercise **3.5.2**), the function

$$f(x) = \begin{cases} e^{-1/x} & x > 0, \\ 0 & x \leq 0, \end{cases}$$

is smooth on $\mathbf{R}$ and satisfies $f^{(n)}(0) = 0$ for all $n \geq 0$. Thus, for all x, the Taylor series centered at zero converges, since it is identically zero. Hence, the Taylor series does not sum to $f(x)$, except when $x \leq 0$.

The above example shows that no growth condition on the sequence $(a_n) = (f^{(n)}(0)/n!)$ can guarantee the analyticity at zero of f. Nevertheless, a necessary condition for the convergence on $(-R,R)$ of the Taylor series centered at c is the boundedness of its terms: $|f^{(n)}(c)|d^n/n! \le C$, $n \ge 1$, for $0 < d < R - |c|$. This is strengthened to a sufficient condition in Exercise **3.5.8**.

To study the convergence of the Taylor series to the function, it makes sense to start with the n-th partial sum. Let f be continuous on (a,b) and assume f is n times differentiable on (a,b). For $n \ge 0$, the *Taylor polynomial of degree n at c* is

$$p_n(x,c) = f(c) + f'(c)(x-c) + \frac{f''(c)}{2!}(x-c)^2 + \cdots + \frac{f^{(n)}(c)}{n!}(x-c)^n.$$

Define the $(n+1)$st *remainder*

$$R_{n+1}(x,c) = f(x) - p_n(x,c), \quad a < x < b;$$

for example, $R_1(x,c) = f(x) - f(c)$. If we write $R_{n+1}(x,c;f)$ to emphasize dependence on f, note that $R'_{n+1}(x,c;f) = R_n(x,c;f')$. Now define $h_{n+1} : (a,b) \to \mathbf{R}$ by

$$h_{n+1}(x) = \begin{cases} \dfrac{R_{n+1}(x,c)}{(x-c)^{n+1}/(n+1)!}, & x \ne c, \\ f^{(n+1)}(c), & x = c. \end{cases}$$

Now assume $f^{(n)}$ is differentiable at c, i.e., assume $f^{(n+1)}(c)$ exists. Then h_{n+1} is continuous when $x \ne c$. Applying l'Hôpital's rule n times,

$$\begin{aligned} \lim_{x\to c} h_{n+1}(x) &= \lim_{x\to c} \frac{R_{n+1}(x,c)}{(x-c)^{n+1}/(n+1)!} = \lim_{x\to c} \frac{R'_{n+1}(x,c)}{(x-c)^n/n!} \\ &= \lim_{x\to c} \frac{R_n(x,c;f')}{(x-c)^n/n!} = \lim_{x\to c} \frac{R_{n-1}(x,c;f'')}{(x-c)^{n-1}/(n-1)!} \\ &= \cdots = \lim_{x\to c} \frac{R_1(x,c;f^{(n)})}{x-c} = \lim_{x\to c} \frac{f^{(n)}(x) - f^{(n)}(c)}{x-c} = f^{(n+1)}(c). \end{aligned}$$

Thus, h_{n+1} is continuous on (a,b).

Theorem 3.5.2 (Taylor's Theorem). *Suppose f is continuous on (a,b) and $n \ge 0$. Suppose also f is n times differentiable on (a,b), and let $p_n(x,c)$ denote the Taylor polynomial at c of degree n. If $f^{(n+1)}(c)$ exists at c in (a,b), then there is a continuous function[5] $h_{n+1} : (a,b) \to \mathbf{R}$ satisfying $h_{n+1}(c) = f^{(n+1)}(c)$ and*

$$f(x) = p_n(x,c) + \frac{h_{n+1}(x)}{(n+1)!}(x-c)^{n+1}. \tag{3.5.8}$$

[5] Taylor's theorem in §4.4 gives a useful formula for h_{n+1}.

Moreover, if $f^{(n+1)}$ *exists on all of* (a,b)*, then for some* ξ *between* c *and* x*,* $R_{n+1}(x,c)$ *is given by the Cauchy form*

$$R_{n+1}(x,c) = \frac{h_{n+1}(x)}{(n+1)!}(x-c)^{n+1} = \frac{f^{(n+1)}(\xi)}{n!}(x-\xi)^n(x-c),$$

and, for some η *between* c *and* x*,* $R_{n+1}(x,c)$ *is given by the Lagrange form*

$$R_{n+1}(x,c) = \frac{h_{n+1}(x)}{(n+1)!}(x-c)^{n+1} = \frac{f^{(n+1)}(\eta)}{(n+1)!}(x-c)^{n+1}.$$

The derivation of (3.5.8) is above. To obtain the Cauchy and Lagrange forms, differentiate the expression defining $R_{n+1}(x,c)$ once *with respect to* c. Then the sum collapses to

$$\frac{d}{dc}R_{n+1}(x,c) = -\frac{f^{(n+1)}(c)}{n!}(x-c)^n.$$

Now apply the mean value theorem to $t \mapsto R_{n+1}(x,t)$ on the interval joining c to x. Since $R_{n+1}(x,x) = 0$, this yields the Cauchy form. To obtain the Lagrange form, set $g(t) = (x-t)^{n+1}/(n+1)!$ and fix x and c. Then $g'(t) = -(x-t)^n/n!$, $g(x) = 0$, and (remember $'$ here is with respect to t) $R'_{n+1}(x,t)/g'(t) = f^{(n+1)}(t)$. So by the generalized mean value theorem, there is an η between c and x with

$$h_{n+1}(x) = \frac{R_{n+1}(x,x) - R_{n+1}(x,c)}{g(x) - g(c)} = \frac{R'_{n+1}(x,\eta)}{g'(\eta)} = f^{(n+1)}(\eta).\square$$

In particular, the Lagrange form implies that, for f smooth on $\mathbf{R}$ and any $n \geq 1$, for each $x \in \mathbf{R}$, there is an η between 0 and x with

$$f(x) = f(0) + f'(0)x + \frac{f''(0)}{2!}x^2 + \cdots + \frac{f^{(n)}(0)}{n!}x^n + \frac{f^{(n+1)}(\eta)}{(n+1)!}x^{n+1}. \quad (3.5.9)$$

Thus, a smooth function f can be approximated near 0 by an nth-degree polynomial with an error $R_n(x,0)$ given by a certain expression, for every $n \geq 1$.

Using (3.5.8) with $f(x) = e^x$, we can show that e is irrational. Indeed, suppose that e were rational. Then there would be a natural N, such that $n!e \in \mathbf{N}$ for all $n \geq N$. So choose n greater than 3 and greater than N, and write (3.5.8) for this n and $x = 1$ to get

$$e = 1 + \frac{1}{1!} + \frac{1}{2!} + \cdots + \frac{1}{n!} + \frac{e^\eta}{(n+1)!}.$$

Then

$$n!e = n!\left(1 + \frac{1}{1!} + \frac{1}{2!} + \cdots + \frac{1}{n!}\right) + \frac{e^\eta}{n+1}.$$

So $e^\eta/(n+1)$ is a natural, which is false since $0<\eta<1$ and $n>3$ imply

$$\frac{e^\eta}{n+1} < \frac{e}{4} \le \frac{3}{4}.$$

This contradiction allows us to conclude the following.

Theorem 3.5.3. *e is irrational.* □

Our last topic is to describe Newton's generalization of the binomial theorem (§3.3) to nonnatural exponents. The result is the same, except that the sum proceeds to ∞. For v real and $n \ge 1$, let

$$\binom{v}{n} = \frac{v\cdot(v-1)\cdot\cdots\cdot(v-n+1)}{1\cdot 2\cdot\cdots\cdot n}.$$

Also let $\binom{v}{0}=1$. If v is a natural, then $\binom{v}{n}$ is the binomial coefficient defined previously for $0\le n\le v$ and $\binom{v}{n}=0$ for $n>v$. These binomial coefficients grow at most polynomially as $n\to\infty$.

Theorem 3.5.4. *For v real,*

$$\left|\binom{v}{n}\right| \le (n+|v|)^{|v|}, \qquad n\ge 0.$$

To see this, let N be the integer part of $v+1$, so $N-1\le v<N$, and consider the case $N\ge 1$. Since for $n<N$ we have $0\le v(v-1)\cdots(v-n+1)\le N!\le N^{N-1}\le (v+1)^v\le (n+v)^v$, we may assume $n\ge N$. Then $v(v-1)\cdots(v-n+1)$ is the product of the positive factors $v(v-1)\cdots(v-N+1)$ and the negative factors $(v-N)\cdots(v-n+1)$. Hence, the absolute value of $v(v-1)\cdots(v-n+1)$ is no larger than $N!(n-N)!$; hence, $|\binom{v}{n}|\le N!(n-N)!/n!\le N!\le (n+v)^v$. Now assume $N\le 0$. Then all the factors in $v(v-1)\cdots(v-n+1)$ are negative; hence, its absolute value is no larger than $(n-N)!$. Hence,

$$\left|\binom{v}{n}\right| \le \frac{(n-N)!}{n!} \le (n-N)^{-N} \le (n-v)^{-v}. \square$$

The above estimate is not sharp; for example, one can see directly $|\binom{-1/2}{n}|\le 1$. In fact, $|\binom{-1/2}{n}|\sim 1/\sqrt{\pi n}$ as $n\to\infty$ (Exercise **5.5.3**).

Theorem 3.5.5 (Newton's Binomial Theorem). *For v real and $|b|<|a|$,*

$$(a+b)^v = \sum_{n=0}^{\infty}\binom{v}{n}a^{v-n}b^n.$$

In particular,

$$(1+x)^v = \sum_{n=0}^{\infty} \binom{v}{n} x^n, \qquad -1 < x < 1.$$

By setting $x = b/a$ and multiplying by a^v in the second form, the first form follows from the second. Let $f(x) = (1+x)^v$. Since

$$f^{(n)}(x) = v(v-1)\dots(v-n+1)(1+x)^{v-n},$$

$f^{(n)}(x)/n! = \binom{v}{n}(1+x)^{v-n}$. Hence, using Lagrange's form of the remainder, for some $0 < \eta < x < 1$ (see §3.3 for the limits), we obtain

$$|R_n(x,c)| = \left|\binom{v}{n+1}\right| (1+\eta)^{v-n-1} x^{n+1} \le (n+1+|v|)^{|v|} x^{n+1}$$

which goes to zero as $n \nearrow \infty$.

To establish the result on $(-1,0)$, apply the Cauchy form of the remainder to $f(x) = (1-x)^v$ and $0 < x < 1$, $c = 0$. Then $f^{(n+1)}(x)/n! = (-1)^{n+1}\binom{v}{n+1}(1-x)^{v-n-1}(n+1)$, so for some $0 < \xi < x$, we obtain

$$\begin{aligned}
|R_n(x,c)| &= (n+1)\left|\binom{v}{n+1}\right| (1-\xi)^{v-n-1}(x-\xi)^n(x-0) \\
&= (n+1)x\left|\binom{v}{n+1}\right| \left(\frac{x-\xi}{1-\xi}\right)^n (1-\xi)^{v-1} \\
&\le (n+1)x(n+1+|v|)^{|v|} \left(\frac{x-\xi}{1-\xi}\right)^n (1-\xi)^{v-1}.
\end{aligned}$$

If $v \ge 1$, $(1-\xi)^{v-1} \le 1$. If $v < 1$, $(1-\xi)^{v-1} \le (1-x)^{v-1}$. Hence, in both cases, $(1-\xi)^{v-1}$ is bounded by $[1+(1-x)^{v-1}]$, a fixed quantity independent of n (remember that ξ may depend on n). Moreover, $(x-\xi)/(1-\xi) < (x-0)/(1-0) = x$. Hence,

$$|R_n(x,c)| \le [1+(1-x)^{v-1}](n+1)(n+1+|v|)^{|v|} x^{n+1},$$

which goes to zero as $n \nearrow \infty$. □

Exercises

3.5.1. Suppose that $f : \mathbf{R} \to \mathbf{R}$ is nonnegative, twice differentiable, and $f''(c) \le 1/2$ for all $c \in \mathbf{R}$. Use Taylor's theorem to conclude that $|f'(c)| \le \sqrt{f(c)}$.

3.5.2. Let

$$h(x) = \begin{cases} e^{-1/x}, & x > 0, \\ 0, & x \leq 0. \end{cases}$$

By induction on $n \geq 1$, show that

A. $h^{(n-1)}(x) = h(x)R_n(x)$, $x > 0$, for some rational function R_n, and
B. $h^{(n-1)}(x) = 0$, $x \leq 0$.

Conclude that h (Figure 3.11) is smooth on **R**.

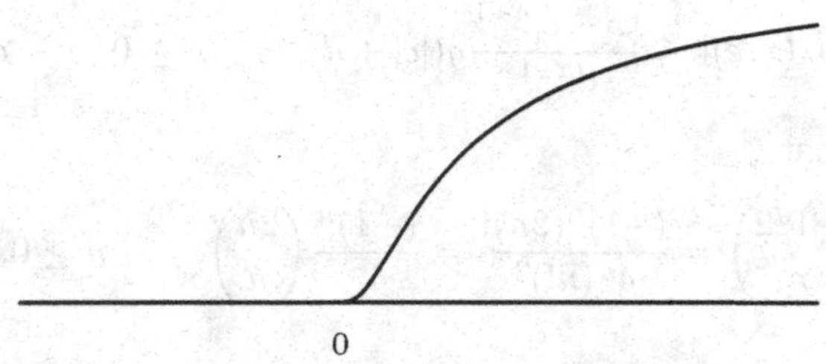

Fig. 3.11 Graph of the function h in Exercise **3.5.2**

3.5.3. Show that

$$\frac{1}{\sqrt{1-x^2}} = 1 + \frac{1}{2} \cdot x^2 + \frac{1}{2} \cdot \frac{3}{4} \cdot x^4 + \frac{1}{2} \cdot \frac{3}{4} \cdot \frac{5}{6} \cdot x^6 + \dots$$

for $|x| < 1$.

3.5.4. Compute the Taylor series of $\log(1+x)$ centered at 0.

3.5.5. Show that

$$\frac{\log(1+x)}{1+x} = x - \left(1 + \frac{1}{2}\right) x^2 + \left(1 + \frac{1}{2} + \frac{1}{3}\right) x^3 - \left(1 + \frac{1}{2} + \frac{1}{3} + \frac{1}{4}\right) x^4 + \dots$$

for $|x| < 1$ by considering the product (§1.7) of the series for $1/(1+x)$ and the series for $\log(1+x)$.

3.5.6. Let $f : \mathbf{R} \to \mathbf{R}$ be twice differentiable with f, f', and f'' continuous. If $f(0) = 1$, $f'(0) = 0$, and $f''(0) = q$, use Taylor's theorem and Exercise **3.2.3** to show that

$$\lim_{n \nearrow \infty} [f(x/\sqrt{n})]^n = \exp(qx^2/2).$$

This result is a key step in the derivation of the *central limit theorem*.

3.5.7. Compute

$$\lim_{t \to 0} \left(\frac{1}{e^t - 1} - \frac{1}{t} \right)$$

by writing $e^t = 1 + t + t^2 h(t)/2$.

3.5.8. Let $f:(-R,R)\to\mathbf{R}$ be smooth and let $|c|<R$, $d<R-|c|$. Suppose there is real $C>0$ with

$$|f^{(n)}(x)|\cdot\frac{d^n}{n!}\le C,\qquad n\ge 1, |x-c|<d.$$

Show that the Taylor series of f centered at c converges to $f(x)$ for $|x-c|<d$.

3.5.9. Suppose $f(x)=\sum a_nx^n$ is a power series on $(-R,R)$ and let $g(x)=\sum|a_n|x^n$. Show that for $|c|<R$ and $d<R-|c|$,

$$|f(x)-p_n(x,c)|\le\frac{|x-c|^{n+1}}{d^{n+1}}g(|c|+d),\qquad n\ge 0, |x-c|<d.$$

3.5.10. Show that

$$\binom{-1/2}{n}=\frac{(-1)^n(2n)!}{4^n(n!)^2}=\frac{(-1)^n}{4^n}\binom{2n}{n},\qquad n\ge 0.$$

3.6 Trigonometry

The trigonometric functions or circular functions sine and cosine may be defined either by power series or by measuring the length of arcs along the unit circle. As the second method involves integration, it cannot be discussed until Chapter 4 (Exercise **4.4.29**). Because of this, we base our development here on power series, which is what we have at hand.

In the previous section, we introduced alternating versions of the even and the odd parts of the exponential series, the *sine function*, and the *cosine function*,

$$\sin x = x-\frac{x^3}{3!}+\frac{x^5}{5!}-\frac{x^7}{7!}+\dots,$$

and

$$\cos x = 1-\frac{x^2}{2!}+\frac{x^4}{4!}-\frac{x^6}{6!}+\dots.$$

Since these functions are *defined* by these convergent power series, they are smooth everywhere and satisfy (§3.4):

$$\begin{aligned}(\sin x)' &= \cos x\\ (\cos x)' &= -\sin x,\end{aligned}$$

$\sin 0=0$, and $\cos 0=1$. The sine function is odd and the cosine function is even,

$$\sin(-x)=-\sin x,$$

and

$$\cos(-x)=\cos x.$$

Since

$$(\sin^2 x + \cos^2 x)' = 2\sin x \cos x + 2\cos x(-\sin x) = 0,$$

$\sin^2 + \cos^2$ is a constant; evaluating $\sin^2 x + \cos^2 x$ at $x = 0$ yields 1; hence,

$$\sin^2 x + \cos^2 x = 1 \tag{3.6.1}$$

for all x. This implies $|\sin x| \le 1$ and $|\cos x| \le 1$ for all x. If a is a critical point of $\sin x$, then $\cos a = 0$; hence, $\sin a = \pm 1$. Hence, $\sin a = 1$ at any positive local maximum a of $\sin x$.

Let f, g be differentiable functions satisfying $f' = g$ and $g' = -f$ on $\mathbf{R}$. Now the derivatives of $f\sin + g\cos$ and $f\cos - g\sin$ vanish; hence,

$$f(x)\sin x + g(x)\cos x = g(0),$$

and

$$f(x)\cos x - g(x)\sin x = f(0)$$

for all x. Multiplying the first equation by $\sin x$ and the second by $\cos x$ and adding, we obtain

$$f(x) = g(0)\sin x + f(0)\cos x.$$

Multiplying the first by $\cos x$ and the second by $-\sin x$ and adding, we obtain

$$g(x) = g(0)\cos x - f(0)\sin x.$$

Fixing y and taking $f(x) = \sin(x+y)$ and $g(x) = \cos(x+y)$, we obtain the identities

$$\begin{aligned} \sin(x+y) &= \sin x \cos y + \cos x \sin y, \\ \cos(x+y) &= \cos x \cos y - \sin x \sin y. \end{aligned} \tag{3.6.2}$$

If we replace y by $-y$ and combine the resulting equations with (3.6.2), we obtain the identities

$$\begin{aligned} \sin(x+y) + \sin(x-y) &= 2\sin x \cos y, \\ \sin(x+y) - \sin(x-y) &= 2\cos x \sin y, \\ \cos(x-y) - \cos(x+y) &= 2\sin x \sin y, \\ \cos(x-y) + \cos(x+y) &= 2\cos x \cos y. \end{aligned} \tag{3.6.3}$$

For $0 \le x \le 3$, the series

$$x - \sin x = \frac{x^3}{3!} - \frac{x^5}{5!} + \frac{x^7}{7!} - \dots$$

is alternating *with decreasing terms* (Exercise **3.6.8**). Hence, by the Leibnitz test (§1.7),

$$x - \sin x \le \frac{x^3}{6}, \qquad 0 \le x \le 3,$$

and

$$x - \sin x \ge \frac{x^3}{6} - \frac{x^5}{120}, \qquad 0 \le x \le 3.$$

Inserting $x = 1$ in the first inequality and $x = 3$ in the second, we obtain

$$\sin 0 = 0, \qquad \sin 1 \ge \frac{5}{6}, \qquad \sin 3 \le \frac{21}{40}.$$

Hence, there is a positive b in $(0,3)$, where $\sin b$ is a positive maximum, which gives $\cos(b) = 0$ and $\sin(b) = 1$. Let $a = \inf\{b > 0 : \sin b = 1\}$. Then the continuity of $\sin x$ implies $\sin a = 1$. Since $\sin 0 = 0$, $a > 0$. Since sin is a specific function, a is a specific real number.

For more than 20 centuries, the real $2a$ has been called "Archimedes' constant." More recently, for the last 200 years, the Greek letter π has been used to denote this real. Thus,

$$\sin\left(\frac{\pi}{2}\right) = 1,$$

and

$$\cos\left(\frac{\pi}{2}\right) = 0.$$

As yet, all we know about π is $0 < \pi/2 < 3$. In §5.2, we address the issue of computing π accurately.

Since the slope of the tangent line of $\sin x$ at $x = 0$ is $\cos 0 = 1$, $\sin x > 0$ for $x > 0$ near 0. Since there are no positive local maxima for $\sin x$ in $(0, \pi/2)$ and $\sin 0 = 0$, we must have $\sin x > 0$ on $(0, \pi/2)$. Hence, $\cos x$ is strictly decreasing on $(0, \pi/2)$. Hence, $\cos x$ is positive on $(0, \pi/2)$. Hence, $\sin x$ is strictly increasing on $(0, \pi/2)$. Moreover, since $(\sin x)'' = -\sin x$ and $(\cos x)'' = -\cos x$, $\sin x$ and $\cos x$ are concave on $(0, \pi/2)$. This justifies the graphs of $\sin x$ and $\cos x$ in the interval $[0, \pi/2]$ (Figure 3.12).

Inserting $y = \pi$ and replacing x by $-x$ in (3.6.2) yields

$$\begin{aligned}\sin(\pi - x) &= \sin x,\\ \cos(\pi - x) &= -\cos x.\end{aligned} \tag{3.6.4}$$

These identities justify the graphs of $\sin x$ and $\cos x$ in the interval $[\pi/2, \pi]$. Hence, the graphs are now justified on $[0, \pi]$. Replacing x by $-x$ in (3.6.4), we obtain

$$\sin(x + \pi) = -\sin x,$$

and

$$\cos(x + \pi) = -\cos x.$$

These identities justify the graphs of $\sin x$ and $\cos x$ on $[0, 2\pi]$. Repeating this reasoning once more,

$$\sin(x + 2\pi) = \sin(x + \pi + \pi) = -\sin(x + \pi) = \sin x,$$

and

$$\cos(x + 2\pi) = \cos(x + \pi + \pi) = -\cos(x + \pi) = \cos x,$$

showing that 2π is a *period* of $\sin x$ and $\cos x$. In fact, repeating this reasoning,

$$\sin(x + 2\pi n) = \sin x, \qquad n \in \mathbf{Z},$$

and

$$\cos(x + 2\pi n) = \cos x, \qquad n \in \mathbf{Z},$$

showing that every integer multiple of 2π is a period of $\sin x$ and $\cos x$. If $\sin x$ or $\cos x$ had any other period p, then by subtracting from p an appropriate integral multiple of 2π, we would obtain a period in $(0, 2\pi)$, contradicting the graphs. Hence, $2\pi\mathbf{Z}$ is the set of periods of $\sin x$ and of $\cos x$.

If we set $x = y$ in (3.6.2), we obtain

$$\sin(2x) = 2 \sin x \cos x,$$

and

$$\cos(2x) = \cos^2 x - \sin^2 x.$$

By (3.6.1), the second identity implies

$$\cos^2 x = \frac{1 + \cos(2x)}{2},$$

and

$$\sin^2 x = \frac{1 - \cos(2x)}{2}.$$

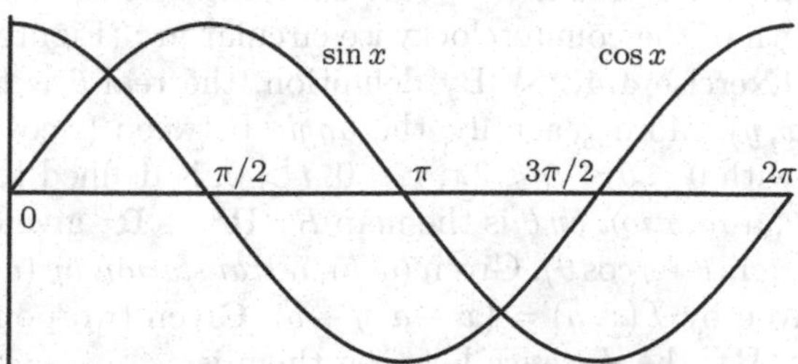

Fig. 3.12 The graphs of sine and cosine

By the inverse function theorem, $\sin x$ has an inverse on $[-\pi/2, \pi/2]$, and $\cos x$ has an inverse on $[0, \pi]$. These inverses are $\arcsin : [-1, 1] \to [-\pi/2, \pi/2]$

and arccos : $[-1,1] \to [0,\pi]$. Since $\cos x > 0$ on $(-\pi/2,\pi/2)$, by (3.6.1), $\cos(\arcsin x) = \sqrt{1-x^2}$ on $[-1,1]$. Similarly, since $\sin x > 0$ on $(0,\pi)$, $\sin(\arccos x) = \sqrt{1-x^2}$ on $[-1,1]$. Thus, the derivatives of the inverse functions are given by

$$(\arcsin x)' = \frac{1}{\cos(\arcsin x)} = \frac{1}{\sqrt{1-x^2}}, \qquad -1 < x < 1,$$

and

$$(\arccos x)' = \frac{1}{-\sin(\arccos x)} = \frac{-1}{\sqrt{1-x^2}}, \qquad -1 < x < 1.$$

As an application,

$$\left(2\arcsin\sqrt{x}\right)' = 2\cdot\frac{1}{\sqrt{1-\sqrt{x}^2}}\cdot\frac{1}{2\sqrt{x}} = \frac{1}{\sqrt{x(1-x)}}, \qquad 0 < x < 1.$$

Now we make the connection with the unit circle. The *unit circle* is the subset $\{(x,y) : x^2+y^2 = 1\}$ of $\mathbf{R}^2$. The interior $\{(x,y) : x^2+y^2 < 1\}$ of the unit circle is the *(open) unit disk*.

Theorem 3.6.1. *If (x,y) is a point on the unit circle, there is a real θ with $(x,y) = (\cos\theta,\sin\theta)$. If ϕ is any other such real, then $\theta-\phi$ is in $2\pi\mathbf{Z}$.*

Since $x^2+y^2 = 1$, $|x| \le 1$. Let $\theta = \arccos x$. Then $\sin^2\theta + \cos^2\theta = 1$ implies $y = \pm\sin\theta$. If $y = \sin\theta$, we have found a real θ, as required. Otherwise, replace θ by $-\theta$. This does not change $\cos\theta = \cos(-\theta)$, but changes $\sin\theta = -\sin(-\theta)$. For the second statement, suppose that $(\cos\theta,\sin\theta) = (\cos\phi,\sin\phi)$. Then

$$\cos(\theta-\phi) = \cos\theta\cos\phi + \sin\theta\sin\phi = 1.$$

Hence, $\theta-\phi$ is an integer multiple of 2π. □

By the Theorem, we may subtract an integer multiple of 2π from θ without affecting (x,y) and thus assume $0 \le \theta < 2\pi$. In this case, it turns out that θ equals the length of the counterclockwise circular arc (Figure 3.13) joining $(1,0)$ to (x,y) (Exercise **4.4.28**). By definition, the real θ is the *angle* corresponding to (x,y). More generally, the *angle* between $(r\cos\theta, r\sin\theta)$ and $(t\cos\phi, t\sin\phi)$, with $0 \le \theta \le \phi < 2\pi$, $r > 0$, $t > 0$, is defined to equal $\phi-\theta$.

Given a real θ, a *rotation by* θ is the map $R : \mathbf{R}^2 \to \mathbf{R}^2$ given by $R(x,y) = (x\cos\theta - y\sin\theta, x\sin\theta + y\cos\theta)$. Given (a,b), a *translation by* (a,b) is the map $T : \mathbf{R}^2 \to \mathbf{R}^2$ given by $T(x,y) = (x+a, y+b)$. Given two points $A = (a,b)$ and $B = (c,d)$ in $\mathbf{R}^2$, the *distance* between them is

$$d(A,B) = \sqrt{(a-c)^2 + (b-d)^2}.$$

This is also (by definition) the *length of the line segment* joining the two points.

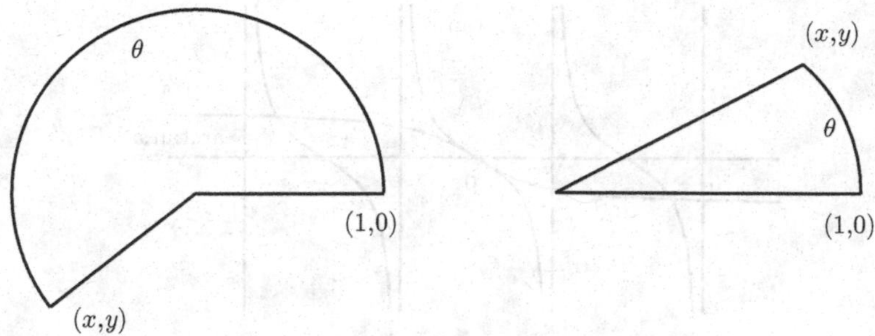

Fig. 3.13 The angle θ corresponding to (x, y)

It then follows that distance is translation and rotation invariant (Exercise **3.6.2**). From this follows the *Pythagoras theorem* for right triangles (Exercise **3.6.4**).

In this section, we defined sin and cos as power series; then we established the connection with the unit circle. It is possible to reverse this point of view and first define sin and cos from the unit circle and then derive their properties and Taylor expansions (see Exercise **4.4.29**). Because of this, the trigonometric functions are also called the *circular functions*.

Going back to the main development, the *tangent function*, $\tan x = \sin x / \cos x$, is smooth everywhere except at odd multiples of $\pi/2$ where the denominator vanishes. Moreover, $\tan x$ is an odd function, and

$$\tan(x + \pi) = \frac{\sin(x + \pi)}{\cos(x + \pi)} = \frac{-\sin x}{-\cos x} = \tan x.$$

So π is the period for $\tan x$. By the quotient rule

$$(\tan x)' = \frac{(\sin x)' \cos x - \sin x(\cos x)'}{(\cos x)^2} = \frac{1}{\cos^2 x}, \qquad -\pi/2 < x < \pi/2.$$

Thus, $\tan x$ is strictly increasing on $(-\pi/2, \pi/2)$. Moreover,

$$\begin{aligned} \tan(\pi/2-) &= \infty, \\ \tan(-\pi/2+) &= -\infty, \end{aligned}$$

and

$$(\tan x)'' = \left(\frac{1}{\cos^2 x}\right)' = 2\frac{\tan x}{\cos^2 x}.$$

Thus, $\tan x$ is convex on $(0, \pi/2)$ and concave on $(-\pi/2, 0)$. The graph is as shown in Figure 3.14.

By the inverse function theorem, $\tan x$ has an inverse on $(-\pi/2, \pi/2)$. This inverse, $\arctan : (-\infty, \infty) \to (-\pi/2, \pi/2)$, is smooth, and

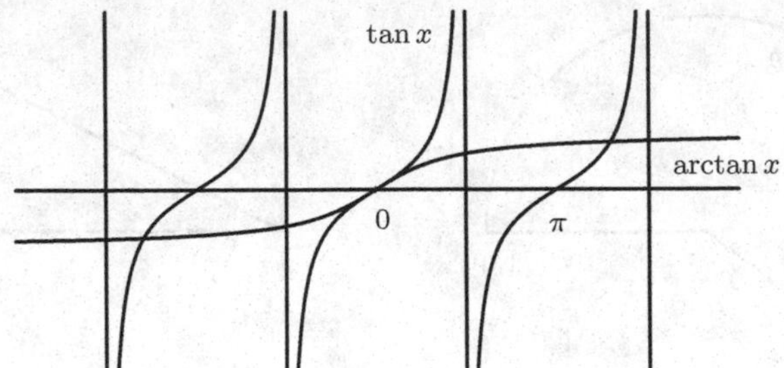

Fig. 3.14 The graphs of $\tan x$ and $\arctan x$

$$(\arctan x)' = 1/\cos(\arctan x)^{-2} = \cos^2(\arctan x).$$

Since $\cos x$ is positive on $(-\pi/2, \pi/2)$, dividing (3.6.1) by $\cos^2 x$, we have $\tan^2 x + 1 = 1/\cos^2 x$. Hence, $\cos^2 x = 1/(1+\tan^2 x)$. Thus,

$$(\arctan x)' = \frac{1}{1+x^2}.$$

It follows that $\arctan x$ is strictly increasing on $\mathbf{R}$. Since $(\arctan x)'' = -2x/(1+x^2)^2$, $\arctan x$ is convex for $x < 0$ and concave for $x > 0$. Moreover, $\arctan \infty = \pi/2$ and $\arctan(-\infty) = -\pi/2$. The graph is as shown in Figure 3.14.

Often, we will use the convenient abbreviations $\sec x = 1/\cos x$, $\csc x = 1/\sin x$, and $\cot x = 1/\tan x$. These are the *secant*, *cosecant*, and *cotangent* functions. For example, $(\tan x)' = \sec^2 x$, and $(\cot x)' = -\csc^2 x$.

If $t = \tan(\theta/2)$, we have the *half-angle formulas*

$$\sin\theta = \frac{2t}{1+t^2},$$

$$\cos\theta = \frac{1-t^2}{1+t^2},$$

and

$$\tan\theta = \frac{2t}{1-t^2}.$$

Exercises

3.6.1. Derive the half-angle formulas.

3.6.2. Let T be a translation, R a rotation, and $A = (a, b)$ and $B = (c, d)$ points in the plane. Show that $d(A, B) = d(T(A), T(B))$ and $d(A, B) = d(R(A), R(B))$.

3.6.3. With $R(x, y) = (x\cos\theta - y\sin\theta, x\sin\theta + y\cos\theta)$, show that

$$R(\cos\phi, \sin\phi) = (\cos(\phi + \theta), \sin(\phi + \theta)),$$

justifying the term "rotation."

3.6.4. Let A, B, C be three vertices of a triangle in the plane. Then the *angle at* C is that obtained after translating the vertex C to the origin. The triangle is a right triangle if the angle at C is $\pi/2$. Suppose a, b, c are the lengths of the sides of the triangle formed by A, B, C, with c the length of the hypotenuse, i.e., the length of the side across from C. Show that $a^2+b^2 = c^2$.

3.6.5. Let $f(x) = x\sin(1/x)$, $x \neq 0$, $f(0) = 0$. Show that f is continuous at all points, but is not of bounded variation on any interval (a, b) containing 0.

3.6.6. Let $f(x) = x^2\sin(1/x)$, $x \neq 0$, $f(0) = 0$. Show that f is differentiable at all points with $|f'(x)| \leq 1 + 2|x|$. Show that f is of bounded variation on any bounded interval.

3.6.7. Let $f(x) = x + 2x^2\sin(1/x)$, $x \neq 0$, and $f(0) = 0$. Show that f is not injective on $(0, \epsilon)$ for any $\epsilon > 0$, but $f'(0) = 1$. This shows the conclusion of the IFT does not hold with the weakened hypothesis $f'(0) \neq 0$.

3.6.8. Show that, for $0 \leq x \leq 3$, the series

$$\frac{x^3}{3!} - \frac{x^5}{5!} + \frac{x^7}{7!} - \dots$$

has decreasing terms.

3.6.9. Use the trigonometric identities to compute the sine, cosine, and tangent of $\pi/6$, $\pi/4$, and $\pi/3$.

3.6.10. Show that $\cos(\pi/9)\cos(2\pi/9)\cos(4\pi/9) = 1/8$.

3.6.11. Use the trigonometric identities to show $2\cos(\pi/5)$ is the golden mean $(1+\sqrt{5})/2$ (Exercise **1.7.8**).

3.6.12. Use (3.6.2) and induction to show that

$$1 + 2\cos x + 2\cos(2x) + \cdots + 2\cos(nx) = \frac{\sin\left[\left(n+\frac{1}{2}\right)x\right]}{\sin\left(\frac{1}{2}x\right)}$$

for $n \geq 1$ and $x \notin 2\pi\mathbf{Z}$.

3.6.13. Show that $2\cot(2x) = \cot x - \tan x$.

3.6.14. Show that

$$\left(x^2 - 2x\cos\theta + 1\right) \cdot \left(x^2 - 2x\cos(\pi - \theta) + 1\right) = \left(x^4 - 2x^2\cos(2\theta) + 1\right).$$

Use this to derive the identity

$$x^{2n} - 1 = (x^2 - 1) \cdot \prod_{k=1}^{n-1} \left(x^2 - 2x\cos(k\pi/n) + 1\right) \tag{3.6.5}$$

for $n = 2, 4, 8, 16, \dots$.

3.6.15. Use the Dirichlet test (§1.7) to show that $\sum_{n=1}^{\infty} \cos(nx)/n$ is convergent for $x \notin 2\pi\mathbf{Z}$.

3.6.16. Let $f : \mathbf{R} \to \mathbf{R}$ be continuous. A *period* of f is a real $p > 0$ satisfying $f(x+p) = f(x)$ for all x. Let P be the set of periods. *The period* of f is by definition $p^* = \inf P$. For example, the period of sin and cos is 2π. Show that $p^* = 0$ implies f is constant.

3.7 Primitives

Let f be defined on (a, b). A differentiable function F is a *primitive* of f if

$$F'(x) = f(x), \qquad a < x < b.$$

For example, $f(x) = x^3$ has the primitive $F(x) = x^4/4$ on $\mathbf{R}$ since $(x^4/4)' = (4x^3)/4 = x^3$.

Not every function has a primitive on a given open interval (a, b). Indeed, if f has a primitive F on (a, b), then by Exercise **3.2.8**, $f = F'$ satisfies the intermediate value property. Hence, $f((a, b))$ must be an interval.

Moreover, Exercise **3.1.6** shows that the presence of a jump discontinuity in f at a single point in (a, b) is enough to prevent the existence of a primitive F on (a, b). In other words, if f is defined on (a, b) and $f(c+)$, $f(c-)$ exist but are not both equal to $f(c)$, for some $c \in (a, b)$, then f has no primitive on (a, b).

Later (§4.4), we see that *every continuous function has a primitive* on any open interval of definition.

Now we investigate the converse of the last statement: To what extent does the existence of a primitive F of f determine the continuity of f? To begin, it is possible (Exercise **3.7.7**) for a function f to have a primitive and to be discontinuous at some points, so the converse is, in general, false. However, the previous paragraph shows that such discontinuities cannot be jump discontinuities but must be wild, in the terminology of §2.3. In fact, it turns out that, wherever f is of bounded variation, the existence of a primitive forces the continuity of f (Exercise **3.7.8**). Thus, a function f that has a primitive on (a, b) and is discontinuous at a particular point $c \in (a, b)$ must have unbounded variation near c, i.e., must be similar to the example in Exercise **3.7.7**.

From the mean value theorem, we have the following simple but fundamental fact.

Theorem 3.7.1. *Any two primitives of f differ by a constant.*

Indeed, if F and G are primitives of f, then $H = F - G$ is a primitive of zero, i.e., $H'(x) = (F(x)-G(x))' = 0$ for all $a < x < b$. Hence, $H(x)-H(y) = H'(c)(x-y) = 0$ for $a < x < y < b$, i.e., H is a constant. □

For example, all the primitives of $f(x) = x^3$ are $F(x) = x^4/4 + C$ with C a real constant. Sometimes, F is called the antiderivative or the indefinite integral of f. We shall use only the term *primitive*, and symbolically, we write

$$F(x) = \int f(x)\,dx \tag{3.7.1}$$

to mean—no more, no less—$F'(x) = f(x)$ on the interval under consideration. The reason for the unusual notation (3.7.1) is due to the connection between the primitive and the integral. This is explained in §4.4. With this notation,

$$\int f'(x)\,dx = f(x)$$

is a tautology. As a mnemonic device, we sometimes write $d[f(x)] = f'(x)dx$. With this notation, $\int$ and d "cancel": $\int d[f(x)] = f(x)$.

Based on the derivative formulas in Chapter 3, we can list the primitives known to us at this point. These identities are valid on any open interval of definition. These formulas, like any formula involving primitives, can be checked by differentiation.

$$\begin{aligned}
\int x^a\,dx &= \frac{x^{a+1}}{a+1}, \qquad a \neq -1,\\
\int \frac{1}{x}\,dx &= \log|x|,\\
\int a^x\,dx &= \frac{1}{\log a}a^x, \qquad a > 0,\\
\int \cos x\,dx &= \sin x,\\
\int \sin x\,dx &= -\cos x,\\
\int \sec^2 x\,dx &= \tan x,\\
\int \csc^2 x\,dx &= -\cot x,\\
\int \frac{1}{\sqrt{1-x^2}}\,dx &= \arcsin x,
\end{aligned}$$

and

$$\int \frac{1}{1+x^2}\,dx = \arctan x.$$

From the linearity of derivatives, we obtain the following.

Theorem 3.7.2. *If f and g have primitives on (a,b) and k is a real, so do $f+g$ and kf, and*

$$\int [f(x)+g(x)]\,dx = \int f(x)\,dx + \int g(x)\,dx,$$

and

$$\int kf(x)\,dx = k\int f(x)\,dx.\square$$

From the product rule, we obtain the following analog for primitives of summation by parts (§1.7).

Theorem 3.7.3 (Integration By Parts). *If f and g are differentiable on (a,b) and $f'g$ has a primitive on (a,b), then so does fg', and*

$$\int f(x)g'(x)\,dx = f(x)g(x) - \int f'(x)g(x)\,dx.$$

To see this, let F be a primitive for $f'g$. Then $(fg-F)' = f'g + fg' - F' = fg'$, so $fg - F$ is a primitive for fg'. $\square$ We caution the reader that integration is taken up in §4.3. Here this last result is called *integration by parts* because of its usefulness for computing integrals in the next chapter. *There are no integrals in this section.*

From the chain rule, we obtain the following.

Theorem 3.7.4 (Substitution). *If g is differentiable on (a,b), $g[(a,b)] \subset (c,d)$ and $\int f(x)\,dx = F(x)$ on (c,d), then*

$$\int f[g(x)]g'(x)\,dx = F[g(x)], \qquad a < x < b.\square$$

We work out some examples. It is convenient to allow the undefined symbol dx to enter into the expression for f and to write, e.g., $\int dx$ instead of $\int 1\,dx$ and $\int dx/x$ instead of $\int (1/x)dx$.

Substitution is often written as $\int f[g(x)]g'(x)\,dx = \int f(u)du$, $u = g(x)$. For example,

$$\begin{aligned}\int \cot x\,dx &= \int \frac{\cos x}{\sin x}\,dx \\ &= \int \frac{d\sin x}{\sin x} = \int \frac{du}{u} = \log|u| = \log|\sin x|.\end{aligned}$$

Similarly,

$$\begin{aligned}\int \frac{2x+1}{1+x^2}\,dx &= \int \frac{2x}{1+x^2}\,dx + \int \frac{1}{1+x^2}\,dx \\ &= \int \frac{d(1+x^2)}{1+x^2} + \arctan x \\ &= \log(1+x^2) + \arctan x.\end{aligned}$$

Particularly useful special cases of substitution are

$$\int f(x)\,dx = F(x) \quad \text{implying} \int f(x+a)\,dx = F(x+a),$$

and

$$\int f(x)\,dx = F(x) \quad \text{implying} \int f(ax)\,dx = \frac{1}{a}F(ax), \qquad a \neq 0.$$

We call these the *translation* and *dilation* properties. Thus, for example,

$$\int e^{ax} dx = \frac{1}{a} e^{ax}, \qquad a \neq 0.$$

Integration by parts is often written as $\int u dv = uv - \int v du$. If we take $u = \log x$, $dv = dx$, then $du = u' dx = dx/x$, $v = x$, so

$$\int \log x\,dx = x \log x - \int x(dx/x) = x \log x - \int dx = x \log x - x.$$

Similarly,

$$\begin{aligned}\int x \log x\,dx &= \frac{x^2}{2} \log x - \int \frac{x^2}{2}(dx/x) \\ &= \frac{x^2}{2} \log x - \int \frac{x}{2}\,dx = \frac{x^2}{2} \log x - \frac{x^2}{4}.\end{aligned}$$

By a trigonometric formula (§3.6),

$$\begin{aligned}\int \cos^2 x\,dx &= \int \frac{1+\cos(2x)}{2}\,dx = \frac{x}{2} + \frac{1}{2}\int \cos(2x)\,dx \\ &= \frac{x}{2} + \frac{\sin(2x)}{4} = \frac{2x+\sin(2x)}{4}.\end{aligned}$$

Since $4x(1-x) = 4x - 4x^2 = 1 - (2x-1)^2$,

$$\int \frac{dx}{\sqrt{x(1-x)}} = \int \frac{2dx}{\sqrt{4x-4x^2}} = \int \frac{d(2x-1)}{\sqrt{1-(2x-1)^2}} = \arcsin(2x-1)$$

by translation and dilation. Of course, we already know (§3.6) that another primitive is $2\arcsin\sqrt{x}$, so the two primitives must differ by a constant (Exercise **3.7.9**). The reduction $4x - 4x^2 = 1 - (2x-1)^2$ is the usual technique of *completing the square.*

To compute $\int \sqrt{1-x^2}\,dx$, let $x = \sin\theta$, $dx = \cos\theta\,d\theta$. Then

$$\begin{aligned}\int \sqrt{1-x^2}\,dx &= \int \sqrt{1-\sin^2\theta}\cos\theta\,d\theta\\ &= \int \cos^2\theta\,d\theta\\ &= \frac{1}{4}[2\theta + \sin(2\theta)]\\ &= \frac{1}{2}(\theta + \sin\theta\cos\theta)\\ &= \frac{1}{2}\left(\arcsin x + x\sqrt{1-x^2}\right).\end{aligned}$$

Alternatively, let $u = \sqrt{1-x^2}$, $dv = dx$. Then $du = -xdx/\sqrt{1-x^2}$, $v = x$. So

$$\begin{aligned}\int \sqrt{1-x^2}\,dx &= x\sqrt{1-x^2} - \int x(-xdx/\sqrt{1-x^2})\\ &= x\sqrt{1-x^2} + \int \frac{x^2dx}{\sqrt{1-x^2}}\\ &= x\sqrt{1-x^2} + \int \frac{(x^2-1)dx}{\sqrt{1-x^2}} + \int \frac{dx}{\sqrt{1-x^2}}\\ &= x\sqrt{1-x^2} - \int \sqrt{1-x^2}\,dx + \arcsin x.\end{aligned}$$

Moving the second term on the right to the left side, we obtain the same result.

To compute $\displaystyle\int \frac{dx}{1-x^2}$, write

$$\frac{1}{1-x^2} = \frac{1}{2}\left(\frac{1}{1+x} + \frac{1}{1-x}\right)$$

to get

$$\int \frac{dx}{1-x^2} = \frac{1}{2}\left[\log(1+x) - \log(1-x)\right] = \frac{1}{2}\log\left(\frac{1+x}{1-x}\right).$$

If f is given as a power series, the primitive is easily found as another power series.

Theorem 3.7.5. *If $R > 0$ is the radius of convergence of*

$$f(x) = a_0 + a_1x + a_2x^2 + \dots,$$

then R is the radius of convergence of $\sum a_n x^{n+1}/(n+1)$, and

$$\int f(x)\,dx = a_0x + \frac{a_1x^2}{2} + \frac{a_2x^3}{3} + \dots \tag{3.7.2}$$

on $(-R, R)$.

To see this, one first checks that the radius of convergence of the series in (3.7.2) is also R using $n^{1/n} \to 1$, as in the previous section. Now differentiate the series in (3.7.2) obtaining $f(x)$. Hence, the series in (3.7.2) is a primitive. □

For example, using the geometric series with $-x$ replacing x,

$$\frac{1}{1+x} = 1 - x + x^2 - x^3 + \dots, \qquad |x| < 1.$$

Hence, by the theorem,

$$\log(1+x) = x - \frac{x^2}{2} + \frac{x^3}{3} - \dots, \qquad |x| < 1. \tag{3.7.3}$$

Indeed, both sides are primitives of $1/(1+x)$, and both sides equal to zero at $x = 0$. Similarly, using the geometric series with $-x^2$ replacing x,

$$\frac{1}{1+x^2} = 1 - x^2 + x^4 - x^6 + \dots, \qquad |x| < 1.$$

Hence, by the theorem,

$$\arctan x = x - \frac{x^3}{3} + \frac{x^5}{5} - \frac{x^7}{7} + \dots, \qquad |x| < 1. \tag{3.7.4}$$

This follows since both sides are primitives of $1/(1+x^2)$ and both sides vanish at $x = 0$. To obtain the sum of the series (1.7.5), we seek to insert $x = 1$ in (3.7.4). We cannot do this directly since (3.7.4) is valid only for $|x| < 1$. Instead, we let $s_n(x)$ denote the nth partial sum of the series in (3.7.4). Since this series is alternating with decreasing terms (§1.7) when $0 < x < 1$,

$$s_{2n}(x) \le \arctan x \le s_{2n-1}(x), \qquad n \ge 1.$$

In this last inequality, the number of terms in the partial sums is finite. Letting $x \nearrow 1$, we obtain

$$s_{2n}(1) \le \arctan 1 \le s_{2n-1}(1), \qquad n \ge 1.$$

Now letting $n \nearrow \infty$ and recalling $\arctan 1 = \pi/4$, we arrive at the sum of the *Leibnitz series*

$$\frac{\pi}{4} = 1 - \frac{1}{3} + \frac{1}{5} - \frac{1}{7} + \dots,$$

first discussed in §1.7. In particular, we conclude that $8/3 < \pi < 4$. In §5.2, we obtain the sum of the Leibnitz series by a procedure that will be useful in many other situations.

The Leibnitz series is "barely convergent" and is not useful for computing π. To compute π, the traditional route is to insert $x = 1/5$ and $x = 1/239$ in (3.7.4) and to use Machin's 1706 formula

$$\frac{\pi}{4} = 4 \arctan \frac{1}{5} - \arctan \frac{1}{239}.$$

Exercises

3.7.1. Compute $\int e^x \cos x \, dx$.

3.7.2. Compute $\int e^{\arcsin x} \, dx$.

3.7.3. Compute $\int \frac{x+1}{\sqrt{1-x^2}} dx$.

3.7.4. Compute $\int \frac{\arctan x}{1+x^2} dx$.

3.7.5. Compute $\int x^2 (\log x)^2 dx$.

3.7.6. Compute $\int \sqrt{1 - e^{-2x}} dx$.

3.7.7. Let $F(x) = x^2 \sin(1/x)$, $x \neq 0$, and let $F(0) = 0$. Show that the derivative $F'(x) = f(x)$ exists for all x, but f is not continuous at $x = 0$.

3.7.8. If f is of bounded variation and $F' = f$ on (a, b), then f is continuous.

3.7.9. Show directly (i.e., without derivatives) $2 \arcsin \sqrt{x} = \arcsin(2x - 1) + \pi/2$.

3.7.10. Show

$$\arcsin x = x + \frac{1}{2} \cdot \frac{x^3}{3} + \frac{1}{2} \cdot \frac{3}{4} \cdot \frac{x^5}{5} + \frac{1}{2} \cdot \frac{3}{4} \cdot \frac{5}{6} \cdot \frac{x^7}{7} + \dots, \qquad |x| < 1.$$

3.7.11. If f is a polynomial and $F(x) = f(x) - f''(x) + f^{(4)}(x) - \dots$, then

$$\int f(x) \sin x \, dx = F'(x) \sin x - F(x) \cos x.$$

3.7.12. Show that

$$\tan(a+b) = \frac{\tan a + \tan b}{1 - \tan a \tan b}.$$

Use this to derive Machin's formula.

3.7.13. Use Machin's formula and (3.7.4) to obtain $\pi = 3.14\dots$ to within an error of 10^{-2}.

3.7.14. Simplify $\arcsin(\sin 100)$.

3.7.15. Compute $\displaystyle\int \frac{-4x}{1-x^2} dx$.

3.7.16. Compute $\displaystyle\int \frac{4\sqrt{2} - 4x}{x^2 - \sqrt{2}x + 1} dx$.

3.7.17. Show that

$$\log 2 = 1 - \frac{1}{2} + \frac{1}{3} - \frac{1}{4} + \dots$$

following the procedure discussed above for the Leibnitz series.

Chapter 4
Integration

4.1 The Cantor Set

The subject of this chapter is the measurement of the areas of subsets of the plane $\mathbf{R}^2 = \mathbf{R} \times \mathbf{R}$. The areas of elementary geometric figures, such as squares, rectangles, and triangles, are already known to us. By *known to us* we mean that, e.g., by defining the area of a rectangle to be the product of the lengths of its sides, we obtain quantities that agree with our intuition. Since every right-angle triangle is half a rectangle, the areas of right-angle triangles are also known to us. Similarly, we can obtain the area of a general triangle.

How does one approach the problem of measuring the area of an unfamiliar figure or subset of $\mathbf{R}^2$, say a subset that cannot be broken up into triangles? For example, how does one measure the area of the unit disk

$$D = \{(x, y) : x^2 + y^2 < 1\}?$$

One solution is to arbitrarily define the area of D to equal whatever one feels is right. The Egyptian book of Ahmes ($\sim$ 3,900 years ago) states that the area of D is $(16/9)^2$. In the Indian Śulbastras (written down $\sim$ 2,500 years ago), the area of D is taken to equal $(26/15)^2$. Albrecht Dürer ($\sim$ 500 years ago) of Nuremburg solved a related problem which amounted to taking the area of D to equal $25/8$.

Which of these answers should we accept as the area of D? If we treat these answers as *estimates* of the area of D, then in our minds, we must have the presumption that such a quantity—the area of D—has a meaningful existence. In that case, we have no way of judging the merit of an estimate except by the quality of the reasoning leading to it.

O. Hijab, *Introduction to Calculus and Classical Analysis*, Undergraduate Texts in Mathematics, DOI 10.1007/978-3-319-28400-2_4

Realizing this, by reasoning that remains perfectly valid today, Archimedes ($\sim$ 2,250 years ago) carefully established

$$\frac{223}{71} < \text{area}\,(D) < \frac{22}{7}.$$

In §4.4, we show that area $(D) = \pi$, where π, "Archimedes' constant," is the real number defined in §3.6.

At the basis of the Greek mathematicians' computations of area was the *method of exhaustion*. This asserted that the area of a set $A \subset \mathbf{R}^2$ could be computed as the limit of areas of a sequence of inscribed sets (A_n) that filled out more and more of A as $n \nearrow \infty$ (Figure 4.1). Nevertheless the Greeks were apparently uncomfortable with the concept of infinity and never used this method as stated. Instead, for example, in dealing with D, Archimedes used inscribed and circumscribed polygons with 96 sides to obtain the above result. He never explicitly passed to the limit. It turns out, however, that the method of exhaustion is so important to integration that in §4.5 we give a careful derivation of it.

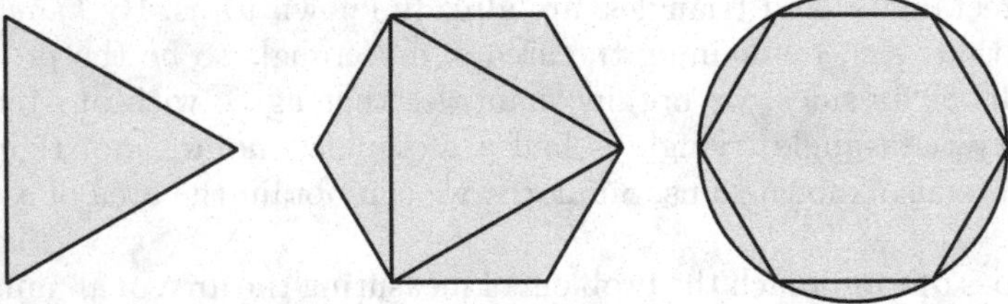

Fig. 4.1 The method of exhaustion

Now the unit disk is not a totally unfamiliar set to the reader. But, if we are presented with some genuinely unfamiliar subset C, the situation changes, and we may no longer have any clear conception of the area of C. If we are unable to come up with a procedure leading us to the area, then we may be forced to reexamine our intuitive notion of area. In particular, we may be led to the conclusion that the "true area" of C may have no meaning. Let us describe such a subset.

Let C_0 denote the compact unit square $[0,1] \times [0,1]$. Divide C_0 into nine equal subsquares and take out from C_0 all but the four compact corner subsquares. Let C_1 be the remainder, i.e., the union of the four remaining compact subsquares. Repeat this process with each of the four subsquares. Divide each subsquare into nine equal compact subsubsquares and take out, in each subsquare, all but the four compact corner subsubsquares. Call the union of the remaining sixteen subsubsquares C_2. Continuing in this manner yields a sequence

$$C_0 \supset C_1 \supset C_2 \supset \dots .$$

The *Cantor set* is the common part (Figure 4.2) of all these sets, i.e., their intersection

$$C = \bigcap_{n=1}^{\infty} C_n.$$

At first glance, it is not clear that C is not empty. But $(0,0) \in C$! Moreover, the sixteen corners of the set C_1 are in C. Similarly, any corner of any sub-square, at any level, lies in C. But the set of such points is countable (§1.7), and it turns out that there is much more: There are as many points in C as there are in the unit square C_0. In particular C is uncountable.

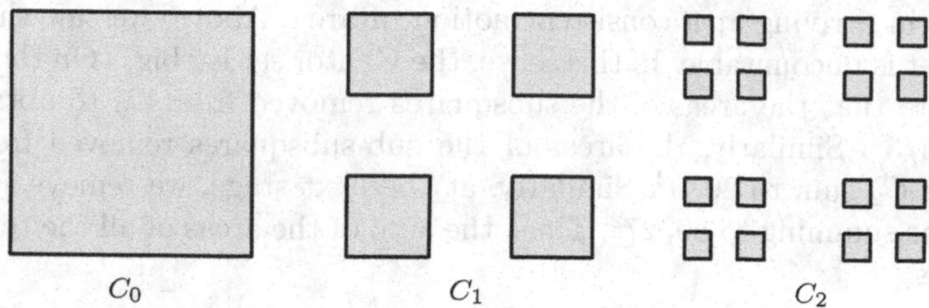

Fig. 4.2 The Cantor set

To see this, recall the concept of ternary expansions (§1.6). Let $a \in [0,1]$. We say that

$$a = .a_1a_2\ldots$$

is the ternary expansion of a if the naturals a_n are ternary digits 0, 1, 2, and

$$a = \sum_{n=1}^{\infty} a_n 3^{-n}.$$

Now let $(a,b) \in C_0$, and let

$$a = .a_1a_2\ldots$$

and

$$b = .b_1b_2\ldots$$

be ternary expansions of a and b. If $a_1 \neq 1$ and $b_1 \neq 1$, then (a,b) is in C_1. Similarly, in addition, if $a_2 \neq 1$ and $b_2 \neq 1$, $(a,b) \in C_2$. Continuing in this manner, we see that, if $a_n \neq 1$ and $b_n \neq 1$ for all $n \geq 1$, $(a,b) \in C$. Conversely, $(a,b) \in C$ implies that there are ternary expansions of a and b as stated. Thus, $(a,b) \in C$ iff a and b have ternary expansions in which the digits are equal to 0 or 2.

Now although some reals may have more than one ternary expansion, a real a cannot have more than one ternary expansion $.a_1a_2\ldots$ where $a_n \neq 1$ for all $n \geq 1$ because any two ternary expansions yielding the same real must have their nth digits differing by 1 for some $n \geq 1$ (Exercise **1.6.2** treats the decimal case). Thus, the mapping

$$(a,b) = (.a_1a_2a_3\ldots,.b_1b_2b_3\ldots) \mapsto \left(\sum_{n=1}^{\infty} a'_n 2^{-n}, \sum_{n=1}^{\infty} b'_n 2^{-n}\right),$$

where $a'_n = a_n/2$, $b'_n = b_n/2$, $n \geq 1$, is well defined. Since any real in $[0,1]$ has a binary expansion, this mapping is a surjection of the Cantor set C onto the unit square C_0. Since C_0 is uncountable (Exercise **1.7.4**), we conclude that C is uncountable (§1.7).

The difficulty of measuring the size of the Cantor set underscores the difficulty in arriving at a consistent notion of area. Above, we saw that the Cantor set is uncountable. In this sense, the Cantor set is "big." On the other hand, note that the areas of the subsquares removed from C_0 to obtain C_1 sum to $5/3^2$. Similarly, the areas of the sub-subsquares removed from C_1 to obtain C_2 sum to $20/9^2$. Similarly, at the next stage, we remove squares with areas summing to $80/27^2$. Thus, the sum of the areas of all the removed squares is

$$\frac{5}{9} + \frac{20}{9^2} + \frac{80}{27^2} + \cdots = \frac{5}{9}\left(1 + \frac{4}{9} + \left(\frac{4}{9}\right)^2 + \ldots\right) = \frac{5}{9} \cdot \frac{1}{1 - \frac{4}{9}} = 1.$$

Since C is the complement of all these squares in C_0 and C_0 has area 1, the area of C is $1 - 1 = 0$. Thus, in the sense of area, the Cantor set is "small."

This argument is perfectly reasonable, except for one aspect. We are assuming that areas can be added and subtracted in the usual manner, even when there are *infinitely* many sets involved. In §4.2, we show that, with an appropriate definition of *area*, this argument can be modified to become correct, and the area of C is in fact zero.

Another indication of the smallness of C is the fact that C has no interior. To explain this, given any set $A \subset \mathbf{R}^2$, let us say that A *has interior* if we can fit some rectangle Q within A, i.e., $Q \subset A$. If we cannot fit a (nontrivial) rectangle, no matter how small, within A, then we say that A *has no interior*. For example, the unit disk has interior, but a line segment has no interior. The Cantor set C has no interior, because there is a point in every rectangle whose coordinate ternary expansions contain at least one digit 1. Alternatively, if C contained a rectangle Q, then the area of C would be at least as much as the area of Q, which is positive. But we saw above that the area of C equals zero.

Since this reasoning applies to any set, we see that if $A \subset \mathbf{R}^2$ has interior, then the area of A is positive. The surprising fact is that the converse of this statement is false. There are sets $A \subset \mathbf{R}^2$ that have positive area but have no interior. Such a set C^α is described in Exercise **4.1.2**. To add to the confusion, even though C^α has no interior, there is (Exercise **4.2.15**) some rectangle Q such that $\operatorname{area}(C^\alpha \cap Q)$ is at least $.99 \operatorname{area}(Q)$.

These issues are discussed to point out the existence of unavoidable phenomena involving area where things do not behave as simply as triangles. In the first three decades of this century, these issues were finally settled. The solution to *the problem of area*, analyzed extensively by Archimedes more than two thousand years ago, can now be explained in a few pages. Why did it take so long for the solution to be discovered? It should not be too surprising that one missing ingredient was the completeness property of the set of real numbers, the importance of which was not fully realized until the nineteenth century.

Exercises

4.1.1. Let $C_0 = [0,1] \times [0,1]$ denote the unit square, and let C'_1 be obtained by throwing out from C_0 the middle subrectangle $(1/3, 2/3) \times [0,1]$ of width $1/3$ and height 1. Then C'_1 consists of two compact subrectangles. Let C'_2 be obtained from C'_1 by throwing out, in each of the subrectangles, the middle sub-subrectangles $(1/9, 2/9) \times [0,1]$ and $(7/9, 8/9) \times [0,1]$, each of width $1/3^2$ and height 1. Then C'_2 consists of four compact sub-subrectangles. Similarly C'_3 consists of eight compact sub-subsubrectangles, obtained by throwing out from C'_2 the middle sub-subsubrectangles of width $1/3^3$ and height 1. Continuing in this manner, we have $C'_1 \supset C'_2 \supset C'_3 \supset \dots$. Let $C' = \bigcap_{n=1}^{\infty} C'_n$. Show that area $(C') = 0$ and C' has no interior.

4.1.2. Fix a real $0 < \alpha < 1$ (e.g., $\alpha = .7$) and let $C_0 = [0,1] \times [0,1]$ be the unit square. Let C_1^α be obtained from C_0 by throwing out the middle subrectangle of width $\alpha/3$ and height 1. Then C_1^α consists of two subrectangles. Let C_2^α be obtained from C_1^α by throwing out, in each of the subrectangles, the *middle* sub-subrectangles of width $\alpha/3^2$ and height 1. Then C_2^α consists of four sub-subrectangles. Similarly C_3^α consists of eight sub-subsubrectangles, obtained by throwing out from C_2^α the middle sub-subsubrectangles of width $\alpha/3^3$ and height 1. Continuing in this manner, we have $C_1^\alpha \supset C_2^\alpha \supset C_3^\alpha \supset \dots$. Let

$$C^\alpha = \bigcap_{n=1}^{\infty} C_n^\alpha.$$

Show that area $(C^\alpha) > 0$, but C^α has no interior.

4.1.3. For $A \subset \mathbf{R}^2$, let

$$A + A = \{(x + x', y + y') : (x, y) \in A, (x', y') \in A\}$$

be the *set of sums*. Show that $C + C = [0,2] \times [0,2]$ (see Exercise **1.6.5**).

4.2 Area

Let I and J be intervals, i.e., subsets of $\mathbf{R}$ of the form (a,b), $[a,b]$, $(a,b]$, or $[a,b)$. As usual, we allow the endpoints a or b to equal $\pm\infty$ when they are not included within the interval. A *rectangle* is a subset of $\mathbf{R}^2 = \mathbf{R}\times\mathbf{R}$ (Figure 4.3) of the form $I \times J$. A rectangle $Q = I \times J$ is *open* if I and J are both open intervals, *closed* if I and J are both closed intervals, and *compact* if I and J are both compact intervals. For example, the plane $\mathbf{R}^2$ and the upper-half plane $\mathbf{R}\times(0,\infty)$ are open rectangles. We say that a rectangle $Q = I \times J$ is *bounded* if I and J are bounded subsets of $\mathbf{R}$. For example, the vertical line segment $\{a\}\times[c,d]$ is a compact rectangle. A single point is a compact rectangle. If I is a bounded interval, let $\bar{I}$ denote the compact interval with the same endpoints, and let I° denote the open interval with the same endpoints. If $Q = I \times J$ is a bounded rectangle, the compact rectangle $\bar{Q} = \bar{I}\times\bar{J}$ is its *compactification.* If $Q = I \times J$ is any rectangle, then the open rectangle $Q^\circ = I^\circ \times J^\circ$ is its *interior.* Note that $Q^\circ \subset Q \subset \bar{Q}$ and $\bar{Q}\setminus Q^\circ$ is a subset of the sides of Q, for any rectangle Q. Note that an open rectangle may be empty, for example, $(a,a)\times(c,d)$ is empty.

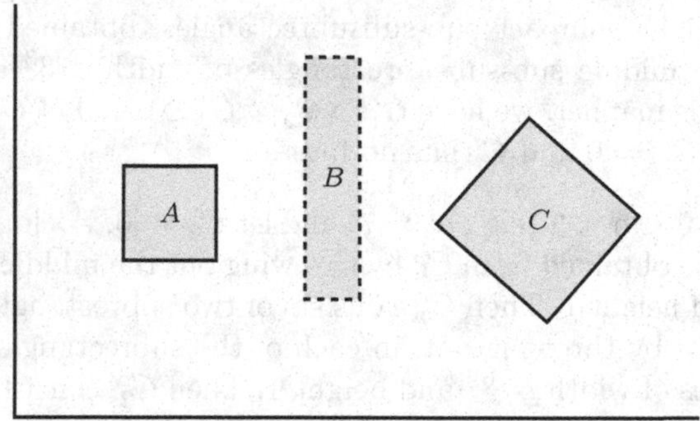

Fig. 4.3 A and B are rectangles, but C is not

Let A be a subset of $\mathbf{R}^2$. A *countable cover* of A is a sequence of sets (A_n) such that A is contained in their union,[1]

$$A \subset \bigcup_{n=1}^{\infty} A_n.$$

In a given cover, the sets (A_n) may overlap, i.e., intersect. If, for some N, $A_n = \emptyset$ for $n > N$, we say that $(A_1,\dots,A_N)$ is a *finite cover* (Figure 4.4).

A *paving of* A is a countable cover (Q_n), where the sets Q_n, $n \ge 1$, are *rectangles.* A *finite paving* is a finite cover that is also a paving (Figure 4.5).

[1] $\bigcup_{n=1}^{\infty} A_n \equiv \bigcup\{A_n : n \in \mathbf{N}\}$ and $\bigcap_{n=1}^{\infty} A_n \equiv \bigcap\{A_n : n \in \mathbf{N}\}$, see §2.1.

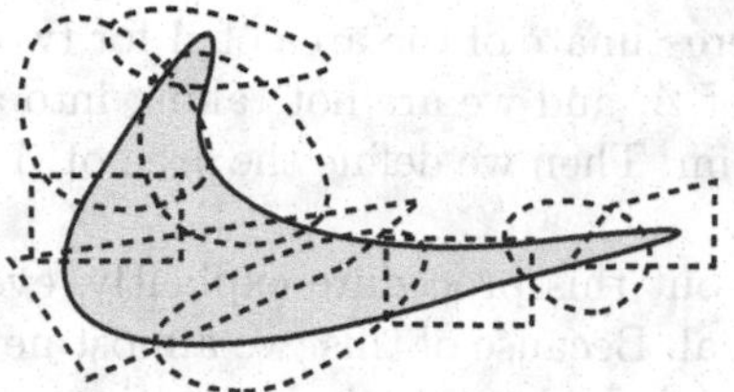

Fig. 4.4 A finite cover

Every subset $A \subset \mathbf{R}^2$ has at least one (not very interesting) paving $Q_1 = \mathbf{R}^2$, $Q_2 = \emptyset$, $Q_3 = \emptyset$, ...

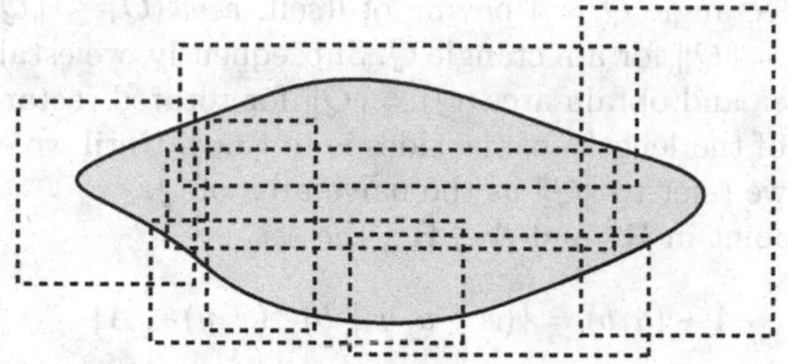

Fig. 4.5 A finite paving

For any interval I as above, let $|I| = b - a$ denote the *length* of I. For any rectangle $Q = I \times J$, let $\|Q\| = |I| \cdot |J|$. More generally, for any rotated rectangle, let $\|Q\|$ denote the product of the lengths of its sides (length is defined in §3.6). Then $\|Q\|$, the traditional high-school formula for the area of a rectangle, is a positive real or 0 or ∞. We also take $\|\emptyset\| = 0$. For reasons discussed below, we call $\|Q\|$ the *naive area* of Q.

Let A be a subset of $\mathbf{R}^2$. The *area*[2] of A is defined by

$$\operatorname{area}(A) = \inf \left\{ \sum_{n=1}^{\infty} \|Q_n\| : \text{ all pavings } (Q_n) \text{ of } A \right\}. \tag{4.2.1}$$

This definition of area is at the basis of all that follows. It is necessarily complicated because it applies to all subsets A of $\mathbf{R}^2$. As an immediate consequence of the definition, area$(\emptyset) = 0$. Similarly, the area of a finite vertical line segment A is zero since A can be covered by a thin rectangle of arbitrarily small naive area.

In words, the definition says that to find the area of a set A, we cover A by a sequence $Q_1, Q_2, \ldots$ of rectangles, measure the sum of their naive areas, and take this sum as an estimate for the area of A. Of course, we expect that

[2] The usual terminology is *two-dimensional Lebesgue measure*.

this sum will be an overestimate of the area of A for two reasons. The paving may cover a superset of A, and we are not taking into account any overlaps when computing the sum. Then we define the area of A to be the inf of these sums.

Of course, carrying out this procedure explicitly, even for simple sets A, is completely impractical. Because of this, we almost never use the definition directly to compute areas. Instead, as is typical in mathematics, we derive the elementary properties of area from the definition, and we use them to compute areas.

Whether or not we can compute the area of a given set, the above definition applies consistently to every subset A. In particular, this is so whether A is a rectangle, a triangle, a smooth graph, or the Cantor set C. Let us now derive the properties of area that follow immediately from the definition.

Since every rectangle Q is a paving of itself, $\text{area}\,(Q) \leq \|Q\|$. Below, we obtain $\text{area}\,(Q) = \|Q\|$ for a rectangle Q. Subsequently, we establish rotation invariance of area and obtain $\text{area}\,(Q) = \|Q\|$ for rotated rectangles Q (rotation invariance of the lengths of the sides is in §3.6). Until we establish this, we repeat that we refer to $\|Q\|$ as the naive area of Q.

If (a, b) is a point in $\mathbf{R}^2$ and $A \subset \mathbf{R}^2$, the set

$$A + (a, b) = \{(x + a, y + b) : (x, y) \in A\}$$

is the *translate* of A by (a, b). Then $[A + (a, b)] + (c, d) = A + (a + c, b + d)$ and, for any rectangle Q, $Q + (a, b)$ is a rectangle and $\|Q + (a, b)\| = \|Q\|$. From this follows the *translation invariance of area*,

$$\text{area}\,[A + (a, b)] = \text{area}\,(A)\,, \qquad A \subset \mathbf{R}^2.$$

To see this, let (Q_n) be a paving of A. Then $(Q_n + (a, b))$ is a paving of $A + (a, b)$, so

$$\text{area}\,[A + (a, b)] \leq \sum_{n=1}^{\infty} \|Q_n + (a, b)\| = \sum_{n=1}^{\infty} \|Q_n\|.$$

Since $\text{area}\,(A)$ is the inf of the sums on the right, $\text{area}\,[A + (a, b)] \leq \text{area}\,(A)$. Now in this last inequality, replace, in order, (a, b) by $(-a, -b)$ and A by $A + (a, b)$. We obtain $\text{area}\,(A) \leq \text{area}\,[A + (a, b)]$. Hence, $\text{area}\,(A) = \text{area}\,[A + (a, b)]$, establishing translation invariance (Figure 4.6).

If $k > 0$ is real and $A \subset \mathbf{R}^2$, the set

$$kA = \{(kx, ky) : (x, y) \in A\}$$

is the *dilate* of A by k. Then $k(cA) = (kc)A$ for k and c positive, kQ is a rectangle, and $\|kQ\| = k^2\|Q\|$ for every rectangle Q. From this follows the *dilation invariance of area*,

$$\text{area}\,(kA) = k^2 \cdot \text{area}\,(A)\,, \quad A \subset \mathbf{R}^2.$$

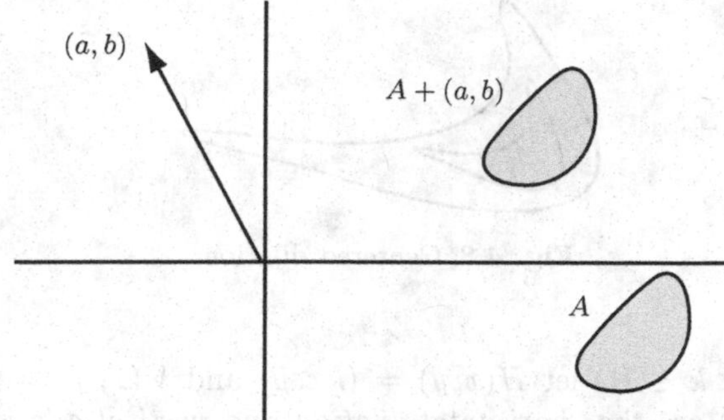

Fig. 4.6 area $(A) = $ area $[A + (a, b)]$

To see this, let (Q_n) be a paving of A. Then (kQ_n) is a paving of kA so

$$\text{area}\,(kA) \leq \sum_{n=1}^{\infty} \|kQ_n\| = k^2 \left(\sum_{n=1}^{\infty} \|Q_n\| \right).$$

Since area (A) is the inf of the sums on the right, we obtain area $(kA) \leq k^2 \cdot$ area (A). Now in this last inequality, replace, in order, k by $1/k$ and A by kA. We obtain $k^2 \cdot$ area $(A) \leq$ area (kA). Hence, area $(kA) = k^2 \cdot$ area (A), establishing dilation invariance (Figure 4.7).

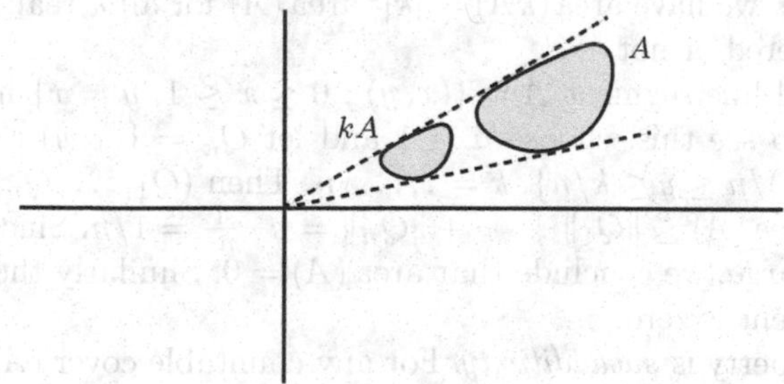

Fig. 4.7 area $(kA) = k^2 \cdot$ area (A)

Instead of dilation from the origin, we can dilate from any point in $\mathbf{R}^2$. In particular, by elementary geometry, certain subsets such as rectangles, triangles, and parallelograms have well-defined centers, and we can dilate from these centers. Given a subset A and an arbitrary point (a, b), its *centered dilation* kA (from (a, b)) is the set (Figure 4.8) obtained by translating (a, b) to $(0, 0)$, dilating as above by the factor $k > 0$ and then translating $(0, 0)$ back to (a, b). Then area $(kA) = k^2 \cdot$ area (A) for centered dilations as well. For example, if $k < 1$, kA is A shrunk towards (a, b).

Fig. 4.8 Centered dilation

Similarly, for $k \in \mathbf{R}$, let $H(x,y) = (kx,y)$ and $V(x,y) = (x,ky)$; then $H(A)$ and $V(A)$ are the *horizontal dilation* and *vertical dilation* of A, and $\text{area}\,(V(A)) = |k|\,\text{area}\,(A)$ and $\text{area}\,(H(A)) = |k|\,\text{area}\,(A)$ for every set A (Exercise **4.2.6**). Note these dilations incorporate reflections when $k < 0$.

The map $F(x,y) = (y,x)$ is a reflection across the line $y = x$; if Q is a rectangle, then so is $F(Q)$, and $\|F(Q)\| = \|Q\|$. It follows as above that the *flip* F preserves area, $\text{area}\,(F(A)) = \text{area}\,(A)$.

As with dilation, setting $-A = \{(-x,-y) : (x,y) \in A\}$, we have *reflection invariance of area,*

$$\text{area}\,(-A) = \text{area}\,(A)\,, \qquad A \subset \mathbf{R}^2,$$

and *monotonicity,*

$$\text{area}\,(A) \le \text{area}\,(B)\,, \qquad A \subset B \subset \mathbf{R}^2.$$

From the above, we have $\text{area}\,(kA) = |k|^2\,\text{area}\,(A)$ for all k real, whether the dilation is centered or not.

The diagonal line segment $A = \{(x,y) : 0 \le x \le 1,\ y = x\}$ has zero area (Figure 4.9). To see this, choose $n \ge 1$ and let $Q_k = \{(x,y) : (k-1)/n \le x \le k/n,\ (k-1)/n \le y \le k/n\}$, $k = 1,\dots,n$. Then $(Q_1,\dots,Q_n)$ is a paving of A. Hence, $\text{area}\,(A) \le \|Q_1\| + \dots + \|Q_n\| = n \cdot \frac{1}{n^2} = 1/n$. Since $n \ge 1$ may be arbitrarily large, we conclude that $\text{area}\,(A) = 0$. Similarly the area of any finite line segment is zero.

Another property is *subadditivity.* For any countable cover (A_n) of a given set A,

$$\text{area}\,(A) \le \sum_{n=1}^{\infty} \text{area}\,(A_n)\,. \tag{4.2.2}$$

Here the sets A_n, $n \ge 1$, need not be rectangles. For future reference, we call the sum on the right side of (4.2.2) the *area of the cover* (A_n).

In particular, since (A,B) is a cover of $A \cup B$,

$$\text{area}\,(A \cup B) \le \text{area}\,(A) + \text{area}\,(B)\,.$$

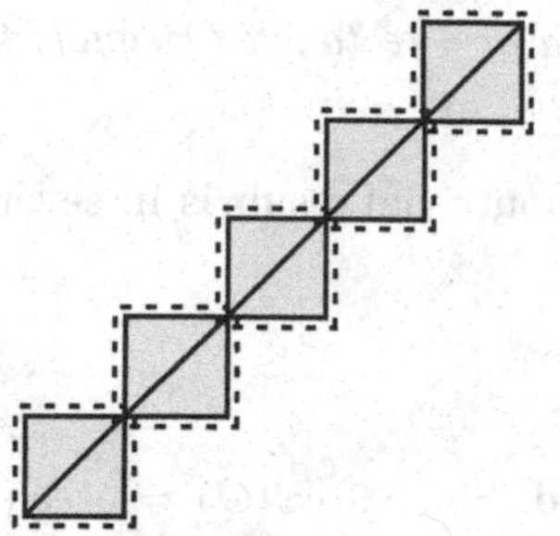

Fig. 4.9 Area of a diagonal line segment

Similarly, $(A_1, A_2, \dots, A_n)$ is a cover of $A_1 \cup A_2 \cup \dots \cup A_n$,[3] so

$$\text{area}(A_1 \cup A_2 \cup \dots \cup A_n) \le \text{area}(A_1) + \text{area}(A_2) + \cdots + \text{area}(A_n).$$

To obtain subadditivity, note that, if the right side of (4.2.2) is ∞, there is nothing to show, since in that case (4.2.2) is true. Hence, we may safely assume $\text{area}(A_n) < \infty$ for all $n \ge 1$. Let $\epsilon > 0$, and, for each $k \ge 1$, choose[4] a paving $(Q_{k,n})$ of A_k satisfying

$$\sum_{n=1}^{\infty} \|Q_{k,n}\| < \text{area}(A_k) + \epsilon 2^{-k}. \tag{4.2.3}$$

This is possible since $\text{area}(A_k)$ is the inf of sums of the form $\sum_{n=1}^{\infty} \|Q_n\|$. Then the double sequence $(Q_{k,n})$ is a cover by rectangles, hence a paving of A. Summing (4.2.3) over $k \ge 1$, we obtain

$$\begin{aligned} \text{area}(A) &\le \sum_{k=1}^{\infty} \left(\sum_{n=1}^{\infty} \|Q_{k,n}\| \right) \\ &\le \sum_{k=1}^{\infty} \left(\text{area}(A_k) + \epsilon 2^{-k}\right) = \left(\sum_{k=1}^{\infty} \text{area}(A_k) \right) + \epsilon. \end{aligned}$$

Since $\epsilon > 0$ is arbitrary, subadditivity follows.

Now let Q be any bounded rectangle and let $\bar{Q}$ and Q° denote its compactification and its interior respectively. We claim $\text{area}(\bar{Q}) = \text{area}(Q^\circ)$. To see this, by monotonicity, we have $\text{area}(\bar{Q}) \ge \text{area}(Q^\circ)$. Conversely, let $t > 1$ and let tQ° denote the centered dilation of Q°. Then $tQ^\circ \supset \bar{Q}$ so by monotonicity and dilation, $\text{area}(\bar{Q}) \le \text{area}(tQ^\circ) = t^2\,\text{area}(Q^\circ)$. Since $t > 1$ is arbitrary, we have $\text{area}(\bar{Q}) \le \text{area}(Q^\circ)$; hence, the claim follows. As a consequence, we see that for any bounded rectangle Q, we have $\text{area}(Q) = \text{area}(\bar{Q})$.

---

[3] $\bigcup_{k=1}^{n} A_k$, $\bigcap_{k=1}^{n} A_k$ are defined by induction; see §2.1.

[4] This uses the axiom of countable choice.

Theorem 4.2.1. *The area of a rectangle* Q *equals the product of the lengths of its sides,* area $(Q) = \|Q\|$.

The derivation of this nontrivial result is in several steps.

Step 1

Assume first Q is bounded. Since area $(Q) =$ area $(\bar{Q})$ and $\|Q\| = \|\bar{Q}\|$, we may assume without loss of generality that Q is a compact rectangle. Since we already know area $(Q) \leq \|Q\|$, we need only derive $\|Q\| \leq$ area (Q). By the definition of area (4.2.1), this means we need to show that

$$\|Q\| \leq \sum_{n=1}^{\infty} \|Q_n\| \tag{4.2.4}$$

for every paving (Q_n) of Q.

Let (Q_n) be a paving of Q. We say (Q_n) is an *open paving* if every rectangle Q_n is open. Suppose we established (4.2.4) for every open paving. Let $t > 1$. Then for any paving (Q_n) (open or not), (tQ_n°) is an open paving since $Q_n \subset tQ_n^\circ$ for $n \geq 1$; under the assumption that (4.2.4) is valid for open pavings, we would have

$$\|Q\| \leq \sum_{n=1}^{\infty} \|tQ_n^\circ\| = t^2 \sum_{n=1}^{\infty} \|Q_n^\circ\| = t^2 \sum_{n=1}^{\infty} \|Q_n\|. \tag{4.2.5}$$

Since $t > 1$ in (4.2.5) is arbitrary, we would then obtain (4.2.4) for the arbitrary paving (Q_n). Thus, it is enough to establish (4.2.4) when the paving (Q_n) is open and Q is a compact rectangle.

Step 2

Assume now Q is a compact rectangle and the paving (Q_n) is open. We show there is an $N > 0$ such that Q is contained in the finite union $Q_1 \cup Q_2 \cup \ldots \cup Q_N$. We argue by contradiction: Suppose that there is no such N. Then for each $n \geq 1$, we may select[5] $(x_n, y_n) \in Q \setminus Q_1 \cup \ldots \cup Q_n$. By Exercise **2.1.1**, (x_n, y_n) subconverges to some $(x, y) \in Q$. Now select $N \geq 1$ such that $(x, y) \in Q_N$. Then $(x_n, y_n) \in Q_N$ for infinitely many n, contradicting the construction of the sequence (x_n, y_n).

---

[5] The axiom of countable choice may be avoided as in the proof of Theorem 2.1.3.

Step 3

We are reduced to establishing

$$\|Q\| \leq \|Q_1\| + \cdots + \|Q_N\| \tag{4.2.6}$$

whenever $Q \subset Q_1 \cup \ldots \cup Q_N$ and Q is a compact rectangle and $Q_1, \ldots, Q_N$ are open.

Since area (Q°) = area $(\bar{Q})$, a moment's thought shows that we may assume both Q and $Q_1, \ldots, Q_N$ are compact rectangles.

Moreover, since $\|Q \cap Q_n\| \leq \|Q_n\|$ and $(Q \cap Q_n)$ is a paving of Q, by replacing (Q_n) by $(Q \cap Q_n)$, we may additionally assume $Q = Q_1 \cup \ldots \cup Q_N$.

This now is a combinatorial, or counting, argument. Write $Q = I \times J$ and $Q_n = I_n \times J_n$, $n = 1, \ldots, N$. Let $c_0 < c_1 < \cdots < c_r$ denote the distinct left and right endpoints of $I_1, \ldots, I_N$, *arranged in increasing order*, and set $I'_i = [c_{i-1}, c_i]$, $i = 1, \ldots, r$. Let $d_0 < d_1 < \cdots < d_s$ denote the distinct left and right endpoints of $J_1, \ldots, J_N$, *arranged in increasing order*, and set $J'_j = [d_{j-1}, d_j]$, $j = 1, \ldots, s$. Let $Q'_{ij} = I'_i \times J'_j$, $i = 1, \ldots, r$, $j = 1, \ldots, s$. Then:

A. The rectangles Q'_{ij} intersect at most along their edges,
B. The union of all the Q'_{ij}, $i = 1, \ldots, r$, $j = 1, \ldots, s$, equals Q,
C. the union of all the Q'_{ij} contained in a fixed Q_n equals Q_n.

Let c_{ijn} equal 1 or 0 according to whether $Q'_{ij} \subset Q_n$ or not. Then

$$\|Q\| = \sum_{i,j} \|Q'_{ij}\| \tag{4.2.7}$$

and

$$\|Q_n\| = \sum_{i,j} c_{ijn} \|Q'_{ij}\|, \qquad 1 \leq n \leq N, \tag{4.2.8}$$

since both sums are telescoping. Combining (4.2.8) and (4.2.7) and interchanging the order of summation, we get

$$\|Q\| = \sum_{i,j} \|Q'_{ij}\| \leq \sum_{i,j} \sum_{n} c_{ijn} \|Q'_{ij}\| = \sum_{n} \sum_{i,j} c_{ijn} \|Q'_{ij}\| = \sum_{n} \|Q_n\|.$$

This establishes (4.2.6).

Step 4

Thus, we have established (4.2.4) for finite pavings, hence by Step 2, for all pavings. Taking the inf over all pavings (Q_n) of Q in (4.2.4), we obtain $\|Q\| \leq$ area (Q); hence, area $(Q) = \|Q\|$ for Q a compact rectangle. As mentioned in Step 1, this implies the result for every bounded rectangle Q. When Q

is unbounded, Q contains bounded subrectangles of arbitrarily large area; hence, the result follows in this case as well. □

The reader may be taken aback by the complications involved in establishing this intuitively obvious result. Why is the derivation so complicated? The answer is that this complication is the price we have to pay if we are to stick to our definition (4.2.1) of area. The fact that we will obtain many powerful results—in a straightforward fashion—easily offsets the seemingly excessive complexity of the above result. In part, we are able to derive the powerful results in the rest of this Chapter and in Chapters 5 and 6, because of our decision to define area as in (4.2.1). The utility of such a choice can only be assessed in terms of the ease with which we obtain our results in what follows.

We now return to the main development.

We can compute the area of a triangle A with a horizontal base of length b and height h by constructing a cover of A consisting of thin horizontal strips (Figure 4.10). Let $\|A\|$ denote the naive area of A, i.e., $\|A\| = hb/2$. By reflection invariance, we may assume A lies above its base. Let us first assume the two base angles are non-obtuse, i.e., are at most $\pi/2$. Since every triangle may be rotated into one whose base angles are such, this restriction may be removed after we establish rotation invariance below:

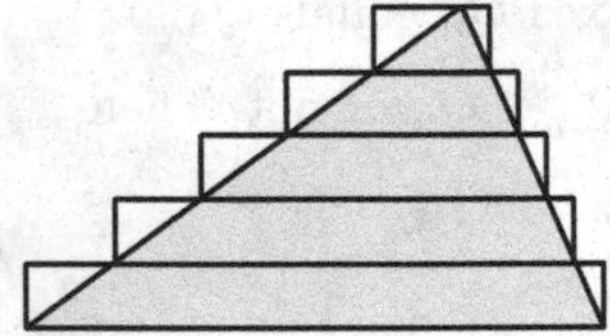

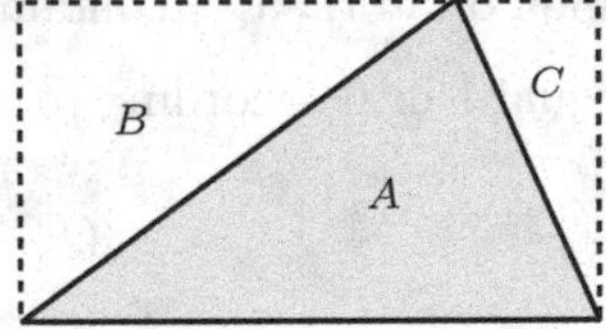

Fig. 4.10 Cover of a triangle

Divide A into n horizontal strips of height h/n as in Figure 4.10. Then the length of the base of each strip is b/n shorter than the length of the base of the strip below it, so (Exercise **1.3.12**)

$$\begin{aligned}\text{area}\,(A) &\leq \frac{h}{n}\,(b + (b - b/n) + (b - 2b/n) + \dots)\\ &= \frac{bh}{n^2}(n + (n-1) + \cdots + 1) = \frac{bh}{n^2} \cdot \frac{n(n+1)}{2} = \frac{bh(n+1)}{2n}.\end{aligned}$$

Now let $n \nearrow \infty$ to obtain area $(A) \leq (hb/2) = \|A\|$.

To obtain the reverse inequality, draw two other triangles B, C with horizontal bases, such that $A \cup B \cup C$ is a rectangle and A, B, and C intersect only along their edges. Then by simple arithmetic, the sum of the naive areas of A, B, and C equals the naive area of $A \cup B \cup C$, so by subadditivity of area,

$$\begin{aligned}\|A\| + \|B\| + \|C\| &= \|A \cup B \cup C\| \\ &= \text{area}\,(A \cup B \cup C) \\ &\le \text{area}\,(A) + \text{area}\,(B) + \text{area}\,(C) \\ &\le \text{area}\,(A) + \|B\| + \|C\|.\end{aligned}$$

Cancelling $\|B\|$, $\|C\|$, we obtain the reverse inequality $\|A\| \le \text{area}\,(A)$.

Theorem 4.2.2. *The area of a triangle equals half the product of the lengths of its base and its height.* □

We have derived this theorem assuming that the base of the triangle is horizontal. The general case follows from rotation invariance, which we do below.

Our next item is additivity. In general, we do not expect area $(A \cup B) =$ area $(A) +$ area (B) because A and B may overlap, i.e., intersect. If A and B are disjoint, one expects to have additivity. To what extent this is true leads to measurable sets (§4.5). Here we establish additivity only for the case when A and B are well separated. Exercises **4.5.12** and **4.5.13** discuss a broader case.

If $A \subset \mathbf{R}^2$ and $B \subset \mathbf{R}^2$, set

$$d(A,B) = \inf \sqrt{(a-c)^2 + (b-d)^2},$$

where the inf is over all points $(a,b) \in A$ and points $(c,d) \in B$. We say A and B are *well separated* if $d(A,B)$ is positive (Figure 4.11). For example, although $\{(2,0)\}$ and the unit disk are well separated, $\mathbf{Q} \times \mathbf{Q}$ and $\{(\sqrt{2},0)\}$ are disjoint but not well separated. Note that, since $\inf \emptyset = \infty$, A empty implies $d(A,B) = \infty$. Hence, the empty set is well separated from any subset of $\mathbf{R}^2$.

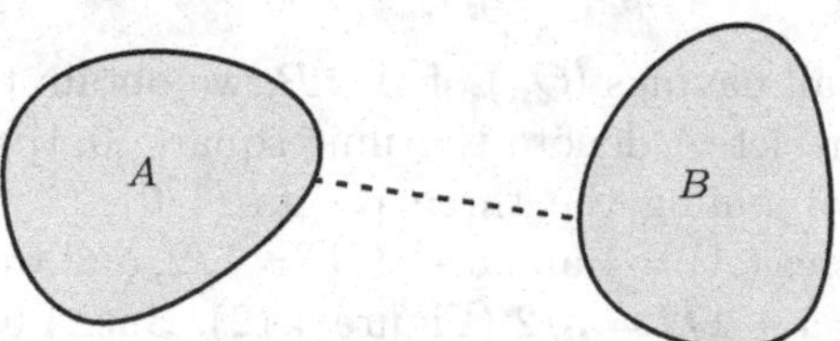

Fig. 4.11 Well-separated sets

If the lengths of the sides of a rectangle Q are a and b, by the *diameter* of Q, we mean the length of the diagonal $\sqrt{a^2 + b^2}$.

Theorem 4.2.3 (Well-Separated Additivity). *If A and B are well separated, then*[6]

$$\text{area}(A \cup B) = \text{area}(A) + \text{area}(B).$$

By subadditivity, $\text{area}(A \cup B) \le \text{area}(A) + \text{area}(B)$, so we need show only that

$$\text{area}(A \cup B) \ge \text{area}(A) + \text{area}(B) \tag{4.2.9}$$

If $\text{area}(A \cup B) = \infty$, (4.2.9) is true, so assume $\text{area}(A \cup B) < \infty$. In this case, to compute the area of $A \cup B$, we need consider only pavings involving bounded rectangles, since the sum of the areas of rectangles with at least one unbounded rectangle is ∞. Let $\epsilon = d(A, B) > 0$. If (Q_n) is a paving of $A \cup B$ with bounded rectangles Q_n, $n \ge 1$, divide each Q_n into subrectangles all with diameter less than ϵ. Since $\|Q_n\|$ equals the sum of the areas of its subrectangles, by replacing each Q_n by its subrectangles, we obtain a paving (Q'_n) of $A \cup B$ by rectangles Q'_n of diameter less than ϵ and

$$\sum_{n=1}^{\infty} \|Q'_n\| = \sum_{n=1}^{\infty} \|Q_n\|.$$

Thus, for each $n \ge 1$, Q'_n intersects A or B or neither but not both. Let (Q_n^A) denote those rectangles in (Q'_n) intersecting A, and let (Q_n^B) denote those rectangles in (Q'_n) intersecting B. Because no Q'_n intersects both A and B, (Q_n^A) is a paving of A and (Q_n^B) is a paving of B. Hence, by subadditivity,

$$\begin{aligned} \text{area}(A) + \text{area}(B) &\le \sum_{n=1}^{\infty} \|Q_n^A\| + \sum_{n=1}^{\infty} \|Q_n^B\| \\ &\le \sum_{n=1}^{\infty} \|Q'_n\| = \sum_{n=1}^{\infty} \|Q_n\|. \end{aligned}$$

Taking the inf over all pavings (Q_n) of $A \cup B$, we obtain the result. □

As an application, let A denote the unit square $[0,1] \times [0,1]$ and B the triangle obtained by joining the three points $(1,0)$, $(1,1)$, and $(2,1)$. We already know that $\text{area}(A) = 1$ and $\text{area}(B) = 1/2$, and we want to conclude that $\text{area}(A \cup B) = 1 + 1/2 = 3/2$ (Figure 4.12). But A and B are not well separated, so we do not have additivity directly. Instead, we dilate A by a factor $0 < \alpha < 1$ towards its center. Then the shrunken set αA and B are well separated. Moreover, $\alpha A \subset A$, so

$$\begin{aligned} \text{area}(A \cup B) &\ge \text{area}((\alpha A) \cup B) = \text{area}(\alpha A) + \text{area}(B) \\ &= \alpha^2 \cdot \text{area}(A) + \text{area}(B) = \alpha^2 + 1/2. \end{aligned}$$

[6] In Caratheodory's terminology [6], this says area is a *metric outer measure.*

Since α can be arbitrarily close to 1, we conclude that area $(A \cup B) \geq 3/2$. Since subadditivity yields area $(A \cup B) \leq$ area (A)+area $(B) = 1+1/2 = 3/2$, we obtain the result we seek, area $(A \cup B) = 3/2$.

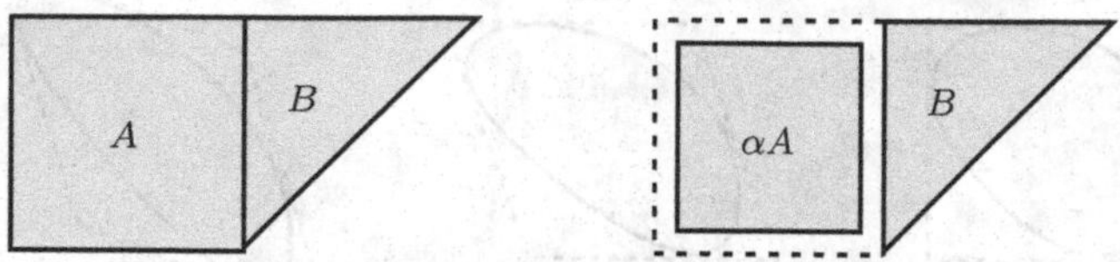

Fig. 4.12 Area of $A \cup B$

More generally, let P be a parallelogram with horizontal or vertical base and let $\|P\|$ denote its naive area, i.e., the product of the length of its base and its height. Then we leave it as an exercise to show that area $(P) = \|P\|$.

Theorem 4.2.4. *The area of a parallelogram equals the product of the lengths of its base and height.* □

We have derived this result assuming the base of the parallelogram is horizontal or vertical. The general case follows after we establish below the rotation invariance of area.

Now we turn to the rotation invariance of area or, more generally, the affine invariance of area.

A *linear map* is a map $T : \mathbf{R}^2 \to \mathbf{R}^2$ given by $T(x,y) = (ax+by, cx+dy)$. Examples of linear maps are

- Centered dilations $D(x,y) = (kx, ky)$,
- Horizontal dilations $H(x,y) = (kx, y)$,
- Vertical dilations $V(x,y) = (x, ky)$,
- Flips $F(x,y) = (y,x)$,
- Horizontal shears $S(x,y) = (x+ty, y)$,
- Vertical shears $S(x,y) = (x, y+tx)$,
- Rotations $R(x,y) = (x\cos\theta - y\sin\theta, x\sin\theta + y\cos\theta)$,
- Upper-triangular maps $U(x,y) = (ax+by, dy)$, and
- Lower-triangular maps $L(x,y) = (ax, cx+dy)$.

The *determinant* of a linear map $T(x,y) = (ax+by, cx+dy)$ is the real $\det(T) = ad - bc$. For example, we have $\det(D) = k^2$, $\det(H) = \det(V) = k$, $\det(F) = -1$, $\det(S) = 1$, and $\det(R) = 1$. A linear map T is *affine* if $\det(T) = \pm 1$.

Theorem 4.2.5. *[Affine Invariance of Area] Let $A \subset \mathbf{R}^2$. If T is a linear map, then*

$$\text{area}\,(T(A)) = |\det(T)|\,\text{area}\,(A)\,.$$

In particular, note that this establishes the rotation invariance of area (Figure 4.13).

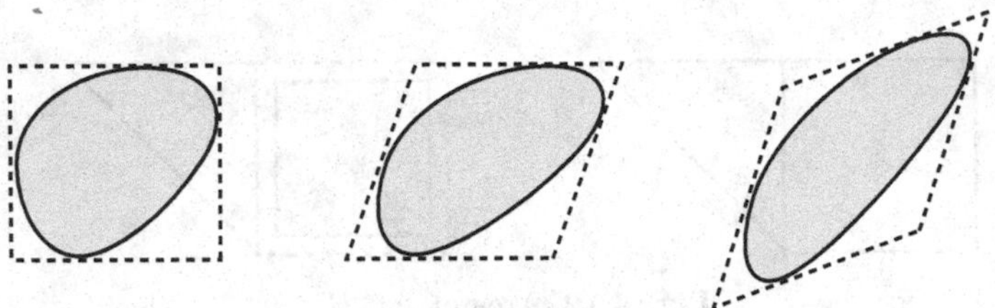

Fig. 4.13 Affine invariance of area

This result has already been established when T is a flip or a dilation. To establish the result when $T = S$ is a shear, note that $S(Q)$ is a parallelogram P with a horizontal or vertical base, when Q is a rectangle. Since we know $\text{area}\,(P) = \|P\|$, it follows that $\text{area}\,(T(A)) = \text{area}\,(A)$ when T is a shear and A is a rectangle. The case of a general set A is now derived as before by the use of pavings.

If (Q_n) is a paving of A, then $(S(Q_n))$ is a cover of $S(A)$; hence,

$$\text{area}\,(S(A)) \le \sum_{n=1}^{\infty} \text{area}\,(S(Q_n)) = \sum_{n=1}^{\infty} \|Q_n\|.$$

Taking the inf over all pavings of A, we obtain $\text{area}\,(S(A)) \le \text{area}\,(A)$. Replacing S by its inverse $T(x,y) = (x-ty, y)$ (or $T(x,y) = (x, y-tx)$ respectively), and A by $S(A)$, yields $\text{area}\,(A) \le \text{area}\,(S(A))$. Thus, $\text{area}\,(S(A)) = \text{area}\,(A)$.

Thus, the result is established for flips, shears, and dilations. The derivation of the general result now depends on the following basic facts[7]:

A. The composition (§1.1) $T_1 \circ T_2$ of linear maps T_1, T_2 is a linear map.
B. If Theorem 4.2.5 holds for linear maps T_1 and T_2, then it holds for their composition $T_1 \circ T_2$.

For example, since upper-triangular maps and lower-triangular maps are compositions of shears, horizontal dilations, and vertical dilations (Exercise **4.2.9**), the result holds for upper-triangular and lower-triangular maps. Moreover, since every linear map T is either of the form $L \circ U$ or of the form $F \circ U$ (Exercise **4.2.9**), Theorem 4.2.5 follows.

Because the derivations of **A** and **B** above play no role in the subsequent development, we relegate them to the exercises. □

By induction, well-separated additivity of area holds for several sets. *If* $A_1, A_2, \dots, A_n$ *are pairwise well-separated subsets of* $\mathbf{R}^2$, *then*

[7] These *group properties* are basic to much of mathematics.

$$\text{area}\,(A_1 \cup \ldots \cup A_n) = \text{area}\,(A_1) + \cdots + \text{area}\,(A_n)\,. \tag{4.2.10}$$

To see this, (4.2.10) is trivially true for $n = 1$, so assume (4.2.10) is true for a particular $n \geq 1$, and let $A_1, \ldots, A_{n+1}$ be pairwise well separated. Let $\epsilon_j = d(A_j, A_{n+1}) > 0$, $j = 1, \ldots, n$. Since

$$d(A_1 \cup \ldots \cup A_n, A_{n+1}) = \min(\epsilon_1, \ldots, \epsilon_n) > 0,$$

A_{n+1} and $A_1 \cup \ldots \cup A_n$ are well separated. Hence, by the inductive hypothesis,

$$\begin{aligned}\text{area}\,(A_1 \cup \ldots \cup A_{n+1}) &= \text{area}\,(A_1 \cup \ldots \cup A_n) + \text{area}\,(A_{n+1}) \\ &= \text{area}\,(A_1) + \cdots + \text{area}\,(A_n) + \text{area}\,(A_{n+1})\,.\end{aligned}$$

By induction, this establishes (4.2.10) for all $n \geq 1$.

More generally, *if (A_n) is a sequence of pairwise well-separated sets, then*

$$\text{area}\left(\bigcup_{n=1}^{\infty} A_n\right) = \sum_{n=1}^{\infty} \text{area}\,(A_n)\,. \tag{4.2.11}$$

To see this, subadditivity yields

$$\text{area}\left(\bigcup_{n=1}^{\infty} A_n\right) \leq \sum_{n=1}^{\infty} \text{area}\,(A_n)\,.$$

For the reverse inequality, apply (4.2.10) and monotonicity to the first N sets, yielding

$$\text{area}\left(\bigcup_{n=1}^{\infty} A_n\right) \geq \text{area}\left(\bigcup_{n=1}^{N} A_n\right) = \sum_{n=1}^{N} \text{area}\,(A_n)\,.$$

Now let $N \nearrow \infty$, obtaining

$$\text{area}\left(\bigcup_{n=1}^{\infty} A_n\right) \geq \sum_{n=1}^{\infty} \text{area}\,(A_n)\,.$$

This establishes (4.2.11).

As an application of (4.2.11), we can now compute the area of the Cantor set C. The Cantor set is constructed by removing, at successive stages, smaller and smaller open subsquares of $C_0 = [0,1] \times [0,1]$. Denote these subsquares $Q_1, Q_2, \ldots$ (at what stage they are removed is not important). Then for each n, C and Q_n are disjoint, so for $0 < \alpha < 1$, C and the centered dilations αQ_n are well separated. Moreover, for each m, n, the centered dilations αQ_n and αQ_m are well separated. But the union of C with all the squares αQ_n, $n \geq 1$, lies in the unit square C_0. Hence, by (4.2.11),

$$\text{area}(C) + \sum_{n=1}^{\infty} \text{area}(\alpha Q_n) = \text{area}\left(C \cup \left(\bigcup_{n=1}^{\infty} \alpha Q_n\right)\right) \leq \text{area}(C_0) = 1.$$

In the previous section, we obtained, $\sum_{n=1}^{\infty} \text{area}(Q_n) = 1$. By dilation invariance, this implies $\text{area}(C)+\alpha^2 \leq 1$. Letting $\alpha \nearrow 1$, we obtain $\text{area}(C)+1 \leq 1$ or $\text{area}(C) = 0$.

Theorem 4.2.6. *The area of the Cantor set is zero.* □

Exercises

4.2.1. Establish reflection invariance and monotonicity of area.

4.2.2. Show that the area of a bounded line segment is zero and the area of any line is zero.

4.2.3. Let P be a parallelogram with a horizontal or vertical base, and let $\|P\|$ denote the product of the length of its base and its height. Then $\text{area}(P) = \|P\|$.

4.2.4. Compute the area of a trapezoid.

4.2.5. If A and B are rectangles, then $\text{area}(A \cup B) = \text{area}(A) + \text{area}(B) - \text{area}(A \cap B)$.

4.2.6. For $k \in \mathbf{R}$, define $H : \mathbf{R}^2 \to \mathbf{R}^2$ and $V : \mathbf{R}^2 \to \mathbf{R}^2$ by $H(x,y) = (kx, y)$ and $V(x,y) = (x, ky)$. Then $\text{area}[V(A)] = |k| \cdot \text{area}(A) = \text{area}[H(A)]$ for every $A \subset \mathbf{R}^2$.

4.2.7. A mapping $T : \mathbf{R}^2 \to \mathbf{R}^2$ is *linear* if it is of the form $T(x,y) = (ax + by, cx + dy)$ with $a, b, c, d \in \mathbf{R}$. The *determinant* of T is the real $\det(T) = ad - bc$. If T and T' are linear, show that $T \circ T'$ is linear and $\det(T \circ T') = \det(T)\det(T')$.

4.2.8. Show that if affine invariance of area holds for T and for T', then it holds for $T \circ T'$.

4.2.9. Show that every upper-triangular map U may be written as a composition $V \circ S \circ H$. Similarly, every lower-triangular map satisfies $L = H \circ S \circ V$ for some H, S, V. If $T(x,y) = (ax+by, cx+dy)$ satisfies $a \neq 0$, show $T = L \circ U$ for some L, U. If $a = 0$, show $T = F \circ U$ for some U.

4.2.10. Show that $\{(\sqrt{2}, 0)\}$ and the unit disk are well separated, but $\{(\sqrt{2}, 0)\}$ and $\mathbf{Q} \times \mathbf{Q}$ are not.

4.2.11. Let D be the unit disk, and let $D^+ = \{(x,y) : x^2+y^2 < 1 \text{ and } y > 0\}$. Show that $\text{area}(D) = 2 \cdot \text{area}(D^+)$.

4.2.12. Compute the area of the sets C' and C^α described in Exercises **4.1.1** and **4.1.2**, using the properties of area.

4.2.13. Let $P_k = (\cos(2\pi k/n), \sin(2\pi k/n))$, $k = 0, 1, \dots, n$. Then P_0, $P_1, \dots,$ P_n are evenly spaced points on the unit circle with $P_n = P_0$. Let D_n denote the n-sided polygon obtained by joining the points P_k. Compute area (D_n).

4.2.14. Let $A \subset \mathbf{R}^2$. A *triangular paving* of A is a cover (T_n) of A where each T_n, $n \geq 1$, is a triangle (oriented arbitrarily). With area (A) as defined previously, show that

$$\text{area}\,(A) = \inf\left\{\sum_{n=1}^{\infty} \|T_n\| : \text{ all triangular pavings } (T_n) \text{ of } A\right\}.$$

Here $\|T\|$ denotes the naive area of the triangle T, i.e., half the product of the length of the base times the height.

4.2.15. Let $A \subset \mathbf{R}^2$. If area $(A) > 0$ and $0 < \alpha < 1$, there is *some* rectangle Q, such that area $(Q \cap A) > \alpha \cdot$ area (Q). (Argue by contradiction, and use the definition of area.)

4.3 The Integral

Let f be defined on an open interval (a, b), where, as usual, a may equal $-\infty$ or b may equal ∞. We say f is *bounded* if $|f(x)| \leq M$, $a < x < b$, for some real M. If f is *nonnegative*, i.e., if $f(x) \geq 0$, $a < x < b$, the *subgraph of f over* (a, b) is the set (Figure 4.14)

$$G = \{(x, y) : a < x < b, 0 < y < f(x)\} \subset \mathbf{R}^2.$$

Note that the inequalities in this definition are strict.

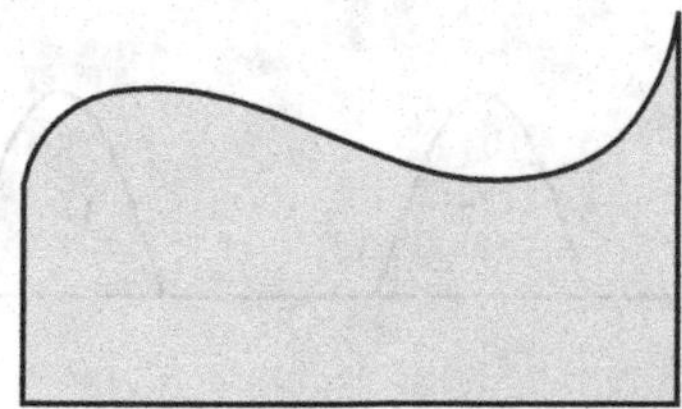

Fig. 4.14 Subgraphs of nonnegative functions

For nonnegative f, we define the *integral*[8] *of f from a to b* to be the area of its subgraph G,

[8] The usual terminology is *Lebesgue integral*.

$$\int_a^b f(x)\,dx = \text{area}\,(G)\,.$$

Then the integral is simply a quantity that is either 0, a positive real, or ∞. The reason for the unusual notation $\int_a^b f(x)\,dx$ for this quantity is explained below.

Thus, according to our definition, *every nonnegative function has an integral*, and integrals of nonnegative functions are areas—nothing more, nothing less—of certain subsets of $\mathbf{R}^2$.

Since the empty set has zero area, we always have $\int_a^a f(x)\,dx = 0$. For each $k \geq 0$, the subgraph of $f(x) = k$, $a < x < b$, over (a, b) is an open rectangle, so

$$\int_a^b k\,dx = k(b-a).$$

Since the area is monotone, so is the integral: If $0 \leq f \leq g$ on (a, b),

$$\int_a^b f(x)\,dx \leq \int_a^b g(x)\,dx.$$

In particular, $0 \leq f \leq M$ on (a, b) implies $0 \leq \int_a^b f(x)\,dx \leq M(b-a)$.

A nonnegative function f is *integrable* over (a, b) if $\int_a^b f(x)\,dx < \infty$. For example, we have just seen that every bounded nonnegative f is integrable over a bounded interval (a, b). Now we discuss the integral of a *signed* function, i.e., a function that takes on positive and negative values.

Given a function $f : (a, b) \to \mathbf{R}$, we set

$$f^+(x) = \max[f(x), 0],$$

and

$$f^-(x) = \max[-f(x), 0].$$

These are (Figure 4.15) the *positive part* and the *negative part* of f, respectively. Note that $f^+ - f^- = f$ and $f^+ + f^- = |f|$.

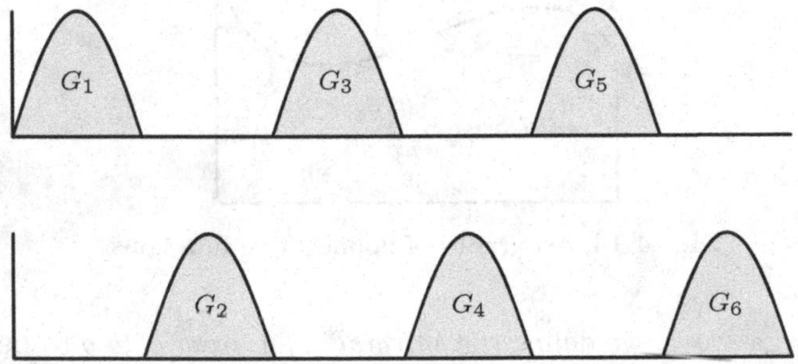

Fig. 4.15 Positive and negative parts of $\sin x$

We say a signed function f is *integrable over* (a,b) if

$$\int_a^b |f(x)|\,dx < \infty.$$

In this case, $\int_a^b f^\pm(x)\,dx \le \int_a^b |f(x)|\,dx$ are both finite. For integrable f, we define the *integral* $\int_a^b f(x)\,dx$ *of* f *from* a *to* b by

$$\int_a^b f(x)\,dx = \int_a^b f^+(x)\,dx - \int_a^b f^-(x)\,dx.$$

From this follows

$$\int_a^b [-f(x)]\,dx = -\int_a^b f(x)\,dx$$

for every integrable function f, since $g = -f$ implies $g^+ = f^-$ and $g^- = f^+$.

We warn the reader that, although $\int_a^b f(x)\,dx = \int_a^b f^+(x)\,dx - \int_a^b f^-(x)\,dx$ is a definition, we have not verified the identity $\int_a^b |f(x)|\,dx = \int_a^b f^+(x)\,dx + \int_a^b f^-(x)\,dx$ for general integrable f. For more on this, see §6.1.

Also from the above discussion, we see that *every bounded (signed) function is integrable over a bounded interval.* For example, $\sin x$ and $\sin x/x$ are integrable over $(0,\pi)$. In fact, both functions are integrable over $(0,b)$ for any finite b, and hence, $\int_0^b \sin x\,dx$ and $\int_0^b (\sin x/x)\,dx$ are defined.

It is reasonable to expect that $\sin x$ is not integrable over $(0,\infty)$. Indeed the subgraph of $|\sin x|$ consists of a union of sets G_n, $n \ge 1$, where each G_n denotes the subgraph over $((n-1)\pi, n\pi)$. By translation invariance, the sets G_n, $n \ge 1$, have the same positive area and the sets $G_1, G_3, G_5, \dots$, are well separated. Hence, we obtain

$$\int_0^\infty (\sin x)^+\,dx = \operatorname{area}\left(\bigcup_{n=1}^\infty G_{2n-1}\right) = \sum_{n=1}^\infty \operatorname{area}(G_{2n-1}) = \infty.$$

By considering, instead, $G_2, G_4, G_6, \dots$, we obtain $\int_0^\infty (\sin x)^-\,dx = \infty$. Thus,

$$\int_0^\infty (\sin x)^+\,dx - \int_0^\infty (\sin x)^-\,dx = \infty - \infty.$$

Hence, $\int_0^\infty \sin x\,dx$ cannot be defined as a difference of two areas. The trick of considering every other set in a sequence to force well-separatedness is generalized in the proof of Theorem 4.3.3.

It turns out that $\sin x/x$ is also not integrable over $(0,\infty)$. To see this, let G_n denote the subgraph of $|\sin x/x|$ over $((n-1)\pi, n\pi)$, $n \ge 1$ (Figure 4.16). In each G_n, we can insert a rectangle Q_n of area $\sqrt{2}/(4n-1)$ so that the rectangles are well separated (select Q_n to have base the open interval obtained by translating $(\pi/4, 3\pi/4)$ by $(n-1)\pi$ and height as large as possible—see Figure 4.16). By additivity, then we obtain

$$\int_0^\infty \frac{|\sin x|}{x}\,dx \geq \operatorname{area}\left(\bigcup_{n=1}^\infty Q_n\right) = \sum_{n=1}^\infty \operatorname{area}(Q_n) = \sum_{n=1}^\infty \frac{\sqrt{2}}{4n-1} = \infty,$$

by comparison with the harmonic series. Thus, $\sin x/x$ is not integrable over $(0,\infty)$. More explicitly, this reasoning also shows that

$$\int_0^\infty \left(\frac{\sin x}{x}\right)^+ dx \geq \operatorname{area}(Q_1) + \operatorname{area}(Q_3) + \operatorname{area}(Q_5) + \cdots = \infty$$

and

$$\int_0^\infty \left(\frac{\sin x}{x}\right)^- dx \geq \operatorname{area}(Q_2) + \operatorname{area}(Q_4) + \operatorname{area}(Q_6) + \cdots = \infty.$$

Thus,

$$\int_0^\infty \left(\frac{\sin x}{x}\right)^+ dx - \int_0^\infty \left(\frac{\sin x}{x}\right)^- dx = \infty - \infty;$$

hence $\int_0^\infty \sin x/x\,dx$ also cannot be defined as a difference of two areas.

To summarize, *the integral of an integrable function is the area of the subgraph of its positive part minus the area of the subgraph of its negative part.* Every property of $\int_a^b f(x)\,dx$ ultimately depends on a corresponding property of area.

Frequently, one checks integrability of a given f by first applying one or more of the properties below to the nonnegative function $|f|$. For example, consider the function $g(x) = 1/x^2$ for $x > 1$, and, for each $n \geq 1$, let G_n denote the compact rectangle $[n, n+1] \times [0, 1/n^2]$. Then (G_n) is a cover of the subgraph of g over $(1,\infty)$ (Figure 4.17). Hence,

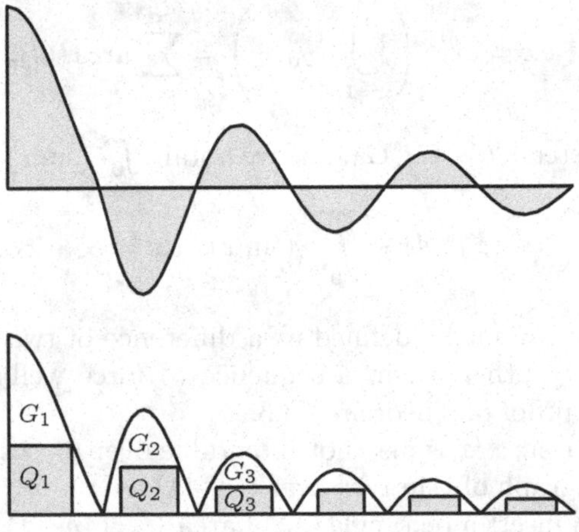

Fig. 4.16 The graphs of $\sin x/x$ and $|\sin x|/x$

$$\int_1^\infty \frac{1}{x^2}\,dx \le \sum_{n=1}^\infty \frac{1}{n^2},$$

which is finite (§1.6). Thus, g is integrable over $(1,\infty)$. Since the signed function $f(x) = \cos x/x^2$ satisfies $|f(x)| \le g(x)$ for $x > 1$, by monotonicity, we conclude that

$$\int_1^\infty \left|\frac{\cos x}{x^2}\right| dx < \infty. \tag{4.3.1}$$

Hence, $\cos x/x^2$ is integrable over $(1,\infty)$.

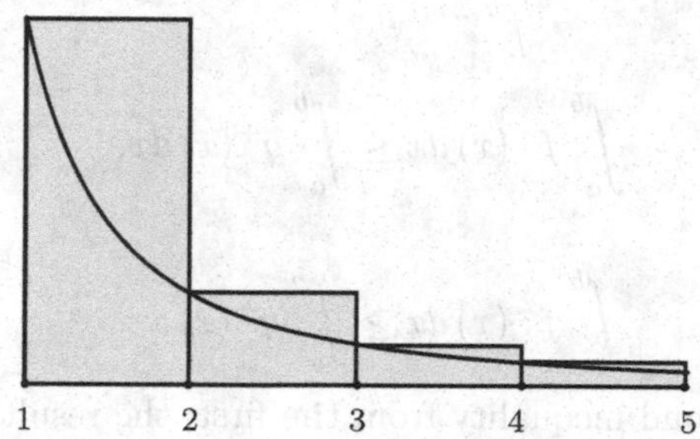

Fig. 4.17 A cover of the subgraph of $1/x^2$ over $(1,\infty)$

Of course, functions may be unbounded and integrable. For example, the function $f(x) = 1/\sqrt{x}$ is integrable over $(0,1)$. To see this, let G_n denote the compact rectangle $[1/(n+1)^2, 1/n^2]\times[0,n+1]$. Then (G_n) is a cover of the subgraph of f over $(0,1)$ (Figure 4.18). Hence,

$$\int_0^1 \frac{1}{\sqrt{x}}\,dx \le \sum_{n=1}^\infty (n+1)\left(\frac{1}{n^2} - \frac{1}{(n+1)^2}\right) = \sum_{n=1}^\infty \frac{2n+1}{n^2(n+1)} \le 2\sum_{n=1}^\infty \frac{1}{n^2},$$

which is finite. Thus, f is integrable over (0,1).

Theorem 4.3.1 (Monotonicity). *Suppose that f and g are both nonnegative or both integrable on (a,b). If $f \le g$ on (a,b), then*

$$\int_a^b f(x)\,dx \le \int_a^b g(x)\,dx.$$

If $0 \le f \le g$, we already know this. For the integrable case, note that $f \le g$ implies $f^+ = \max(f,0) \le \max(g,0) = g^+$ and $g^- = \max(-g,0) \le \max(-f,0) = f^-$ on (a,b). Hence,

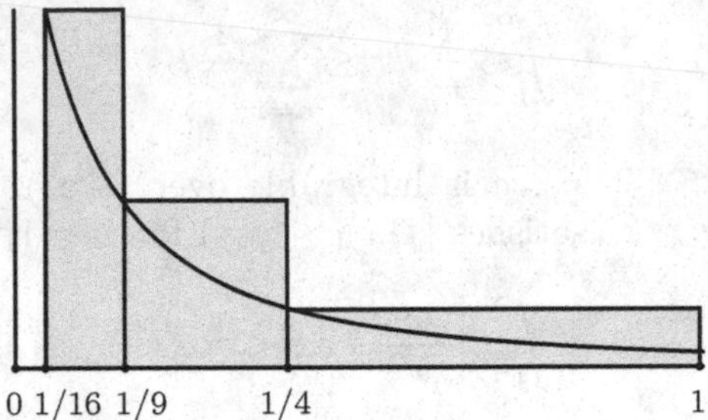

Fig. 4.18 A cover of the subgraph of $1/\sqrt{x}$ over $(0,1)$

$$\int_a^b f^+(x)\,dx \le \int_a^b g^+(x)\,dx,$$

and

$$\int_a^b f^-(x)\,dx \ge \int_a^b g^-(x)\,dx.$$

Subtracting the second inequality from the first, the result follows. □

Since $\pm f \le |f|$, the theorem implies

$$\pm \int_a^b f(x)\,dx = \int_a^b \pm f(x)\,dx \le \int_a^b |f(x)|\,dx$$

which yields

$$\left| \int_a^b f(x)\,dx \right| \le \int_a^b |f(x)|\,dx$$

for every integrable f.

Theorem 4.3.2 (Translation and Dilation Invariance). *Let f be nonnegative or integrable on (a,b). Choose $c \in \mathbf{R}$ and $k > 0$. Then*

$$\int_a^b f(x+c)\,dx = \int_{a+c}^{b+c} f(x)\,dx,$$

$$\int_a^b kf(x)\,dx = k\int_a^b f(x)\,dx,$$

and

$$\int_a^b f(kx)\,dx = \frac{1}{k}\int_{ka}^{kb} f(x)\,dx.$$

If f is nonnegative, let G denote the subgraph of $f(x+c)$ over (a,b) (Figure 4.19). Then the translate $G + (c,0)$ equals

$$\{(x,y) : a+c < x < b+c, 0 < y < f(x)\},$$

which is the subgraph of $f(x)$ over the interval $(a+c, b+c)$. By translation invariance of area, we obtain translation invariance of the integral in the nonnegative case. If f is integrable, by the nonnegative case,

$$\int_a^b f^+(x+c)\,dx = \int_{a+c}^{b+c} f^+(x)\,dx,$$

and

$$\int_a^b f^-(x+c)\,dx = \int_{a+c}^{b+c} f^-(x)\,dx.$$

Now if $g(x) = f(x+c)$, then $g^+(x) = f^+(x+c)$ and $g^-(x) = f^-(x+c)$. So subtracting the last equation from the previous one, we obtain translation invariance in the integrable case.

For the second equation and f nonnegative, recall that from the previous section, the dilation mapping $V(x,y) = (x, ky)$, and let G denote the subgraph of f over (a,b). Then $V(G) = \{(x,y) : a < x < b, 0 < y < kf(x)\}$. Hence, $\operatorname{area}(V(G)) = \int_a^b kf(x)\,dx$. Now dilation invariance of the area (Exercise **4.2.6**) yields $\int_a^b kf(x)\,dx = k\int_a^b f(x)\,dx$ for f nonnegative. For integrable f, the result follows by applying, as above, the nonnegative case to f^+ and f^-.

For the third equation, let $H(x,y) = (kx, y)$, and let G denote the subgraph of $f(kx)$ over (a,b). Then $H(G) = \{(x,y) : ka < x < kb, 0 < y < f(x)\}$. The third equation now follows, as before, by dilation invariance. For integrable f, the result follows by applying the nonnegative case to $f^\pm$. □

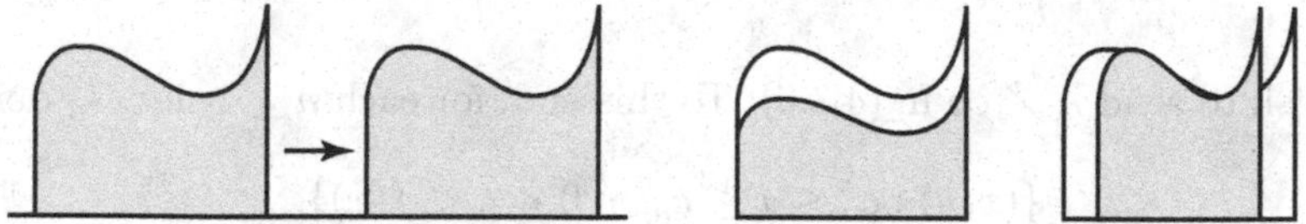

Fig. 4.19 Translation and dilation invariance of integrals

By similar reasoning, one can also derive (Figure 4.20)

$$\int_{-b}^{-a} f(-x)\,dx = \int_a^b f(x)\,dx,$$

valid for f nonnegative or integrable over (a,b).

The next property is additivity.

Theorem 4.3.3 (Additivity). *Suppose that f is nonnegative or integrable over (a,b), and choose $a < c < b$. Then*

$$\int_a^b f(x)\,dx = \int_a^c f(x)\,dx + \int_c^b f(x)\,dx.$$

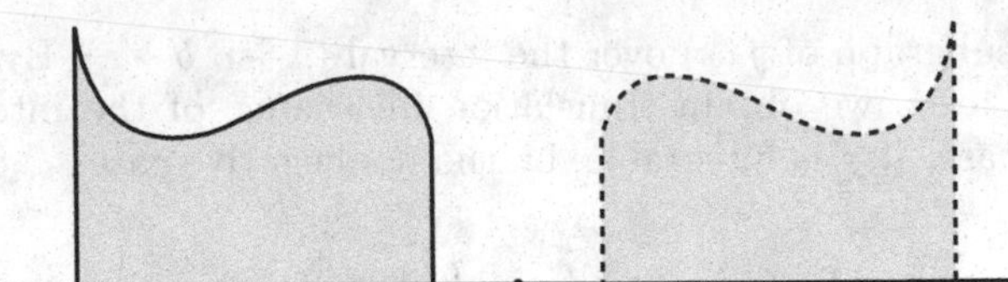

Fig. 4.20 Reflection invariance of integrals

To see this, first, assume that f is nonnegative. Since the vertical line $x = c$ has zero area, subadditivity yields

$$\int_a^b f(x)\,dx \le \int_a^c f(x)\,dx + \int_c^b f(x)\,dx.$$

So we need only show that

$$\int_a^b f(x)\,dx \ge \int_a^c f(x)\,dx + \int_c^b f(x)\,dx. \tag{4.3.2}$$

If f is not integrable, (4.3.2) is immediate since the left side is infinite, so assume f is nonnegative and integrable. Now, choose any strictly increasing sequence $a < c_1 < c_2 < \dots$ converging to c. Then for $n \ge 1$, the subgraph of f over (a, c_n) and the subgraph of f over (c, b) are well separated (Figure 4.21). So by monotonicity and well-separated additivity,

$$\int_a^b f(x)\,dx \ge \int_a^{c_n} f(x)\,dx + \int_c^b f(x)\,dx. \tag{4.3.3}$$

We wish to send $n \nearrow \infty$ in (4.3.3). To this end, for each $n \ge 1$, let G_n denote

$$\{(x, y) : c_n \le x \le c_{n+1}, 0 < y < f(x)\}.$$

Since $G_2, G_4, G_6, \dots$, are pairwise well separated,

$$\text{area}\,(G_2) + \text{area}\,(G_4) + \text{area}\,(G_6) + \dots \le \int_a^c f(x)\,dx < \infty.$$

Since $G_1, G_3, G_5, \dots$, are pairwise well separated,

$$\text{area}\,(G_1) + \text{area}\,(G_3) + \text{area}\,(G_5) + \dots \le \int_a^c f(x)\,dx < \infty.$$

Adding the last two inequalities yields the convergence of $\sum_{n=1}^{\infty} \text{area}\,(G_n)$. Hence, the tail (§1.6) goes to zero:

$$\lim_{n \nearrow \infty} \sum_{k=n}^{\infty} \text{area}\,(G_k) = 0.$$

Since the subgraph of f over (c_n, c) is contained in $G_n \cup G_{n+1} \cup G_{n+2} \cup \dots$, monotonicity and subadditivity imply

$$0 \le \int_{c_n}^{c} f(x)\, dx \le \sum_{k=n}^{\infty} \text{area}\,(G_k), \qquad n \ge 1.$$

Hence, we obtain

$$\lim_{n \nearrow \infty} \int_{c_n}^{c} f(x)\, dx = 0. \tag{4.3.4}$$

Since by monotonicity and subadditivity, again,

$$\int_a^{c_n} f(x)\, dx \le \int_a^{c} f(x)\, dx \le \int_a^{c_n} f(x)\, dx + \int_{c_n}^{c} f(x)\, dx, \quad n \ge 1,$$

we conclude that

$$\lim_{n \nearrow \infty} \int_a^{c_n} f(x)\, dx = \int_a^{c} f(x)\, dx.$$

Now sending $n \nearrow \infty$ in (4.3.3) yields (4.3.2). Hence, the result for f nonnegative. If f is integrable, apply the nonnegative case to f^+ and f^-. Then

$$\int_a^b f^+(x)\, dx = \int_a^c f^+(x)\, dx + \int_c^b f^+(x)\, dx,$$

and

$$\int_a^b f^-(x)\, dx = \int_a^c f^-(x)\, dx + \int_c^b f^-(x)\, dx.$$

Subtracting the second equation from the first, we obtain the result in the integrable case. □

The trick in the above proof of considering every other set in a sequence to force well-separatedness is generalized in the proof of Theorem 4.5.4.

The bulk of the derivation above involves establishing (4.3.4). If f is bounded, say by M, then the integral in (4.3.4) is no more than $M(c - c_n)$, hence trivially, goes to zero. The delicacy is necessary to handle unbounded situations.

The main point in the derivation is that, although the subgraph G of f over (a, c) and the subgraph G' of f over (c, b) are not well separated, we still have additivity, because we know something—the existence of the vertical edges—about the geometry of G and G'.

In the previous section, when we wanted to apply additivity to several sets (e.g., when we computed the area of the Cantor set) that were not well separated, we dilated them by a factor $0 < \alpha < 1$ and then applied additivity to the shrunken sets.

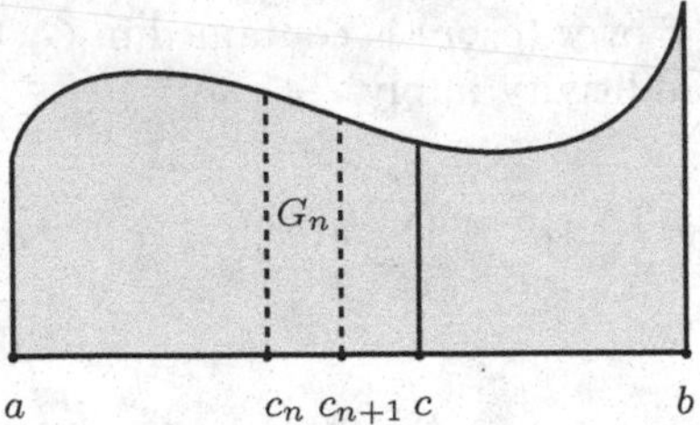

Fig. 4.21 Additivity of integrals

Why don't we use the same trick here for G or G'? The reason is that if the graph of f is sufficiently "jagged" (Figure 4.22), we do not have $\alpha G \subset G$, a necessary step in applying the shrinking trick of the previous section.

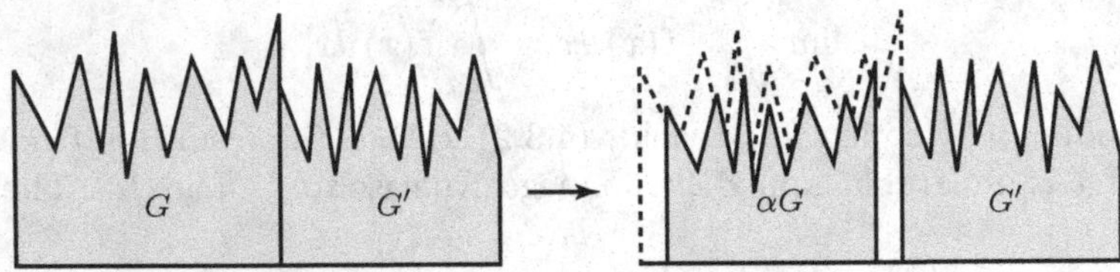

Fig. 4.22 A "jagged" function

By induction, additivity holds for a partition (§2.2) of (a, b): *If* $a = x_0 < x_1 < \cdots < x_n = b$ *and* f *is nonnegative or integrable over* (a, b), *then*

$$\int_a^b f(x)\,dx = \sum_{k=1}^{n} \int_{x_{k-1}}^{x_k} f(x)\,dx. \tag{4.3.5}$$

Since the right side does not involve the values of f at the points defining the partition, we conclude that *the integrals of two functions* $f : (a, b) \to \mathbf{R}$ *and* $g : (a, b) \to \mathbf{R}$ *are equal, whenever they differ only on finitely many points* $a < x_1 < \cdots < x_{n-1} < b$.

Another application of additivity is to piecewise constant functions. A function $f : (a, b) \to \mathbf{R}$ is *piecewise constant* if there is a partition $a = x_0 < x_1 < \cdots < x_n = b$, such that f, restricted to each open subinterval (x_{i-1}, x_i), $i = 1, \ldots, n$, is constant. (Note that the values of a piecewise constant function at the partition points x_i, $1 \le i \le n-1$, are not restricted in any way.) In this case, additivity implies

$$\int_a^b f(x)\,dx = \sum_{i=1}^{n} c_i \Delta x_i,$$

where $\Delta x_i = x_i - x_{i-1}$, $i = 1, \ldots, n$. Since a continuous function can be closely approximated by a piecewise constant function (§2.3), the integral should be thought of as a sort of sum with Δx_i "infinitely small," hence the notation dx replacing Δx_i and $\int$ replacing $\sum$.

This view is supported by Exercise **4.3.3**. Indeed, by defining integrals as areas of subgraphs, we capture the intuition that integrals are approximately sums of areas of rectangles in *any* paving and not just finite vertical pavings as given by the "Riemann sums" of Exercise **4.3.3**.

Also since the integral is, by definition, a combination of certain areas and the notation $\int_a^b f(x)\,dx$ is just a mnemonic device, the variable inside the integral sign is a "dummy" variable, i.e., $\int_a^b f(x)\,dx = \int_a^b f(t)\,dt$. Nevertheless, the interpretation of the integral as a "continuous sum" is basic, useful, and important.

Let us go back to the integrals of $\sin x$ and $\sin x/x$ over $(0,\infty)$. Above, we saw that these functions were not integrable over $(0,\infty)$, and so $\int_0^\infty \sin x\,dx$ and $\int_0^\infty \sin x/x\,dx$ could not be defined as the difference of the areas of the positive and the negative parts. An alternate approach is to consider $F(b) = \int_0^b \sin x\,dx$ and to take the limit $F(\infty) = \lim_{b\to\infty} F(b)$. However, since the areas of the sets G_n, $n \ge 1$, are equal, by additivity, $F(n\pi) = \text{area}\,(G_1) - \text{area}\,(G_2) + \cdots \pm \text{area}\,(G_n)$ equals $\text{area}\,(G_1)$ or zero according to whether n is odd or even. Thus, the limit $F(\infty)$ does not exist, and this approach fails for $\sin x$.

For $\sin x/x$, however, it is a different story. Let $F(b) = \int_0^b \sin x/x\,dx$, and let G_n denote the subgraph of $|\sin x|/x$ over $((n-1)\pi, n\pi)$ for each $n \ge 1$. Then by additivity $F(n\pi) = \text{area}\,(G_1) - \text{area}\,(G_2) + \cdots \pm \text{area}\,(G_n)$. Hence,

$$\lim_{n\nearrow\infty} \int_0^{n\pi} \frac{\sin x}{x}\,dx = \text{area}\,(G_1) - \text{area}\,(G_2) + \text{area}\,(G_3) - \dots .$$

But this last series has a finite sum since it is alternating with decreasing terms! Thus,

$$\int_0^\infty \frac{\sin x}{x}\,dx \neq \lim_{n\nearrow\infty} \int_0^{n\pi} \frac{\sin x}{x}\,dx \tag{4.3.6}$$

since the left side is not defined and the right side is a well defined, finite real.

More generally (Exercise **4.3.14**) $F(\infty) = \lim_{b\to\infty} F(b)$ exists and $F(\infty)$ is computed in Exercise **5.4.12**. Even more generally, there is an integral version (Exercise **4.4.31**) of the Dirichlet test for series: If f is smooth and decreasing to zero as $x \to \infty$, with $f(x)\sin x$ and $f'(x)(1-\cos x)$ integrable over $(0,b)$ for all $b > 0$, then

$$\lim_{b\to\infty} \int_0^b f(x)\sin x\,dx$$

exists.

On the other hand, when f is nonnegative or integrable, its integral over an interval (a,b) *can* be obtained as a limit of integrals over subintervals (a_n,b_n) (Figure 4.23), and the behavior (4.3.6) does not occur.

Theorem 4.3.4 (Continuity at the Endpoints). *If f is nonnegative or integrable on (a,b) and $a_n \to a+$, $b_n \to b-$, then*

$$\int_a^b f(x)\,dx = \lim_{n\nearrow\infty} \int_{a_n}^{b_n} f(x)\,dx. \tag{4.3.7}$$

If f is integrable on (a,b) and $a_n \to a+$, $b_n \to b-$, then in addition,

$$\lim_{n\nearrow\infty} \int_a^{a_n} f(x)\,dx = 0, \tag{4.3.8}$$

and

$$\lim_{n\nearrow\infty} \int_{b_n}^{b} f(x)\,dx = 0. \tag{4.3.9}$$

To see this, first assume that f is nonnegative and $b_n \nearrow b$, and fix $a < c < b$. Since area is monotone, the sequence $\int_c^{b_n} f(x)\,dx$, $n \geq 1$, is increasing and

$$\lim_{n\nearrow\infty} \int_c^{b_n} f(x)\,dx \leq \int_c^b f(x)\,dx.$$

For the reverse inequality, let G_n denote the subgraph of f over (b_n, b_{n+1}), $n \geq 1$. By additivity,

$$\int_c^{b_n} f(x)\,dx = \int_c^{b_1} f(x)\,dx + \sum_{k=1}^{n-1} \text{area}\,(G_k)\,,$$

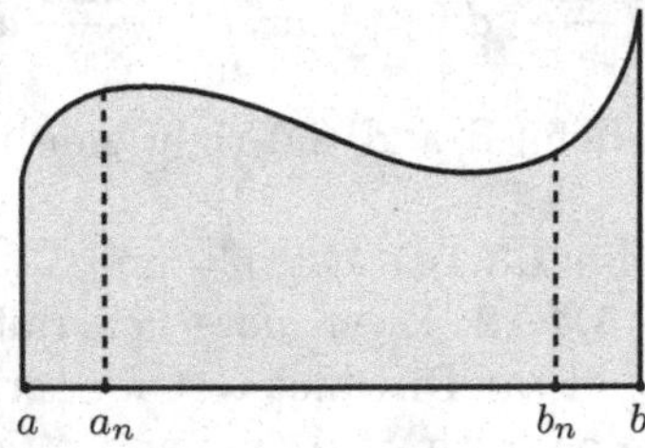

Fig. 4.23 Continuity at the endpoints

so taking the limit and using subadditivity,

$$\lim_{n\nearrow\infty} \int_c^{b_n} f(x)\,dx = \int_c^{b_1} f(x)\,dx + \sum_{k=1}^{\infty} \text{area}\,(G_k)$$

$$\geq \int_c^{b_1} f(x)\,dx + \int_{b_1}^b f(x)\,dx \geq \int_c^b f(x)\,dx.$$

Hence,

$$\lim_{n\nearrow\infty} \int_c^{b_n} f(x)\,dx = \int_c^b f(x)\,dx. \tag{4.3.10}$$

In general, if $b_n \to b-$, then $b_{n*} \nearrow b$ (§1.5), and $b_{n*} \leq b_n < b$. Hence,

$$\int_c^{b_{n*}} f(x)\,dx \leq \int_c^{b_n} f(x)\,dx \leq \int_c^b f(x)\,dx, \qquad n \geq 1,$$

which implies (4.3.10), for general $b_n \to b-$. Since

$$\int_{a_n}^c f(x)\,dx = \int_{-c}^{-a_n} f(-x)\,dx,$$

applying what we just learned to $f(-x)$ yields

$$\lim_{n\nearrow\infty} \int_{a_n}^c f(x)\,dx = \int_a^c f(x)\,dx.$$

Hence,

$$\begin{aligned}
\int_a^b f(x)\,dx &= \int_a^c f(x)\,dx + \int_c^b f(x)\,dx \\
&= \lim_{n\nearrow\infty} \int_{a_n}^c f(x)\,dx + \lim_{n\nearrow\infty} \int_c^{b_n} f(x)\,dx \\
&= \lim_{n\nearrow\infty} \int_{a_n}^{b_n} f(x)\,dx.
\end{aligned}$$

For the integrable case, apply (4.3.7) to $f^{\pm}$ to get (4.3.7) for f. Since

$$\int_a^{a_n} f(x)\,dx = \int_a^b f(x)\,dx - \int_{a_n}^b f(x)\,dx,$$

we get (4.3.8). Similarly, we get (4.3.9). □

For example,

$$\int_0^1 x^r\,dx = \lim_{a\to 0+} \int_a^1 x^r\,dx,$$

and

$$\int_1^\infty x^r\,dx = \lim_{b\to\infty} \int_1^b x^r\,dx,$$

both for r real.

When f is integrable, the last theorem can be improved: We have continuity of the integral at every point in (a, b).

Theorem 4.3.5 (Continuity). *Suppose that f is integrable over (a, b), and set*

$$F(t) = \int_a^t f(x)\,dx, \qquad a < t < b.$$

Then F is continuous on (a, b).

To see this, fix $a < c < b$, and let $c_n \to c-$. Applying the previous theorem on (a, c), we obtain $F(c_n) \to F(c)$. Hence, we obtain continuity of F from the left at every real in (a, b).

Now let $g(x) = f(-x)$, $-b < x < -a$, and

$$G(t) = \int_t^{-a} g(x)\,dx, \qquad -b < t < -a.$$

Since, by additivity,

$$G(t) = \int_{-b}^{-a} g(x)\,dx - \int_{-b}^{t} g(x)\,dx, \qquad -b < t < -a,$$

by the previous paragraph applied to g, the function G is continuous from the left at every point in $(-b, -a)$. Thus, the function

$$G(-t) = \int_{-t}^{-a} f(-x)\,dx = \int_a^t f(x)\,dx = F(t), \qquad a < t < b,$$

is continuous from the right at every point in (a, b). This establishes continuity of F on (a, b). □

Our last item is the integral test for positive series.

Theorem 4.3.6 (Integral Test). *Let $f : (0, \infty) \to (0, \infty)$ be decreasing. Then*

$$\gamma = \lim_{n \nearrow \infty} \left[\sum_{k=1}^{n} f(k) - \int_1^{n+1} f(x)\,dx \right] \tag{4.3.11}$$

exists and $0 \le \gamma \le f(1)$. In particular, the integral $\int_1^\infty f(x)\,dx$ is finite iff the sum $\sum_{n=1}^\infty f(n)$ converges.

For each $n \ge 1$, let $B_n = (n, n+1) \times (0, f(n))$, $B_n' = (n, n+1) \times [f(n+1), f(n)]$, and let G_n denote the subgraph of f over $(n, n+1)$ (Figure 4.24). Since f is decreasing, $G_n \subset B_n \subset G_n \cup B_n'$, for all $n \ge 1$. Then the quantity whose limit is the right side of (4.3.11) equals

$$\sum_{k=1}^{n} \left[\text{area}\,(B_k) - \text{area}\,(G_k) \right],$$

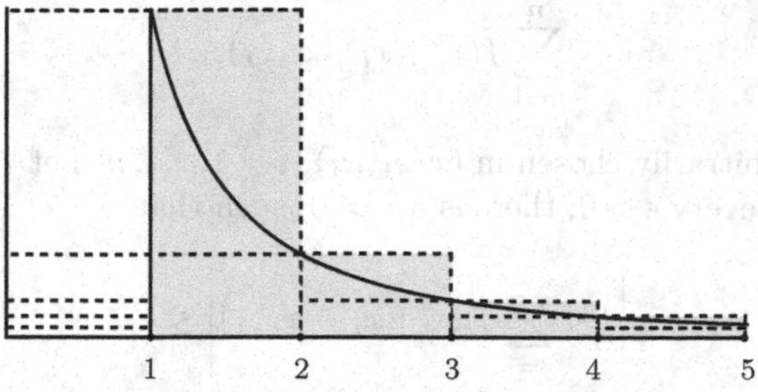

Fig. 4.24 Integral test

which is clearly increasing with n (here, we used additivity). Hence, the limit $\gamma \geq 0$ exists. On the other hand, by subadditivity, we get

$$\text{area}\,(B_k) - \text{area}\,(G_k) \leq \text{area}\,(B'_k) = f(k) - f(k+1).$$

So

$$\gamma = \sum_{n=1}^{\infty} [\text{area}\,(B_k) - \text{area}\,(G_k)] \leq \sum_{k=1}^{\infty} [f(k) - f(k+1)] = f(1).$$

Thus, $\gamma \leq f(1)$. If either $\int_1^\infty f(x)\,dx$ or $\sum_{n=1}^\infty f(n)$ is finite, (4.3.11) simplifies to

$$\gamma = \sum_{n=1}^{\infty} f(n) - \int_1^\infty f(x)\,dx,$$

which shows that the sum is finite iff the integral is finite. □

For $a < b$ we *define*

$$\int_b^a f(x)\,dx = -\int_a^b f(x)\,dx.$$

This is useful in **4.3.13** below and elsewhere.

Exercises

4.3.1. Show that $\int_0^\infty f(kx)x^{-1}\,dx = \int_0^\infty f(x)x^{-1}\,dx$ for $k > 0$ and $f(x)/x$ nonnegative or integrable over $(0, \infty)$.

4.3.2. Show that $\int_{-b}^{-a} f(-x)\,dx = \int_a^b f(x)\,dx$ for f nonnegative or integrable over (a, b).

4.3.3. Let $f : [a, b] \to \mathbf{R}$ be continuous. If $a = x_0 < x_1 < \cdots < x_n = b$ is a partition of $[a, b]$, a *Riemann sum* corresponding to this partition is the real (Figure 4.25)

$$\sum_{i=1}^{n} f(x_i^{\#})(x_i - x_{i-1}),$$

where $x_i^{\#}$ is arbitrarily chosen in (x_{i-1}, x_i), $i = 1, \dots, n$. Let $I = \int_a^b f(x)\,dx$. Show that, for every $\epsilon > 0$, there is a $\delta > 0$, such that

$$\left| I - \sum_{i=1}^{n} f(x_i^{\#})(x_i - x_{i-1}) \right| \leq \epsilon \tag{4.3.12}$$

for any partition $a = x_0 < x_1 < \cdots < x_n = b$ of mesh less than δ and choice of points $x_1^{\#}, \dots, x_n^{\#}$. (Approximate f by a piecewise constant f_ϵ as in §2.3.)

4.3.4. Let $f : (0,1) \to \mathbf{R}$ be given by

$$f(x) = \begin{cases} x & \text{if } x \text{ irrational}, \\ 0 & \text{if } x \text{ rational}. \end{cases}$$

Compute $\int_0^1 f(x)\,dx$.

4.3.5. Let $f : (a,b) \to \mathbf{R}$ be nonnegative, and suppose that $g : (a,b) \to \mathbf{R}$ is nonnegative and piecewise constant. Use additivity to show that

$$\int_a^b [f(x) + g(x)]\,dx = \int_a^b f(x)\,dx + \int_a^b g(x)\,dx.$$

(First, do this for g constant.)

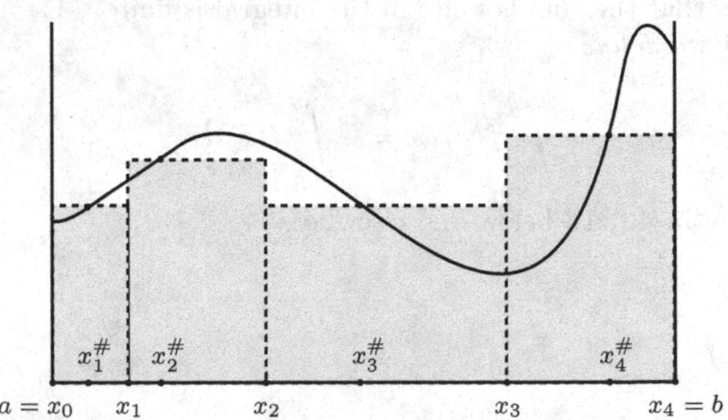

Fig. 4.25 Riemann sums

4.3.6. Let $f : (0,\infty) \to \mathbf{R}$ be nonnegative and equal to a constant c_n on each subinterval $(n-1, n)$ for $n = 1, 2, \dots$. Then

$$\int_0^\infty f(x)\,dx = \sum_{n=1}^{\infty} c_n.$$

Instead, if f is integrable, then $\sum_{n=1}^{\infty} c_n$ is absolutely convergent and the equality holds.

4.3.7. A function $f : (a,b) \to \mathbf{R}$ is *Riemann integrable over* (a,b) if there is a real I satisfying the following property: For all $\epsilon > 0$, there is a $\delta > 0$, such that (4.3.12) holds for any partition $a = x_0 < x_1 < \cdots < x_n = b$ of mesh less than δ and choice of intermediate points $x_1^\#, \dots, x_n^\#$. Thus, Exercise **4.3.3** says every function continuous on a compact interval $[a,b]$ is Riemann integrable over (a,b). Let $f(x) = 0$ for $x \in \mathbf{Q}$ and $f(x) = 1$ for $x \notin \mathbf{Q}$. Show that this f is *not* Riemann integrable over $(0,1)$.

4.3.8. Let $g : (0,\infty) \to (0,\infty)$ be decreasing and bounded. Show that

$$\lim_{\delta \to 0+} \delta \sum_{n=1}^{\infty} g(n\delta) = \int_0^\infty g(x)\, dx.$$

(Apply the integral test to $f(x) = g(x\delta)$.)

4.3.9. Let $f : (-b,b) \to \mathbf{R}$ be nonnegative or integrable. If f is even, then $\int_{-b}^{b} f(x)\, dx = 2\int_0^b f(x)\, dx$. Now let f be integrable. If f is odd, then $\int_{-b}^{b} f(x)\, dx = 0$.

4.3.10. Show that $\int_{-\infty}^{\infty} e^{-a|x|}\, dx < \infty$ for $a > 0$.

4.3.11. If $f : \mathbf{R} \to \mathbf{R}$ is superlinear (Exercise **2.3.20**) and continuous, the *Laplace transform*

$$L(s) = \int_{-\infty}^{\infty} e^{sx} e^{-f(x)}\, dx$$

is finite for all $s \in \mathbf{R}$. (Write $\int_{-\infty}^{\infty} = \int_{-\infty}^{a} + \int_a^b + \int_b^\infty$ for appropriate a, b.)

4.3.12. A function $\delta : \mathbf{R} \to \mathbf{R}$ is a *Dirac delta function* if it is nonnegative and satisfies

$$\int_{-\infty}^{\infty} \delta(x) f(x)\, dx = f(0) \tag{4.3.13}$$

for *all continuous* nonnegative $f : \mathbf{R} \to \mathbf{R}$. Show that there is no such function. (Construct continuous f's which take on the two values 0 or 1 on most or all of $\mathbf{R}$, and insert them into (4.3.13).)

4.3.13. If f is convex on (a,b) and $a < c - \delta < c < c + \delta < b$, then (Exercise **3.3.7**)

$$f(c \pm \delta) - f(c) \ge \pm f'_\pm(c)\delta \ge \int_{c-(\pm\delta)}^{c} f'_\pm(x)\, dx. \tag{4.3.14}$$

Here $\pm$ means there are two cases, either all $+$s or all $-$s. Use this to conclude that if f is convex on an open interval containing $[a,b]$, then

$$f(b) - f(a) = \int_a^b f'_+(x)\, dx = \int_a^b f'_-(x)\, dx.$$

(Break $[a,b]$ into an evenly spaced partition $a = x_0 < x_1 < \cdots < x_n = b$, $x_i - x_{i-1} = \delta$, and apply (4.3.14) at each point $c = x_i$.)

4.3.14. With

$$F(b) = \int_0^b \frac{\sin x}{x}\,dx,$$

show that $F(\infty)$ exists. Conclude that $F : (0,\infty) \to \mathbf{R}$ is bounded. This is a special case of Exercise **4.4.31**.

4.4 The Fundamental Theorems of Calculus

By constructing appropriate covers, Archimedes was able to compute areas and integrals in certain situations. For example, he knew that $\int_0^1 x^2\,dx = 1/3$. On the other hand, Archimedes was also able to compute tangent lines to certain curves and surfaces. However, he apparently had no idea that these two processes were intimately related, through the fundamental theorem of calculus. It was the discovery of the fundamental theorem, in the seventeenth century, that turned the computation of areas from a mystery to a simple and straightforward reality.

In this section, all functions will be continuous. Since we will use f^+ and f^- repeatedly, it is important to note that (§2.3) *a function is continuous iff both its positive and negative parts are continuous.*

Let f be continuous on (a,b), and let $[c,d]$ be a compact subinterval. Since (§2.3) continuous functions map compact intervals to compact intervals, f is bounded on $[c,d]$, hence integrable over (c,d).

Let f be continuous on (a,b), fix $a < c < b$, and set

$$F_c(x) = \begin{cases} \displaystyle\int_c^x f(t)\,dt, & c \le x < b, \\ \displaystyle -\int_x^c f(t)\,dt, & a < x \le c. \end{cases}$$

By the previous paragraph, $F_c(x)$ is finite for all $a < x < b$. From the previous section, we know that F_c is continuous. Here we show that F_c is differentiable and $F_c'(x) = f(x)$ on (a,b) (Figure 4.26). We will need the modulus of continuity μ_x (§2.3) of f at x. To begin, by additivity, $F_c(y) - F_c(x) = F_x(y) - F_x(x)$ for any two points x, y in (a,b), whether they are to the right or the left of c.

Then for $a < x < t < y < b$, $|f(t) - f(x)| \le \mu_x(y-x)$. Thus, $f(t) \le f(x) + \mu_x(y-x)$. Hence,

$$\begin{aligned} \frac{F_c(y) - F_c(x)}{y-x} &= \frac{F_x(y) - F_x(x)}{y-x} \\ &= \frac{1}{y-x}\int_x^y f(t)\,dt \\ &\le \frac{1}{y-x}\int_x^y [f(x) + \mu_x(y-x)]\,dt = f(x) + \mu_x(y-x). \end{aligned}$$

Similarly, since $a < x < t < y < b$ implies $f(x) - \mu_x(y-x) \le f(t)$,

$$\frac{F_c(y) - F_c(x)}{y - x} \ge f(x) - \mu_x(y-x).$$

Combining the last two inequalities, we obtain

$$\left| \frac{F_c(y) - F_c(x)}{y - x} - f(x) \right| \le \mu_x(y-x)$$

for $a < x < y < b$. If $a < y < x < b$, repeating the same steps yields

$$\left| \frac{F_c(y) - F_c(x)}{y - x} - f(x) \right| \le \mu_x(x-y).$$

Hence, if $a < x \neq y < b$,

$$\left| \frac{F_c(y) - F_c(x)}{y - x} - f(x) \right| \le \mu_x(|y-x|),$$

which implies, by continuity of f at x,

$$\lim_{y \to x} \frac{F_c(y) - F_c(x)}{y - x} = f(x).$$

Hence, $F_c'(x) = f(x)$. We have established the following result, first mentioned in §3.7.

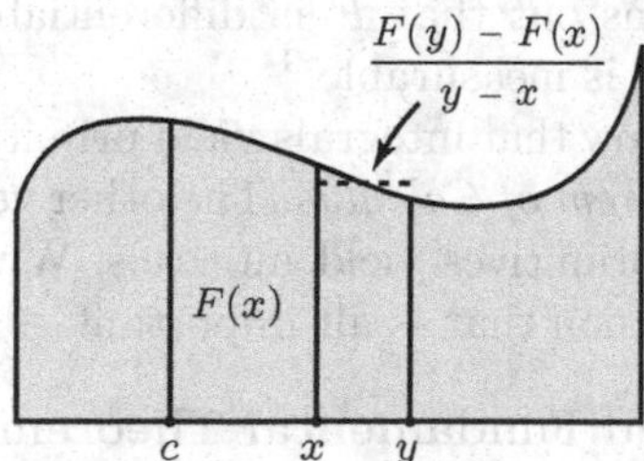

Fig. 4.26 The derivative at x of the integral of f is $f(x)$

Theorem 4.4.1. *Every continuous $f : (a,b) \to \mathbf{R}$ has a primitive on (a,b).* □

When f is continuous and integrable on (a,b), we can do better.

Theorem 4.4.2 (First Fundamental Theorem of Calculus). *Let f be continuous and integrable on (a,b). Then*

$$F(x) = \int_a^x f(t)\,dt, \quad a < x < b,$$

implies

$$F'(x) = f(x), \quad a < x < b,$$

and

$$F(x) = \int_x^b f(t)\,dt, \quad a < x < b,$$

implies

$$F'(x) = -f(x), \quad a < x < b.$$

To see this, for the first implication, write $\int_a^x f(t)\,dt = \int_a^c f(t)\,dt + F_c(x)$, and use $F_c'(x) = f(x)$. Since, by additivity, $\int_a^x f(t)\,dt + \int_x^b f(t)\,dt$ equals the constant $\int_a^b f(x)\,dx$, the second implication follows. □

For example, if

$$F(\theta) = \int_0^{\tan\theta} e^{-t^2}\,dt, \qquad 0 < \theta < \frac{\pi}{2},$$

then $F'(\theta) = e^{-\tan^2\theta}\sec^2\theta$ by the above theorem combined with the chain rule. We will need this in §5.4.

The above result begs an answer to a broader question. Let $f : (a,b) \to \mathbf{R}$ be an arbitrary—not necessarily continuous—integrable function. We know from §4.3 that F is continuous. We also know F is differentiable and $F' = f$ on (a,b) when f is continuous. *Are there any other integrable functions f for which this is so?* It turns out that F is differentiable and $F' = f$ almost everywhere on (a,b) iff f is measurable.[9]

The last two results show that integrals yield primitives. This is one version of the *Fundamental Theorem of Calculus*. The other version of the fundamental theorem states that primitives yield integrals. When one is seeking areas or integrals, it is this version that is all important.

Theorem 4.4.3 (Second Fundamental Theorem of Calculus). *Let f be nonnegative or integrable over (a,b). Suppose f is continuous and let F be any primitive of f on (a,b). Then $F(b-)$ and $F(a+)$ exist, and*

$$\int_a^b f(x)\,dx = F(b-) - F(a+).$$

To see this, first, assume that f is nonnegative. Then F is increasing ($F' = f \ge 0$). Hence, $F(b-)$ and $F(a+)$ exist for any primitive F. In particular, with F_c as above, $F_c(b-)$ and $F_c(a+)$ exist. Since $F_c - F = k$ is a constant, by continuity at the endpoints,

---

[9] This generalization is derived in §6.6.

$$\begin{aligned}\int_a^b f(x)\,dx &= \int_a^c f(x)\,dx + \int_c^b f(x)\,dx \\ &= \lim_{n\nearrow\infty}\int_{a+1/n}^c f(x)\,dx + \lim_{n\nearrow\infty}\int_c^{b-1/n} f(x)\,dx \\ &= -\lim_{n\nearrow\infty} F_c(a+1/n) + \lim_{n\nearrow\infty} F_c(b-1/n) \\ &= F_c(b-) - F_c(a+) \\ &= (F(b-)+k) - (F(a+)+k) = F(b-) - F(a+).\end{aligned}$$

For the integrable case, let $F^\pm$ denote primitives of $f^\pm$ (here, $F^\pm$ are *not* the positive and negative parts of F). Then $F^+ - F^-$ differs from any primitive F of f by a constant k. Since $F^\pm(b-)$ and $F^\pm(a+)$ exist, so do $F(b-)$ and $F(a+)$. Hence,

$$\begin{aligned}\int_a^b f(x)\,dx &= \int_a^b f^+(x)\,dx - \int_a^b f^-(x)\,dx \\ &= F^+(b-) - F^+(a+) - F^-(b-) + F^-(a+) \\ &= F(b-) + k - F(a+) - k = F(b-) - F(a+).\square\end{aligned}$$

Note that, in the second fundamental theorem, as stated above, a or b or $F(a+)$ or $F(b-)$ may be infinite. This result is generalized to noncontinuous functions in §6.3.

When a, b, $F(b-)$, and $F(a+)$ are all finite, the second fundamental theorem simplifies slightly. Indeed, in this case, by defining $F(b) = F(b-)$, $F(a) = F(a+)$, the primitive F extends to a continuous function on the compact interval $[a,b]$ and the fundamental theorem becomes

$$\int_a^b f(x)\,dx = F(b) - F(a).$$

In particular, this simpler form of the fundamental theorem applies when f and (a,b) are both bounded. All primitives displayed below were obtained in §3.7.

For example, $\sin x$ is bounded and has the primitive $-\cos x$ on $(0,\pi)$. So

$$\int_0^\pi \sin x\,dx = (-\cos\pi) - (-\cos 0) = 2.$$

Similarly, x^n, $n \ge 0$, is bounded and has the primitive $x^{n+1}/(n+1)$ over any bounded interval (a,b), so

$$\int_a^b x^n\,dx = \frac{b^{n+1}}{n+1} - \frac{a^{n+1}}{n+1}, \qquad n \ge 0. \tag{4.4.1}$$

In particular, x^{2n} is nonnegative, hence (4.4.1) also holds when a or b are infinite. For example,

$$\int_{-\infty}^{\infty} x^{2n}\,dx = \frac{\infty^{2n+1}}{2n+1} - \frac{(-\infty)^{2n+1}}{2n+1} = \infty$$

is perfectly valid.

Below, it is convenient to denote $F(b) - F(a) = F(x)|_a^b$. Since, in §3.7, a primitive of f was written $\int f(x)\,dx$, the fundamental theorem becomes

$$\int_a^b f(x)\,dx = \left.\int f(x)\,dx\right|_a^b.$$

This explains the notation $\int f(x)\,dx$ for primitives. (The notation $\int_a^b f(x)\,dx$ for integrals was explained in §4.3.)

Also $f(x) = 1/\sqrt{1-x^2} > 0$ has the primitive $F(x) = \arcsin x$ continuous over $[-1,1]$, so

$$\int_{-1}^{1} \frac{dx}{\sqrt{1-x^2}} = \arcsin x|_{-1}^{1} = \arcsin 1 - \arcsin(-1) = \pi.$$

Similarly, since $f(x) = 1/(1+x^2)$ is nonnegative and has the primitive $F(x) = \arctan x$ over $\mathbf{R}$,

$$\int_{-\infty}^{\infty} \frac{dx}{1+x^2} = \arctan x|_{-\infty}^{\infty} = \frac{\pi}{2} - \left(-\frac{\pi}{2}\right) = \pi.$$

The unit disk

$$D = \{(x,y) : x^2 + y^2 < 1\}$$

is the disjoint union of a horizontal line segment and the two half-disks

$$D^{\pm} = \{(x,y) : x^2 + y^2 < 1, \pm y > 0\}.$$

Then area$\,(D) = 2 \cdot$ area$\,(D^+)$ (Exercise **4.2.11**). But D^+ is the subgraph of $f(x) = \sqrt{1-x^2}$ over $(-1,1)$, which has a primitive continuous on $[-1,1]$. Hence,

$$\int_{-1}^{1} \sqrt{1-x^2}\,dx = \frac{1}{2}\left(\arcsin x + x\sqrt{1-x^2}\right)\Big|_{-1}^{1} = \frac{\pi}{2}.$$

This yields the following.

Theorem 4.4.4. *The area of the unit disk is π.* □

Of course, by translation and dilation invariance, the area of any disk of radius $r > 0$ is πr^2. Another integral is

$$\int_0^1 (-\log x)\,dx = (x - x\log x)|_0^1 = 1 + \lim_{x\to 0+} x\log x = 1 + 0 = 1.$$

Our next item is the linearity of the integral.

Theorem 4.4.5 (Linearity). *Suppose that f, g are continuous on (a,b). If f and g are both nonnegative or both integrable over (a,b), then*

$$\int_a^b [f(x)+g(x)]\,dx = \int_a^b f(x)\,dx + \int_a^b g(x)\,dx.$$

To see this, let F and G be primitives corresponding to f and g. Then $f+g = F'+G' = (F+G)'$. So $F+G$ is a primitive of $f+g$. By the fundamental theorem,

$$\begin{aligned}\int_a^b [f(x)+g(x)]dx &= F(b-)+G(b-)-F(a+)-G(a+)\\ &= \int_a^b f(x)\,dx + \int_a^b g(x)\,dx.\square\end{aligned}$$

More generally, linearity is valid when f and g are both nonnegative and f or g is continuous (Exercise **4.4.24**). To what extend is linearity valid when f or g are signed? Since integrals are defined in terms of areas, the scope of the validity of the linearity of integrals is intimately connected with the scope of the validity of additivity of areas

$$\text{area}\,(A\cup B) = \text{area}\,(A) + \text{area}\,(B)$$

for disjoint sets A and B. This leads to measurable sets, discussed in §4.5, and measurable functions, discussed in §6.1.

We say $f:(a,b)\to\mathbf{R}$ is *piecewise continuous* if there is a partition $a=x_0<x_1<\cdots<x_n=b$, such that f is continuous on each subinterval (x_{i-1},x_i), $i=1,\dots,n$. Now by additivity, the integral $\int_a^b$ can be broken up into $\int_{x_{i-1}}^{x_i}$, $i=1,\dots,n$. We conclude that *additivity also holds for piecewise continuous functions.*

By induction, additivity holds for finitely many (piecewise) continuous functions. *If $f_1,\dots,f_n$ are (piecewise) continuous and all nonnegative or all integrable over (a,b), then*

$$\int_a^b \sum_{k=1}^n f_k(x)\,dx = \sum_{k=1}^n \int_a^b f_k(x)\,dx.$$

Since primitives are connected to integrals by the fundamental theorem, there is an integration by parts (§3.7) result for integrals.

Theorem 4.4.6 (Integration by Parts). *Let f and g be differentiable on (a,b) with $f'g$ and fg' continuous. If $f'g$ and fg' are both nonnegative or both integrable, then*

$$\int_a^b f(x)g'(x)\,dx = f(x)g(x)\Big|_{a+}^{b-} - \int_a^b f'(x)g(x)\,dx.$$

This follows by applying the fundamental theorem to $f'g + fg' = (fg)'$ and using linearity. $\square$

Since primitives are connected to integrals by the fundamental theorem, there is a substitution (§3.7) result for integrals. Recall (§2.3) that continuous strictly monotone functions map open intervals to open intervals.

Theorem 4.4.7 (Substitution). *Let g be differentiable and strictly monotone on an interval (a,b) with g' continuous, and let $(m,M) = g[(a,b)]$. Let $f : (m,M) \to \mathbf{R}$ be continuous. If f is nonnegative or integrable over (m,M), then $f[g(t)]|g'(t)|$ is nonnegative or integrable over (a,b), and*

$$\int_m^M f(x)\,dx = \int_a^b f[g(t)]|g'(t)|\,dt. \tag{4.4.2}$$

To see this, first, assume that g is strictly increasing and f is nonnegative; let F be a primitive of f, let $H(t) = F[g(t)]$, and let $h(t) = f[g(t)]g'(t)$. Then $(m,M) = (g(a+), g(b-))$ and $H'(t) = F'[g(t)]g'(t) = f[g(t)]g'(t) = h(t)$ by the chain rule. Hence, H is a primitive for h. Moreover, h is continuous and nonnegative, $F(M-) = H(b-)$, and $F(m+) = H(a+)$. By the fundamental theorem,

$$\begin{aligned}\int_m^M f(x)\,dx &= F(M-) - F(m+) \\ &= H(b-) - H(a+) \\ &= \int_a^b h(t)\,dt \\ &= \int_a^b f[g(t)]g'(t)\,dt.\end{aligned}$$

Since $|g'(t)| = g'(t)$, this establishes the case with g strictly increasing and f nonnegative. If f is integrable, apply the nonnegative case to $f^{\pm}$. Since the positive and negative parts of $f[g(t)]g'(t)$ are $f^{\pm}[g(t)]g'(t)$, the integrable case follows.

If g is strictly decreasing, then $(m,M) = (g(b-), g(a+))$. Now $h(t) = g(-t)$ is strictly increasing, $h((-b,-a)) = (m,M)$, and $h'(-t) = -g'(t) = |g'(t)|$ is nonnegative on (a,b). Applying what we just learned to f and h over $(-b,-a)$ yields

$$\begin{aligned}\int_m^M f(x)\,dx &= \int_{-b}^{-a} f[h(t)]h'(t)\,dt \\ &= \int_a^b f[h(-t)]h'(-t)\,dt \\ &= \int_a^b f[g(t)]|g'(t)|\,dt. \square\end{aligned}$$

If g is not monotone, then (4.4.2) has to be reformulated (Exercise **4.4.25**). To see what happens, let us consider a simple example with $f(x) = 1$. Let $g : (a, b) \to (m, M)$ be *piecewise linear* with line segments inclined at $\pm\pi/4$. By this, we mean g is continuous on (a, b) and the graph of g is a line segment with slope ± 1 on each subinterval (t_{i-1}, t_i), $i = 1, \dots, n$, for some partition $a = t_0 < t_1 < \cdots < t_n = b$ of (a, b) (Figure 4.27). Then $|g'(t)| = 1$ for all but finitely many t, so $\int_a^b |g'(t)|\, dt = b - a$. On the other hand, substituting $f(x) = 1$ in (4.4.2) gives $\int_a^b |g'(t)|\, dt = M - m$. Thus, in such a situation, (4.4.2) cannot be correct unless the domain and the range have the same length, i.e., $M - m = b - a$.

To fix this, we have to take into account the extent to which g is not a bijection. To this end, for each x in (m, M), let $\#(x)$ denote the number of points in the inverse image $g^{-1}(\{x\})$. Since (m, M) is the range of g, $\#(x) \geq 1$ for all $m < x < M$. The correct replacement[10] for (4.4.2) with $f(x) = 1$ is

$$\int_m^M \#(x)\, dx = \int_a^b |g'(t)|\, dt. \tag{4.4.3}$$

This holds as long as g is continuous on (a, b) and there is a partition $a = t_0 < t_1 < \cdots < t_n = b$ of (a, b) with g differentiable, g' continuous, and g strictly monotone on each subinterval (t_{i-1}, t_i), for each $i = 1, \dots, n$ (Exercise **4.4.25**). For example, supposing $g : (a, b) \to (m, M)$ piecewise linear with slopes ± 1 reduces (4.4.3) to $\int_m^M \#(x)\, dx = b - a$. Dividing by $M - m$ yields

$$\frac{1}{M - m} \int_m^M \#(x)\, dx = \frac{b - a}{M - m}. \tag{4.4.4}$$

Now the left side of (4.4.4) may be thought of as the average value of $\#(x)$ over (m, M). We conclude that, for a piecewise linear g with slopes ± 1, the average value of the number of inverse images equals the ratio of the lengths of the domain over the range.

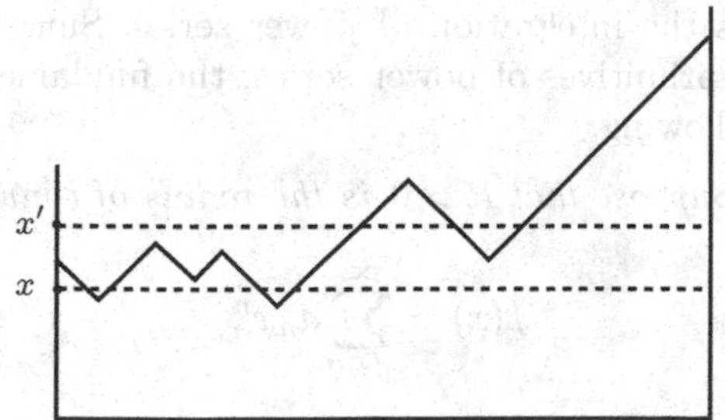

Fig. 4.27 Piecewise linear: $\#(x) = 4$, $\#(x') = 3$

[10] This is generalized to any *continuous* g in Theorem 6.6.4.

Now we derive the integral version of

Theorem 4.4.8 (Taylor's Theorem). *Let $n \geq 0$ and suppose that f is $(n+1)$ times differentiable on (a,b), with $f^{(n+1)}$ continuous on (a,b). Suppose that $f^{(n+1)}$ is nonnegative or integrable over (a,b), and fix $a < c < x < b$. Then*

$$f(x) = f(c) + f'(c)(x-c) + \frac{f''(c)}{2!}(x-c)^2 + \dots$$
$$\dots + \frac{f^{(n)}(c)}{n!}(x-c)^n + \frac{h_{n+1}(x)}{(n+1)!}(x-c)^{n+1},$$

where

$$h_{n+1}(x) = (n+1)\int_0^1 (1-s)^n f^{(n+1)}[c + s(x-c)]\, ds.$$

To see this, recall, in §3.5, that we obtained $R_{n+1}(x,x) = 0$ and (here, $'$ denotes derivative with respect to t)

$$R'_{n+1}(x,t) = -\frac{f^{(n+1)}(t)}{n!}(x-t)^n.$$

Now apply the fundamental theorem to $-R'_{n+1}(x,t)$ and substitute $t = c + s(x-c)$, $dt = (x-c)ds$, obtaining

$$\begin{aligned} R_{n+1}(x,c) &= \frac{1}{n!}\int_c^x f^{(n+1)}(t)(x-t)^n\, dt \\ &= \frac{(x-c)^{n+1}}{n!}\int_0^1 f^{(n+1)}(c + s(x-c))(1-s)^n\, ds \\ &= \frac{(x-c)^{n+1}}{(n+1)!} h_{n+1}(x). \square \end{aligned}$$

In contrast with the Lagrange and Cauchy forms (§3.5) of the remainder, here, we need continuity and nonnegativity or integrability of $f^{(n+1)}$.

Our last item is the integration of power series. Since we already know (§3.7) how to find primitives of power series, the fundamental theorem and (4.4.1) yield the following.

Theorem 4.4.9. *Suppose that $R > 0$ is the radius of convergence of*

$$f(x) = \sum_{n=0}^{\infty} a_n x^n.$$

If $[a,b] \subset (-R,R)$, then

$$\int_a^b f(x)\, dx = \sum_{n=0}^{\infty} \int_a^b a_n x^n\, dx. \quad \square \tag{4.4.5}$$

For example, substituting $-x^2$ for x in the exponential series,

$$e^{-x^2} = 1 - x^2 + \frac{x^4}{2!} - \frac{x^6}{3!} + \frac{x^8}{4!} - \dots .$$

Integrating this over $(0,1)$, we obtain

$$\int_0^1 e^{-x^2}\,dx = 1 - \frac{1}{1!3} + \frac{1}{2!5} - \frac{1}{3!7} + \frac{1}{4!9} - \dots .$$

This last result is, in general, false if $a = -R$ or $b = R$. For example, with $f(x) = e^{-x} = \sum_{n=0}^\infty (-1)^n x^n/n!$ and $(a,b) = (0,\infty)$, (4.4.5) reads $1 = \infty - \infty + \infty - \infty + \dots$. Under additional assumptions, however, (4.4.5) is true, even in these cases (see §5.2).

Exercises

4.4.1. Compute $\int_0^\infty e^{-sx}\,dx$ for $s > 0$.

4.4.2. Compute $\int_0^1 x^{r-1}\,dx$ and $\int_1^\infty x^{r-1}\,dx$ and $\int_0^\infty x^{r-1}\,dx$ for r real. (There are three cases, $r < 0$, $r = 0$, and $r > 0$.)

4.4.3. Suppose that f is continuous over (a,b), and let F be any primitive. If f and (a,b) are both bounded, then f is integrable, and $F(a+)$ and $F(b-)$ are finite.

4.4.4. Let f be continuous on $[1,\infty)$ and differentiable on $(1,\infty)$ with f' continuous and nonnegative over $(1,\infty)$. Show $f(\infty)$ exists iff f' is integrable over $(1,\infty)$.

4.4.5. Let f be continuous on $[1,\infty)$ and differentiable on $(1,\infty)$ with f' continuous, decreasing, and nonnegative over $(1,\infty)$. Show that

$$\sum_{n=1}^\infty f'(n) < \infty$$

iff

$$\sum_{n=1}^\infty \frac{f'(n)}{f(n)} < \infty.$$

(Use the integral test.)

4.4.6. Let $f(x) = \sin x/x$, $x > 0$, and let $F(b) = \int_0^b f(x)\,dx$, $b > 0$. Show that $F(\infty) = \lim_{b\to\infty} F(b)$ exists and is finite. (Write $F(b) = \int_0^1 f(x)\,dx + \int_1^b f(x)\,dx$, integrate the second integral by parts, and use (4.3.1). This limit is computed in Exercise **5.4.12**.)

4.4.7. For f continuous and nonnegative or integrable over $(0,1)$,

$$\int_0^1 f(x)\,dx = \int_0^\infty e^{-t} f(e^{-t})\,dt.$$

4.4.8. Compute $\int_0^\infty e^{-sx}x^n\,dx$ for $s > 0$ and $n \ge 0$. (Integration by parts.)

4.4.9. Compute $\int_0^\infty e^{-nx}\sin(sx)\,dx$ and $\int_0^\infty e^{-nx}\cos(sx)\,dx$ for $n \ge 1$. (Integration by parts.)

4.4.10. Show that $\int_0^\infty e^{-t^2/2}t^x\,dt = (x-1)\int_0^\infty e^{-t^2/2}t^{x-2}\,dt$ for $x > 1$. Use this to derive

$$\int_0^\infty e^{-t^2/2}t^{2n+1}\,dt = 2^n n!, \qquad n \ge 0.$$

(Integration by parts.)

4.4.11. Compute $\int_0^1 (1-t)^n t^{x-1}\,dt$ for $x > 0$ and $n \ge 1$. (Integration by parts.)

4.4.12. Compute $\int_0^1 (-\log x)^n\,dx$.

4.4.13. Show that

$$\int_{-1}^1 (x^2-1)^n\,dx = (-1)^n \frac{2n\cdot(2n-2)\cdot\dots\cdot 2}{(2n+1)\cdot(2n-1)\cdot\dots\cdot 3}\cdot 2.$$

(Integrate by parts.)

4.4.14. For $n \ge 0$, the *Legendre polynomial* P_n (of degree n) is given by $P_n(x) = f^{(n)}(x)/2^n n!$, where $f(x) = (x^2-1)^n$. Show that

$$\int_{-1}^1 P_n(x)^2\,dx = \frac{2}{2n+1}.$$

4.4.15. If f is a polynomial, show

$$\int_0^\pi f(x)\sin x\,dx = F(0) + F(\pi)$$

where $F(x) = f(x) - f''(x) + f^{(4)}(x) - \dots$ (Exercise **3.7.11**).

4.4.16. If $\pi = a/b$ is rational, let

$$I_n = \int_0^\pi g_n(x)\sin x\,dx, \qquad n \ge 1,$$

where g_n is defined in Exercise **3.3.30**. Show that:

- $I_n > 0$
- I_n is an integer
- $I_n \to 0$ as $n \to \infty$.

Conclude that π is irrational.

4.4.17. Use the integral test (§4.3) to show that

$$\zeta(s) = \sum_{n=1}^{\infty} \frac{1}{n^s}, \qquad s > 1,$$

converges.

4.4.18. Use the integral test (§4.3) to show that

$$\gamma = \lim_{n \nearrow \infty} \left(1 + \frac{1}{2} + \frac{1}{3} + \cdots + \frac{1}{n} - \log n \right)$$

exists and $0 < \gamma < 1$. This particular real γ is *Euler's constant.*

4.4.19. Compute $\int_{-\pi}^{\pi} x \cos(nx)\, dx$ and $\int_{-\pi}^{\pi} x \sin(nx)\, dx$ for $n \geq 0$. (Integration by parts.)

4.4.20. Compute $\int_{-\pi}^{\pi} f(nx) g(mx)\, dx$, $n, m \geq 0$, with $f(x)$ and $g(x)$ equal to $\sin x$ or $\cos x$ (three possibilities—use (3.6.3)).

4.4.21. If $f, g : (a, b) \to \mathbf{R}$ are nonnegative and continuous, derive the *Cauchy–Schwarz inequality*

$$\left[\int_a^b f(x) g(x)\, dx \right]^2 \leq \left[\int_a^b f(x)^2\, dx \right] \cdot \left[\int_a^b g(x)^2\, dx \right].$$

(Use the fact that $q(t) = \int_a^b [f(x) + t g(x))]^2\, dx$ is a nonnegative quadratic polynomial and Exercise **1.4.5**.)

4.4.22. For $n \geq 1$, show that

$$\int_0^n \frac{1 - (1 - t/n)^n}{t}\, dt = 1 + \frac{1}{2} + \cdots + \frac{1}{n}.$$

4.4.23. We say $f : (a, b) \to \mathbf{R}$ is *piecewise differentiable* if there is a partition $a = x_0 < x_1 < \cdots < x_n = b$, such that f restricted to (x_{i-1}, x_i) is differentiable for $i = 1, \ldots, n$. Let $f : (a, b) \to \mathbf{R}$ be piecewise continuous and integrable. Show that $F(x) = \int_a^x f(t)\, dt$, $a < x < b$, is continuous on (a, b) and piecewise differentiable on (a, b).

4.4.24. If $f:(a,b)\to\mathbf{R}$ is nonnegative and $g:[a,b]\to\mathbf{R}$ is nonnegative and continuous, then

$$\int_a^b [f(x)+g(x)]\,dx = \int_a^b f(x)\,dx + \int_a^b g(x)\,dx.$$

(Use Exercise **4.3.5** and approximate g by a piecewise constant g_ϵ as in §2.3. Since f is arbitrary, linearity may not be used directly.)

4.4.25. Suppose that $g:(a,b)\to(m,M)$ is continuous, and suppose that there is a partition $a=t_0<t_1<\cdots<t_n=b$ of (a,b), such that g is differentiable, g' is continuous, and g is strictly monotone on each subinterval (t_{i-1},t_i), for each $i=1,\dots,n$. For each x in (m,M), let $\#(x)$ denote the number of points in the inverse image $g^{-1}(\{x\})$. Also let $f:(m,M)\to\mathbf{R}$ be continuous and nonnegative. Then[11]

$$\int_m^M f(x)\#(x)\,dx = \int_a^b f[g(t)]|g'(t)|\,dt. \tag{4.4.6}$$

(Use additivity on the integral $\int_a^b$.)

4.4.26. Let f be differentiable with f' continuous on (a,b). Show (Exercise **2.2.4**) $v_f(a,b)$ is bounded by $\int_a^b |f'(x)|dx$. Use Exercise **4.3.3** to show

$$v_f(a,b)=\int_a^b |f'(x)|\,dx.$$

(Restrict to a compact subinterval $[c,d]\subset(a,b)$ and rewrite the variation of f over a given partition as a Riemann sum for $|f'|$.)

4.4.27. Let f be a differentiable function on (a,b). The *length of the graph of f over* (a,b) is defined by the formula

$$\int_a^b \sqrt{1+f'(x)^2}\,dx. \tag{4.4.7}$$

Apply this formula to show that the length p of the upper-half unit circle $y=\sqrt{1-x^2}$, $-1<x<1$, equals

$$\int_{-1}^1 \frac{dx}{\sqrt{1-x^2}}=\pi.$$

Show directly, without using any trigonometry, that the integral is finite and in fact between 0 and 4. (Use $x^2<x$ on $(0,1)$).

4.4.28. Let $(x,y)=(\cos\theta,\sin\theta)$ be on the unit circle and assume $0\le\theta<2\pi$. If $y\ge 0$, the length L of the counterclockwise circular arc joining $(1,0)$ to

[11] (4.4.6) is actually valid under general conditions; see §6.6.

(x, y) is defined by (4.4.7). If $y < 0$, given that the length of the upper-half unit circle is π, it is natural to define the length L of the counterclockwise circular arc joining $(1, 0)$ to (x, y) as $\pi + L'$, where L' is the length of the counterclockwise circular arc joining $(-1, 0)$ to (x, y). Show that in either case $L = \theta$ (Figure 3.13). In particular, this shows the length of the unit circle equals 2π (Figure 4.28).

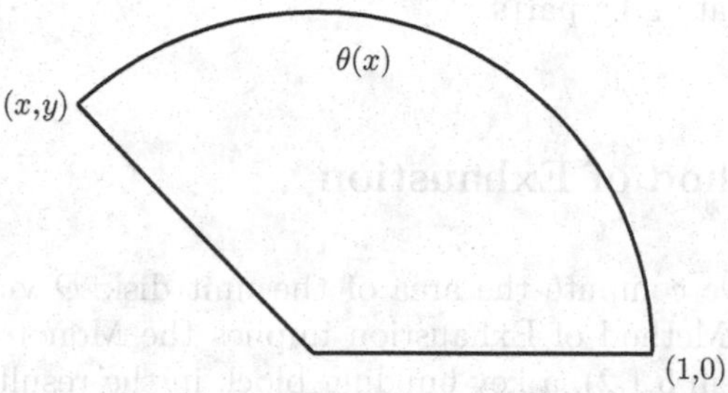

Fig. 4.28 Geometric definition of the inverse cosine function

4.4.29. This problem shows how one can define π, sin, cos directly and geometrically from the unit circle. Given a point (x, y), $y \geq 0$, on the upper-half unit circle $x^2 + y^2 = 1$, we define the *angle* corresponding to (x, y) to be the length of the circular arc lying above the interval $(x, 1)$,

$$\theta(x) = \int_x^1 \frac{dt}{\sqrt{1 - t^2}}, \quad -1 \leq x \leq 1.$$

By Exercise **4.4.28**, this agrees with the definition in §3.6. Note that $\theta(x) \leq \theta(-1) = p$ is less than 4 (Exercise **4.4.27**).

A. Show that $\theta'(x) = -1/\sqrt{1 - x^2}$, $-1 < x < 1$.
B. Show that $\theta : [-1, 1] \to [0, p]$ is continuous on $[-1, 1]$ and strictly decreasing, hence by the IFT, has an inverse.
C. Let $c : [0, p] \to [-1, 1]$ be the inverse of θ. Show that $c' = -\sqrt{1 - c^2}$ on $(0, p)$.
D. Let $s = \sqrt{1 - c^2}$. Show that $s' = c$ and $c' = -s$ on $(0, p)$.

Of course, p equals π and the functions c, s are the functions cos, sin, and this approach may be used as the basis for §3.6.

4.4.30. Let $f : (-R, R) \to \mathbf{R}$ be smooth and suppose there is an integrable $g : (-R, R) \to \mathbf{R}$ such that for $|c| < R$ and $d < R - |c|$,

$$|f^{(n)}(x)| \cdot \frac{d^n}{n!} \leq g(x) \qquad n \geq 1, |x - c| < d.$$

Show that the Taylor series of f centered at c converges to f on $|x - c| < d$. Compare with Exercise **3.5.8**.

4.4.31. Let $f : (0,\infty) \to \mathbf{R}$ be differentiable with $f' : (0,\infty) \to \mathbf{R}$ continuous. Assume f is decreasing to zero as $x \to \infty$. Let $g : (0,\infty) \to \mathbf{R}$ be continuous with bounded primitive G on $(0,\infty)$. If both fg and $f'G$ are integrable over $(0,b)$ for all $b > 0$, then

$$\lim_{b\to\infty} \int_0^b f(x)g(x)\,dx$$

exists. Use integration by parts.

4.5 The Method of Exhaustion

In this section, we compute the area of the unit disk D via the method of exhaustion. The Method of Exhaustion implies the Monotone Convergence Theorem (Theorem 5.1.2), a key building block in the results of Chapter 5.

For $n \geq 3$, let $P_k = (\cos(2\pi k/n), \sin(2\pi k/n))$, $0 \leq k \leq n$. Then the points P_k are evenly spaced about the unit circle $\{(x,y) : x^2+y^2 = 1\}$, and $P_n = P_0$. Let $D_n \subset D$ be the interior of the inscribed regular n-sided polygon obtained by joining the points $P_0, P_1, \dots, P_n$ (we do not include the edges of D_n in the definition of D_n). Then (Exercise **4.2.13**),

$$\operatorname{area}(D_n) = \frac{n}{2}\sin(2\pi/n) = \pi \cdot \frac{\sin(2\pi/n)}{2\pi/n}.$$

Since $\lim_{x\to 0} \dfrac{\sin x}{x} = \sin' 0 = \cos 0 = 1$, we obtain

$$\lim_{n\nearrow\infty} \operatorname{area}(D_n) = \pi. \tag{4.5.1}$$

Since

$$D_4 \subset D_8 \subset D_{16} \subset \dots, \qquad \text{and} \qquad D = \bigcup_{n=2}^{\infty} D_{2^n}, \tag{4.5.2}$$

it is reasonable to make the guess that

$$\operatorname{area}(D) = \lim_{n\nearrow\infty} \operatorname{area}(D_{2^n}), \tag{4.5.3}$$

and hence conclude that $\operatorname{area}(D) = \pi$. The reasoning that leads from (4.5.2) to (4.5.3) is generally correct. The result is called the *method of exhaustion.*

Although $\operatorname{area}(D)$ was computed in the previous section using the fundamental theorem, in Chapter 5 we will need the method to compute other areas.

We say that a sequence of sets (A_n) is *increasing* (Figure 4.29) if $A_1 \subset A_2 \subset A_3 \subset \dots$.

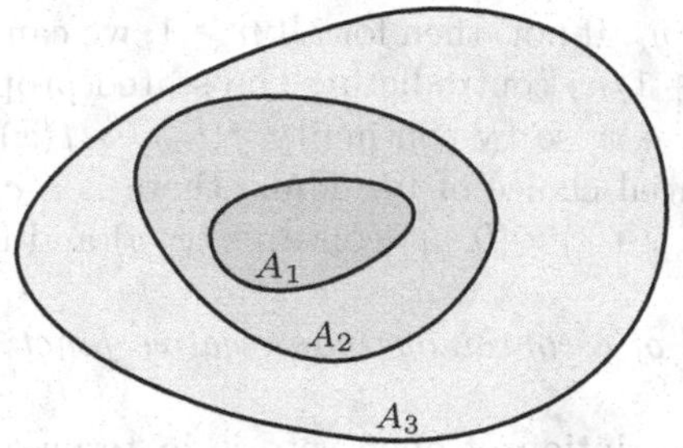

Fig. 4.29 An increasing sequence of sets

Theorem 4.5.1 (Method of Exhaustion). *If $A_1 \subset A_2 \subset \dots$ is an increasing sequence of subsets of* $\mathbf{R}^2$, *then*

$$\text{area}\left(\bigcup_{n=1}^{\infty} A_n\right) = \lim_{n\nearrow\infty} \text{area}\,(A_n).$$

We warn the reader that the result is false, in general, for decreasing sequences. For example, take $A_n = (n,\infty) \times (-\infty,\infty)$, $n \geq 1$. Then $\text{area}\,(A_n) = \infty$ for all $n \geq 1$, but $\bigcap_{n=1}^{\infty} A_n = \emptyset$ so $\text{area}\,(\bigcap_{n=1}^{\infty} A_n) = 0$. This lack of symmetry between increasing and decreasing sequences is a reflection of the lack of symmetry in the definition of area: area (A) is defined as an inf of overestimates, not as a sup of underestimates.

The derivation of the method is not as compelling as the applications in Chapter 5. Moreover, the techniques used in this derivation are not used elsewhere in the text. Because of this, the reader may wish to skip the derivation on first reading and come back to it after progressing further.

The method is established in three stages: first, when (definitions below) the sets A_n, $n \geq 1$, are open; then when the sets A_n, $n \geq 1$, are interopen; and, finally, for arbitrary sets A_n, $n \geq 1$. *Open* and *interopen* are structural properties of sets that we describe below.

We call a set $G \subset \mathbf{R}^2$ *open* if every point $(a,b) \in G$ can be surrounded by a nonempty, open rectangle wholly contained in G. For example, an open rectangle is an open set, but a compact rectangle Q is not, since no point on the edges of Q can be surrounded by a rectangle wholly contained in Q. The n-sided polygon D_n, considered above, is an open set as is the unit disk D. Since there are no points in $\emptyset$ for which the open criterion fails, $\emptyset$ is open.

For our purposes, the most important example of an open set is given by the following.

Theorem 4.5.2. *If $f \geq 0$ is continuous on (a,b), its subgraph is an open subset of* $\mathbf{R}^2$.

To see this, pick x and y with $a < x < b$ and $0 < y < f(x)$. We have to find a rectangle Q containing (x,y) and contained in the subgraph. Pick $y < y_1 < f(x)$. We claim there is a $c > 0$, such that $|t - x| < c$ implies

$a < t < b$ and $f(t) > y_1$. If not, then for all $n \geq 1$, we can find a real t_n in the interval $(x - 1/n, x + 1/n)$ contradicting the stated property, i.e., satisfying $f(t_n) \leq y_1$. Then $t_n \to x$, so by continuity $f(t_n) \to f(x)$. Hence, $f(x) \leq y_1$, contradicting our initial choice of y_1. Thus, there is a $c > 0$, such that the rectangle $Q = (x - c, x + c) \times (0, y_1)$ contains (x, y) and lies in the subgraph. □

Thus, *the integral of a continuous nonnegative function is the area of an open set.*

An alternative description of open sets is in terms of distance. If (a, b) is a point and A is a set, then the distance $d((a, b), A)$ between the point (a, b) and A, by definition, is the distance between the set $\{(a, b)\}$ and the set A (§4.2). For example, if Q is an open rectangle and $(a, b) \in Q$, then $d((a, b), Q^c)$ is positive. Here and below, $A^c = \mathbf{R}^2 \setminus A$.

Theorem 4.5.3. *A set G is open iff $d((a, b), G^c) > 0$ for all points $(a, b) \in G$.*

Indeed, if $(a, b) \in G$ and $Q \subset G$ contains (a, b), then $Q^c \supset G^c$, so $d((a, b), G^c) \geq d((a, b), Q^c) > 0$. Conversely, if $d = d((a, b), G^c) > 0$, then the disk R of radius $d/2$ and center (a, b) lies wholly in G. Now choose any rectangle Q in R containing (a, b). □

If G, G' are open subsets, so are $G \cup G'$ and $G \cap G'$. In fact, if (G_n) is a sequence of open sets, then $G = \bigcup_{n=1}^\infty G_n$ is open. To see this, if $(a, b) \in G$, then $(a, b) \in G_n$ for some specific n. Since the specific G_n is open, there is a rectangle Q with $(a, b) \in Q \subset G_n \subset G$. Hence, G is open. Thus, an infinite union of open sets is open. If $G_1, \dots, G_n$ are finitely many open sets, then $G = G_1 \cap G_2 \cap \dots \cap G_n$ is open. To see this, if $(a, b) \in G$, then $(a, b) \in G_k$ for all $1 \leq k \leq n$, so there are open rectangles Q_k with $(a, b) \in Q_k \subset G_k$, $1 \leq k \leq n$. Hence, $Q = \bigcap_{k=1}^n Q_k$ is an open rectangle containing (a, b) and contained in G (a finite intersection of open rectangles is an open rectangle). Thus, a finite intersection of open sets is open. However, an infinite intersection of open sets need not be open.

If $A \subset \mathbf{R}^2$ is any set and $\epsilon > 0$, by definition of area, we can find an open set G containing A and satisfying $\text{area}\,(G) \leq \text{area}\,(A) + \epsilon$ (Exercise **4.5.6**). If we had additivity and $\text{area}\,(A) < \infty$, writing $\text{area}\,(G) = \text{area}\,(A) + \text{area}\,(G \setminus A)$, we would conclude that $\text{area}\,(G \setminus A) \leq \epsilon$. Conversely, if we are seeking properties of sets that guarantee additivity, we may, instead, focus on those sets M in $\mathbf{R}^2$ satisfying the above approximability condition: *For all $\epsilon > 0$, there is an open superset G of M, such that* $\text{area}\,(G \setminus M) \leq \epsilon$. Instead of doing this, however, it will be quicker for us to start with an alternate equivalent (Exercise **4.5.15**) formulation.

We say a set $M \subset \mathbf{R}^2$ is *measurable* if

$$\text{area}\,(A) = \text{area}\,(A \cap M) + \text{area}\,(A \cap M^c)\,, \qquad \text{for all } A \subset \mathbf{R}^2. \qquad (4.5.4)$$

For example, the empty set is measurable, and M is measurable iff M^c is measurable. Below, we show that every open set is measurable. Measurability may be looked upon as a strengthened form of additivity, since the equality in (4.5.4) is required to hold for *every* $A \subset \mathbf{R}^2$. Note that the trick, below, of summing alternate areas $C_1, C_3, C_5, \dots$ was already used in derivating additivity in Theorem 4.3.3. Compare the next derivation with that derivation!

In §4.2, we established additivity when the sets were well separated. Now we establish a similar result involving open sets.

Theorem 4.5.4. *If G is open, then G is measurable.*

To see this, we need show only that

$$\text{area}(A) \geq \text{area}(A \cap G) + \text{area}(A \cap G^c) \tag{4.5.5}$$

for every $A \subset \mathbf{R}^2$, since the reverse inequality follows by subadditivity. Let $A \subset \mathbf{R}^2$ be arbitrary. If area $(A) = \infty$, (4.5.5) is immediate, so let us assume that area $(A) < \infty$. Let G_n be the set of points in G whose distance from G^c is at least $1/n$. Since $A \cap G_n$ and $A \cap G^c$ are well separated (Figure 4.30),

$$\text{area}(A) \geq \text{area}(A \cap G_n) + \text{area}(A \cap G^c).$$

By subadditivity,

$$\text{area}(A \cap G) \leq \text{area}(A \cap G_n) + \text{area}(A \cap G \cap G_n^c).$$

Combining the last two inequalities, we obtain

$$\text{area}(A) \geq \text{area}(A \cap G) + \text{area}(A \cap G^c) - \text{area}(A \cap G \cap G_n^c). \tag{4.5.6}$$

Thus, if we show that

$$\lim_{n \nearrow \infty} \text{area}(A \cap G \cap G_n^c) = 0, \tag{4.5.7}$$

letting $n \nearrow \infty$ in (4.5.6), we obtain (4.5.5), hence the result.

To obtain (4.5.7), let C_n be the set of points (a, b) in G satisfying $1/(n+1) \leq d((a,b), G^c) < 1/n$. Since G is open, $d((a,b), G^c) > 0$ for every point in G. Thus,

$$G \cap G_n^c = C_n \cup C_{n+1} \cup C_{n+2} \cup \dots .$$

But the sets C_n, C_{n+2}, C_{n+4}, ..., are well separated. Hence,

$$\text{area}(A \cap C_n) + \text{area}(A \cap C_{n+2}) + \text{area}(A \cap C_{n+4}) + \dots \leq \text{area}(A \cap G).$$

Since C_{n+1}, C_{n+3}, C_{n+5}, ..., are well separated,

$$\text{area}(A \cap C_{n+1}) + \text{area}(A \cap C_{n+3}) + \text{area}(A \cap C_{n+5}) + \dots \leq \text{area}(A \cap G).$$

Adding the last two inequalities, by subadditivity, we obtain

$$\text{area}(A \cap G \cap G_n^c) \leq \sum_{k=n}^{\infty} \text{area}(A \cap C_k) \leq 2\,\text{area}(A \cap G) < \infty. \tag{4.5.8}$$

Now (4.5.8) with $n = 1$ shows that the series $\sum_{k=1}^{\infty} \text{area}\,(A \cap C_k)$ converges. Thus, the tail series, starting from $k = n$ in (4.5.8), approaches zero, as $n \nearrow \infty$. This establishes (4.5.7). □

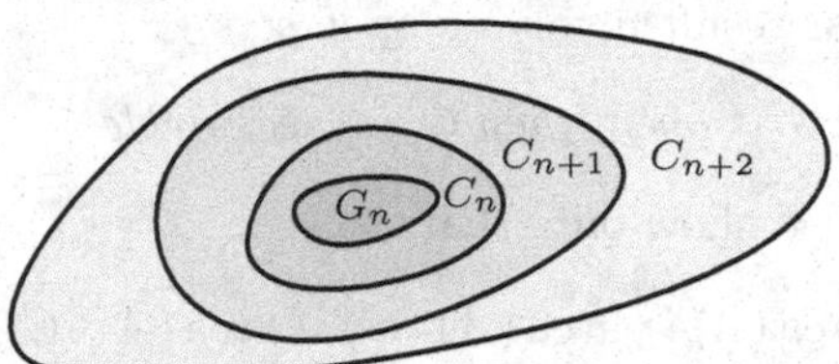

Fig. 4.30 An open set is measurable

Now we establish the method for measurable, hence for open, sets. In fact, we need to establish a strengthened form of the method for measurable sets.

Theorem 4.5.5 (Measurable Method of Exhaustion). *If $M_1 \subset M_2 \subset \dots$ is an increasing sequence of measurable subsets of $\mathbf{R}^2$ and $A \subset \mathbf{R}^2$ is arbitrary, then*

$$\text{area}\left[A \cap \left(\bigcup_{n=1}^{\infty} M_n\right)\right] = \lim_{n \nearrow \infty} \text{area}\,(A \cap M_n)\,.$$

To derive this, let $M_\infty = \bigcup_{n=1}^{\infty} M_n$. Since $A \cap M_n \subset A \cap M_\infty$, by monotonicity, the sequence $(\text{area}\,(A \cap M_n))$ is increasing and bounded above by $\text{area}\,(A \cap M_\infty)$. Thus,

$$\lim_{n \nearrow \infty} \text{area}\,(A \cap M_n) \le \text{area}\,(A \cap M_\infty)\,.$$

To obtain the reverse inequality, apply (4.5.4) with M and A, there, replaced by M_1 and $A \cap M_2$ respectively, obtaining

$$\text{area}\,(A \cap M_2) = \text{area}\,(A \cap M_2 \cap M_1) + \text{area}\,(A \cap M_2 \cap M_1^c)\,.$$

Since $A \cap M_2 \cap M_1 = A \cap M_1$, this implies that

$$\text{area}\,(A \cap M_2) = \text{area}\,(A \cap M_1) + \text{area}\,(A \cap M_2 \cap M_1^c)\,.$$

Now apply (4.5.4) with M and A, there, replaced by M_2 and $A \cap M_3$ respectively, obtaining

$$\begin{aligned}\text{area}\,(A \cap M_3) &= \text{area}\,(A \cap M_2) + \text{area}\,(A \cap M_3 \cap M_2^c) \\ &= \text{area}\,(A \cap M_1) + \text{area}\,(A \cap M_2 \cap M_1^c) \\ &\quad + \text{area}\,(A \cap M_3 \cap M_2^c)\,.\end{aligned}$$

Proceeding in this manner, we obtain

$$\operatorname{area}(A \cap M_n) = \operatorname{area}(A \cap M_1) + \sum_{k=2}^{n} \operatorname{area}\left(A \cap M_k \cap M_{k-1}^c\right).$$

Sending $n \nearrow \infty$, we obtain

$$\lim_{n \nearrow \infty} \operatorname{area}(A \cap M_n) = \operatorname{area}(A \cap M_1) + \sum_{k=2}^{\infty} \operatorname{area}\left(A \cap M_k \cap M_{k-1}^c\right).$$

Since

$$M_1 \cup (M_2 \cap M_1^c) \cup (M_3 \cap M_2^c) \cup \cdots = M_\infty,$$

subadditivity implies that

$$\operatorname{area}(A \cap M_\infty) \leq \operatorname{area}(A \cap M_1) + \sum_{k=2}^{\infty} \operatorname{area}\left(A \cap M_k \cap M_{k-1}^c\right).$$

Hence, we obtain the reverse inequality

$$\lim_{n \nearrow \infty} \operatorname{area}(A \cap M_n) \geq \operatorname{area}(A \cap M_\infty) . \square$$

By choosing $A = \mathbf{R}^2$, we conclude that the method is valid for measurable, hence open sets. *This completes stage one of the derivation of the method.*

Next we establish the method for interopen sets. A set $I \subset \mathbf{R}^2$ is *interopen* if I is the infinite intersection of a sequence of open sets (G_n), $I = \bigcap_{n=1}^\infty G_n$. Of course, every open set is interopen. Also every compact rectangle is interopen (Exercise **4.5.5**). The key feature of interopen sets is that any set A can be covered by some interopen set I, $A \subset I$, having the same area, $\operatorname{area}(A) = \operatorname{area}(I)$ (Exercise **4.5.7**).

Theorem 4.5.6. *If (M_n) is a sequence of measurable sets, then $\bigcap_{n=1}^\infty M_n$ is measurable.*

To derive this theorem, we start with two measurable sets M, N, and we show that $M \cap N$ is measurable. First, note that

$$(M \cap N)^c = (M \cap N^c) \cup (M^c \cap N) \cup (M^c \cap N^c). \tag{4.5.9}$$

Let $A \subset \mathbf{R}^2$ be arbitrary. Since N is measurable, write (4.5.4) with $A \cap M$ and N replacing A and M, respectively, obtaining

$$\operatorname{area}(A \cap M) = \operatorname{area}(A \cap M \cap N) + \operatorname{area}(A \cap M \cap N^c).$$

Now write (4.5.4) with $A \cap M^c$ and N replacing A and M respectively, obtaining

$$\operatorname{area}(A \cap M^c) = \operatorname{area}(A \cap M^c \cap N) + \operatorname{area}(A \cap M^c \cap N^c).$$

Now insert the last two equalities in (4.5.4). By (4.5.9) and subadditivity, we obtain

$$\begin{aligned}\text{area}\,(A) &= \text{area}\,(A\cap M) + \text{area}\,(A\cap M^c)\\ &= \text{area}\,(A\cap(M\cap N)) + \text{area}\,(A\cap(M\cap N^c))\\ &\quad + \text{area}\,(A\cap(M^c\cap N)) + \text{area}\,(A\cap(M^c\cap N^c))\\ &\geq \text{area}\,(A\cap(M\cap N)) + \text{area}\,(A\cap(M\cap N)^c)\,.\end{aligned}$$

Hence,

$$\text{area}\,(A) \geq \text{area}\,(A\cap(M\cap N)) + \text{area}\,(A\cap(M\cap N)^c)\,.$$

Since the reverse inequality is an immediate consequence of subadditivity, we conclude that $M\cap N$ is measurable.

Now let (M_n) be a sequence of measurable sets and set $N_n = \bigcap_{k=1}^n M_k$, $n\geq 1$. Then N_n, $n\geq 1$, are measurable. Indeed $N_1 = M_1$ is measurable. For the inductive step, suppose that N_n is measurable. Since $N_{n+1} = N_n\cap M_{n+1}$, we conclude that N_{n+1} is measurable. Hence, by induction, N_n is measurable for all $n\geq 1$. Now $M_\infty = \bigcap_{n=1}^\infty M_n = \bigcap_{n=1}^\infty N_n$ and $N_1 \supset N_2 \supset \dots$. Hence, $N_1^c \subset N_2^c \subset \dots$, so by the measurable method, we obtain

$$\begin{aligned}\text{area}\,(A\cap M_\infty^c) &= \text{area}\left[A\cap\left(\bigcap_{n=1}^\infty N_n\right)^c\right] \qquad (4.5.10)\\ &= \text{area}\left(A\cap\bigcup_{n=1}^\infty N_n^c\right) = \lim_{n\nearrow\infty}\text{area}\,(A\cap N_n^c)\,.\end{aligned}$$

Here we used De Morgan's law (§1.1). Now for each $n\geq 1$,

$$\begin{aligned}\text{area}\,(A) &= \text{area}\,(A\cap N_n) + \text{area}\,(A\cap N_n^c)\\ &\geq \text{area}\,(A\cap M_\infty) + \text{area}\,(A\cap N_n^c)\,. \qquad (4.5.11)\end{aligned}$$

Sending $n\nearrow\infty$ in (4.5.11) and using (4.5.11) yields

$$\text{area}\,(A) \geq \text{area}\,(A\cap M_\infty) + \text{area}\,(A\cap M_\infty^c)\,.$$

Since the reverse inequality follows from subadditivity, we conclude that $M_\infty = \bigcap_{n=1}^\infty M_n$ is measurable. □

By choosing (M_n) in the theorem to consist of open sets, we see that every interopen set is measurable. Hence, we conclude that the method is valid for interopen sets. *This completes stage two of the derivation of the method.*

The third and final stage of the derivation of the method is to establish it for an increasing sequence of arbitrary sets. To this end, let $A_1 \subset A_2 \subset \dots$ be an arbitrary increasing sequence of sets. For each $n\geq 1$, by Exercise **4.5.7**, choose an interopen set I_n containing A_n and having the same area: $I_n \supset A_n$ and area $(I_n) =$ area (A_n). For each $n\geq 1$, let

$$J_n = I_n\cap I_{n+1}\cap I_{n+2}\cap\dots\,.$$

Then J_n is interopen, $A_n \subset J_n \subset I_n$, and area $(J_n) = \text{area}(A_n)$, for all $n \geq 1$. Moreover, $J_n = I_n \cap J_{n+1}$. Hence, (and this is the reason for introducing the sequence (J_n)), the sequence (J_n) is increasing. Thus, by applying the method for interopen sets,

$$\begin{aligned}\lim_{n\nearrow\infty} \text{area}(A_n) &= \lim_{n\nearrow\infty} \text{area}(J_n) \\ &= \text{area}\left(\bigcup_{n=1}^{\infty} J_n\right) \geq \text{area}\left(\bigcup_{n=1}^{\infty} A_n\right). \end{aligned} \tag{4.5.12}$$

On the other hand, by monotonicity, the sequence (area (A_n)) is increasing and bounded above by area $(\bigcup_{n=1}^{\infty} A_n)$. Hence,

$$\lim_{n\nearrow\infty} \text{area}(A_n) \leq \text{area}\left(\bigcup_{n=1}^{\infty} A_n\right).$$

Combining this with (4.5.12), we conclude that

$$\lim_{n\nearrow\infty} \text{area}(A_n) = \text{area}\left(\bigcup_{n=1}^{\infty} A_n\right).$$

This completes stage three, hence the derivation of the method. □

We end by describing the connection between the areas of the inscribed and circumscribed polygons of the unit disk D, as the number of sides doubles. Let

$$P_k = \left(\frac{\cos(2\pi k/n)}{\cos(\pi/n)}, \frac{\sin(2\pi k/n)}{\cos(\pi/n)}\right), \qquad 0 \leq k \leq n.$$

Then the points P_k are evenly spaced about the circle $\{(x,y) : x^2 + y^2 = \sec^2(\pi/n)\}$, and $P_n = P_0$. Let D_n' denote the interior of the regular n-sided polygon obtained by joining the points $P_0, \ldots, P_n$ by line segments. Then $D_n' \supset D$ and $D_n' = c \cdot D_n$ with $c = \sec(\pi/n)$. Hence, by dilation invariance, we obtain

$$\text{area}(D_n') = c^2 \cdot \text{area}(D_n) = n\tan(\pi/n)$$

which also goes to π as $n \nearrow \infty$.

Let a_n, a_n' denote the areas of the inscribed and circumscribed n-sided polygons D_n, D_n', respectively. Then using trigonometry, one obtains (Exercise **4.5.11**)

$$a_{2n} = \sqrt{a_n a_n'} \tag{4.5.13}$$

and

$$\frac{1}{a_{2n}'} = \frac{1}{2}\left(\frac{1}{a_{2n}} + \frac{1}{a_n'}\right). \tag{4.5.14}$$

Since $a_4 = 2$ and $a_4' = 4$, we obtain $a_8 = 2\sqrt{2}$ and $a_8' = 8(\sqrt{2}-1)$. Thus,

$$2\sqrt{2} < \pi < 8(\sqrt{2}-1).$$

Continuing in this manner, one obtains approximations to π. These identities are very similar to those leading to Gauss's *arithmetic–geometric mean*, which we discuss in §5.3.

Exercises

4.5.1. If Q is an open rectangle and $(x,y) \in Q$, then $d((x,y),Q^c) > 0$.

4.5.2. Find a sequence (A_n) of open sets, such that $\bigcap_{n=1}^\infty A_n$ is not open.

4.5.3. A set A is *closed* if A^c is open. Show that a compact rectangle is closed, an infinite intersection of closed sets is closed, and a finite union of closed sets is closed. Find a sequence (A_n) of closed sets, such that $\bigcup_{n=1}^\infty A_n$ is not closed. (You will need De Morgan's law (§1.1).)

4.5.4. Given a real a, let L_a denote the vertical infinite line through a, $L_a = \{(x,y) : x = a, y \in \mathbf{R}\}$. Also set $L_\infty = L_{-\infty} = \emptyset$. Let f be nonnegative and continuous on (a,b). Show that

$$C = \{(x,y) : a < x < b, 0 \le y \le f(x)\} \cup L_a \cup L_b$$

is a closed set and

$$\int_a^b f(x)\,dx = \text{area}\,(C)\,.$$

This shows that the integral of a continuous nonnegative function is also the area of a closed set. (Compare C with the subgraph of $f(x) + \epsilon/(1+x^2)$ for $\epsilon > 0$ small.)

4.5.5. Show that C is closed iff

$$d((x,y),C) = 0 \qquad \Longleftrightarrow \qquad (x,y) \in C.$$

If C is closed and $G_n = \{(x,y) : d((x,y),C) < 1/n\}$, then G_n is open and $C = \bigcap_{n=1}^\infty G_n$. Thus, every closed set is interopen.

4.5.6. Let $A \subset \mathbf{R}^2$ be arbitrary. Use the definition of area$\,(A)$ to show: For all $\epsilon > 0$, there is an open superset G of A satisfying area$\,(G) \le$ area$\,(A) + \epsilon$. Conclude that

$$\text{area}\,(A) = \inf\{\text{area}\,(G) : A \subset G,\, G \text{ open}\}.$$

4.5.7. Let $A \subset \mathbf{R}^2$ be arbitrary. Show that there is an interopen set I containing A and having the same area as A (use Exercise **4.5.6**).

4.5.8. Show that M is measurable iff there is an interopen superset $I \supset M$ satisfying area $(I - M) = 0$. (use Exercise **4.5.7**).

4.5.9. If (M_n) is a sequence of measurable sets, then $\bigcup_{n=1}^{\infty} M_n$ is measurable.

4.5.10. Show that $D_n' \supset D$.

4.5.11. Derive (4.5.13) and (4.5.14).

4.5.12. If A and B are disjoint and A is measurable, then area $(A \cup B) =$ area $(A) +$ area (B).

4.5.13. If (A_n) is a sequence of disjoint measurable sets, then

$$\text{area}\left(\bigcup_{n=1}^{\infty} A_n\right) = \sum_{n=1}^{\infty} \text{area}\,(A_n)\,.$$

4.5.14. If A and B are measurable, then area $(A \cup B) =$ area $(A) +$ area $(B) -$ area $(A \cap B)$.

4.5.15. Show that M is measurable iff, for all $\epsilon > 0$, there is an open superset G of M, such that area $(G \setminus M) \leq \epsilon$.

4.5.16. Let $A \subset \mathbf{R}^2$ be measurable. If area $(A) > 0$, there is an $\epsilon > 0$, such that area $(A \cap A') > 0$ for all translates $A' = A + (a, b)$ of A with $|a| < \epsilon$ and $|b| < \epsilon$. (Start with A a rectangle, and use Exercise **4.2.15**.)

4.5.17. $N \subset \mathbf{R}^2$ is *negligible* if area $(N) = 0$. If N is negligible, then N is measurable.

4.5.18. If $A \subset \mathbf{R}^2$ is measurable and area $(A) > 0$, let

$$A - A = \{(x - x', y - y') : (x, y) \text{ and } (x', y') \in A\}$$

be the *set of differences*. Note that $A - A$ contains the origin. Then for some $\epsilon > 0$, $A - A$ must contain the open rectangle $Q_\epsilon = (-\epsilon, \epsilon) \times (-\epsilon, \epsilon)$. (Use Exercise **4.5.16**.)

Chapter 5
Applications

5.1 Euler's Gamma Function

In this section, we derive the formula

$$\int_0^1 \frac{dx}{x^x} = \sum_{n=1}^{\infty} \frac{1}{n^n}$$
$$= \frac{1}{1^1} + \frac{1}{2^2} + \frac{1}{3^3} + \dots. \qquad (5.1.1)$$

Along the way, we will meet Euler's gamma function and the monotone convergence theorem, both of which play roles in subsequent sections.

The gamma function is defined by

$$\Gamma(x) = \int_0^{\infty} e^{-t} t^{x-1}\, dt, \qquad x > 0.$$

Clearly $\Gamma(x)$ is positive for $x > 0$. Below we see that the gamma function is finite, and, in the next section, we see that it is continuous. Since

$$\int_0^{\infty} e^{-at} dt = \frac{1}{-a} e^{-at}\bigg|_0^{\infty} = -\frac{1}{a}\left(e^{-a\infty} - e^{-a0}\right) = \frac{1}{a}, \qquad a > 0, \quad (5.1.2)$$

we have $\Gamma(1) = 1$. Below, we use the convention $0! = 1$.

Theorem 5.1.1. *The gamma function* $\Gamma(x)$ *is positive, finite, and* $\Gamma(x+1) = x\Gamma(x)$ *for* $x > 0$*. Moreover,* $\Gamma(n) = (n-1)!$ *for* $n \geq 1$*.*

O. Hijab, *Introduction to Calculus and Classical Analysis*, Undergraduate Texts in Mathematics, DOI 10.1007/978-3-319-28400-2_5

To derive the first identity, use integration by parts with $u = t^x$, $dv = e^{-t}dt$. Then $v = -e^{-t}$, and $du = xt^{x-1}dt$. Hence, we obtain the following equality between primitives:

$$\int e^{-t}t^x dt = -e^{-t}t^x + x\int e^{-t}t^{x-1}dt.$$

Since $e^{-t}t^x$ vanishes at $t = 0$ and $t = \infty$ for $x > 0$ fixed, and the integrands are positive, by the fundamental theorem, we obtain

$$\begin{aligned}\Gamma(x+1) &= \int_0^\infty e^{-t}t^x dt \\ &= -e^{-t}t^x\big|_0^\infty + x\int_0^\infty e^{-t}t^{x-1}dt \\ &= 0 + x\Gamma(x) = x\Gamma(x).\end{aligned}$$

Note that this identity is true whether or not $\Gamma(x)$ is finite. We derive $\Gamma(n) = (n-1)!$ by induction. The statement is true for $n = 1$ since $\Gamma(1) = 1$ from above. Assuming the statement is true for n, $\Gamma(n+1) = n\Gamma(n) = n(n-1)! = n!$. Hence, the statement is true for all $n \geq 1$. Now we show that $\Gamma(x)$ is finite for all $x > 0$. Since the integral $\int_0^1 e^{-t}t^{x-1}\,dt \leq \int_0^1 t^{x-1}\,dt = 1/x$ is finite for $x > 0$, it is enough to verify integrability of $e^{-t}t^{x-1}$ over $(1, \infty)$. Over this interval, $e^{-t}t^{x-1}$ increases with x; hence, $\int_1^\infty e^{-t}t^{x-1}\,dt \leq \int_1^\infty e^{-t}t^{n-1}\,dt \leq \Gamma(n)$ for any natural $n \geq x$. But we already know that $\Gamma(n) = (n-1)! < \infty$, hence the result. □

Because of this result, we *define* $x! = \Gamma(x+1)$ for $x > -1$. For example, in Exercise **5.4.1**, we obtain $(1/2)! = \sqrt{\pi}/2$.

We already know (linearity §4.4) that the integral of a finite sum of continuous functions is the sum of their integrals. To obtain linearity for infinite sums, we first derive the following.

Theorem 5.1.2 (Monotone Convergence Theorem). *Let* $0 \leq f_1 \leq f_2 \leq f_3 \leq \dots$ *be an increasing sequence of nonnegative functions,*[1] *all defined on an interval* (a, b). *If*

$$\lim_{n\nearrow\infty} f_n(x) = f(x), \qquad a < x < b,$$

then

$$\lim_{n\nearrow\infty}\int_a^b f_n(x)\,dx = \int_a^b \lim_{n\nearrow\infty} f_n(x)\,dx = \int_a^b f(x)\,dx.$$

We caution that this result may be false when the sequence (f_n) is not increasing (Exercise **5.1.1**). Nevertheless, one can still obtain roughly half this result for any sequence (f_n) of nonnegative functions (Exercise **5.1.2**).

[1] These functions need not be continuous, they may be *arbitrary*.

To see this, let G_n denote the subgraph of f_n over (a,b), G the subgraph of f over (a,b). Then $G_n \subset G_{n+1}$ since $y < f_n(x)$ implies $y < f_{n+1}(x)$. Moreover, $y < f(x)$ iff $y < f_n(x)$ for some $n \geq 1$; hence, $G = \bigcup_{n=1}^{\infty} G_n$. The result now follows from the method of exhaustion and the definition of the integral of a nonnegative function. □

In the next section, we show that Γ is continuous; in Exercise **5.1.7**, we show that Γ is convex. Later (§5.4), we show that Γ is strictly convex. Since $\Gamma(x) = \Gamma(x+1)/x$ for $x > 0$ and $\Gamma(1) = 1$, it follows that $\Gamma(0+) = \infty$. Also for $x \geq n$, $\Gamma(x) \geq \int_1^\infty e^{-t}t^{x-1}\,dt \geq \int_1^\infty e^{-t}t^{n-1}\,dt \geq (n-1)! - 1/n$; hence, $\Gamma(\infty) = \infty$. Putting all of this together, we conclude that Γ has exactly one global positive minimum and the graph for $x > 0$ is as in Figure 5.1. Later (§5.8), we will extend the domain of Γ to negative reals.

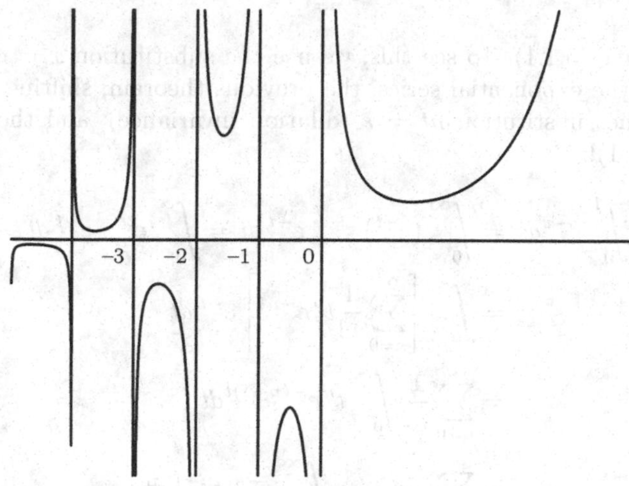

Fig. 5.1 The gamma function

Theorem 5.1.3 (Summation Under the Integral Sign: Positive Case). *Let f_n, $n \geq 1$, be a sequence of nonnegative functions on (a,b). If f_n, $n \geq 1$ are continuous, then*

$$\int_a^b \left[\sum_{n=1}^{\infty} f_n(x)\right] dx = \sum_{n=1}^{\infty} \int_a^b f_n(x)\,dx.$$

A key aspect of the proof is the use of the linearity of the integral, as derived in §4.4 for continuous functions.[2]

For alternating series, a different version of this result is needed (next section).

[2] The general version is in Chapter 6.

To derive this, use linearity and the monotone convergence theorem (Theorem 5.1.2) to obtain

$$\begin{aligned}\int_a^b \left[\sum_{n=1}^\infty f_n(x)\right] dx &= \int_a^b \lim_{n\nearrow\infty} \left[\sum_{k=1}^n f_k(x)\right] dx \\ &= \lim_{n\nearrow\infty} \int_a^b \left[\sum_{k=1}^n f_k(x)\right] dx \\ &= \lim_{n\nearrow\infty} \sum_{k=1}^n \int_a^b f_k(x)\, dx \\ &= \sum_{n=1}^\infty \int_a^b f_n(x)\, dx. \quad \square\end{aligned}$$

Now we derive (5.1.1). To see this, we use the substitution $x = e^{-t}$ (Exercise **4.4.7**), the exponential series, the previous theorem, shifting the index n by one, the substitution $nt = s$ (dilation invariance), and the property $\Gamma(n) = (n-1)!$:

$$\begin{aligned}\int_0^1 x^{-x}\, dx &= \int_0^\infty \left(e^{-t}\right)^{-e^{-t}} e^{-t}\, dt = \int_0^\infty e^{te^{-t}} e^{-t}\, dt \\ &= \int_0^\infty \left[\sum_{n=0}^\infty \frac{1}{n!} t^n e^{-nt}\right] e^{-t}\, dt \\ &= \sum_{n=0}^\infty \frac{1}{n!} \int_0^\infty t^n e^{-(n+1)t} dt \\ &= \sum_{n=1}^\infty \frac{1}{(n-1)!} \int_0^\infty e^{-nt} t^{n-1}\, dt \\ &= \sum_{n=1}^\infty \frac{1}{(n-1)!} n^{-n} \int_0^\infty e^{-s} s^{n-1}\, ds \\ &= \sum_{n=1}^\infty \frac{1}{(n-1)!} n^{-n} \Gamma(n) = \sum_{n=1}^\infty n^{-n}. \quad \square\end{aligned}$$

We end with a special case of the monotone convergence theorem.

Theorem 5.1.4 (Monotone Convergence Theorem (For Series)). *Let (a_{nj}), $n \geq 1$, be a sequence of sequences, and let (a_j) be a given sequence. Suppose that $0 \leq a_{1j} \leq a_{2j} \leq a_{3j} \leq \dots$ for all $j \geq 1$. If*

$$\lim_{n\nearrow\infty} a_{nj} = a_j, \qquad j \geq 1,$$

then

$$\lim_{n\nearrow\infty}\sum_{j=1}^{\infty}a_{nj}=\sum_{j=1}^{\infty}\lim_{n\nearrow\infty}a_{nj}=\sum_{j=1}^{\infty}a_j.$$

To see this, define piecewise constant functions $f_n(x)=a_{nj}$, $j-1<x\le j$, $j\ge 1$, and $f(x)=a_j$, $j-1<x\le j$, $j\ge 1$. Then (f_n) is nonnegative on $(0,\infty)$ and increasing to f. Now apply the monotone convergence theorem for integrals, and use Exercise **4.3.6**. □

Using this theorem, one can derive an analog of summation under the integral sign involving series ("summation under the summation sign") rather than integrals. But we already did this in §1.7.

Exercises

5.1.1. Find a sequence $f_1\ge f_2\ge f_3\ge\cdots\ge 0$ of nonnegative functions, such that $f_n(x)\to 0$ for all $x\in\mathbf{R}$, but $\int_{-\infty}^{\infty}f_n(x)\,dx=\infty$ for all $n\ge 1$. This shows that the monotone convergence theorem is false for decreasing sequences.

5.1.2. (Fatou's lemma) Let f_n, $n\ge 1$, be nonnegative functions, all defined on (a,b), and suppose that $f_n(x)\to f(x)$ for all x in (a,b). Then the lower limit of the sequence $\left(\int_a^b f_n(x)\,dx\right)$ is greater or equal to $\int_a^b f(x)\,dx$,

$$\int_a^b f(x)\,dx\le\liminf_{n\to\infty}\int_a^b f_n(x)\,dx.$$

(For each x, let $(g_n(x))$ equal the lower sequence (§1.5) of the sequence $(f_n(x))$ and apply the monotone convergence theorem.)

5.1.3. Let $f_0(x)=1-x^2$ for $|x|<1$ and $f_0(x)=0$ for $|x|\ge 1$, and let $f_n(x)=f_0(x-n)$ for $-\infty<x<\infty$ and $n\ge 1$. Compute $f(x)=\lim_{n\nearrow\infty}f_n(x)$, $-\infty<x<\infty$, $\int_{-\infty}^{\infty}f_n(x)\,dx$, $n\ge 1$, and $\int_{-\infty}^{\infty}f(x)\,dx$ (Figure 5.2). Conclude that, for this example, the inequality in Fatou's lemma is strict.

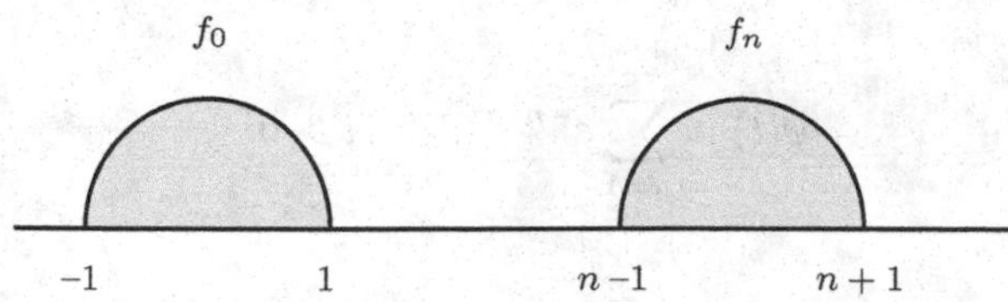

Fig. 5.2 Exercise **5.1.3**

5.1.4. Show that

$$\Gamma(x) = \lim_{n\nearrow\infty} \int_0^n \left(1-\frac{t}{n}\right)^n t^{x-1}\,dt, \qquad x>0.$$

(Use Exercise **3.2.4.**)

5.1.5. Use substitution $t = ns$, and integrate by parts to get

$$\int_0^n \left(1-\frac{t}{n}\right)^n t^{x-1}\,dt = \frac{n^x n!}{x(x+1)\dots(x+n)}, \qquad x>0,$$

for $n \ge 1$. Conclude that

$$\Gamma(x) = \lim_{n\nearrow\infty} \frac{n^x n!}{x(x+1)\dots(x+n)}, \qquad x>0, \tag{5.1.3}$$

and

$$x! = \lim_{n\nearrow\infty} \frac{n^x n!}{(x+1)\dots(x+n)}, \qquad x>-1. \tag{5.1.4}$$

5.1.6. We say that a function $f:(a,b)\to\mathbf{R}^+$ is *log-convex* if $\log f$ is convex (§3.3). Show that the right side of (5.1.3) is log-convex on $(0,\infty)$. Suppose that $f_n:(a,b)\to\mathbf{R}$, $n\ge 1$, is a sequence of convex functions, and $f_n(x)\to f(x)$, as $n\nearrow\infty$, for all x in (a,b). Show that f is convex on (a,b). Conclude that the gamma function is log-convex on $(0,\infty)$.

5.1.7. Show that Γ is convex on $(0,\infty)$. (Consider $\Gamma = \exp(\log\Gamma)$ and use Exercise **3.3.3.**)

5.1.8. Let $s_n(t)$ denote the nth partial sum of

$$\frac{1}{e^t-1} = e^{-t}+e^{-2t}+e^{-3t}+\dots, \qquad t>0.$$

Use $s_n(t)$ to derive

$$\int_0^\infty \frac{t^{x-1}}{e^t-1}\,dt = \Gamma(x)\zeta(x), \qquad x>1,$$

where $\zeta(x) = \sum_{n=1}^\infty n^{-x}$, $x>1$.

5.1.9. Let

$$\psi(t) = \sum_{n=1}^\infty e^{-n^2\pi t}, \qquad t>0.$$

Show that

$$\int_0^\infty \psi(t)t^{x/2-1}dt = \pi^{-x/2}\Gamma(x/2)\zeta(x), \qquad x>1,$$

where ζ is as in the previous exercise.

5.1.10. Show that $\int_0^1 t^{x-1}(-\log t)^{n-1}\,dt = \Gamma(n)/x^n$ for $x > 0$ and $n \ge 1$ (Exercise **4.4.7**).

5.1.11. Show that

$$\int_0^\infty e^{-t}t^{x-1}|\log t|^{n-1}\,dt \le \frac{\Gamma(n)}{x^n} + \Gamma(x+n-1)$$

for $x > 0$ and $n \ge 1$. (Break up the integral into $\int_0^1 + \int_1^\infty$ and use $\log t \le t$ for $t \ge 1$.)

5.1.12. Use the monotone convergence theorem for series to compute $\zeta(1+)$ and $\psi(0+)$.

5.1.13. With $\tau(t) = t/(1-e^{-t})$, show that

$$\int_0^\infty e^{-xt}\tau(t)\,dt = \sum_{n=0}^\infty \frac{1}{(x+n)^2}, \qquad x > 0.$$

(Compare with Exercise **5.1.8**.)

5.1.14. Use the monotone convergence theorem to derive continuity at the endpoints (§4.3): If $f : (a,b) \to \mathbf{R}$ is nonnegative and $a_n \searrow a$, $b_n \nearrow b$, then $\int_{a_n}^{b_n} f(x)\,dx \to \int_a^b f(x)\,dx$.

5.1.15. Show that

$$\exp\left(\int_s^{s+1} \log\Gamma(x)\,dx\right) = \text{constant} \times \left(\frac{s}{e}\right)^s, \qquad s > 0.$$

The left side is the *geometric mean* of Γ over the interval $(s, s+1)$. The constant is evaluated in Exercise **5.5.9**.

5.2 The Number π

In this section, we discuss several formulas for the irrational (Exercise **1.3.18**) number π, namely:

- The *Leibnitz series*

$$\frac{\pi}{4} = 1 - \frac{1}{3} + \frac{1}{5} - \frac{1}{7} + \dots, \tag{5.2.1}$$

- The *Wallis product*

$$\frac{\pi}{2} = \lim_{n\nearrow\infty} \frac{2\cdot 2\cdot 4\cdot 4\cdot 6\cdot 6\cdot \dots \cdot 2n\cdot 2n}{1\cdot 3\cdot 3\cdot 5\cdot 5\cdot 7\cdot \dots \cdot (2n-1)\cdot(2n+1)}, \tag{5.2.2}$$

- The *Vieta formula*

$$\frac{2}{\pi} = \sqrt{\frac{1}{2}} \cdot \sqrt{\frac{1}{2} + \frac{1}{2}\sqrt{\frac{1}{2}}} \cdot \sqrt{\frac{1}{2} + \frac{1}{2}\sqrt{\frac{1}{2} + \frac{1}{2}\sqrt{\frac{1}{2}}}} \cdots, \tag{5.2.3}$$

- the *continued fraction expansion*

$$\frac{\pi}{4} = \cfrac{1}{1 + \cfrac{1}{2 + \cfrac{9}{2 + \cfrac{25}{2 + \cfrac{49}{2 + \dots}}}}}, \tag{5.2.4}$$

- The *Bailey–Borwein–Plouffe series*

$$\pi = \sum_{n=0}^{\infty} \frac{1}{16^n} \left(\frac{4}{8n+1} - \frac{2}{8n+4} - \frac{1}{8n+5} - \frac{1}{8n+6} \right). \tag{5.2.5}$$

Along the way, we will meet the dominated convergence theorem, and we also compute the Laplace transform of the Bessel function of order zero.

It is one thing to derive these remarkable formulas and quite another to *discover* them. We begin by rederiving the Leibnitz series for π by an alternate method to that in §3.7.

Start with the power series expansion

$$\frac{1}{1+x^2} = 1 - x^2 + x^4 - x^6 + \dots, \qquad 0 < x < 1. \tag{5.2.6}$$

We seek to integrate (5.2.6), term by term, as in §4.4. Since $\arctan 1 = \pi/4$ and $\arctan x$ is a primitive of $1/(1+x^2)$, we seek to integrate (5.2.6) over the interval $(0,1)$. However, since the radius of convergence of (5.2.6) is 1, the result in §4.4 is not applicable. On the other hand, the theorem in §5.1 allows us to integrate, term by term, any series of nonnegative functions. Since (5.2.6) is alternating, again this is not applicable.

If we let $s_n(x)$ denote the nth partial sum in (5.2.6), then by the Leibnitz test (§1.7),

$$0 < s_n(x) < 1, \qquad 0 < x < 1, n \geq 1. \tag{5.2.7}$$

It turns out that (5.2.7) allows us to integrate (5.2.6) over the interval (0,1), term by term. This is captured in the following theorem.

Theorem 5.2.1 (Dominated Convergence Theorem). *Let f_n, $n \geq 1$, be a sequence of functions defined on (a,b). Suppose there is a function g integrable on (a,b) satisfying $|f_n(x)| \leq g(x)$ for all x in (a,b) and all $n \geq 1$. If*

$$\lim_{n\nearrow\infty} f_n(x) = f(x), \qquad a < x < b,$$

then f and f_n, $n \geq 1$, are integrable on (a,b). If g, f, and f_n, $n \geq 1$, are continuous, then

$$\lim_{n\nearrow\infty} \int_a^b f_n(x)\,dx = \int_a^b \lim_{n\nearrow\infty} f_n(x)\,dx = \int_a^b f(x)\,dx. \tag{5.2.8}$$

Note that (5.2.8) says we can switch the limit and the integral, exactly as in the monotone convergence theorem. This theorem takes its name from the hypothesis $|f_n(x)| \leq g(x)$, $a < x < b$, which is read *f_n is dominated by g over (a,b)*. The point of this hypothesis is the existence of a *single integrable* g that *dominates all* the f_n's.

The two results, the monotone convergence theorem and the dominated convergence theorem, are used throughout analysis to justify the interchange of integrals and limits. Which theorem is applied when depends on which hypothesis is applicable to the problem at hand. When trigonometric or more general oscillatory functions are involved, the monotone convergence theorem is not applicable. In these cases, it is the dominated convergence theorem that saves the day.

When $(a,b) = (0,\infty)$ and the functions f_n, $n \geq 1$, f, g, are piecewise constant, the dominated convergence theorem reduces to a theorem about series, which we discuss at the end of the section. Also one can allow the interval (a_n, b_n) to vary with $n \geq 1$ (Exercise **5.2.15**). We defer the derivation of the dominated convergence theorem to the end of the section.

Going back to the derivation of (5.2.1), since the nth partial sum s_n converges to $f(x) = 1/(1+x^2)$ and $|s_n(x)| \leq 1$ by (5.2.7), we can choose $g(x) = 1$ which is integrable on $(0,1)$. Hence, applying the fundamental theorem and the dominated convergence theorem yields

$$\begin{aligned}
\frac{\pi}{4} &= \arctan 1 - \arctan 0 = \int_0^1 \frac{1}{1+x^2}\,dx = \int_0^1 \left[\sum_{n=0}^{\infty}(-1)^n x^{2n}\right] dx \\
&= \int_0^1 \lim_{N\nearrow\infty} s_N(x)\,dx = \lim_{N\nearrow\infty} \int_0^1 s_N(x)\,dx \\
&= \lim_{N\nearrow\infty} \sum_{n=0}^{N}(-1)^n \int_0^1 x^{2n}\,dx = \sum_{n=0}^{\infty}(-1)^n \int_0^1 x^{2n}\,dx \\
&= \sum_{n=0}^{\infty}(-1)^n \frac{1}{2n+1} = 1 - \frac{1}{3} + \frac{1}{5} - \frac{1}{7} + \dots .
\end{aligned}$$

This completes the derivation of (5.2.1).

The idea behind this derivation of (5.2.1) can be carried out more generally.

Theorem 5.2.2 (Summation Under the Integral Sign: Alternating Case). *Let f_n, $n \geq 1$, be a decreasing sequence of nonnegative functions on (a, b), and suppose that f_1 is integrable on (a, b). Then f_n, $n \geq 1$, and $\sum_{n=1}^{\infty}(-1)^{n-1} f_n$ are integrable on (a, b). If f_n, $n \geq 1$, and $\sum_{n=1}^{\infty}(-1)^{n-1} f_n$ are continuous, then*

$$\int_a^b \left[\sum_{n=1}^{\infty}(-1)^{n-1} f_n(x)\right] dx = \sum_{n=1}^{\infty}(-1)^{n-1} \int_a^b f_n(x)\, dx.$$

To derive this, we need only note that the nth partial sum s_n is nonnegative and no greater than $g = f_1$, which is integrable. Hence, we may apply the dominated convergence theorem, as above, to the sequence (s_n) of partial sums. □

For example, using this theorem to integrate the geometric series $1/(1 + x) = 1 - x + x^2 - x^3 + \dots$ over $(0, 1)$, we obtain

$$\log 2 = 1 - \frac{1}{2} + \frac{1}{3} - \frac{1}{4} + \dots .$$

Now we discuss the general case.

Theorem 5.2.3 (Summation Under the Integral Sign: Absolute Case). *Let f_n, $n \geq 1$, be a sequence of functions on (a, b), and suppose that there is a function g integrable on (a, b) and satisfying $\sum_{n=1}^{\infty} |f_n(x)| \leq g(x)$ for all x in (a, b). Then f_n, $n \geq 1$, and $\sum_{n=1}^{\infty} f_n$ are integrable. If g, f_n, $n \geq 1$, and $\sum_{n=1}^{\infty} f_n$ are continuous, then*

$$\int_a^b \left[\sum_{n=1}^{\infty} f_n(x)\right] dx = \sum_{n=1}^{\infty} \int_a^b f_n(x)\, dx.$$

To derive this, we need only note that $|s_n(x)| \leq |f_1(x)| + \dots + |f_n(x)| \leq g(x)$, which is integrable. Hence, we may apply the dominated convergence theorem, as above, to the sequence (s_n) of partial sums. □

The *Bessel function of order zero* is defined by

$$J_0(x) = \sum_{n=0}^{\infty}(-1)^n \frac{x^{2n}}{4^n (n!)^2}, \qquad -\infty < x < \infty.$$

To check the convergence, rewrite the series using Exercise **3.5.10** obtaining

$$J_0(x) = \sum_{n=0}^{\infty}\binom{-1/2}{n}\frac{x^{2n}}{(2n)!}, \qquad -\infty < x < \infty.$$

Now use the definition of $\binom{v}{n}$ (§3.5) to check the inequality $|\binom{-1/2}{n}| \le 1$ for all $n \ge 0$. Hence,

$$|J_0(x)| \le \sum_{n=0}^{\infty} \left| \binom{-1/2}{n} \frac{x^{2n}}{(2n)!} \right| \le \sum_{n=0}^{\infty} \frac{x^{2n}}{(2n)!} \le e^{|x|}, \quad -\infty < x < \infty. \quad (5.2.9)$$

This shows that the series J_0 converges absolutely for all x real. Since J_0 is a convergent power series, J_0 is a smooth function on $\mathbf{R}$. We wish to use summation under the integral sign to obtain

$$\int_0^{\infty} e^{-sx} J_0(x)\, dx = \frac{1}{\sqrt{1+s^2}}, \qquad s > 1. \qquad (5.2.10)$$

The left side of (5.2.10), by definition, is the *Laplace transform* of J_0. Thus, (5.2.10) exhibits the Laplace transform of the Bessel function J_0. In Exercise **5.2.2**, you are asked to derive the Laplace transform of $\sin x/x$.

To obtain (5.2.10), fix $s > 1$, and set $f_n(x) = e^{-sx}\binom{-1/2}{n} x^{2n}/(2n)!$, $n \ge 0$. Then by (5.2.9), we may apply summation under the integral sign with $g(x) = e^{-sx}e^{x}$, $x > 0$, which is positive, continuous, and integrable (since $s > 1$). Hence,

$$\int_0^{\infty} e^{-sx} J_0(x)\, dx = \sum_{n=0}^{\infty} \binom{-1/2}{n} \frac{1}{(2n)!} \int_0^{\infty} e^{-sx} x^{2n} dx.$$

Inserting the substitution $x = t/s$, $dx = dt/s$, and recalling Newton's generalization of the binomial theorem (§3.5) yields

$$\begin{aligned}\int_0^{\infty} e^{-sx} J_0(x)\, dx &= \sum_{n=0}^{\infty} \binom{-1/2}{n} \frac{1}{(2n)! s^{2n+1}} \int_0^{\infty} e^{-t} t^{2n} dt \\ &= \sum_{n=0}^{\infty} \binom{-1/2}{n} \frac{\Gamma(2n+1)}{(2n)! s^{2n+1}} \\ &= \frac{1}{s} \sum_{n=0}^{\infty} \binom{-1/2}{n} \left(\frac{1}{s^2}\right)^n \\ &= \frac{1}{s} \cdot \frac{1}{\sqrt{1+(1/s)^2}} = \frac{1}{\sqrt{1+s^2}}.\end{aligned}$$

This establishes (5.2.10).

Now let

$$F(x) = \int_0^{\infty} \frac{\sin(xt)}{1+t^2}\, dt, \qquad -\infty < x < \infty.$$

We use the dominated convergence theorem to show that F is continuous on $\mathbf{R}$. To show this, fix $x \in \mathbf{R}$, and let $x_n \to x$; we have to show that $F(x_n) \to F(x)$. Set $f_n(t) = \sin(x_n t)/(1+t^2)$, $f(t) = \sin(xt)/(1+t^2)$, and

$g(t) = 1/(1+t^2)$. Then $f_n(t)$, $n \ge 1$, $f(t)$, and $g(t)$ are continuous; all the $f_n(t)$'s are dominated by $g(t)$ over $(0,\infty)$, $f_n(t) \to f(t)$ for all $t > 0$, and $g(t)$ is integrable over $(0,\infty)$ since $\int_0^\infty g(t)\,dt = \pi/2$. Hence, the theorem applies, and

$$\begin{aligned}\lim_{n\nearrow\infty} F(x_n) &= \lim_{n\nearrow\infty}\int_0^\infty \frac{\sin(x_n t)}{1+t^2}\,dt \\ &= \int_0^\infty \lim_{n\nearrow\infty}\frac{\sin(x_n t)}{1+t^2}\,dt \\ &= \int_0^\infty \frac{\sin(xt)}{1+t^2}\,dt = F(x).\end{aligned}$$

This establishes the continuity of F.

Similarly, one can establish the continuity of the gamma function on $(0,\infty)$. To this end, choose $0 < a < x < b < \infty$, and let $x_n \to x$ with $a < x_n < b$. We have to show $\Gamma(x_n) \to \Gamma(x)$. Now $f_n(t) = e^{-t}t^{x_n-1}$ satisfies

$$|f_n(t)| \le \begin{cases} e^{-t}t^{b-1}, & 1 \le t < \infty, \\ e^{-t}t^{a-1}, & 0 < t \le 1. \end{cases}$$

If we call the right side of this inequality $g(t)$, we see that $f_n(t)$, $n \ge 1$, are all dominated by $g(t)$ over $(0,\infty)$. Moreover, $g(t)$ is continuous (especially at $t=1$) and integrable over $(0,\infty)$, since $\int_0^\infty g(t)\,dt \le \Gamma(a) + \Gamma(b)$. Also the functions $f_n(t)$, $n \ge 1$, and $f(t) = e^{-t}t^{x-1}$ are continuous, and $f_n(t) \to f(t)$ for all $t > 0$. Thus, the dominated convergence theorem applies. Hence,

$$\begin{aligned}\lim_{n\nearrow\infty} \Gamma(x_n) &= \lim_{n\nearrow\infty}\int_0^\infty e^{-t}t^{x_n-1}dt \\ &= \int_0^\infty \lim_{n\nearrow\infty} e^{-t}t^{x_n-1}dt \\ &= \int_0^\infty e^{-t}t^{x-1}dt = \Gamma(x).\end{aligned}$$

Hence, Γ is continuous on (a,b). Since $0 < a < b$ are arbitrary, Γ is continuous on $(0,\infty)$.

Next, we derive Wallis' product (5.2.2). Begin with integrating by parts to obtain

$$\int \sin^n x\,dx = -\frac{1}{n}\sin^{n-1}x\cos x + \frac{n-1}{n}\int \sin^{n-2}x\,dx, \qquad n \ge 2. \quad (5.2.11)$$

Evaluating at 0 and $\pi/2$ yields

$$\int_0^{\pi/2} \sin^n x\,dx = \frac{n-1}{n}\int_0^{\pi/2} \sin^{n-2}x\,dx, \qquad n \ge 2.$$

Since $\int_0^{\pi/2} \sin^0 x\,dx = \pi/2$ and $\int_0^{\pi/2} \sin^1 x\,dx = 1$, by the last equation and induction,

$$I_{2n} = \int_0^{\pi/2} \sin^{2n} x\,dx = \frac{(2n-1)\cdot(2n-3)\cdot\dots\cdot 1}{2n\cdot(2n-2)\cdot\dots\cdot 2}\cdot\frac{\pi}{2},$$

and

$$I_{2n+1} = \int_0^{\pi/2} \sin^{2n+1} x\,dx = \frac{2n\cdot(2n-2)\cdot\dots\cdot 2}{(2n+1)\cdot(2n-1)\cdot\dots\cdot 3}\cdot 1,$$

for $n \geq 1$. Since $0 < \sin x < 1$ on $(0, \pi/2)$, the integrals I_n are decreasing in n. But, by the formula for I_n with n odd,

$$1 \leq \frac{I_{2n-1}}{I_{2n+1}} \leq 1 + \frac{1}{2n} \qquad n \geq 1.$$

Thus

$$1 \leq \frac{I_{2n}}{I_{2n+1}} \leq \frac{I_{2n-1}}{I_{2n+1}} \leq 1 + \frac{1}{2n}, \qquad n \geq 1,$$

or $I_{2n}/I_{2n+1} \to 1$, as $n \nearrow \infty$. Since

$$\frac{I_{2n}}{I_{2n+1}} = \frac{(2n+1)\cdot(2n-1)\cdot(2n-1)\cdot\dots\cdot 3\cdot 3\cdot 1}{2n\cdot 2n\cdot(2n-2)\cdot\dots\cdot 4\cdot 2\cdot 2}\cdot\frac{\pi}{2},$$

we obtain (5.2.2).

A derivation of Vieta's formula (5.2.3) starts with the identity

$$\frac{\sin\theta}{2^n \sin(\theta/2^n)} = \cos\left(\frac{\theta}{2}\right)\cos\left(\frac{\theta}{2^2}\right)\dots\cos\left(\frac{\theta}{2^n}\right) \tag{5.2.12}$$

which follows by multiplying both sides by $\sin(\theta/2^n)$ and using the double-angle formula $\sin(2x) = 2\sin x \cos x$ repeatedly. Now insert in (5.2.12) $\theta = \pi/2$, and use, repeatedly, the formula $\cos(\theta/2) = \sqrt{(1+\cos\theta)/2}$. This yields

$$\frac{2}{\pi}\cdot\frac{\pi/2^{n+1}}{\sin(\pi/2^{n+1})} = \sqrt{\frac{1}{2}}\cdot\sqrt{\frac{1}{2}+\frac{1}{2}\sqrt{\frac{1}{2}}}\cdot\sqrt{\frac{1}{2}+\frac{1}{2}\sqrt{\frac{1}{2}+\frac{1}{2}\sqrt{\frac{1}{2}}}}\cdot\dots$$

$$\dots\cdot\sqrt{\frac{1}{2}+\frac{1}{2}\sqrt{\frac{1}{2}+\frac{1}{2}\sqrt{\frac{1}{2}+\dots}}},$$

where the last (nth) factor involves n square roots. Letting $n \nearrow \infty$ yields (5.2.3) since $\sin x/x \to 1$ as $x \to 0$.

To derive the continued fraction expansion (5.2.4), first, we must understand what it means. To this end, introduce the *convergents*

$$c_n = \cfrac{1}{1+\cfrac{1}{2+\cfrac{9}{2+\cfrac{25}{2+\cfrac{49}{\ddots\, 2+\cfrac{(2n-1)^2}{2}}}}}}. \tag{5.2.13}$$

Then we take (5.2.4) to mean that the sequence (c_n) converges to $\pi/4$. To derive this, it is enough to show that c_n equals the nth partial sum

$$s_n = 1 - \frac{1}{3} + \frac{1}{5} - \cdots \pm \frac{1}{2n+1}$$

of the Leibniz series (5.2.1) for all $n \geq 1$.

Given reals $a_1, \ldots, a_n$, let

$$s_n^* = a_1 + a_1 a_2 + a_1 a_2 a_3 + \cdots + a_1 a_2 \ldots a_n.$$

Then $s_n^* = s_n^*(a_1, \ldots, a_n)$ is a function of the n variables $a_1, \ldots, a_n$. Later, we will make a judicious choice of $a_1, \ldots, a_n$. Note that

$$a_1 + a_1 s_n^*(a_2, \ldots, a_{n+1}) = s_{n+1}^*(a_1, a_2, \ldots, a_{n+1}).$$

Let $f(x, y) = x/(1 + x - y)$, and let

$$c_1^*(a_1) = f(a_1, 0) = \frac{a_1}{1+a_1},$$

$$c_2^*(a_1, a_2) = f(a_1, f(a_2, 0)) = \cfrac{a_1}{1 + a_1 - \cfrac{a_2}{1+a_2}},$$

$$c_3^*(a_1, a_2, a_3) = f(a_1, f(a_2, f(a_3, 0))) = \cfrac{a_1}{1 + a_1 - \cfrac{a_2}{1 + a_2 - \cfrac{a_3}{1+a_3}}},$$

and so on. More systematically, define $c_n^*(a_1, \ldots, a_n)$ inductively by setting $c_1^*(a_1) = a_1/(1 + a_1)$ and

$$c_{n+1}^*(a_1, a_2, \ldots, a_{n+1}) = \frac{a_1}{1 + a_1 - c_n^*(a_2, \ldots, a_{n+1})}, \qquad n \geq 1.$$

We claim that:

$$c_n^* = \frac{s_n^*}{1+s_n^*}, \qquad n \geq 1, \tag{5.2.14}$$

and we verify this by induction. Here $c_n^* = c_n^*(a_1, \ldots, a_n)$ and $s_n^* = s_n^*(a_1, \ldots, a_n)$. Clearly $c_1^* = s_1^*/(1+s_1^*)$ since $s_1^* = a_1$. Now assume $c_n^* = s_n^*/(1+s_n^*)$. Replacing $a_1, \ldots, a_n$ by $a_2, \ldots, a_{n+1}$ yields

$$c_n^*(a_2, \ldots, a_{n+1}) = s_n^*(a_2, \ldots, a_{n+1})/[1 + s_n^*(a_2, \ldots, a_{n+1})].$$

Then

$$\begin{aligned} c_{n+1}^*(a_1, a_2, \ldots, a_{n+1}) &= \frac{a_1}{1 + a_1 - c_n^*(a_2, \ldots, a_{n+1})} \\ &= \frac{a_1}{1 + a_1 - \dfrac{s_n^*(a_2, \ldots, a_{n+1})}{1 + s_n^*(a_2, \ldots, a_{n+1})}} \\ &= \frac{s_{n+1}^*(a_1, a_2, \ldots, a_{n+1})}{1 + s_{n+1}^*(a_1, a_2, \ldots, a_{n+1})}. \end{aligned}$$

Thus, $c_{n+1}^* = s_{n+1}^*/(1+s_{n+1}^*)$. Hence, by induction, the claim is true.

Solving for $1 + s_n^*$ in (5.2.14) and multiplying by a_0 yield

$$a_0 + a_0 s_n^* = \frac{a_0}{1 - c_n^*}.$$

We arrive at *Euler's continued fraction formula.*

Theorem 5.2.4. *For $a_0, a_1, a_2, \ldots, a_n$ real,*

$$a_0 + a_0 a_1 + \cdots + a_0 a_1 \ldots a_n = \cfrac{a_0}{1 - \cfrac{a_1}{1 + a_1 - \cfrac{a_2}{1 + a_2 - \cfrac{\ddots}{\ddots \cfrac{a_{n-1}}{1 + a_{n-1} - \cfrac{a_n}{1 + a_n}}}}}}. \qquad \square$$

Now choose

$$a_0 = 1, a_1 = -\frac{1}{3}, a_2 = -\frac{3}{5}, a_3 = -\frac{5}{7}, \ldots, a_n = -\frac{2n-1}{2n+1}.$$

Then the left side of Euler's continued fraction formula is s_n, and the right side is

$$\cfrac{1}{1+\cfrac{1/3}{1-1/3+\cfrac{3/5}{1-3/5+\cfrac{5/7}{1-7/9+ \ddots +\cfrac{(2n-1)/(2n+1)}{1-(2n-1)/(2n+1)}}}}}. \tag{5.2.15}$$

Now multiply the top and bottom of the first fraction by 1, and then the top and bottom of the second fraction by 3, and then the top and bottom of the third fraction by 5, and so on. Then (5.2.15) becomes c_n. Since $s_n \to \pi/4$, we conclude that $c_n \to \pi/4$. This completes the derivation of (5.2.4).

The series (5.2.5) is remarkable not only because of its rapid convergence, but because it can be used to compute specific digits in the hexadecimal (base 16, see §1.6) expansion of π, without computing all previous digits ([5]).

To obtain (5.2.5), check that

$$\frac{4\sqrt{2}-8x^3-4\sqrt{2}x^4-8x^5}{1-x^8} = \frac{4\sqrt{2}-4x}{x^2-\sqrt{2}x+1} - \frac{4x}{1-x^2}$$

using $x^8-1=(x^4-1)(x^4+1)$ and $x^4+1=(x^2+\sqrt{2}x+1)(x^2-\sqrt{2}x+1)$. Hence, (Exercises **3.7.15** and **3.7.16**),

$$\begin{aligned}&\int \frac{4\sqrt{2}-8x^3-4\sqrt{2}x^4-8x^5}{1-x^8}\,dx \\ &= 4\arctan(\sqrt{2}x-1) - 2\log(x^2-\sqrt{2}x+1) + 2\log(1-x^2).\end{aligned}$$

Evaluating at 0 and $1/\sqrt{2}$ yields

$$\pi = \int_0^{1/\sqrt{2}} \frac{4\sqrt{2}-8x^3-4\sqrt{2}x^4-8x^5}{1-x^8}\,dx. \tag{5.2.16}$$

To see the equivalence of (5.2.16) and (5.2.5), note that

$$\begin{aligned}\int_0^{1/\sqrt{2}} \frac{x^{k-1}}{1-x^8}dx &= \int_0^{1/\sqrt{2}} \left(\sum_{n=0}^{\infty} x^{k-1+8n}\right) dx \\ &= \sum_{n=0}^{\infty} \int_0^{1/\sqrt{2}} x^{k-1+8n}\,dx\end{aligned}$$

$$= \sum_{n=0}^{\infty} \frac{1}{(k+8n)(\sqrt{2})^{k+8n}}$$
$$= \frac{1}{\sqrt{2}^k} \sum_{n=0}^{\infty} \frac{1}{16^n(k+8n)}.$$

Now use this with k equal 1, 4, 5, and 6, and insert the resulting four series in (5.2.16). You obtain (5.2.5).

To derive the dominated convergence theorem, we will need *Fatou's lemma* which is Exercise **5.1.2**. This states that, for any sequence $f_n : (a,b) \to \mathbf{R}$, $n \geq 1$, of nonnegative functions satisfying $f_n(x) \to f(x)$ for all x in (a,b), the lower limit of the sequence $\left(\int_a^b f_n(x)\,dx\right)$ is greater or equal to $\int_a^b f(x)\,dx$. Although shelved as an exercise, we caution the reader that Fatou's lemma is so frequently useful that it rivals the monotone convergence theorem and the dominated convergence theorem in importance.

Let I^* and I_* denote the upper and lower limits of the sequence $(I_n) = \left(\int_a^b f_n(x)\,dx\right)$, and let $I = \int_a^b f(x)\,dx$. It is enough to show that

$$I \leq I_* \leq I^* \leq I, \tag{5.2.17}$$

since this implies the convergence of (I_n) to I.

If f_n, $n \geq 1$, are as given, then $\pm f_n(x) \leq g(x)$. Hence, $g(x) - f_n(x)$ and $g(x) + f_n(x)$, $n \geq 1$, are nonnegative and converge to $g(x) - f(x)$ and $g(x) + f(x)$, respectively, for all x in (a,b).

Apply Fatou's lemma to the sequence $(g + f_n)$. Then

$$\begin{aligned}\int_a^b g(x)\,dx + I &= \int_a^b [g(x) + f(x)]\,dx \leq \liminf_{n\to\infty} \int_a^b [g(x) + f_n(x)]\,dx \\ &= \int_a^b g(x)\,dx + \liminf_{n\to\infty} I_n = \int_a^b g(x)\,dx + I_*;\end{aligned}$$

hence, $I \leq I_*$ which is half of (5.2.17). Here we are justified in using linearity (Theorem 4.4.5) since the functions f, g, and f_n, $n \geq 1$, are continuous.

Now apply Fatou's lemma to the sequence $(g - f_n)$. Then

$$\begin{aligned}\int_a^b g(x)\,dx - I &= \int_a^b [g(x) - f(x)]\,dx \leq \liminf_{n\to\infty} \int_a^b [g(x) - f_n(x)]\,dx \\ &= \int_a^b g(x)\,dx - \limsup_{n\to\infty} I_n = \int_a^b g(x)\,dx - I^* ;\end{aligned}$$

hence, $I \geq I^*$ which is the other half of (5.2.17). □

A useful consequence of the dominated convergence theorem is the following.

Theorem 5.2.5 (Continuity Under the Integral Sign). *Let $f : (a,b) \times (c,d) \to \mathbf{R}$ be such that $f(\cdot,t)$ is continuous for all $c < t < d$. Suppose there is an integrable $g : (c,d) \to \mathbf{R}$ satisfying $|f(x,t)| \le g(t)$ for $a < x < b$ and $c < t < d$. Then $f(x,\cdot)$ is integrable on (c,d) for $a < x < b$ and*

$$F(x) = \int_c^d f(x,t)\,dt, \qquad a < x < b, \tag{5.2.18}$$

is well defined. If g and $f(x,\cdot)$, $a < x < b$, are continuous, then F is continuous.

Note that the domination hypothesis guarantees that F is well defined. To establish this, fix x in (a,b) and let $x_n \to x$. We have to show that $F(x_n) \to F(x)$. Let $k_n(t) = f(x_n,t)$, $c < t < d$, $n \ge 1$, and $k(t) = f(x,t)$, $c < t < d$. Then $k_n(t)$ and $k(t)$ are continuous on (c,d) and $k_n(t) \to k(t)$ for $c < t < d$. By the domination hypothesis, $|k_n(t)| \le g(t)$. Thus, the dominated convergence theorem applies, and

$$F(x_n) = \int_c^d k_n(t)\,dt \to \int_c^d k(t)\,dt = F(x). \quad \square$$

For example, the continuity of the gamma function on (a,b), $0 < a < b < \infty$, follows by choosing, as in the beginning of the section,

$$g(t) = \begin{cases} e^{-t}t^{b-1}, & 1 \le t < \infty, \\ e^{-t}t^{a-1}, & 0 < t \le 1. \end{cases}$$

Moreover, the separate continuity of f in (x,t) is immediate because f is the product of two continuous functions, one of x and one of t. Because continuity is established in the same manner in all our examples below, we will usually omit this step.

In fact, continuity under the integral sign is nothing but a packaging of the derivation of continuity of Γ, presented earlier.

Let us go back to the statement of the dominated convergence theorem. When $(a,b) = (0,\infty)$ and the functions (f_n), f, and g are piecewise constant, the dominated convergence theorem reduces to the following.

Theorem 5.2.6 (Dominated Convergence Theorem (for Series). *Let (a_{nj}), $n \ge 1$, be a sequence of sequences, and let (a_j) be a given sequence. Also suppose that there is a convergent positive series $\sum_{j=1}^\infty g_j$ satisfying $|a_{nj}| \le g_j$ for all $j \ge 1$ and $n \ge 1$. If*

$$\lim_{n \nearrow \infty} a_{nj} = a_j, \qquad j \ge 1,$$

then

$$\lim_{n\nearrow\infty}\sum_{j=1}^{\infty}a_{nj}=\sum_{j=1}^{\infty}\lim_{n\nearrow\infty}a_{nj}=\sum_{j=1}^{\infty}a_j.$$

To see this, for $j-1<x\le j$ set $f_n(x)=a_{nj}$, $n\ge 1$, $f(x)=a_j$, and $g(x)=g_j$, $j=1,2,\dots$, and use Exercise **4.3.6**. □

Let us use the dominated convergence theorem for series to show[3]

$$\lim_{x\to 1}\left(1-\frac{1}{2^x}+\frac{1}{3^x}-\frac{1}{4^x}+\dots\right)=1-\frac{1}{2}+\frac{1}{3}-\frac{1}{4}+\dots, \tag{5.2.19}$$

which sums to $\log 2$. For this, by the mean value theorem, $(2j-1)^{-x}-(2j)^{-x}=x(2j-t)^{-x-1}$ for some $0<t<1$. Hence, $(2j-1)^{-x}-(2j)^{-x}\le 2(2j-1)^{-3/2}$ when $1/2<x<2$. Now let $x_n\to 1$ with $1/2<x_n<2$, and set $a_{nj}=(2j-1)^{-x_n}-(2j)^{-x_n}$, $a_j=(2j-1)^{-1}-(2j)^{-1}$, $g_j=2(2j-1)^{-3/2}$ for $j\ge 1$, $n\ge 1$. Then $a_{nj}\to a_j$ and $|a_{nj}|\le g_j$ for all $j\ge 1$. Hence, the theorem applies, and, since the sequence (x_n) is arbitrary, we obtain (5.2.19). Note how, here, we are not choosing a_{nj} as the individual terms but as pairs of terms, producing an absolutely convergent series out of a conditionally convergent one (cf. the Dirichlet test (§1.7)).

Just as we used the dominated convergence theorem for integrals to obtain continuity under the integral sign, we can use the theorem for series to obtain the following.

Theorem 5.2.7 (Continuity Under the Summation Sign). *Let (f_n) be a sequence of continuous functions defined on (a,b), and suppose that there is a convergent positive series $\sum_{n=1}^{\infty}g_n$ of numbers satisfying $|f_n(x)|\le g_n$ for $n\ge 1$ and $a<x<b$. If*

$$F(x)=\sum_{n=1}^{\infty}f_n(x),\qquad a<x<b,$$

then $F:(a,b)\to\mathbf{R}$ is continuous. □

For example,

$$\zeta(x)=\sum_{n=1}^{\infty}\frac{1}{n^x}$$

is continuous on (a,∞) for $a>1$, since $1/n^x\le 1/n^a$ for $x\ge a$ and $\sum g_n=\sum 1/n^a$ converges. Since $a>1$ is arbitrary, ζ is continuous on $(1,\infty)$.

[3] This series converges for $x>0$ by the Leibnitz test.

Exercises

5.2.1. Derive

$$\int_0^1 \frac{t^4(1-t)^4}{1+t^2}\,dt = \frac{22}{7} - \pi.$$

Thus, $\pi \neq 22/7$.

5.2.2. Use

$$\frac{\sin x}{x} = 1 - \frac{x^2}{3!} + \frac{x^4}{5!} - \dots$$

to derive the Laplace transform

$$\int_0^\infty e^{-sx}\frac{\sin x}{x}\,dx = \arctan\left(\frac{1}{s}\right), \qquad s > 1.$$

5.2.3. Suppose that f_n, $n \geq 1$, f, and g are as in the dominated convergence theorem. Show that f is integrable over (a, b).

5.2.4. Show that $x^2 J_0''(x) + xJ_0'(x) + x^2 J_0(x) = 0$ for all x.

5.2.5. Derive (5.2.11) by integrating by parts.

5.2.6. Show that

$$\sin x/x = \cos(x/2)\cos(x/4)\cos(x/8)\dots.$$

5.2.7. This is an example where switching the integral and the series changes the answer. Show that

$$\sum_{n=0}^\infty \frac{(-1)^n}{n!}\int_0^\infty e^{-x}x^n\,dx \neq \int_0^\infty e^{-x}\left[\sum_{n=0}^\infty \frac{(-x)^n}{n!}\right]dx.$$

5.2.8. Show that the *Fourier transform* (compare with Exercise **5.1.8**)

$$\int_0^\infty \frac{\sin(sx)}{e^x - 1}\,dx = \sum_{n=1}^\infty \frac{s}{n^2 + s^2}, \qquad -\infty < s < \infty.$$

5.2.9. Show that

$$\int_0^\infty \frac{\sinh(sx)}{e^x - 1}\,dx = \sum_{n=1}^\infty \frac{s}{n^2 - s^2}, \qquad |s| < 1.$$

5.2.10. Let s_n, $n \geq 0$, be the nth partial sum of the Bailey–Borwein–Plouffe series. Show

$$s_n < \pi < s_n + \frac{1}{4(n+1)^2 16^{n+1}} \equiv S_n, \qquad n \geq 0.$$

5.2.11. With s_n and S_n as in the previous exercise, $n \geq 0$, write computer code to show

$$s_5 = 4041374233034931 6707/1286409372291563 5200,$$
$$S_5 = 6207550822759532098 6877/1975924795839841 5667200.$$

Following Exercise **1.3.18** show the continued fraction expansions of s_5 and S_5 both start out as $3 + [7, 15, 1, 292, \dots]$. Conclude (Exercise **1.3.23**)

$$\pi = 3 + \cfrac{1}{7 + \cfrac{1}{15 + \cfrac{1}{1 + \cfrac{1}{292 + \dots}}}}$$

leading to the convergents $22/7 = 3 + [7]$ and $355/113 = 3 + [7, 15, 1]$.

5.2.12. Show that the νth *Bessel function*

$$J_\nu(x) = \frac{1}{\pi} \int_0^\pi \cos(\nu t - x \sin t)\, dt, \qquad -\infty < x < \infty,$$

is continuous. Here ν is any real.

5.2.13. Show that $\psi(x) = \sum_{n=1}^\infty e^{-n^2 \pi x}$, $x > 0$, is continuous.

5.2.14. Let $f_n, f, g : (a, b) \to \mathbf{R}$ be as in the dominated convergence theorem, and suppose that $a_n \to a+$ and $b_n \to b-$. Suppose that we have domination $|f_n(x)| \leq g(x)$ only on (a_n, b_n), $n \geq 1$. Show that

$$\lim_{n \nearrow \infty} \int_{a_n}^{b_n} f_n(x)\, dx = \int_a^b f(x)\, dx.$$

5.2.15. Show J_0 in the text is the same as J_ν in Exercise **5.2.12** with $\nu = 0$.

5.2.16. Use Exercise **4.4.22** to show that Euler's constant satisfies

$$\gamma = \lim_{n \nearrow \infty} \left[\int_0^1 \frac{1 - (1 - t/n)^n}{t}\, dt - \int_1^n \frac{(1 - t/n)^n}{t}\, dt \right].$$

Use the dominated convergence theorem to conclude that

$$\gamma = \int_0^1 \frac{1 - e^{-t}}{t}\, dt - \int_1^\infty \frac{e^{-t}}{t}\, dt.$$

(For the second part, first show $0 \leq [1 - (1 - t/n)^n]/t \leq 1$.)

5.2.17. Use Euler's continued fraction formula to derive

$$\arctan x = \cfrac{x}{1 + \cfrac{x^2}{3 - x^2 + \cfrac{(3x)^2}{5 - 3x^2 + \cfrac{(5x)^2}{7 - 5x^2 + \cfrac{(7x)^2}{\ddots}}}}}.$$

5.2.18. Use Euler's continued fraction formula to derive

$$e^x = 1 + \cfrac{x}{1 - \cfrac{x}{2 + x - \cfrac{2x}{3 + x - \cfrac{3x}{4 + x - \cfrac{4x}{\ddots}}}}}.$$

5.3 Gauss' Arithmetic–Geometric Mean

Given $a > b > 0$, their *arithmetic mean* is given by

$$a' = \frac{a+b}{2},$$

and their *geometric mean* by

$$b' = \sqrt{ab}.$$

Since

$$a' - b' = \frac{a+b}{2} - \sqrt{ab} = \frac{1}{2}\left(\sqrt{a} - \sqrt{b}\right)^2 > 0, \tag{5.3.1}$$

these equations transform the pair (a, b), $a > b > 0$, into a pair (a', b'), $a' > b' > 0$. Gauss discovered that iterating this transformation leads to a limit with striking properties.

To begin, since a is the larger of a and b and a' is their arithmetic mean, $a' < a$. Similarly, since b is the smaller of a and b, $b' > b$. Thus, $b < b' < a' < a$.

Set $a_0 = a$ and $b_0 = b$, and define the iteration

$$a_{n+1} = \frac{a_n + b_n}{2}, \tag{5.3.2}$$

$$b_{n+1} = \sqrt{a_n b_n}, \qquad n \geq 0. \tag{5.3.3}$$

By the previous paragraph, for $a > b > 0$, this gives a strictly decreasing sequence (a_n) and a strictly increasing sequence (b_n) with all the a's greater than all the b's. Thus, both sequences converge (Figure 5.3) to finite positive limits a_*, b^* with $a_* \geq b^* > 0$.

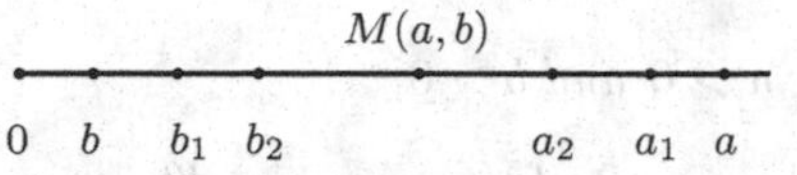

$M(a,b)$

$0 \quad b \quad b_1 \quad b_2 \qquad\qquad a_2 \quad a_1 \quad a$

Fig. 5.3 The AGM iteration

Letting $n \nearrow \infty$ in (5.3.2), we see that $a_* = (a_* + b^*)/2$ which yields $a_* = b^*$. Thus, both sequences converge to a common limit, the *arithmetic–geometric mean (AGM) of (a,b)*, which we denote

$$M(a,b) = \lim_{n \nearrow \infty} a_n = \lim_{n \nearrow \infty} b_n.$$

If (a'_n), (b'_n) are the sequences associated with $a' = ta$ and $b' = tb$, then from (5.3.2) and (5.3.3), $a'_n = ta_n$, and $b'_n = tb_n$, $n \geq 1$, $t > 0$. This implies that M is homogeneous in (a,b),

$$M(ta, tb) = t \cdot M(a,b), \quad t > 0.$$

The convergence of the sequences (a_n), (b_n) to the real $M(a,b)$ is *quadratic* in the following sense. The differences $a_n - M(a,b)$ and $M(a,b) - b_n$ are no more than $2c_{n+1}$, where

$$c_{n+1} = \frac{a_n - b_n}{2}, \qquad n \geq 0. \tag{5.3.4}$$

By (5.3.1),

$$0 < c_{n+1} = \frac{1}{4}\left(\sqrt{a_{n-1}} - \sqrt{b_{n-1}}\right)^2 = \frac{c_n^2}{\left(\sqrt{a_{n-1}} + \sqrt{b_{n-1}}\right)^2} \leq \frac{1}{4b} c_n^2.$$

Iterating the last inequality yields

$$0 < a_n - b_n \leq 8b \left(\frac{a-b}{8b}\right)^{2^n}, \qquad n \geq 1. \tag{5.3.5}$$

This shows that each additional iteration roughly doubles the number-of-decimal-place agreement, at least if $(a-b)/8b < 1$. For a general pair (a,b), eventually, $(a_N - b_N)/8b_N < 1$. After this point, we have the rapid convergence (5.3.5). In (5.3.15) below, we improve (5.3.5) from an inequality to an asymptotic equality. In Exercise **5.7.5**, we further improve this to an actual equality.

For future reference, note that

$$a_n^2 = b_n^2 + c_n^2, \qquad n \geq 1.$$

The following remarkable formula is due to Gauss.

Theorem 5.3.1. *For $a > 0$ and $b > 0$,*

$$\frac{1}{M(a,b)} = \frac{2}{\pi} \int_0^{\pi/2} \frac{d\theta}{\sqrt{a^2 \cos^2 \theta + b^2 \sin^2 \theta}}.$$

Gauss was initially guided to this formula by noting both sides agreed to eleven decimal places when $(a,b) = (1, 1/\sqrt{2})$. We compute $M(1, 1/\sqrt{2})$ explicitly in the next section (see (5.4.5)).

The derivation is best understood within the context of complex numbers.[4] We present this proof cosmetically altered to remain within the real domain.

To derive the formula, let

$$I(a,b) = \frac{1}{\pi} \int_0^{\pi} \frac{d\theta}{\sqrt{a^2 \cos^2 \theta + b^2 \sin^2 \theta}} = \frac{2}{\pi} \int_0^{\pi/2} \frac{d\theta}{\sqrt{a^2 \cos^2 \theta + b^2 \sin^2 \theta}}.$$

Note $I(a,b) = I(b,a)$ and $I(m,m) = 1/m$. The main step is to establish invariance of I under the AGM iteration

$$I(a,b) = I\left(\frac{a+b}{2}, \sqrt{ab}\right) = I(a', b'). \qquad (5.3.6)$$

The result follows from (5.3.6) by iteration $I(a_n, b_n) = I(a,b)$ followed by passing to the limit $n \to \infty$ (Exercise **5.3.2**). Since $(a_n, b_n) \to (m,m)$, $m = M(a,b)$, the result follows.

The invariance (5.3.6) is established by the substitution

$$\theta' = \arccos\left(\frac{a\cos^2\theta - b\sin^2\theta}{a\cos^2\theta + b\sin^2\theta}\right).$$

This map $\theta \mapsto \theta'$ is smooth on $(0, \pi/2)$ and is a continuous bijection of $[0, \pi/2]$ onto $[0, \pi]$ (Exercise **5.3.3**).

To compute the derivative $d\theta'/d\theta$, we use the fact (§3.6) that (x,y) is on the unit circle iff $(x,y) = (\cos\theta, \sin\theta)$ and define

$$x' = \frac{ax^2 - by^2}{ax^2 + by^2}, \qquad y' = \frac{2\sqrt{ab}xy}{ax^2 + by^2}.$$

Then $(x,y) = (\cos\theta, \sin\theta)$ on $[0, \pi/2]$ iff $(x', y') = (\cos\theta', \sin\theta')$ on $[0, \pi]$.

[4] Via the unit circle map $x' + iy' = (\sqrt{a}x + i\sqrt{b}y)/(\sqrt{a}x - i\sqrt{b}y)$.

Now let $\lambda = 1/(ax^2 + by^2)$. Then

$$xdy - ydx = \cos\theta\cos\theta\, d\theta - \sin\theta(-\sin\theta)\, d\theta = (\cos^2\theta + \sin^2\theta)d\theta = d\theta$$

and $(x', y') = (\lambda(ax^2 - by^2), 2\lambda b' xy)$ so

$$\begin{aligned} dx' &= \frac{d\lambda}{\lambda}x' + \lambda\,(2axdx - 2bydy)\,, \\ dy' &= \frac{d\lambda}{\lambda}y' + \lambda 2b'\,(xdy + ydx)\,; \end{aligned}$$

hence,

$$d\theta' = x'dy' - y'dx' = 2b'\lambda^2(ax^2 + by^2)(xdy - ydx) = 2b'\lambda d\theta.$$

Now

$$\begin{aligned} b'^2x'^2 + a'^2y'^2 &= \lambda^2 b'^2(ax^2 - by^2)^2 + \lambda^2(a+b)^2 b'^2 x^2 y^2 \\ &= \lambda^2 b'^2\left(a^2x^2y^2 + b^2x^2y^2 + b^2y^4 + a^2x^4\right) \\ &= \lambda^2 b'^2(x^2 + y^2)(a^2x^2 + b^2y^2). \end{aligned}$$

Dividing the last two equations yields

$$\frac{d\theta'}{\sqrt{b'^2\cos^2\theta' + a'^2\sin^2\theta'}} = 2\frac{d\theta}{\sqrt{a^2\cos^2\theta + b^2\sin^2\theta}}$$

which implies

$$\int_0^\pi \frac{d\theta'}{\sqrt{b'^2\cos^2\theta' + a'^2\sin^2\theta'}} = \int_0^{\pi/2} 2\frac{d\theta}{\sqrt{a^2\cos^2\theta + b^2\sin^2\theta}}.$$

Thus, $I(a', b') = I(b', a') = I(a, b)$. □

Next, we look at the behavior of $M(1, x)$, as $x \to 0+$. When $a_0 = 1$ and $b_0 = 0$, the arithmetic–geometric iteration yields $a_n = 2^{-n}$ and $b_n = 0$ for all $n \geq 1$. Hence, $M(1, 0) = 0$. This leads us to believe that $M(1, x) \to 0$, as $x \to 0+$, or, what is the same, $1/M(1, x) \to \infty$, as $x \to 0+$. Exactly at what speed this happens leads us to another formula for π.

Theorem 5.3.2.

$$\lim_{x\to 0+}\left[\frac{\pi/2}{M(1,x)} - \log\left(\frac{4}{x}\right)\right] = 0. \tag{5.3.7}$$

To derive this, from Exercise **5.3.4**,

$$\frac{\pi/2}{M(1,x)} = \int_0^\infty \frac{dt}{\sqrt{(1+t^2)(x^2+t^2)}}. \tag{5.3.8}$$

By Exercise **5.3.7**, this equals

$$\frac{\pi/2}{M(1,x)} = 2\int_0^{1/\sqrt{x}} \frac{dt}{\sqrt{(1+t^2)(1+(xt)^2)}}. \tag{5.3.9}$$

Now call the right side of (5.3.9) $I(x)$. Thus, the result will follow if we show that

$$\lim_{x\to 0+}\left[I(x) - \log\left(\frac{4}{x}\right)\right] = 0. \tag{5.3.10}$$

To derive (5.3.10), note that

$$\begin{aligned} J(x) = 2\int_0^{1/\sqrt{x}} \frac{dt}{\sqrt{1+t^2}} &= 2\log\left(t+\sqrt{1+t^2}\right)\Big|_0^{1/\sqrt{x}} \\ &= 2\log\left(1+\sqrt{x+1}\right) + \log\left(\frac{1}{x}\right) \\ &= 2\log\left(\frac{1}{2}+\frac{1}{2}\sqrt{x+1}\right) + \log\left(\frac{4}{x}\right) \end{aligned}$$

and, so $\log(4/x) - J(x) \to 0$ as $x \to 0+$. Thus, it is enough to show that

$$\lim_{x\to 0+}\left[I(x) - J(x)\right] = 0. \tag{5.3.11}$$

But, for $xt > 0$,

$$0 \le 1 - \frac{1}{\sqrt{1+(xt)^2}} \le 1 - \frac{1}{1+xt} = \frac{xt}{1+xt} \le xt.$$

So

$$\begin{aligned} 0 \le J(x) - I(x) &= 2\int_0^{1/\sqrt{x}} \frac{1}{\sqrt{1+t^2}}\left[1 - \frac{1}{\sqrt{1+(xt)^2}}\right] dt \\ &\le 2\int_0^{1/\sqrt{x}} \frac{xt}{\sqrt{1+t^2}}\, dt \\ &= 2x\sqrt{1+t^2}\Big|_0^{1/\sqrt{x}} = 2\sqrt{x(1+x)} - 2x, \end{aligned}$$

which clearly goes to zero as $x \to 0+$. □

Our next topic is the functional equation. First and foremost, since the AGM limit starting from (a_0, b_0) is the same as that starting from (a_1, b_1),

$$M(a,b) = M\left(\frac{a+b}{2}, \sqrt{ab}\right).$$

Below, given $0 < x < 1$, we let $x' = \sqrt{1 - x^2}$ be the *complementary variable.* For example, $(x')' = x$ and $k = 2\sqrt{x}/(1 + x)$ implies $k' = (1 - x)/(1 + x)$ since

$$\left(\frac{2\sqrt{x}}{1+x}\right)^2 + \left(\frac{1-x}{1+x}\right)^2 = 1.$$

Also with a_n, b_n, c_n, $n \geq 1$, as above, $(b_n/a_n)' = c_n/a_n$. The functional equation we are after is best expressed in terms of the function

$$Q(x) = \frac{M(1,x)}{M(1,x')}, \qquad 0 < x < 1.$$

Note that $Q(x') = 1/Q(x)$.

Theorem 5.3.3 (AGM Functional Equation).

$$Q(x) = 2Q\left(\frac{1-x'}{1+x'}\right), \qquad 0 < x < 1. \tag{5.3.12}$$

To see this, note that $M(1 + x', 1 - x') = M(1, x)$. So

$$M(1,x) = M(1+x', 1-x') = (1+x')M\left(1, \frac{1-x'}{1+x'}\right). \tag{5.3.13}$$

Here we used homogeneity of M. On the other hand,

$$\begin{aligned} M(1,x') = M\left[(1+x')/2, \sqrt{x'}\right] &= \frac{(1+x')}{2} M\left(1, \frac{2\sqrt{x'}}{1+x'}\right) \\ &= \frac{(1+x')}{2} M\left[1, \left(\frac{1-x'}{1+x'}\right)'\right]. \end{aligned} \tag{5.3.14}$$

Here again, we used homogeneity of M. Dividing (5.3.13) by (5.3.14), the result follows. □

If (a_n) and (b_n) are positive sequences, we say that (a_n) *and* (b_n) *are asymptotically equal*, and we write $a_n \sim b_n$, as $n \nearrow \infty$, if $a_n/b_n \to 1$, as $n \nearrow \infty$. Note that (a_n) and (b_n) are asymptotically equal iff $\log(a_n) - \log(b_n) \to 0$. Now we combine the last two results to obtain the following improvement of (5.3.5).

Theorem 5.3.4. *Let $a > b > 0$, and let a_n, b_n, $n \geq 1$, be as in* (5.3.2),(5.3.3). *Then*

$$a_n - b_n \sim 8M(a,b) \cdot q^{2^n}, \qquad n \nearrow \infty, \tag{5.3.15}$$

where $q = e^{-\pi Q(b/a)}$.

To derive this, use (5.3.4) and $a_n \to M(a, b)$ to check that (5.3.15) is equivalent to

$$\frac{c_n}{4a_n} \sim \left(e^{-\pi Q(b/a)/2}\right)^{2^n}, \qquad n \nearrow \infty. \tag{5.3.16}$$

Now let $x_n = c_n/a_n$. Then $x_n \to 0$, as $n \nearrow \infty$. By taking the log of (5.3.16), it is enough to show that

$$\log\left(\frac{4}{x_n}\right) - 2^n \frac{\pi}{2} Q\left(\frac{b}{a}\right) \to 0, \qquad n \nearrow \infty. \tag{5.3.17}$$

By (5.3.7), (5.3.17) is implied by

$$\frac{1}{M(1, x_n)} - 2^n Q\left(\frac{b}{a}\right) \to 0, \qquad n \nearrow \infty. \tag{5.3.18}$$

By Exercise **5.1.6**, (5.3.18) is implied by

$$\frac{1}{Q(x_n)} - 2^n Q\left(\frac{b}{a}\right) \to 0, \qquad n \nearrow \infty. \tag{5.3.19}$$

In fact, we will show that the left side of (5.3.19) is zero for all $n \geq 1$. To this end, since $c_n/a_n = (b_n/a_n)'$,

$$\begin{aligned} \frac{c_{n+1}}{a_{n+1}} &= \frac{a_n - b_n}{a_n + b_n} \\ &= \frac{1 - (b_n/a_n)}{1 + (b_n/a_n)} \\ &= \frac{1 - (c_n/a_n)'}{1 + (c_n/a_n)'}. \end{aligned}$$

Hence, by the functional equation,

$$Q\left(c_{n+1}/a_{n+1}\right) = Q\left(\frac{1 - (c_n/a_n)'}{1 + (c_n/a_n)'}\right) = \frac{1}{2} Q\left(c_n/a_n\right).$$

Iterating this down to $n = 1$, we obtain

$$\begin{aligned} Q(c_n/a_n) &= 2^{-1} Q(c_{n-1}/a_{n-1}) = 2^{-2} Q(c_{n-2}/a_{n-2}) = \dots \\ &\dots = 2^{-(n-1)} Q(c_1/a_1) = 2^{-n} Q((b/a)') = 2^{-n}/Q(b/a), \quad n \geq 1. \end{aligned}$$

This shows that $1/Q(x_n) = 2^n Q(b/a)$. □

Dividing by 2^n in (5.3.17), we obtain

$$\lim_{n \nearrow \infty} 2^{-n} \log\left(\frac{a_n}{c_n}\right) = \frac{\pi}{2} Q\left(\frac{b}{a}\right), \tag{5.3.20}$$

which we will need in §5.7. Note that we have discarded the 4 since $2^{-n} \log 4 \to 0$. In the exercises below, the AGM is generalized from two variables (a, b) to d variables $(x_1, \dots, x_d)$.

Exercises

5.3.1. Fix $0 < t < 1$. For $0 < b < a$, let

$$a' = (1-t)a + tb, \qquad b' = a^{1-t}b^t.$$

Define an iteration by $a_{n+1} = a'_n$, $b_{n+1} = b'_n$. Show that (a_n), (b_n) converge to a common limit.

5.3.2. Use the dominated convergence theorem to show that $a_n \to a$ and $b_n \to b$, $a > b > 0$, implies $I(a_n, b_n) \to I(a, b)$.

5.3.3. Show that the map $\theta' = G(\theta)$ is a continuous bijection from $[0, \pi/2]$ to $[0, \pi]$.

5.3.4. Use $t = b\tan\theta$ to show

$$I(a,b) = \frac{2}{\pi}\int_0^\infty \frac{dt}{\sqrt{(a^2+t^2)(b^2+t^2)}}.$$

5.3.5. Show that

$$\frac{1}{M(1+x, 1-x)} = \frac{2}{\pi}\int_0^{\pi/2} \frac{d\theta}{\sqrt{1 - x^2\sin^2\theta}}, \qquad 0 < x < 1.$$

5.3.6. Show that

$$\frac{1}{M(1+x, 1-x)} = \sum_{n=0}^\infty 16^{-n}\binom{2n}{n}^2 x^{2n}, \qquad 0 < x < 1,$$

by using the binomial theorem to expand the square root and integrating term by term.

5.3.7. Show that

$$\frac{1}{M(1,x)} = \frac{4}{\pi}\int_0^{\sqrt{x}} \frac{dt}{\sqrt{(1+t^2)(x^2+t^2)}} = \frac{4}{\pi}\int_0^{1/\sqrt{x}} \frac{dt}{\sqrt{(1+(xt)^2)(1+t^2)}}.$$

(Break (5.3.8) into $\int_0^{\sqrt{x}} + \int_{\sqrt{x}}^\infty$, and substitute $t = x/s$ in the second piece.)

5.3.8. With $x' = \sqrt{1-x^2}$, show that $M(1+x, 1-x) = M((1+x')/2, \sqrt{x'})$.

5.3.9. Show that

$$\left|\frac{1}{M(1,x)} - \frac{1}{Q(x)}\right| \le x, \qquad 0 < x < 1.$$

5.3.10. Show that

$$Q(x) = \frac{1}{2} Q\left(\frac{2\sqrt{x}}{1+x}\right), \qquad 0 < x < 1.$$

5.3.11. Show that $M(1, \cdot) : (0,1) \to (0,1)$ and $Q : (0,1) \to (0,\infty)$ are strictly increasing, continuous bijections.

5.3.12. Show that for each $a > 1$, there exists a unique $1 > b = f(a) > 0$, such that $M(a,b) = 1$.

5.3.13. With f as in the previous Exercise, use (5.3.7) to show

$$f(a) \sim 4ae^{-\pi a/2}, \qquad a \to \infty.$$

(Let $x = b/a = f(a)/a$ and take logs of both sides.)

5.3.14. Given reals $a_1, \dots, a_d$, let $p_1, \dots, p_d$ be given by

$$(x + a_1)(x + a_2) \dots (x + a_d) = x^d + \binom{d}{1} p_1 x^{d-1} + \dots + \binom{d}{d-1} p_{d-1} x + p_d.$$

Then $p_1, \dots, p_d$ are polynomials in $a_1, \dots, a_d$, the so-called elementary symmetric polynomials.[5] Show that

$$p_k(1, 1, \dots, 1) = 1, \qquad 1 \le k \le d,$$

p_1 is the arithmetic mean $(a_1 + \dots + a_d)/d$ and p_d is the product $a_1 a_2 \dots a_d$. For $a_1, \dots, a_d$ positive, conclude (Exercise **3.3.28**) the *arithmetic and geometric mean inequality*

$$\frac{a_1 + \dots + a_d}{d} \ge (a_1 a_2 \dots a_d)^{1/d},$$

with equality iff all the a_j's are equal.

5.3.15. Given $a_1 \ge a_2 \ge \dots \ge a_d > 0$, let $a_1' = p_1(a_1, \dots, a_d)$ be their arithmetic mean and let $a_d' = p_d(a_1, \dots, a_d)^{1/d}$ be their geometric mean. Use Exercise **3.2.10** to show

$$\left(\frac{a_1'}{a_d'} - 1\right) \le \left(\frac{d-1}{d}\right)^2 \left(\frac{a_1}{a_d} - 1\right).$$

5.3.16. Given $a_1 \ge a_2 \ge \dots \ge a_d > 0$, let

$$(a_1', a_2', \dots, a_d') = G(a_1, a_2, \dots, a_d) = (p_1, p_2^{1/2}, \dots, p_d^{1/d}),$$

$$(a_1'', a_2'', \dots, a_d'') = G^2(a_1, a_2, \dots, a_d) = G(a_1', a_2', \dots, a_d'),$$

[5] More accurately, the *normalized* elementary symmetric polynomials

and so on. This defines a sequence

$$(a_1^{(n)}, a_2^{(n)}, \dots, a_d^{(n)}) = G^n(a_1, a_2, \dots, a_d), \qquad n \geq 0.$$

Show that $(a_1^{(n)})$ is decreasing, $(a_d^{(n)})$ is increasing, and

$$\frac{a_1^{(n)}}{a_d^{(n)}} - 1 \leq \left(\frac{d-1}{d}\right)^{2n} \left(\frac{a_1}{a_d} - 1\right), \qquad n \geq 0.$$

Conclude that there is a positive real m such that

$$a_j^{(n)} \to m \qquad n \to \infty, 1 \leq j \leq d,$$

(Exercise **3.3.28** and Exercise **5.3.15**). If we set $m = M(a_1, \dots, a_d)$, show that

$$M(a_1, a_2, \dots, a_d) = M\left(p_1, p_2^{1/2}, \dots, p_d^{1/d}\right).$$

An integral formula for $M(a_1, a_2, \dots, a_d)$ analogous to that of Theorem 5.3.1 is not known.

5.4 The Gaussian Integral

In this section, we derive the *Gaussian integral*

$$\int_{-\infty}^{\infty} e^{-x^2/2}\, dx = \sqrt{2\pi}. \tag{5.4.1}$$

This formula is remarkable because the primitive of $e^{-x^2/2}$ cannot be expressed in terms of the elementary functions (i.e., the functions studied in Chapter 3). Nevertheless the area (Figure 5.4) of the (total) subgraph of $e^{-x^2/2}$ is explicitly computable. Because of (5.4.1), the *Gaussian function* $g(x) = e^{-x^2/2}/\sqrt{2\pi}$ has total area under its graph equal to 1.

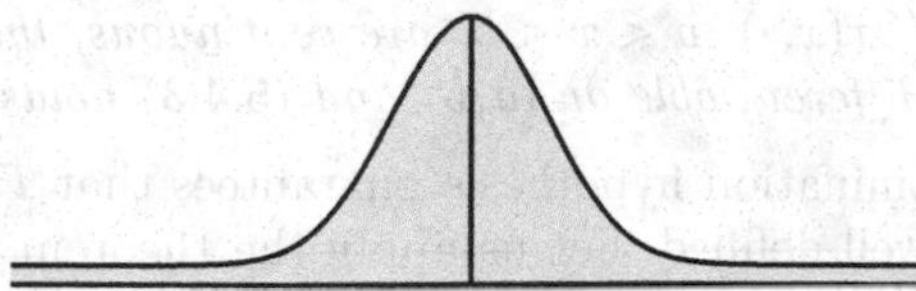

Fig. 5.4 The Gaussian function

The usual derivation of (5.4.1) involves changing variables from Cartesian coordinates (x, y) to polar coordinates in a double integral. How to do this is a two-variable result. Here we give an elementary derivation that uses only the one-variable material we have studied so far. To derive (5.4.1), we will, however, need to know how to "differentiate under an integral sign."

To explain this, consider the integral

$$F(x) = \int_c^d f(x,t)\,dt, \qquad a < x < b, \tag{5.4.2}$$

where $f(x,t) = 3(2x+t)^2$ and $a < b$, $c < d$ are reals. We wish to differentiate F. There are two ways we can do this. The first method is to evaluate the integral obtaining $F(x) = (2x+d)^3 - (2x+c)^3$ and then to differentiate to get $F'(x) = 6(2x+d)^2 - 6(2x+c)^2$. The second method is to differentiate the integrand $f(x,t) = 3(2x+t)^2$ with respect to x, obtaining $12(2x+t)$ and, then to evaluate the integral $\int_c^d 12(2x+t)\,dt$, obtaining $6(2x+d)^2 - 6(2x+c)^2$. Since both methods yield the same result, for $f(x,t) = 3(2x+t)^2$, we conclude that

$$F'(x) = \int_c^d \frac{\partial f}{\partial x}(x,t)\,dt, \qquad a < x < b, \tag{5.4.3}$$

where the *partial derivative* $\partial f/\partial x(x,t)$ is the derivative with respect to x,

$$\frac{\partial f}{\partial x}(x,t) = \lim_{x' \to x} \frac{f(x',t) - f(x,t)}{x' - x}, \qquad a < x < b.$$

It turns out that (5.4.2) implies (5.4.3) in a wide variety of cases.

Theorem 5.4.1 (Differentiation Under the Integral Sign). *Let* $f : (a,b) \times (c,d) \to \mathbf{R}$ *be a function of two variables* (x,t)*, such that*

$$\frac{\partial f}{\partial x}(x,t), \qquad a < x < b, c < t < d,$$

exists. Suppose there is an integrable function $g : (c,d) \to \mathbf{R}$*, such that*

$$|f(x,t)| + \left|\frac{\partial f}{\partial x}(x,t)\right| \le g(t), \qquad a < x < b, c < t < d.$$

Then $f(x,\cdot)$ *and* $\partial f/\partial x(x,\cdot)$ *are integrable for all* $a < x < b$*. If* g, $f(x,\cdot)$, $a < x < b$*, and* $\partial f/\partial x(x,\cdot)$, $a < x < b$*, are continuous, then* $F : (a,b) \to \mathbf{R}$ *given by* (5.4.2) *is differentiable on* (a,b) *and* (5.4.3) *holds.*

Note that the domination hypothesis guarantees that $F(x)$ and the right side of (5.4.3) are well defined. Let us apply the theorem right away to obtain (5.4.1).

To this end, let $I = \int_0^\infty e^{-s^2/2}\,ds$ be half the integral in (5.4.1). Since $(s-1)^2 \ge 0$, $-s^2/2 \le (1/2) - s$. Hence, $I \le \int_0^\infty e^{(1/2)-s}\,ds = \sqrt{e}$. Thus, I is finite and

$$I^2 = I \int_0^\infty e^{-t^2/2}\,dt = \int_0^\infty e^{-t^2/2} I\,dt. \tag{5.4.4}$$

Now set

$$f(\theta,t) = e^{-t^2/2}\int_0^{t\cdot\tan\theta} e^{-s^2/2}\,ds, \qquad 0<t<\infty, 0<\theta<\pi/2.$$

Since $\tan(\pi/2-)=\infty$, by continuity at the endpoints,

$$f(\pi/2-,t) = e^{-t^2/2}I, \qquad t>0.$$

Now let

$$F(\theta) = \int_0^\infty f(\theta,t)\,dt, \qquad 0<\theta<\pi/2.$$

Since $f(\theta,t) \le Ie^{-t^2/2}$ and $g(t) = Ie^{-t^2/2}$ is integrable by (5.4.4), by the dominated convergence theorem, we obtain

$$F(\pi/2-) = \lim_{\theta\to\pi/2-}\int_0^\infty f(\theta,t)\,dt = \int_0^\infty f(\pi/2-,t)\,dt = I^2.$$

Thus, to evaluate I^2, we need to compute $F(\theta)$. Although $F(\theta)$ is not directly computable from its definition, it turns out that $F'(\theta)$ is, using differentiation under the integral sign (Figure 5.5).

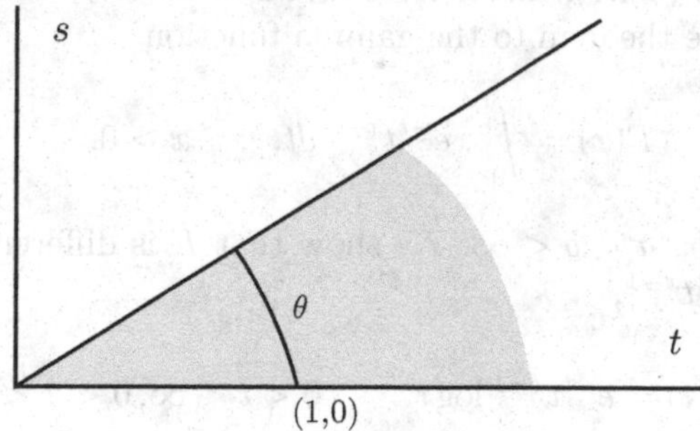

Fig. 5.5 The region of integration defining $F(\theta)$

To motivate where the formula for F comes from, note that the formula for I^2 can be thought of as a double integral over the first quadrant $0<s<\infty$, $0<t<\infty$, in the st-plane, and the formula for $F(\theta)$ can be thought of as a double integral over the triangular sector $0<s<t\cdot\tan\theta$, $0<t<\infty$, in the st-plane. As the angle θ opens up to $\pi/2$, the triangular sector fills the quadrant. Of course, we do not actually use double integrals in the derivation of (5.4.1).

Now by the fundamental theorem and the chain rule,

$$\frac{\partial f}{\partial\theta}(\theta,t) = e^{-t^2(1+\tan^2\theta)/2}t\sec^2\theta = e^{-t^2\sec^2\theta/2}t\sec^2\theta.$$

We verify the hypotheses of the theorem on $(0,b)\times(0,\infty)$, where $0<b<\pi/2$ is fixed. Note, first, that $f(\theta,t)$ and $\partial f/\partial\theta$ are continuous in (θ,t). Moreover,

$$0 \le f(\theta,t) \le Ie^{-t^2/2}, \qquad 0 \le \partial f/\partial\theta(\theta,t) \le e^{-t^2/2} t\sec^2 b$$

($\sec^2\theta \ge 1$ is increasing on $(0,\pi/2)$). So we may take $g(t) = e^{-t^2/2}(I + t\sec^2 b)$, which is integrable.[6] This verifies all the hypotheses. Applying the theorem yields

$$F'(\theta) = \int_0^\infty e^{-t^2\sec^2\theta/2} t\sec^2\theta\,dt = \int_0^\infty e^{-u}\,du = 1, \qquad 0<\theta<b.$$

Here we used the substitution $u = t^2\sec^2\theta/2$, $du = t\sec^2\theta\,dt$. Since $0<b<\pi/2$ is arbitrary, $F'(\theta)=1$ is valid on $(0,\pi/2)$.

Thus, $F(\theta) = \theta + \text{constant}$ on $(0,\pi/2)$. To evaluate the constant, note that $f(0+,t)=0$ for all $0<t<\infty$, by continuity at the endpoints. Then since $f(\theta,t)\le Ie^{-t^2/2}$, we can apply the dominated convergence theorem to get

$$F(0+) = \lim_{\theta\to 0+}\int_0^\infty f(\theta,t)\,dt = \int_0^\infty f(0+,t)\,dt = 0.$$

This shows that $F(\theta)=\theta$, so $F(\pi/2-)=\pi/2$. Hence, $I^2=\pi/2$. Since I is half the integral in (5.4.1), this derives (5.4.1). □

Let us apply the theorem to the gamma function

$$\Gamma(x) = \int_0^\infty e^{-t}t^{x-1}\,dt, \qquad x>0.$$

To this end, fix $0<a<b<\infty$. We show that Γ is differentiable on (a,b). With $f(x,t)=e^{-t}t^{x-1}$,

$$\frac{\partial f}{\partial x}(x,t) = e^{-t}t^{x-1}\log t, \qquad 0<t<\infty, 0<x<\infty.$$

Then f and $\partial f/\partial x$ are continuous on $(a,b)\times(0,\infty)$. Since $|f|+|\partial f/\partial x| \le g(t)$ on $(a,b)\times(0,\infty)$, where

$$g(t) = \begin{cases} e^{-t}t^{a-1}(|\log t|+1), & 0<t\le 1, \\ e^{-t}t^{b-1}(|\log t|+1), & 1\le t, \end{cases}$$

and g is continuous and integrable over $(0,\infty)$ (Exercise **5.1.11**), the domination hypothesis of the theorem is verified. Thus, we can apply the theorem to obtain

$$\Gamma'(x) = \int_0^\infty e^{-t}t^{x-1}\log t\,dt, \qquad a<x<b.$$

---

[6] $\int_0^\infty g(t)\,dt = I^2 + \sec^2 b$.

Since $0 < a < b$ are arbitrary, this shows that Γ is differentiable on $(0,\infty)$. Since this argument can be repeated,

$$\Gamma''(x) = \int_0^\infty e^{-t}t^{x-1}(\log t)^2\,dt, \qquad x > 0.$$

Since this last quantity is positive, we see that Γ is strictly convex on $(0,\infty)$ (§3.3). Differentiating repeatedly we obtain $\Gamma^{(n)}(x)$ for all $n \ge 1$. Hence, the gamma function is smooth on $(0,\infty)$.

In Exercise **5.2.2**, the Laplace transform

$$F(x) = \int_0^\infty e^{-xt}\frac{\sin t}{t}\,dt = \arctan\left(\frac{1}{x}\right)$$

is computed for $x > 1$ by expanding $\sin t/t$ in a series. Now we compute $F(x)$ for $x > 0$ by using differentiation under the integral sign. In Exercise **5.4.12**, we need to know this for $x > 0$; $x > 1$ is not enough. Note that, to compute $F(x)$ for $x > 0$, it is enough to compute $F(x)$ for $x > a$, where $a > 0$ is arbitrarily small.

First, $\partial f/\partial x = -e^{-xt}\sin t$, so f and $\partial f/\partial x$ are continuous on $(a,\infty)\times(0,\infty)$. Since $\sin t$ and $\sin t/t$ are bounded by 1, $|f(x,t)| + |\partial f/\partial x|$ is dominated by $2e^{-at}$ on $(a,\infty)\times(0,\infty)$. Applying the theorem and Exercise **4.4.9** yields

$$F'(x) = -\int_0^\infty e^{-xt}\sin t\,dt = -\frac{1}{1+x^2}, \qquad x > a.$$

Now by the dominated convergence theorem, $F(\infty) = \lim_{x\to\infty} F(x) = 0$. So

$$\begin{aligned} F(x) &= F(x) - F(\infty) = -\int_x^\infty F'(t)\,dt \\ &= \arctan t\big|_x^\infty = \pi/2 - \arctan x = \arctan\left(\frac{1}{x}\right), \qquad x > a. \end{aligned}$$

Since $a > 0$ is arbitrarily small, this is what we wanted to show.

Now we derive the theorem. To this end, fix $a < x < b$, and let $x_n \to x$, with $x_n \ne x$ for all $n \ge 1$. We have to show that

$$\frac{F(x_n) - F(x)}{x_n - x} \to \int_c^d \frac{\partial f}{\partial x}(x,t)\,dt.$$

Let

$$k_n(t) = \frac{f(x_n,t) - f(x,t)}{x_n - x}, \qquad c < t < d, n \ge 1,$$

and

$$k(t) = \frac{\partial f}{\partial x}(x,t), \qquad c < t < d.$$

Then $k_n(t)$, $n \geq 1$, and $k(t)$ are continuous on (c, d). By the mean value theorem

$$k_n(t) = \frac{\partial f}{\partial x}(x'_n, t), \qquad c < t < d, n \geq 1,$$

for some x'_n[7] between x_n and x. By the domination hypothesis, we see that $|k_n(t)| \leq g(t)$. Thus, we can apply the dominated convergence theorem, which yields

$$\frac{F(x_n) - F(x)}{x_n - x} = \int_c^d k_n(t)\,dt \to \int_c^d k(t)\,dt = \int_c^d \frac{\partial f}{\partial x}(x,t)\,dt.$$

This establishes (5.4.3). □

Now we compute $M\left(1, 1/\sqrt{2}\right)$.

Theorem 5.4.2.

$$M\left(1, \frac{1}{\sqrt{2}}\right) = \frac{\Gamma(3/4)}{\Gamma(1/4)}\sqrt{2\pi}. \tag{5.4.5}$$

To this end, bring in the *beta function*[8]

$$B(x,y) = \int_0^1 t^{x-1}(1-t)^{y-1}\,dt, \qquad x > 0, y > 0. \tag{5.4.6}$$

The next result shows that $1/B(x,y)$ extends the binomial coefficient $\binom{x+y}{x}$ to nonnatural x and y.

Theorem 5.4.3. *For all $a > 0$ and $b > 0$,*

$$B(a,b) = \frac{\Gamma(a)\Gamma(b)}{\Gamma(a+b)}. \tag{5.4.7}$$

We derive this following the method used to obtain (5.4.1). First, write

$$\Gamma(b)\Gamma(a) = \int_0^\infty \Gamma(b)e^{-t}t^{a-1}\,dt \tag{5.4.8}$$

and

$$\begin{aligned}
\Gamma(b)e^{-t}t^{a-1} &= e^{-t}t^{a-1}\int_0^\infty e^{-s}s^{b-1}\,ds \\
&= \int_0^\infty e^{-s-t}s^{b-1}t^{a-1}\,ds \\
&= \int_t^\infty e^{-r}(r-t)^{b-1}t^{a-1}\,dr, \qquad t > 0.
\end{aligned}$$

[7] x'_n also depends on t.

[8] $B(x,y)$ is finite by (5.4.7).

Here we substituted $r = s + t$, $dr = ds$. Now set

$$h(t,r) = e^{-r}(r-t)^{b-1}t^{a-1}, i$$

$$f(x,t) = \int_{t/x}^{\infty} h(t,r)\,dr, \qquad t > 0, 0 < x < 1,$$

and

$$F(x) = \int_0^{\infty} f(x,t)\,dt, \qquad 0 < x < 1.$$

By continuity at the endpoints (the integrand is nonnegative), $f(1-,t) = \int_t^{\infty} h(t,r)\,dr = e^{-t}t^{a-1}\Gamma(b)$. Then (5.4.8) says

$$\int_0^{\infty} f(1-,t)\,dt = \Gamma(a)\Gamma(b).$$

Since $f(x,t) \le f(1-,t)$ for $0 < x < 1$ and $f(1-,t)$ is integrable, the dominated convergence theorem applies, and we conclude that $F(1-) = \Gamma(a)\Gamma(b)$.

Moreover, $F(0+) = 0$. To see this, note, by continuity at the endpoints (the integrand is integrable), that we have $f(0+,t) = 0$ for all $t > 0$. By the dominated convergence theorem, again, it follows that $F(0+) = 0$.

Now by the fundamental theorem and the chain rule,

$$\begin{aligned}\frac{\partial f}{\partial x}(x,t) &= -e^{-t/x}t^{a-1}(t/x - t)^{b-1}\left(-\frac{t}{x^2}\right) \\ &= \left(\frac{t}{x}\right)^{a+b-1} e^{-t/x}x^{a-1}(1-x)^{b-1}\cdot\frac{1}{x}. \qquad (5.4.9)\end{aligned}$$

hence $(0 < x < 1)$

$$\left|\frac{\partial f}{\partial x}(x,t)\right| \le \frac{e^{-t}t^{a+b-1}}{x(1-x)}\cdot\left(\frac{1}{x} - 1\right)^b. \qquad (5.4.10)$$

Fix $0 < \epsilon < 1$ and suppose $\epsilon \le x \le 1 - \epsilon$. Then the function $x(1-x)$ is minimized on $\epsilon \le x \le 1 - \epsilon$ at the endpoints, so its minimum value is $\epsilon(1-\epsilon)$. The maximum value of the factor $((1/x) - 1)^b$ is attained at $x = \epsilon$ and equals $((1/\epsilon) - 1)^b$. Hence if we set $C_\epsilon = ((1/\epsilon) - 1)^b/\epsilon(1-\epsilon)$, we obtain

$$\left|\frac{\partial f}{\partial x}(x,t)\right| \le C_\epsilon e^{-t}t^{a+b-1}, \quad \epsilon < x < 1 - \epsilon, t > 0.$$

Thus, the domination hypothesis is verified on $(\epsilon, 1-\epsilon) \times (0,\infty)$ with[9] $g(t) = f(1-,t) + C_\epsilon e^{-t}t^{a+b-1}$. Differentiating under the integral sign and substituting $t/x = u$, $dt/x = du$,

[9] $\int_0^{\infty} g(t)\,dt = \Gamma(a)\Gamma(b) + C_\epsilon\Gamma(a+b)$.

$$F'(x) = \int_0^\infty \left(\frac{t}{x}\right)^{a+b-1} e^{-t/x} x^{a-1}(1-x)^{b-1} \cdot \frac{1}{x}\,dt$$
$$= \int_0^\infty u^{a+b-1} e^{-u} x^{a-1}(1-x)^{b-1}\,du = x^{a-1}(1-x)^{b-1}\Gamma(a+b),$$

valid on $(\epsilon, 1-\epsilon)$. Since $\epsilon > 0$ is arbitrary, we obtain

$$F'(x) = x^{a-1}(1-x)^{b-1}\Gamma(a+b), \qquad 0 < x < 1.$$

Integrating, we arrive at

$$\Gamma(a)\Gamma(b) = F(1-) - F(0+) = \int_0^1 F'(x)\,dx$$
$$= \Gamma(a+b)\int_0^1 x^{a-1}(1-x)^{b-1}\,dx = \Gamma(a+b)B(a,b),$$

which is (5.4.7). □

To derive (5.4.5), we use (5.4.7) and a sequence of substitutions. From §5.3,

$$\frac{\pi/2}{M(1,1/\sqrt{2})} = \int_0^{\pi/2} \frac{d\theta}{\sqrt{1-\frac{1}{2}\sin^2\theta}}.$$

Substituting $\sin\theta = t$, we obtain

$$\frac{\pi/2}{M(1,1/\sqrt{2})} = \sqrt{2}\int_0^1 \frac{dt}{\sqrt{(1-t^2)(2-t^2)}}.$$

Now substitute $x^2 = t^2/(2-t^2)$ to obtain

$$\frac{\pi/2}{M(1,1/\sqrt{2})} = \sqrt{2}\int_0^1 \frac{dx}{\sqrt{1-x^4}}.$$

Now substitute $u = x^4$ to get

$$\frac{\pi/2}{M(1,1/\sqrt{2})} = \frac{\sqrt{2}}{4}\int_0^1 u^{1/4-1}(1-u)^{1/2-1}\,du = \frac{\sqrt{2}}{4}B\left(\frac{1}{4},\frac{1}{2}\right).$$

Since

$$B\left(\frac{1}{4},\frac{1}{2}\right) = \frac{\Gamma(1/4)\Gamma(1/2)}{\Gamma(3/4)}$$

and (Exercise **5.4.1**) $\Gamma(1/2) = \sqrt{\pi}$, we obtain (5.4.5). □

We end the section with an important special case of the theorem. Suppose that $(c,d) = (0,\infty)$ and $f(x,t)$ is piecewise constant in t, i.e., suppose that $f(x,t) = f_n(x)$, $a < x < b$, $n-1 < t \le n$, $n \ge 1$. Then the integral in (5.4.2) reduces to an infinite series. Hence, the theorem takes the following form.

Theorem 5.4.4 (Differentiation Under the Summation Sign). *Let $f_n : (a,b) \to \mathbf{R}$, $n \geq 1$, be a sequence of differentiable functions. Suppose that there is a convergent positive series $\sum g_n$ of numbers, such that*

$$|f_n(x)| + |f_n'(x)| \leq g_n, \qquad a < x < b, n \geq 1.$$

If

$$F(x) = \sum_{n=1}^{\infty} f_n(x), \qquad a < x < b,$$

then F is differentiable on (a,b), $F' : (a,b) \to \mathbf{R}$ is continuous, and

$$F'(x) = \sum_{n=1}^{\infty} f_n'(x), \qquad a < x < b.$$

To derive this, one, of course, applies the dominated convergence theorem for series instead of the theorem for integrals. □

Let $f : (a,b) \times (c,d) \to \mathbf{R}$ be a function of two variables (x,y), and suppose that $\partial f/\partial x$ exists. If $\partial f/\partial x$ is differentiable with respect to x, we denote its derivative by

$$\frac{\partial}{\partial x}\left(\frac{\partial f}{\partial x}\right) = \frac{\partial^2 f}{\partial x^2}.$$

This is needed in Exercise **5.7.6**.

Exercises

5.4.1. Use the substitution $x = \sqrt{2t}$ in (5.4.1) to obtain $\Gamma(1/2) = \sqrt{\pi}$. Conclude that $(1/2)! = \sqrt{\pi}/2$.

5.4.2. Show that the Laplace transform

$$L(s) = \int_{-\infty}^{\infty} e^{sx} e^{-x^2/2}\, dx, \qquad -\infty < s < \infty,$$

is given by $L(s) = \sqrt{2\pi} e^{s^2/2}$. (Complete the square in the exponent, and use translation invariance.)

5.4.3. Compute $L^{(2n)}(0)$ with L as in the previous Exercise, to obtain

$$\int_{-\infty}^{\infty} e^{-x^2/2} x^{2n}\, dx = \sqrt{2\pi} \cdot \frac{(2n)!}{2^n n!}, \qquad n \geq 0.$$

(Writing out the power series of L yields $L^{(2n)}(0)$.)

5.4.4. Show that the Fourier transform

$$F(s) = \int_{-\infty}^{\infty} e^{-x^2/2} \cos(sx)\, dx, \qquad -\infty < s < \infty,$$

is finite and differentiable on $(-\infty, \infty)$. Differentiate under the integral sign, and integrate by parts to show that $F'(s)/F(s) = -s$ for all s. Integrate this equation over $(0, s)$, and use $F(0) = \sqrt{2\pi}$ to obtain

$$F(s) = \sqrt{2\pi} e^{-s^2/2}.$$

5.4.5. Derive the *Hecke integral*

$$H(a) = \int_0^{\infty} e^{-x-a/x} \frac{dx}{\sqrt{x}} = \sqrt{\pi} e^{-2\sqrt{a}}, \qquad a > 0, \tag{5.4.11}$$

by differentiating under the integral sign and substituting $x = a/t$ to obtain $H'(a)/H(a) = -1/\sqrt{a}$. Integrate this equation over $(0, a)$, and use $H(0) = \Gamma(1/2) = \sqrt{\pi}$ to obtain (5.4.11).

5.4.6. Show that

$$\int_{-\infty}^{\infty} e^{-x^2/2q}\, dx = \sqrt{2\pi q}, \qquad q > 0.$$

5.4.7. Let $\psi(t) = \sum_{n=1}^{\infty} e^{-n^2\pi t}$, $t > 0$. Use the integral test (Exercise **4.3.8**) to show that

$$\lim_{t \to 0+} \sqrt{t} \cdot \psi(t) = \frac{1}{2}.$$

5.4.8. Show that $\zeta(x) = \sum_{n=1}^{\infty} 1/n^x$, $x > 1$, is smooth (differentiation under the summation sign).

5.4.9. Show that $\psi(t) = \sum_{n=1}^{\infty} e^{-n^2\pi t}$, $t > 0$, is smooth.

5.4.10. Show that the Bessel function J_ν (Exercise **5.2.12**) is smooth. If ν is an integer, show that J_ν satisfies *Bessel's equation*

$$x^2 J_\nu''(x) + x J_\nu'(x) + (x^2 - \nu^2) J_\nu(x) = 0, \qquad -\infty < x < \infty.$$

(Differentiation under the integral sign and integration by parts.)

5.4.11. Suppose that $f : \mathbf{R} \to \mathbf{R}$ is nonnegative, superlinear, and continuous, and let

$$F(s) = \int_{-\infty}^{\infty} e^{sx} e^{-f(x)}\, dx, \qquad -\infty < s < \infty,$$

denote the Laplace transform of e^{-f} (Exercise **4.3.11**). Show that F is smooth, and compute $(\log F)''$. Use the Cauchy–Schwarz inequality (Exercise **4.4.21**) to conclude that $\log F$ is convex.

5.4.12. Let $F(b) = \int_0^b \sin x/x\,dx$, $b > 0$. From Exercise **4.3.14**, we know F is bounded and $F(\infty)$ exists. Integrate by parts to show that

$$\int_0^b e^{-sx}\frac{\sin x}{x}\,dx = e^{-sb}F(b) + s\int_0^b e^{-sx}F(x)\,dx, \qquad s > 0.$$

Let $b \to \infty$, change variables on the right, and let $s \to 0+$ to get

$$\lim_{s\to 0+}\int_0^\infty e^{-sx}\frac{\sin x}{x}\,dx = \lim_{b\to\infty}\int_0^b \frac{\sin x}{x}\,dx.$$

Conclude that $F(\infty) = \pi/2$.

5.5 Stirling's Approximation

The main purpose of this section is to derive Stirling's approximation to $n!$. If (a_n) and (b_n) are positive sequences, we say that (a_n) *and* (b_n) *are asymptotically equal*, and we write $a_n \sim b_n$ as $n \nearrow \infty$, if $a_n/b_n \to 1$ as $n \nearrow \infty$. Note that $a_n \sim b_n$ as $n \nearrow \infty$ iff $\log a_n - \log b_n \to 0$ as $n \nearrow \infty$.

Theorem 5.5.1. *If x is any real, then*

$$\Gamma(x+n) \sim n^{x+n-1/2}e^{-n}\sqrt{2\pi}, \qquad n \nearrow \infty. \tag{5.5.1}$$

In particular, if $x = 1$, we have Stirling's approximation

$$n! \sim n^{n+1/2}e^{-n}\sqrt{2\pi}, \qquad n \nearrow \infty.$$

A consequence of Stirling's approximation is Raabe's formula (Exercise **5.5.9**) which yields this *Stirling's identity*

$$\exp\left(\int_s^{s+1}\log\Gamma(x)\,dx\right) = s^s e^{-s}\sqrt{2\pi}, \qquad s > 0. \tag{5.5.2}$$

Here the left side is the *geometric mean* of Γ over $(s, s+1)$. The subtlety here is the exact constant $\sqrt{2\pi}$. Apart from this, this identity is an immediate consequence of the definition of Γ (Exercise **5.1.15**).

Note that $\Gamma(x+n)$ is defined, as soon as $n > -x$. By taking the log of both sides, (5.5.1) is equivalent to

$$\lim_{n\nearrow\infty} \log\Gamma(x+n) - \left[\left(x+n-\frac{1}{2}\right)\log n - n\right] = \frac{1}{2}\log(2\pi).$$

To derive (5.5.1), recall that

$$\Gamma(x+n) = \int_0^\infty e^{-t} t^{x+n-1} dt, \qquad x > 0. \tag{5.5.3}$$

Since this integral is the area of the subgraph of $e^{-t}t^{x+n-1}$ and all we want is an approximation, not an exact evaluation, of this integral, let us check where the integrand is maximized, as this will tell us where the greatest contribution to the area is located. A simple computation shows that the integrand is maximized at $t = x + n - 1$, which goes to infinity with n. To get a handle on this region of maximum area, perform the change of variable $t = ns$. This leads to

$$\Gamma(x+n) = n^{x+n} \int_0^\infty e^{-ns} s^{x+n-1} ds = n^{x+n} \int_0^\infty e^{nf(s)} s^{x-1}\, ds, \tag{5.5.4}$$

where

$$f(s) = \log s - s, \qquad s > 0.$$

Now the varying part $e^{nf(s)}$ of the integrand is maximized at the maximum of f, which occurs at $s = 1$, since $f(0+) = -\infty$, $f(\infty) = -\infty$. Since the maximum value of f at $s = 1$ is -1, the maximum value of the integrand is roughly $e^{nf(1)} = e^{-n}$. By analogy with sums (Exercise **5.5.1**), we expect the limiting behavior of the integral in (5.5.4) to involve the maximum value of the integrand. Let us pause in the derivation of Stirling's formula, and turn to the study of the limiting behavior of such integrals, in general.

Theorem 5.5.2. *Suppose that $f : (a,b) \to \mathbf{R}$ is continuous and bounded above on a bounded interval (a,b). Then*

$$\lim_{n \nearrow \infty} \frac{1}{n} \log \left[\int_a^b e^{nf(x)}\, dx \right] = \sup\{f(x) : a < x < b\}. \tag{5.5.5}$$

To see this (Figure 5.6), let I_n denote the integral, and let $M = \sup\{f(x) : a < x < b\}$. Then M is finite since f is bounded above. Given $\epsilon > 0$, choose $c \in (a,b)$ with $f(c) > M - \epsilon$, and, by continuity, choose $\delta > 0$, such that $f(x) > f(c) - \epsilon$ on $(c-\delta, c+\delta)$. Then $f(x) > M - 2\epsilon$ on $(c-\delta, c+\delta)$, and

$$(b-a)e^{nM} \ge I_n \ge \int_{c-\delta}^{c+\delta} e^{nf(x)}\, dx \ge \int_{c-\delta}^{c+\delta} e^{n(M-2\epsilon)}\, dx = 2\delta e^{n(M-2\epsilon)}.$$

Now take the log of this last inequality, and divide by n to obtain

$$\frac{1}{n}\log(b-a) + M \ge \frac{1}{n}\log(I_n) \ge \frac{1}{n}\log(2\delta) + M - 2\epsilon.$$

Sending $n \nearrow \infty$, the upper and lower limits of the sequence $((1/n)\log(I_n))$ lie between M and $M - 2\epsilon$. Since $\epsilon > 0$ is arbitrary, $(1/n)\log(I_n) \to M$. $\square$

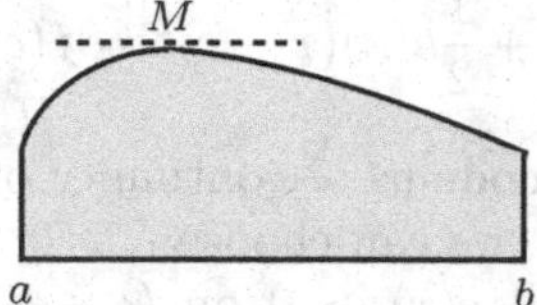

Fig. 5.6 The global max is what counts

Although a good start, this result is not quite enough to obtain Stirling's approximation. The exact form of the limiting behavior, due to Laplace, is given by the following.

Theorem 5.5.3 (Laplace's Theorem). *Let $f : (a,b) \to \mathbf{R}$ be differentiable and assume f is concave. Suppose that f has a global maximum at $c \in (a,b)$ with f twice differentiable at c and $f''(c) < 0$. Suppose that $g : (a,b) \to \mathbf{R}$ is continuous with polynomial growth and $g(c) > 0$. Then*

$$I_n = \int_a^b e^{nf(x)} g(x)\, dx \sim e^{nf(c)} g(c) \sqrt{\frac{2\pi}{-nf''(c)}}, \qquad n \nearrow \infty. \tag{5.5.6}$$

This result is motivated by the fact that when $g(x) = 1$, $a = -\infty$, $b = \infty$, and $f(x)$ is a quadratic polynomial, (5.5.6) is an equality.

By *polynomial growth*, we mean that $|g(x)| \le A + B|x|^p$, $a < x < b$, for some constants A, B, p. Before we derive this theorem, let us apply it to obtain the asymptotic behavior of (5.5.4) to complete the derivation of (5.5.1).

In the case of (5.5.4), $f'(s) = 1/s - 1$, and $f''(s) = -1/s^2$, so f is strictly concave, has a global maximum $f(1) = -1$ at $c = 1$, and $f''(1) = -1 < 0$. Since $g(s) = s^{x-1}$ has polynomial growth (in s), the integral in (5.5.4) is asymptotic to $e^{-n}\sqrt{2\pi/n}$, which yields (5.5.1). □

Now we derive Laplace's theorem. We write $I_n = I_n^- + I_n^0 + I_n^+$, where

$$\begin{aligned} I_n^- &= \int_a^{c-\delta} e^{nf(x)} g(x)\, dx, \\ I_n^0 &= \int_{c-\delta}^{c+\delta} e^{nf(x)} g(x)\, dx, \\ I_n^+ &= \int_{c+\delta}^{b} e^{nf(x)} g(x)\, dx. \end{aligned}$$

Since c is a maximum, $f'(c) = 0$. Since $f''(c)$ exists, by Taylor's theorem (§3.5), there is a continuous function $h : (a,b) \to \mathbf{R}$ satisfying $h(c) = f''(c)$, and

$$f(x) = f(c) + f'(c)(x-c) + \frac{1}{2}h(x)(x-c)^2 = f(c) + \frac{1}{2}h(x)(x-c)^2. \quad (5.5.7)$$

If we let $\mu_c(\delta)$ denote the modulus of continuity of h at c and let $\epsilon = \mu_c(\delta)$, then $\epsilon \to 0$ as $\delta \to 0$. Thus, we can choose $\delta > 0$, such that $h(x) \le f''(c) + \mu_c(\delta) = f''(c) + \epsilon < 0$ and $g(x) > 0$ on $(c-\delta, c+\delta)$. Now substituting $x = c + t/\sqrt{n}$ in I_n^0, $dx = dt/\sqrt{n}$, and inserting (5.5.7),

$$\begin{aligned} I_n^0 &= \int_{-\delta}^{\delta} e^{nf(c)+nh(x)(x-c)^2/2} g(x)\,dx \\ &= \frac{e^{nf(c)}}{\sqrt{n}} \int_{-\delta\sqrt{n}}^{\delta\sqrt{n}} e^{h(c+t/\sqrt{n})t^2/2} g(c+t/\sqrt{n})\,dt. \end{aligned}$$

But $g(x)$ is bounded on $(c-\delta, c+\delta)$. Hence,

$$e^{h(c+t/\sqrt{n})t^2/2}|g(c+t/\sqrt{n})| \le Ce^{(f''(c)+\epsilon)t^2/2}, \qquad |t| < \delta\sqrt{n},$$

which is integrable[10] over $(-\infty, \infty)$. Thus, the dominated convergence theorem applies, and[11]

$$\begin{aligned} e^{-nf(c)} I_n^0 \sqrt{n} &= \int_{-\delta\sqrt{n}}^{\delta\sqrt{n}} e^{h(c+t/\sqrt{n})t^2/2} g(c+t/\sqrt{n})\,dt \\ &\to \int_{-\infty}^{\infty} e^{f''(c)t^2/2} g(c)\,dt = g(c)\sqrt{\frac{2\pi}{-f''(c)}}, \end{aligned}$$

by Exercise **5.4.6**.

We conclude that

$$I_n^0 \sim e^{nf(c)} g(c) \sqrt{\frac{2\pi}{-nf''(c)}}. \quad (5.5.8)$$

To finish the derivation, it is enough to show that $I_n \sim I_n^0$ as $n \nearrow \infty$.

To derive $I_n \sim I_n^0$, it is enough to obtain $I_n^+/I_n^0 \to 0$ and $I_n^-/I_n^0 \to 0$, since

$$\frac{I_n}{I_n^0} = \frac{I_n^-}{I_n^0} + 1 + \frac{I_n^+}{I_n^0}, \qquad n \ge 1.$$

To obtain $I_n^+/I_n^0 \to 0$, we use convexity. Since f is concave, $-f$ is convex. Hence, the graph of $-f$ on $(c+\delta, b)$ lies above its tangent line at $c+\delta$ (Exercise **3.3.7**). Thus,

$$f(x) \le f(c+\delta) + f'(c+\delta)(x-c-\delta), \qquad a < x < b.$$

[10] The integral is $C\sqrt{2\pi/(-f''(c)-\epsilon)}$.

[11] By Exercise **5.2.14**

Since f is strictly concave at c, $f'(c+\delta) < 0$ and $f(c+\delta) < f(c)$. Inserting this in the definition for I_n^+ and substituting $x = t + c + \delta$,

$$|I_n^+| \le e^{nf(c+\delta)} \int_{c+\delta}^{b} e^{nf'(c+\delta)(x-c-\delta)} |g(x)|\, dx \tag{5.5.9}$$

$$\le e^{nf(c+\delta)} \int_0^{b-c-\delta} e^{nf'(c+\delta)t}(A + B|t+c+\delta|^p)\, dt \tag{5.5.10}$$

$$\le e^{nf(c+\delta)} \int_0^{\infty} e^{f'(c+\delta)t}(A + B|t+c+\delta|^p)\, dt \tag{5.5.11}$$

$$= Ce^{nf(c+\delta)}, \tag{5.5.12}$$

where C denotes the (finite) integral in (5.5.9). Now divide this last expression by the expression in (5.5.8), obtaining

$$\left|\frac{I_n^+}{I_n^0}\right| \le \frac{Ce^{nf(c+\delta)}}{I_n^0} \sim \frac{Ce^{nf(c+\delta)}}{e^{nf(c)}g(c)\sqrt{\frac{2\pi}{-nf''(c)}}} = \text{constant} \cdot \sqrt{n} \cdot e^{-n(f(c)-f(c+\delta))},$$

which goes to zero as $n \nearrow \infty$ since the exponent is negative. Since I_n^-/I_n^0 is similar, this completes the derivation. □

Since Stirling's approximation provides a manageable expression for $n!$, it is natural to use it to derive the asymptotics of the binomial coefficient

$$\binom{n}{k} = \frac{n!}{k!(n-k)!}, \qquad 0 \le k \le n.$$

Actually, of more interest is the binomial coefficient divided by 2^n, since this is the probability of obtaining k heads in n tosses of a fair coin.

To this end, suppose $0 < t < 1$ and let (k_n) be a sequence of naturals such that $k_n/n \to t$ as $n \to \infty$. Applying Stirling to $n!$, $k_n! = (t_n n)!$, and $(n-k_n)! = ((1-t_n)n)!$ and simplifying, we obtain the following:

Theorem 5.5.4. *Fix $0 < t < 1$. If (k_n) is a sequence of naturals such that the ratio $k_n/n \to t$, as $n \nearrow \infty$, then the probabilities $\binom{n}{k}2^{-n}$ of tossing $k = k_n$ heads in n tosses satisfy*

$$\binom{n}{k}2^{-n} \sim \frac{1}{\sqrt{2\pi n}} \cdot \frac{1}{\sqrt{t(1-t)}} \cdot e^{-nH(t)}, \qquad n \nearrow \infty,$$

where

$$H(t) = t\log(2t) + (1-t)\log[2(1-t)], \qquad 0 < t < 1. \quad \square$$

Because the binomial coefficients are so basic, the function H which governs their asymptotic decay must be important. The function H, called the *entropy*, controls the rate of decay of the binomial coefficients. Note that H is convex (Figure 5.7) on $(0,1)$ and has a global minimum of zero at $t = 1/2$ with $H''(1/2) = 4$.

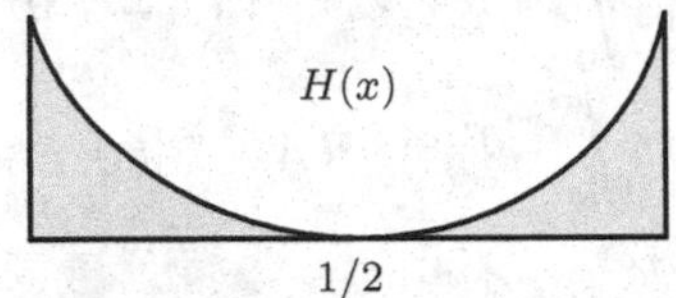

Fig. 5.7 The entropy $H(x)$

We end the section with an application of (5.5.1) to the following formula for the gamma function.

Theorem 5.5.5 (The Duplication Formula). *For* $s > 0$,

$$2^{2s} \cdot \frac{\Gamma(s)\Gamma(s+1/2)}{\Gamma(2s)} = 2\sqrt{\pi}.$$

To derive this, let $f(s)$ denote the left side. Then using $\Gamma(s+1) = s\Gamma(s)$, check that f is periodic of period 1, i.e., $f(s+1) = f(s)$. Hence, $f(s+n) = f(s)$ for all $n \geq 1$. Now inserting the asymptotic (5.5.1) (three times) in the expression for $f(s+n)$ yields $f(s+n) \sim 2\sqrt{\pi}$, as $n \nearrow \infty$. Hence, $f(s) = 2\sqrt{\pi}$, which is the duplication formula. □

Exercises

5.5.1. Show that

$$\lim_{n \nearrow \infty} (a^n + b^n + c^n)^{1/n} = \max(a,b,c), \qquad a,b,c > 0,$$

and

$$\lim_{n \nearrow \infty} \frac{1}{n} \log\left(e^{na} + e^{nb} + e^{nc}\right) = \max(a,b,c), \qquad a,b,c \in \mathbf{R}.$$

Moreover, if $\log(a_n)/n \to A$, $\log(b_n)/n \to B$, and $\log(c_n)/n \to C$, then

$$\lim_{n \to \infty} \frac{1}{n} \log\left(a_n + b_n + c_n\right) = \max(A,B,C).$$

5.5.2. Write computer code to obtain 100! and its Stirling approximation s. Compute the relative error $|100! - s|/100!$.

5.5.3. Show that $\binom{2n}{n}2^{-2n} \sim 1/\sqrt{\pi n}$ as $n \nearrow \infty$.

5.5.4. Apply Stirling to $n!$, $k!$, and $(n-k)!$ to derive the asymptotic for $\binom{n}{k}2^{-n}$ given in Theorem 5.5.4.

5.5.5. Let $0 < p < 1$. Graph

$$H(t,p) = t\log(t/p) + (1-t)\log[(1-t)/(1-p)], \qquad 0 < t < 1.$$

5.5.6. Suppose that a flawed coin is such that the probability of obtaining heads in a single toss is p, where $0 < p < 1$. Let $0 < t < 1$ and let (k_n) be a sequence of naturals satisfying $k_n/n \to t$ as $n \to \infty$. Show that the probabilities $\binom{n}{k}p^k(1-p)^{n-k}$ of obtaining $k = k_n$ heads in n tosses satisfy

$$\binom{n}{k}p^k(1-p)^{n-k} \sim \frac{1}{\sqrt{2\pi n}} \cdot \frac{1}{\sqrt{t(1-t)}} \cdot e^{-nH(t,p)}, \qquad n \nearrow \infty.$$

5.5.7. For $0 < q < 1$ and $0 < a < b < \infty$, let $f(q) = \int_a^b q^{x^2}\,dx$. Compute

$$\lim_{n\nearrow\infty} \frac{1}{n}\log f(q^n).$$

5.5.8. Show that

$$3^{3s} \cdot \frac{\Gamma(s)\Gamma(s+1/3)\Gamma(s+2/3)}{\Gamma(3s)} = 2\pi\sqrt{3}, \qquad s > 0.$$

Generalize to

$$n^{ns} \cdot \frac{\Gamma(s)\Gamma(s+1/n)\cdots\Gamma(s+(n-1)/n)}{\Gamma(ns)} = \sqrt{n}\cdot(2\pi)^{(n-1)/2}, \qquad s > 0.$$

5.5.9. Take the limit of the logarithm in the previous exercise using Riemann sums to get *Raabe's formula*

$$\int_0^1 \log\Gamma(t+s)\,dt = s\log s - s + \frac{1}{2}\log(2\pi), \qquad s > 0,$$

which implies the Stirling identity (5.5.2).

5.5.10. Let $f : \mathbf{R} \to \mathbf{R}$ be superlinear and continuous. Consider the Laplace transforms

$$L_n(y) = \int_{-\infty}^{\infty} e^{xy}e^{-nf(x)}\,dx, \qquad n \geq 1.$$

Show that

$$\lim_{n\nearrow\infty} \frac{1}{n}\log\left[L_n(ny)\right] = g(y),$$

where g is the Legendre transform (3.3.10) of f. (Break $L(ny)$ into three pieces, as in Exercise **4.3.11**, and use Exercise **5.5.1**.)

5.5.11. Differentiate the log of the duplication formula to obtain

$$\frac{\Gamma'(1)}{\Gamma(1)} - \frac{\Gamma'(1/2)}{\Gamma(1/2)} = 2\log 2.$$

5.5.12. Use the duplication formula to get $\Gamma(1/4)\Gamma(3/4) = \pi\sqrt{2}$. Hence,

$$M\left(1, \frac{1}{\sqrt{2}}\right) = \frac{2\pi^{3/2}}{\Gamma^2(1/4)}.$$

5.6 Infinite Products

Given a sequence (a_n), let $p_n = (1+a_1)(1+a_2)\dots(1+a_n)$ denote the *nth partial product*, $n \geq 1$. We say that the *infinite product* $\prod_{n=1}^{\infty}(1+a_n)$ *converges* if there is a *finite* L, such that $p_n \to L$. In this case, we write

$$L = \prod_{n=1}^{\infty}(1+a_n).$$

For example, by induction, check that, for $n \geq 1$,

$$(1+x)\left(1+x^2\right)\left(1+x^4\right)\dots\left(1+x^{2^{n-1}}\right) = 1 + x + x^2 + x^3 + \dots + x^{2^n-1}.$$

If $|x| < 1$, the sum converges. Hence, the product converges to[12]

$$\prod_{n=0}^{\infty}\left(1+x^{2^n}\right) = 1 + x + x^2 + x^3 + \dots = \frac{1}{1-x}, \qquad |x| < 1.$$

If $\prod_{n=1}^{\infty}(1+a_n)$ converges and $L \neq 0$, then $1 + a_n = p_n/p_{n-1} \to L/L = 1$. Hence, *a necessary condition for convergence, when* $L \neq 0$, *is* $a_n \to 0$.

Theorem 5.6.1. *For* $x \neq 0$,

$$\frac{\sinh(\pi x)}{\pi x} = \prod_{n=1}^{\infty}\left(1 + \frac{x^2}{n^2}\right). \tag{5.6.1}$$

[12] This identity is simply a reflection of the fact that every natural has a unique binary expansion (§1.6).

This result provides an "infinite degree" *polynomial factorization* of

$$\frac{\sinh(\pi x)}{\pi x} = \frac{e^{\pi x} - e^{-\pi x}}{2\pi x}.$$

Since for large N (Exercise **3.2.3**) $e^{\pi x}$ is approximated by the polynomial $(1+\pi x/2N)^{2N}$, to derive (5.6.1), it makes sense to first factor the polynomial

$$\left(1 + \frac{\pi x}{2N}\right)^{2N} - \left(1 - \frac{\pi x}{2N}\right)^{2N},$$

which in turn suggests we use

$$X^{2N} - 1 = (X^2 - 1) \cdot \prod_{n=1}^{N-1} \left(X^2 - 2X \cdot \cos(n\pi/N) + 1\right). \tag{5.6.2}$$

This factorization, trivial for $N = 2$, is most easily derived using complex numbers. However, by replacing X by X^2 in (5.6.2) and using the double-angle formula, one obtains (5.6.2) with $2N$ replacing N (Exercise **3.6.14**). Hence, by induction, and without recourse to complex numbers, one obtains (5.6.2) for $N = 2, 4, 8, \dots$. In fact, this is all we need to derive (5.6.1).

Insert $X = a/b$ in (5.6.2) and multiply through by b^{2N}, obtaining

$$a^{2N} - b^{2N} = (a^2 - b^2) \cdot \prod_{n=1}^{N-1} \left[a^2 - 2ab \cdot \cos\left(\frac{n\pi}{N}\right) + b^2\right].$$

Now insert

$$a = \left(1 + \frac{\pi x}{2N}\right), \qquad b = \left(1 - \frac{\pi x}{2N}\right).$$

Simplifying and dividing by $2\pi x$, we obtain

$$\begin{aligned} &\frac{\left(1 + \frac{\pi x}{2N}\right)^{2N} - \left(1 - \frac{\pi x}{2N}\right)^{2N}}{2\pi x} \\ &\quad = \frac{1}{N} \cdot \prod_{n=1}^{N-1} \left\{2\left[1 - \cos\left(\frac{n\pi}{N}\right)\right] + \frac{\pi^2 x^2}{2N^2}\left[1 + \cos\left(\frac{n\pi}{N}\right)\right]\right\} \\ &\quad = \frac{1}{N} \cdot \prod_{n=1}^{N-1} \left[4\sin^2\left(\frac{n\pi}{2N}\right) + \frac{\pi^2 x^2}{N^2}\cos^2\left(\frac{n\pi}{2N}\right)\right], \end{aligned} \tag{5.6.3}$$

where we used the double-angle formula, again. Taking the limit of both sides as $x \to 0$ using l'Hôpital's rule (§3.2), we obtain

$$1 = \frac{1}{N} \cdot \prod_{n=1}^{N-1} \left[4\sin^2\left(\frac{n\pi}{2N}\right)\right]. \tag{5.6.4}$$

Now divide (5.6.3) by (5.6.4), factor by factor, obtaining

$$\frac{\left(1+\frac{\pi x}{2N}\right)^{2N}-\left(1-\frac{\pi x}{2N}\right)^{2N}}{2\pi x}=\prod_{n=1}^{N-1}\left[1+\frac{x^2}{n^2}\cdot f\left(\frac{n\pi}{2N}\right)\right],$$

where $f(x)=x^2\cot^2 x$. To obtain (5.6.1), we wish to take the limit $N\nearrow\infty$. But $(\tan x)'=\sec^2 x\geq 1$. So $\tan x\geq x$, so $f(x)\leq 1$ on $(0,\pi/2)$. Thus,

$$\frac{\left(1+\frac{\pi x}{2N}\right)^{2N}-\left(1-\frac{\pi x}{2N}\right)^{2N}}{2\pi x}\leq\prod_{n=1}^{N-1}\left(1+\frac{x^2}{n^2}\right).$$

Sending $N\nearrow\infty$ through powers of 2, we obtain

$$\frac{\sinh(\pi x)}{\pi x}=\frac{e^{\pi x}-e^{-\pi x}}{2\pi x}\leq\prod_{n=1}^{\infty}\left(1+\frac{x^2}{n^2}\right),$$

which is half of (5.6.1). On the other hand, for $M\leq N$,

$$\frac{\left(1+\frac{\pi x}{2N}\right)^{2N}-\left(1-\frac{\pi x}{2N}\right)^{2N}}{2\pi x}\geq\prod_{n=1}^{M-1}\left[1+\frac{x^2}{n^2}\cdot f\left(\frac{n\pi}{2N}\right)\right].$$

Since $\lim_{x\to 0}f(x)=1$, sending $N\nearrow\infty$ through powers of 2 in this last equation, we obtain

$$\frac{\sinh(\pi x)}{\pi x}\geq\prod_{n=1}^{M-1}\left(1+\frac{x^2}{n^2}\right).$$

Now let $M\nearrow\infty$, obtaining the other half of (5.6.1). □

To give an example of the power of (5.6.1), take the log of both sides to get

$$\log\left(\frac{\sinh(\pi x)}{\pi x}\right)=\sum_{n=1}^{\infty}\log\left(1+\frac{x^2}{n^2}\right),\qquad x\neq 0. \tag{5.6.5}$$

Now differentiate under the summation sign to obtain

$$\pi\coth(\pi x)-\frac{1}{x}=\sum_{n=1}^{\infty}\frac{2x}{n^2+x^2},\qquad x\neq 0. \tag{5.6.6}$$

Here coth = cosh / sinh is the *hyperbolic cotangent*. To justify this, note that $\log(1+t)=\int_0^t ds/(1+s)\leq\int_0^t ds=t$. Hence, $\log(1+t)\leq t$ for $t\geq 0$. Thus, with $f_n(x)=\log(1+x^2/n^2)$,

$$|f_n(x)|+|f_n'(x)|\leq(2b+b^2)/n^2\equiv g_n,\qquad |x|<b,$$

and $\sum g_n < \infty$. Thus, (5.6.6) is valid on $0 < |x| < b$ and hence on $x \neq 0$. Now dividing (5.6.6) by $2x$, letting $x \searrow 0$, and setting $t = \pi x$ yield[13]

$$\begin{aligned}\sum_{n=1}^{\infty} \frac{1}{n^2} &= \lim_{x\searrow 0} \sum_{n=1}^{\infty} \frac{1}{n^2+x^2} \\ &= \lim_{x\searrow 0} \frac{\pi x \coth(\pi x) - 1}{2x^2} \\ &= \pi^2 \cdot \lim_{t\searrow 0} \frac{t\coth t - 1}{2t^2}.\end{aligned}$$

But this last limit can be evaluated as follows. Since

$$\frac{\sinh t}{t} = 1 + \frac{t^2}{3!} + \frac{t^4}{5!} + \dots,$$

and

$$\cosh t = 1 + \frac{t^2}{2!} + \frac{t^4}{4!} + \dots,$$

it follows that

$$\begin{aligned}\frac{t\coth t - 1}{2t^2} &= \frac{1}{2t^2}\left(\frac{\cosh t}{\sinh t/t} - 1\right) = \frac{1}{2t^2}\left(\frac{1+\frac{t^2}{2!}+\frac{t^4}{4!}+\dots}{1+\frac{t^2}{3!}+\frac{t^4}{5!}+\dots} - 1\right) \\ &= \frac{1}{2t^2}\cdot\frac{\left(1+\frac{t^2}{2!}+\frac{t^4}{4!}+\frac{t^6}{6!}+\dots\right)-\left(1+\frac{t^2}{3!}+\frac{t^4}{5!}+\frac{t^6}{7!}+\dots\right)}{1+\frac{t^2}{3!}+\frac{t^4}{5!}+\dots} \\ &= \frac{1}{2t^2}\cdot\frac{\frac{t^2}{3}+\frac{t^4}{3!5}+\frac{t^6}{5!7}+\dots}{1+\frac{t}{3!}+\frac{t^2}{5!}+\dots} \\ &= \frac{\frac{1}{6}+\frac{t^2}{60}+\frac{t^4}{1680}+\dots}{1+\frac{t}{3!}+\frac{t^2}{5!}+\dots}.\end{aligned}$$

Now take the limit, as $t \searrow 0$, obtaining the following:

Theorem 5.6.2.

$$\frac{\pi^2}{6} = \frac{1}{1^2} + \frac{1}{2^2} + \frac{1}{3^2} + \dots . \quad \square$$

---

[13] By the monotone convergence theorem for series.

Recalling the zeta function

$$\zeta(x) = \sum_{n=1}^{\infty} \frac{1}{n^x}, \qquad x > 1,$$

this result says that $\zeta(2) = \pi^2/6$, a result due to Euler. In fact, Euler used (5.6.6) to compute $\zeta(2n)$ for all $n \geq 1$. This computation involves certain rational numbers first studied by Bernoulli.

The *Bernoulli function* is defined by

$$\tau(x) = \begin{cases} \dfrac{x}{1 - e^{-x}}, & x \neq 0, \\ 1, & x = 0. \end{cases}$$

Clearly, τ is a smooth function on $x \neq 0$. The *Bernoulli numbers* B_n, $n \geq 0$, are defined by the *Bernoulli series*

$$\tau(x) = B_0 + B_1 x + \frac{B_2}{2!} x^2 + \frac{B_3}{3!} x^3 + \dots . \tag{5.6.7}$$

Since $1 - e^{-x} = x - x^2/2! + x^3/3! - x^4/4! + \dots$,

$$\frac{1 - e^{-x}}{x} = 1 - x/2! + x^2/3! - x^3/4! + \dots .$$

Hence, to obtain the B_n's, one computes the reciprocal of this last series, which is obtained by setting the Cauchy product (§1.7)

$$\left(1 - \frac{x}{2!} + \frac{x^2}{3!} - \frac{x^3}{4!} + \dots\right)\left(B_0 + B_1 x + \frac{B_2}{2!} x^2 + \frac{B_3}{3!} x^3 + \dots\right) = 1.$$

Multiplying, this leads to $B_0 = 1$ and the recursion formula

$$\frac{B_{n-1}}{(n-1)!1!} - \frac{B_{n-2}}{(n-2)!2!} + \dots + (-1)^{n-1} \frac{B_0}{0!n!} = 0, \qquad n \geq 2. \tag{5.6.8}$$

Computing, we see that each B_n is a rational number with

$$B_1 = \frac{1}{2}, \qquad B_2 = \frac{1}{6}, \qquad B_4 = -\frac{1}{30},$$
$$B_6 = \frac{1}{42}, \qquad B_8 = -\frac{1}{30}, \qquad B_{10} = \frac{5}{66}, \qquad \dots .$$

It turns out (Exercise **5.6.2**) that $|B_n| \leq 2^n n!$. Hence, by the root test, the Bernoulli series (5.6.7) converges, at least, for $|x| < 1/2$. In particular, this shows that τ is smooth near zero. Hence, τ is smooth on $\mathbf{R}$. Let $2\pi\beta > 0$

denote the radius[14] of convergence of (5.6.7). Then (5.6.7) holds for $|x| < 2\pi\beta$. Since

$$\frac{x}{1-e^{-x}} - \frac{x}{2} = \frac{x}{2} \cdot \frac{1+e^{-x}}{1-e^{-x}} = \frac{x}{2} \cdot \frac{e^{x/2}+e^{-x/2}}{e^{x/2}-e^{-x/2}} = \frac{x}{2} \coth\left(\frac{x}{2}\right),$$

subtracting $x/2 = B_1 x$ from both sides of (5.6.7), we obtain

$$\frac{x}{2} \coth\left(\frac{x}{2}\right) = 1 + \sum_{n=2}^{\infty} \frac{B_n}{n!} x^n, \qquad 0 < |x| < 2\pi\beta.$$

But $(x/2)\coth(x/2)$ is even. Hence, $B_3 = B_5 = B_7 = \cdots = 0$, and

$$\frac{x}{2} \coth\left(\frac{x}{2}\right) - 1 = \sum_{n=1}^{\infty} \frac{B_{2n}}{(2n)!} x^{2n}, \qquad 0 < |x| < 2\pi\beta.$$

Now replacing x by $2\pi\sqrt{x}$ and dividing by x,

$$\frac{\pi\sqrt{x}\coth(\pi\sqrt{x}) - 1}{x} = \sum_{n=1}^{\infty} \frac{B_{2n}}{(2n)!} (2\pi)^{2n} x^{n-1}, \qquad 0 < x < \beta^2.$$

Thus, from (5.6.6), we conclude that

$$\frac{1}{2} \sum_{n=1}^{\infty} \frac{B_{2n}}{(2n)!} (2\pi)^{2n} x^{n-1} = \sum_{n=1}^{\infty} \frac{1}{n^2 + x}, \qquad 0 < x < \beta^2.$$

Since the left side is a power series, we may differentiate it term by term (§3.4). On the other hand, the right side may[15] be differentiated under the summation sign. Differentiating both sides $r - 1$ times,

$$\frac{1}{2} \sum_{n=r}^{\infty} \frac{B_{2n}}{(2n)!} (2\pi)^{2n} \cdot \frac{(n-1)!}{(n-r)!} x^{n-r} = \sum_{n=1}^{\infty} \frac{(-1)^{r-1}(r-1)!}{(n^2+x)^r}, \qquad 0 < x < \beta^2.$$

Sending $x \to 0+$, the right side becomes $(-1)^{r-1}(r-1)!\zeta(2r)$, whereas the left side reduces to the first coefficient (that corresponding to $n = r$). We have derived the following.

Theorem 5.6.3. *For all* $n \geq 1$,

$$\zeta(2n) = \frac{(-1)^{n-1}}{2} \cdot \frac{B_{2n}}{(2n)!} \cdot (2\pi)^{2n}. \qquad \square$$

---

[14] In fact, below we see $\beta = 1$ and the radius is 2π.

[15] With $f_n(x) = 1/(n^2+x)$ and $g_n = r!/n^2$, $|f_n(x)| + |f_n'(x)| + \cdots + |f_n^{(r-1)}(x)| \leq g_n$ and $\sum g_n = r!\zeta(2)$.

As an immediate consequence, we obtain the radius of convergence of the Bernoulli series (5.6.7).

Theorem 5.6.4. *The radius of convergence of the Bernoulli series* (5.6.7) *is* 2π.

The derivation is an immediate consequence of the previous theorem, the root test (§3.4), and the fact $\zeta(\infty) = 1$. □

Inserting the formula for $\zeta(2n)$ into the series for $\tau(2\pi x)$ yields

$$\frac{2\pi x}{1 - e^{-2\pi x}} = 1 + \pi x + 2\zeta(2)x^2 - 2\zeta(4)x^4 + 2\zeta(6)x^6 - \dots, \qquad |x| < 1. \quad (5.6.9)$$

This series at once encapsulates our use of the Bernoulli function to compute explicitly $\zeta(2)$, $\zeta(4)$, $\zeta(6)$,... and may be used directly to compute the values of ζ at the even naturals.

A natural question is whether one can find a function whose Taylor series has coefficients involving all of $\zeta(2)$, $\zeta(3)$, $\zeta(4)$,.... It turns out $\log(x!)$ serves this purpose (Exercise **5.6.12**); alas, however, $\log(x!)$ is not tractable enough to compute explicitly any of the odd values $\zeta(2n+1)$, not even $\zeta(3)$.

Above, we saw that relating an infinite series to an infinite product led to some nice results. In particular, we derived the infinite product for $\sinh \pi x / \pi x$, which we rewrite, now, as

$$1 + \frac{\pi^2 x^2}{3!} + \frac{\pi^4 x^4}{5!} + \dots = \prod_{n=1}^{\infty} \left(1 + \frac{x^2}{n^2}\right), \qquad x \neq 0. \qquad (5.6.10)$$

We wish to derive the analog of this result for the sine function, i.e., we want to obtain (5.6.10) with $-x^2$ replacing x^2.

To this end, consider the following identity

$$1 + b_1 x + b_2 x^2 + \dots = \prod_{n=1}^{\infty} (1 + a_n x), \qquad 0 < x < R. \qquad (5.6.11)$$

We seek the relations between (a_n) and (b_n). As a special case, if we suppose that $a_n = 0$ for all $n \geq 3$, (5.6.11) reduces to

$$1 + b_1 x + b_2 x^2 = (1 + a_1 x)(1 + a_2 x)$$

which implies $b_1 = a_1 + a_2$ and $b_2 = a_1 a_2$. Similarly, if we suppose that $a_n = 0$ for all $n \geq 4$, (5.6.11) reduces to

$$1 + b_1 x + b_2 x^2 + b_3 x^3 = (1 + a_1 x)(1 + a_2 x)(1 + a_3 x)$$

which implies

$$b_1 = a_1 + a_2 + a_3 = \sum_i a_i,$$

$$b_2 = a_1 a_2 + a_1 a_3 + a_2 a_3 = \sum_{i<j} a_i a_j,$$

and

$$b_3 = a_1a_2a_3 = \sum_{i<j<k} a_ia_ja_k.$$

Theorem 5.6.5. *Suppose that* (a_n) *and* (b_n) *are positive sequences and the series in* (5.6.11) *converges on* $(-R, R)$. *Suppose also* (5.6.11) *holds; then*

$$1 - b_1x + b_2x^2 - \cdots = \prod_{n=1}^{\infty}(1 - a_nx), \qquad 0 < x < R. \tag{5.6.12}$$

We call (5.6.12) the *alternating version* of (5.6.11). Let us immediately apply this theorem to derive the infinite product for the sine. Replacing x by $\sqrt{x}$ in (5.6.10), we obtain

$$1 + \frac{\pi^2x}{3!} + \frac{\pi^4x^2}{5!} + \cdots = \prod_{n=1}^{\infty}\left(1 + \frac{x}{n^2}\right), \qquad x > 0. \tag{5.6.13}$$

Now the alternating version of (5.6.13) is given by

$$1 - \frac{\pi^2x}{3!} + \frac{\pi^4x^2}{5!} - \cdots = \prod_{n=1}^{\infty}\left(1 - \frac{x}{n^2}\right), \qquad x > 0.$$

Replacing x by x^2 in the last equation leads to

$$1 - \frac{\pi^2x^2}{3!} + \frac{\pi^4x^4}{5!} - \cdots = \prod_{n=1}^{\infty}\left(1 - \frac{x^2}{n^2}\right), \qquad x \neq 0.$$

But this last series is the series for $\sin(\pi x)/\pi x$.

Theorem 5.6.6. *For* $x \neq 0$,

$$\frac{\sin(\pi x)}{\pi x} = \prod_{n=1}^{\infty}\left(1 - \frac{x^2}{n^2}\right). \quad \square \tag{5.6.14}$$

This is the alternating version of (5.6.1).

To derive (5.6.12) from (5.6.11), write the finite version of (5.6.11),

$$1 + b_1^{(N)}x + b_2^{(N)}x^2 + \cdots + b_N^{(N)}x^N = \prod_{n=1}^{N}(1 + a_nx), \tag{5.6.15}$$

and let $b_1^{(N)}$, $b_2^{(N)}$, ..., $b_N^{(N)}$, denote the coefficients obtained by expanding the right side. Then

$$b_1^{(N)} = \sum_{i=1}^{N} a_i \nearrow \sum_{i=1}^{\infty} a_i = b_1^{(\infty)},$$

as $N \nearrow \infty$,

$$b_2^{(N)} = \sum_{1 \le i < j \le N} a_i a_j \nearrow \sum_{1 \le i < j < \infty} a_i a_j = b_2^{(\infty)},$$

as $N \nearrow \infty$, and so on. Here $b_1^{(\infty)}$, $b_2^{(\infty)}$, ..., are *defined* as the positive infinite sums

$$b_1^{(\infty)} = \sum_i a_i, \qquad b_2^{(\infty)} = \sum_{i<j} a_i a_j, \qquad \ldots .$$

We want to show that $\left(b_n^{(\infty)}\right)$ equals the given sequence (b_n). For this, let $N \nearrow \infty$ in (5.6.15). Since x is positive, there is no problem with the limits (everything is increasing), and we get

$$1 + b_1^{(\infty)} x + b_2^{(\infty)} x^2 + \cdots = \prod_{n=1}^{\infty} (1 + a_n x), \qquad 0 < x < R.$$

Since the coefficients of a power series are unique (§3.5), this and (5.6.11) yield $b_n = b_n^{(\infty)}$ for $n \ge 1$. Hence, $b_n^{(N)} \nearrow b_n$ for all $n \ge 1$, as $N \nearrow \infty$.

Now replace x by $-x$ in (5.6.15) to get

$$1 - b_1^{(N)} x + b_2^{(N)} x^2 - \cdots + (-1)^N b_N^{(N)} x^N = \prod_{n=1}^{N} (1 - a_n x), \qquad 0 < x < R.$$

Clearly, as $N \nearrow \infty$, the right side of this last equation decreases to the right side of (5.6.12) ($a_n \to 0$ since $\sum a_n < \infty$ since $\prod(1 + a_n x)$ converges). Thus, to derive the theorem, it is enough to show that

$$\sum_{n=1}^{N} (-1)^n b_n^{(N)} x^n \to \sum_{n=1}^{\infty} (-1)^n b_n x^n, \qquad N \nearrow \infty.$$

But, for $0 < x < R$,

$$\left| \sum_{n=1}^{N} (-1)^n b_n^{(N)} x^n - \sum_{n=1}^{\infty} (-1)^n b_n x^n \right| \le \sum_{n=1}^{N} \left[b_n - b_n^{(N)} \right] x^n + \sum_{n=N+1}^{\infty} b_n x^n.$$

Now the second sum on the right is the tail (§1.6) of a convergent series, hence goes to zero, as $N \nearrow \infty$, whereas the first sum on the right goes to zero by the dominated convergence theorem for series. Indeed, the terms in the first sum on the right are no greater than $g_n = b_n x^n$ with $\sum g_n$ finite by assumption. Thus, we arrive at (5.6.12). □

Exercises

5.6.1. Use (5.6.9) to compute $\zeta(2)$, $\zeta(4)$, $\zeta(6)$, $\zeta(8)$.

5.6.2. Use the recursion (5.6.8) to derive $|B_n| \le 2^n n!$, $n \ge 1$. Conclude that the Bernoulli series (5.6.7) converges on $(-1/2, 1/2)$. Also show that the numbers $(B_2, B_4, B_6, \dots)$, form an alternating sequence $(+,-,+,\dots)$.

5.6.3. If (a_n) is a positive sequence, then

$$\sum_{n=1}^{\infty} a_n \le \prod_{n=1}^{\infty}(1+a_n) \le \exp\left(\sum_{n=1}^{\infty} a_n\right). \tag{5.6.16}$$

Conclude

$$\sum a_n < \infty \quad \text{iff} \quad \prod(1+a_n) < \infty.$$

5.6.4. Use Exercise **5.1.5** to show that

$$\Gamma(x) = \frac{e^{-\gamma x}}{x} \prod_{n=1}^{\infty} \left[\frac{e^{x/n}}{1+\dfrac{x}{n}}\right], \qquad x > 0,$$

and

$$x! = e^{-\gamma x} \prod_{n=1}^{\infty} \left[\frac{e^{x/n}}{1+\dfrac{x}{n}}\right], \qquad x > -1.$$

Here γ is Euler's constant (Exercise **4.4.18**). (Use $1+1/2+\cdots+1/n-\log n \to \gamma$ and $n^x = e^{x\log n}$.)

5.6.5. Use Exercise **5.1.5** applied to $\Gamma(x)$ and $\Gamma(1-x)$ to show that

$$\frac{\pi}{\Gamma(x)\Gamma(1-x)} = \sin(\pi x), \qquad 0 < x < 1.$$

5.6.6. Let

$$B(x) = 1 + \sum_{n=1}^{\infty}(-1)^n \frac{B_{2n}}{(2n)!}x^{2n}, \qquad |x| < 2\pi\beta.$$

Use Exercise **1.7.7** to show

$$(x/2)\cos(x/2) = B(x)\sin(x/2), \qquad \text{for } |x| < 2\pi\beta.$$

Conclude

$$\frac{x}{2}\cot\left(\frac{x}{2}\right) = 1 + \sum_{n=1}^{\infty}(-1)^n \frac{B_{2n}}{(2n)!}x^{2n}, \qquad 0 < |x| < \min(2\pi, 2\pi\beta).$$

5.6.7. Use Exercise **5.6.6** to conclude that $\beta \le 1$, i.e., the radius of convergence of the Bernoulli series (5.6.7) is no more than 2π.

5.6.8. Use (5.6.14) and modify the development leading up to (5.6.6) to obtain

$$\pi\cot(\pi x) - \frac{1}{x} = \sum_{n=1}^{\infty} \frac{2x}{x^2 - n^2}, \qquad 0 < |x| < 1.$$

5.6.9. Use Exercise **5.6.6** above and Exercise **3.6.13** to obtain

$$\tan x = \sum_{n=1}^{\infty} (-1)^{n-1} \frac{B_{2n}}{(2n)!} 2^{2n}(2^{2n} - 1)x^{2n-1}, \qquad |x| < \pi\beta/2.$$

5.6.10. Use Exercise **5.6.4**, and differentiate under the summation sign to get

$$\frac{d}{dx}\log\Gamma(x) = \frac{\Gamma'(x)}{\Gamma(x)} = -\gamma - \frac{1}{x} + \sum_{n=1}^{\infty}\left(\frac{1}{n} - \frac{1}{n+x}\right), \qquad x > 0,$$

and

$$\frac{d}{dx}\log(x!) = -\gamma + \sum_{n=1}^{\infty}\left(\frac{1}{n} - \frac{1}{n+x}\right), \qquad x > -1.$$

$\Psi(x) = \Gamma'(x)/\Gamma(x)$ is the *digamma function.*

5.6.11. Using the previous exercise, show

$$\frac{1}{r!}\frac{d^{r+1}}{dx^{r+1}}\log(x!) = \sum_{n=1}^{\infty} \frac{(-1)^{r+1}}{(n+x)^{r+1}}, \qquad r \ge 1, x > -1,$$

and

$$\left|\frac{1}{r!}\frac{d^{r+1}}{dx^{r+1}}\log(x!)\right| \le \frac{1}{(x+1)^{r+1}} + \zeta(2), \qquad r \ge 1, x > -1.$$

(To justify the differentiation, assume first x is in $[-1+\epsilon, 1/\epsilon]$.)

5.6.12. Derive the zeta series

$$\log(x!) = -\gamma x + \frac{1}{2}\zeta(2)x^2 - \frac{1}{3}\zeta(3)x^3 + \frac{1}{4}\zeta(4)x^4 - \dots, \qquad 1 \ge x > -1.$$

Conclude

$$\gamma = \frac{\zeta(2)}{2} - \frac{\zeta(3)}{3} + \frac{\zeta(4)}{4} - \dots.$$

(Use the previous exercise to estimate the remainder as in Exercise **3.5.8** or Exercise **4.4.30**.)

5.6.13. Use Exercise **5.6.10** to show

$$\Gamma'(1) = -\gamma \qquad \text{and} \qquad \Gamma'(2) = 1 - \gamma.$$

Conclude the global minimum of $\Gamma(x)$, $x > 0$, lies in the interval $(1, 2)$.

5.6.14. Differentiate under the summation sign to obtain the Laplace transform of τ,

$$\frac{d^2}{dx^2} \log \Gamma(x) = \Psi'(x) = \int_0^\infty e^{-xt} \tau(t)\, dt, \qquad x > 0.$$

(Exercise **5.6.10** above and Exercise **5.1.13**.)

5.6.15. Use Exercise **5.6.10** to show

$$\lim_{x \to 1-} \left\{ -\frac{1}{2} \frac{\Gamma'[(1-x)/2]}{\Gamma[(1-x)/2]} + \frac{1}{x-1} \right\} = \frac{1}{2}\gamma.$$

5.7 Jacobi's Theta Functions

The *theta function* is defined by

$$\theta(s) = \sum_{-\infty}^{\infty} e^{-n^2 \pi s} = 1 + 2e^{-\pi s} + 2e^{-4\pi s} + 2e^{-9\pi s} + \dots, \qquad s > 0.$$

This positive sum, over all integers n (positive and negative and zero), converges for $s > 0$ since

$$\begin{aligned}\sum_{-\infty}^{\infty} e^{-n^2 \pi s} &\le \sum_{-\infty}^{\infty} e^{-|n| \pi s} \\ &= 1 + 2 \sum_{n=1}^{\infty} e^{-n \pi s} \\ &= 1 + \frac{2e^{-\pi s}}{1 - e^{-\pi s}} < \infty.\end{aligned}$$

Recall (Exercise **5.1.9**) the function

$$\psi(s) = \sum_{n=1}^{\infty} e^{-n^2 \pi s}.$$

This is related to θ by $\theta = 1 + 2\psi$. The main result in this section is the following remarkable identity, which we need in the next section.

Theorem 5.7.1 (Theta Functional Equation). *For all $s > 0$,*

$$\theta(1/s) = \sqrt{s}\theta(s), \qquad s > 0, \tag{5.7.1}$$

which can be rewritten as

$$\sum_{-\infty}^{\infty} e^{-n^2\pi/s} = \sqrt{s}\sum_{-\infty}^{\infty} e^{-n^2\pi s}, \qquad s > 0. \tag{5.7.2}$$

As one indication of the power of (5.7.2), plug in $s = .01$. Then the series for $\theta(.01)$ converges slowly (the tenth term is $1/e^{\pi}$), whereas the series for $\theta(100)$ converges quickly. In fact, the sum $\theta(100)$ of the series on the left differs from its zeroth term 1 by less than 10^{-100}.

In terms of ψ, the functional equation becomes

$$1 + 2\psi(1/s) = \sqrt{s}[1 + 2\psi(s)]. \tag{5.7.3}$$

To derive (5.7.1), we need to introduce three power series, *Jacobi's theta functions*, and relate them to the arithmetic–geometric mean of §5.3. These functions' most striking property, double-periodicity, does not appear unless one embraces the complex plane. Nevertheless, within the confines of the real line, we shall be able to get somewhere.

The *Jacobi theta functions*, defined for $|q| < 1$, are[16]

$$\theta_0(q) = \sum_{-\infty}^{\infty} q^{n^2} = 1 + 2q + 2q^4 + 2q^9 + \dots,$$

$$\theta_-(q) = \sum_{-\infty}^{\infty} (-1)^n q^{n^2} = 1 - 2q + 2q^4 - 2q^9 + \dots,$$

and

$$\theta_+(q) = \sum_{-\infty}^{\infty} q^{(n+1/2)^2} = 2q^{1/4} + 2q^{9/4} + 2q^{25/4} + \dots.$$

By comparing these series with the geometric series, we see that they all converge for $|q| < 1$.

The simplest properties of these functions depend on parity properties of integers. For example, because n is odd iff n^2 is odd, $\theta_0(-q) = \theta_-(q)$, since

$$\theta_0(-q) = \sum_{-\infty}^{\infty} (-1)^{n^2} q^{n^2} = \sum_{-\infty}^{\infty} (-1)^n q^{n^2} = \theta_-(q).$$

Similarly, since $\theta_-(q)$ is the alternating version (§1.7) of $\theta_0(q)$,

$$\theta_0(q) + \theta_-(q) = 2 \sum_{n \text{ even}} q^{n^2} = 2\sum_{-\infty}^{\infty} q^{(2n)^2} = 2\sum_{-\infty}^{\infty} \left(q^4\right)^{n^2} = 2\theta_0\left(q^4\right). \tag{5.7.4}$$

---

[16] The index notation in θ_0, θ_+, θ_- is not standard.

In the remainder of the section, we restrict q to lie in the interval $(0,1)$. In this case, $\theta_0(q) \geq 1$, $\theta_-(q)$ is bounded in absolute value by 1 by the Leibnitz test, and, hence $\theta_-^2(q) \leq \theta_0^2(q)$.

For $n \geq 0$, let $\sigma(n)$ be the number of ways of writing n as a sum of squares, $n = i^2 + j^2$, with $i, j \in \mathbf{Z}$, where permutations and signs are taken into account. Thus,

$$\begin{aligned}
\sigma(0) &= 1 \quad \text{because} \quad 0 = 0^2 + 0^2, \\
\sigma(1) &= 4 \quad \text{because} \quad 1 = (\pm 1)^2 + 0^2 = 0^2 + (\pm 1)^2, \\
\sigma(2) &= 4 \quad \text{because} \quad 2 = (\pm 1)^2 + (\pm 1)^2, \\
\sigma(3) &= 0, \\
\sigma(4) &= 4 \quad \text{because} \quad 4 = (\pm 2)^2 + 0^2 = 0^2 + (\pm 2)^2, \\
\sigma(5) &= 8 \quad \text{because} \quad 5 = (\pm 2)^2 + (\pm 1)^2 = (\pm 1)^2 + (\pm 2)^2, \\
\sigma(6) &= \sigma(7) = 0, \\
\sigma(8) &= 4 \quad \text{because} \quad 8 = (\pm 2)^2 + (\pm 2)^2, \\
\sigma(9) &= 4 \quad \text{because} \quad 9 = (\pm 3)^2 + 0^2 = 0^2 + (\pm 3)^2, \\
\sigma(10) &= 8 \quad \text{because} \quad 10 = (\pm 1)^2 + (\pm 3)^2 = (\pm 3)^2 + (\pm 1)^2, \\
&\quad \text{etc.}
\end{aligned}$$

Then

$$\begin{aligned}
\theta_0^2(q) &= \left(\sum_{-\infty}^{\infty} q^{n^2} \right)^2 && (5.7.5) \\
&= \left(\sum_{-\infty}^{\infty} q^{i^2} \right) \left(\sum_{-\infty}^{\infty} q^{j^2} \right) \\
&= \sum_{i,j \in \mathbf{Z}} q^{i^2+j^2} && (5.7.6) \\
&= \sum_{n=0}^{\infty} \left(\sum_{i^2+j^2=n} q^n \right) && (5.7.7) \\
&= \sum_{n=0}^{\infty} \sigma(n) q^n && (5.7.8) \\
&= 1 + 4q + 4q^2 + 4q^4 + 8q^5 + 4q^8 + 4q^9 + 8q^{10} + \dots.
\end{aligned}$$

Similarly, since n is even iff n^2 is even,

$$\theta_-^2(q) = \sum_{n=0}^{\infty} (-1)^n \sigma(n) q^n. \tag{5.7.9}$$

Now if $n = i^2 + j^2$, then $2n = (i+j)^2 + (i-j)^2 = k^2 + \ell^2$. Conversely, if $2n = k^2 + \ell^2$, then $n = ((k+\ell)/2)^2 + (k-\ell)/2)^2 = i^2 + j^2$. Thus,

$$\sigma(2n) = \sigma(n), \qquad n \geq 1.$$

For example, $\sigma(1) = \sigma(2) = \sigma(4) = \sigma(8)$. Here we used the fact that $k^2 + \ell^2$ is even iff $k + \ell$ is even iff $k - \ell$ is even. Since the series (5.7.9) is the alternating version of the series (5.7.8),

$$\begin{aligned} \theta_0^2(q) + \theta_-^2(q) &= 2 \sum_{n \text{ even}} \sigma(n) q^n \\ &= 2 \sum_{n=0}^{\infty} \sigma(2n) q^{2n} \\ &= 2 \sum_{n=0}^{\infty} \sigma(n) \left(q^2\right)^n \\ &= 2\theta_0^2\left(q^2\right). \end{aligned} \tag{5.7.10}$$

Now subtract (5.7.10) from the square of (5.7.4). You obtain

$$\begin{aligned} 2\theta_0(q)\theta_-(q) &= \left[\theta_0(q) + \theta_-(q)\right]^2 - \left[\theta_0^2(q) + \theta_-^2(q)\right] \\ &= 4\theta_0^2\left(q^4\right) - 2\theta_0^2\left(q^2\right) = 2\theta_-^2\left(q^2\right), \end{aligned} \tag{5.7.11}$$

where the last equality is by (5.7.10) again. Rewriting (5.7.10) and (5.7.11), we have arrived at the AGM iteration

$$\begin{aligned} \frac{\theta_0^2(q) + \theta_-^2(q)}{2} &= \theta_0^2\left(q^2\right), \\ \sqrt{\theta_0^2(q)\theta_-^2(q)} &= \theta_-^2\left(q^2\right). \end{aligned} \tag{5.7.12}$$

Setting $a_0 = \theta_0^2(q)$ and $b_0 = \theta_-^2(q)$, let (a_n), (b_n), be the AGM iteration (5.3.2),(5.3.3). Iterating (5.7.12), we obtain

$$a_n = \theta_0^2\left(q^{2^n}\right),$$

and

$$b_n = \theta_-^2\left(q^{2^n}\right),$$

$n \geq 0$. Since $\theta_0(0) = 1 = \theta_-(0)$, $q^{2^n} \to 0$, and $a_n \to M(a_0, b_0)$, $b_n \to M(a_0, b_0)$, we arrive at $M(a_0, b_0) = 1$ or $M\left(\theta_0^2(q), \theta_-^2(q)\right) = 1$.

Theorem 5.7.2. *Suppose that (a, b) lies in the first quadrant of the ab-plane with $a > 1 > b$. Then (a, b) lies on the AGM curve*

$$M(a,b) = 1 \qquad \textit{iff} \qquad (a,b) = (\theta_0^2(q), \theta_-^2(q))$$

for a unique $0 < q < 1$. *In particular,*

$$M\left(\theta_0^2(q), \theta_-^2(q)\right) = 1, \qquad 0 < q < 1. \tag{5.7.13}$$

Above we derived (5.7.13). To get the rest, suppose that $a > 1 > b > 0$ and $M(a,b) = 1$. Since $\theta_0^2 : (0,1) \to (1,\infty)$ is a bijection (Exercise **5.7.1**), there is a unique q in $(0,1)$, satisfying $a = \theta_0^2(q)$. Then by (5.7.13), $M\left[a, \theta_-^2(q)\right] = 1 = M(a,b)$. Since $b \mapsto M(a,b)$ is strictly increasing, we must have $b = \theta_-^2(q)$. □

Now let $c_n = \sqrt{a_n^2 - b_n^2}$, $n \ge 0$, be given by (5.3.4). We show that

$$c_n = \theta_+^2\left(q^{2^n}\right), \qquad n \ge 0. \tag{5.7.14}$$

To this end, compute

$$\begin{aligned}
\theta_0^2(q) - \theta_0^2\left(q^2\right) &= \sum_{n=0}^{\infty} \sigma(n) q^n - \sum_{n=0}^{\infty} \sigma(2n) q^{2n} \\
&= \sum_{n \text{ odd}} \sigma(n) q^n \\
&= \sum_{\substack{i,j \in \mathbf{Z} \\ i^2+j^2 \text{ odd}}} q^{i^2+j^2}.
\end{aligned}$$

Now i^2+j^2 is odd iff $i+j$ is odd iff $i-j$ is odd, in which case $k = (j+i-1)/2$ and $\ell = (j-i-1)/2$ are integers. Solving, since $i = k - \ell$, and $j = k + \ell + 1$, the last sum equals

$$\begin{aligned}
\theta_0^2(q) - \theta_0^2\left(q^2\right) &= \sum_{k,\ell \in \mathbf{Z}} q^{(k-\ell)^2 + (k+\ell+1)^2} \\
&= \sum_{k,\ell \in \mathbf{Z}} \left(q^2\right)^{(k^2+k+1/4)+(\ell^2+\ell+1/4)} \\
&= \left[\sum_{-\infty}^{\infty} \left(q^2\right)^{k^2+k+1/4}\right]^2 = \theta_+^2\left(q^2\right).
\end{aligned}$$

Hence,

$$\theta_0^2\left(q^2\right) + \theta_+^2\left(q^2\right) = \theta_0^2(q). \tag{5.7.15}$$

Adding (5.7.10) and (5.7.15) leads to

$$\theta_0^2\left(q^2\right) - \theta_+^2\left(q^2\right) = \theta_-^2\left(q\right).$$

Multiplying the last two equations and recalling (5.7.11) lead to

$$\theta_0^4\left(q^2\right) = \theta_-^4\left(q^2\right) + \theta_+^4\left(q^2\right). \tag{5.7.16}$$

Now replacing q^2 by q^{2^n} in (5.7.16) leads to (5.7.14) since $a_n^2 = b_n^2 + c_n^2$. This establishes (5.7.14). □

From §5.3, we know that $c_n \to 0$. Let us compute the rate at which this happens. It turns out that the decay rate is exponential, in the sense that

$$\lim_{n \nearrow \infty} \frac{1}{2^n} \log(c_n) = \frac{1}{2} \log q. \tag{5.7.17}$$

To see this, let us denote, for clarity, $2^n = N$. Then by (5.7.14),

$$\begin{aligned} \frac{1}{N} \log(c_n) &= \frac{1}{N} \log\left(\theta_+^2\left(q^N\right)\right) \\ &= \frac{2}{N} \log\left(2q^{N/4} + 2q^{9N/4} + 2q^{25N/4} + \dots\right) \\ &= \frac{2}{N} \left[\log 2 + (N/4) \log q + \log\left(1 + q^{2N} + q^{6N} + \dots\right)\right]. \end{aligned}$$

Hence, since $q^N \to 0$, we obtain (5.7.17). This computation should be compared with Exercise **5.5.7**. Now (5.3.20) says

$$\lim_{n \nearrow \infty} \frac{1}{2^n} \log\left(\frac{a_n}{c_n}\right) = \frac{\pi}{2} Q\left(\frac{b}{a}\right).$$

Inserting $a_n \to \theta_0(0) = 1$ and (5.7.17) into this equation and recalling $a = \theta_0^2(q)$, $b = \theta_-^2(q)$, we obtain

$$-\frac{1}{\pi} \log q = Q\left(\frac{\theta_-^2(q)}{\theta_0^2(q)}\right). \tag{5.7.18}$$

Here $Q(x) = M(1,x)/M(1,x')$. Solving for q, we obtain the following sharpening of the previous theorem.

Theorem 5.7.3. *Suppose that* (a,b) *satisfies* $M(a,b) = 1$, $a > 1 > b > 0$, *and let* $q \in (0,1)$ *be such that* $(a,b) = (\theta_0^2(q), \theta_-^2(q))$. *Then*

$$q = e^{-\pi Q(b/a)}. \quad \square \tag{5.7.19}$$

Now go back and look at (5.3.15). In Exercise **5.7.5**, (5.3.15) is improved to an equality.

Now let $q = e^{-\pi s}$, $s > 0$, and set

$$\theta_0(s) = \theta_0\left(e^{-\pi s}\right), \qquad \theta_\pm(s) = \theta_\pm\left(e^{-\pi s}\right).$$

Then (5.7.18) can be written

$$s = Q\left(\frac{\theta_-^2(s)}{\theta_0^2(s)}\right) = \frac{M\left(1, \theta_-^2(s)/\theta_0^2(s)\right)}{M\left(1, (\theta_-^2(s)/\theta_0^2(s))'\right)}, \qquad s > 0. \tag{5.7.20}$$

Replacing s by $1/s$ in this last equation and using $1/Q(x) = Q(x')$, we obtain

$$s = Q\left(\left(\frac{\theta_-^2(1/s)}{\theta_0^2(1/s)}\right)'\right) = Q\left(\frac{\theta_+^2(1/s)}{\theta_0^2(1/s)}\right), \qquad s > 0.$$

Here we use (5.7.16) to show that $(\theta_-^2/\theta_0^2)' = \theta_+^2/\theta_0^2$. Equating the last two expressions for s and using the strict monotonicity of Q (Exercise **5.3.11**), we arrive at

$$\frac{\theta_-^2(s)}{\theta_0^2(s)} = \frac{\theta_+^2(1/s)}{\theta_0^2(1/s)}, \qquad s > 0. \tag{5.7.21}$$

Now we can derive the theta functional equation (5.7.1), as follows:

$$\begin{aligned}
s\theta_0^2(s) &= \frac{s\theta_0^2(s)}{M\left(\theta_0^2(s), \theta_-^2(s)\right)} && [(5.7.13)] \\
&= \frac{s}{M\left(1, \theta_-^2(s)/\theta_0^2(s)\right)} && [\text{homogeneity}] \\
&= \left[M\left(1, \left(\theta_-^2(s)/\theta_0^2(s)\right)'\right)\right]^{-1} && [(5.7.20)] \\
&= \left[M\left(1, \theta_+^2(s)/\theta_0^2(s)\right)\right]^{-1} \\
&= \left[M\left(1, \theta_-^2(1/s)/\theta_0^2(1/s)\right)\right]^{-1} && [(5.7.21)] \\
&= \frac{\theta_0^2(1/s)}{M\left(\theta_0^2(1/s), \theta_-^2(1/s)\right)} && [\text{homogeneity}] \\
&= \theta_0^2(1/s). && [(5.7.13)]
\end{aligned}$$

Since $\theta_0(s) = \theta(s)$, this completes the derivation of (5.7.1). □ Combining (5.7.1) with (5.7.21), we obtain the companion functional equation

$$\sqrt{s}\theta_-(e^{-\pi s}) = \theta_+\left(e^{-\pi/s}\right), \qquad s > 0. \tag{5.7.22}$$

Exercises

5.7.1. Show that θ_0 and θ_+ are strictly increasing functions of $(0,1)$ onto $(1,\infty)$.

5.7.2. Derive (5.7.22).

5.7.3. Show that θ_- is a strictly decreasing function of $(0,1)$ onto $(0,1)$. (Use (5.7.22) to compute $\theta_-(1-)$.)

5.7.4. Compute $\sigma(n)$ for $n = 11, 12, 13, 14, 15$. Show that $\sigma(4n-1) = 0$ for $n \geq 1$.

5.7.5. Let $a > b > 0$, let (a_n) and (b_n) be the AGM iteration, and let q be as in (5.7.19). Show that

$$a_n - b_n = 8M(a,b)q^{2^n} \times \left(1 + 2q^{2^{n+2}} + q^{2^{n+3}} + \dots\right)$$

for $n \geq 0$.

5.7.6. Let

$$\psi(t,x) = \sum_{n=1}^{\infty} e^{-n^2\pi t}\cos(nx), \qquad t > 0, x \in \mathbf{R}.$$

Show ψ satisfies the *heat equation*

$$\frac{\partial\psi}{\partial t} = \pi\frac{\partial^2\psi}{\partial x^2}.$$

5.8 Riemann's Zeta Function

In this section, we study the *Riemann zeta function*

$$\zeta(x) = \sum_{n=1}^{\infty} \frac{1}{n^x} = 1 + \frac{1}{2^x} + \frac{1}{3^x} + \dots, \qquad x > 1,$$

and we discuss

- The behavior of ζ near $x = 1$,
- The extension of the domains of definition of Γ and ζ,
- The functional equation,
- The values of the zeta function at the nonpositive integers,
- The Euler product, and
- Primes in arithmetic progressions.

Most of the results in this section are due to Euler. Nevertheless, ζ is associated with Riemann because, as Riemann showed, the subject really takes off only after x is allowed to range in the complex plane.

We already know that $\zeta(x)$ is smooth for $x > 1$ (Exercise **5.4.8**) and $\zeta(1) = \zeta(1+) = \infty$ (Exercise **5.1.12**). We say that f is *asymptotically equal* to g as $x \to a$, and we write $f(x) \sim g(x)$, as $x \to a$, if $f(x)/g(x) \to 1$, as $x \to a$ (compare with $a_n \sim b_n$ §5.5).

Theorem 5.8.1.

$$\zeta(x) \sim \frac{1}{x-1}, \qquad x \to 1+.$$

We have to show that $(x-1)\zeta(x) \to 1$ as $x \to 1+$. Multiply $\zeta(x)$ by 2^{-x} to get

$$2^{-x}\zeta(x) = \frac{1}{2^x} + \frac{1}{4^x} + \frac{1}{6^x} + \dots, \qquad x > 1. \tag{5.8.1}$$

Then

$$\left(1 - \frac{2}{2^x}\right)\zeta(x) = 1 - \frac{1}{2^x} + \frac{1}{3^x} - \frac{1}{4^x} + \dots, \qquad x > 1. \tag{5.8.2}$$

Now by the Leibniz test, the series in (5.8.2) converges for $x > 0$, equals $\log 2$ at $x = 1$ (Exercise **3.7.17**), and

$$\lim_{x\to 1}\left(1 - \frac{1}{2^x} + \frac{1}{3^x} - \frac{1}{4^x} + \dots\right) = 1 - \frac{1}{2} + \frac{1}{3} - \frac{1}{4} + \dots$$

by the dominated convergence theorem for series (see (5.2.19)). On the other hand, by l'Hôpital's rule,

$$\lim_{x\to 1}\frac{1 - 2^{1-x}}{x-1} = \log 2.$$

Thus,

$$\begin{aligned}\lim_{x\to 1+}(x-1)\zeta(x) &= \lim_{x\to 1+}\frac{x-1}{1-2^{1-x}}\cdot\left(1 - \frac{2}{2^x}\right)\zeta(x)\\ &= \frac{1}{\log 2}\left(1 - \frac{1}{2} + \frac{1}{3} - \frac{1}{4} + \dots\right) = 1. \quad \square\end{aligned}$$

Thus, $\zeta(x)$ and $1/(x-1)$ are asymptotically equal as $x \to 1+$. Nevertheless, it may be possible that the difference $\zeta(x) - 1/(x-1)$ still goes to infinity. For example, x^2 and x^2+x are asymptotically equal as $x \to \infty$, but $(x^2+x) - x^2 \to \infty$, as $x \to \infty$. In fact, for $\zeta(x)$, we show that this does not happen.

Theorem 5.8.2.

$$\lim_{x\to 1+}\left[\zeta(x) - \frac{1}{x-1}\right] = \gamma, \tag{5.8.3}$$

where γ is Euler's constant.

To see this, use Exercise **5.1.8** and $\Gamma(x) = (x-1)\Gamma(x-1)$ to get, for $x > 1$,

$$\begin{aligned}\left[\zeta(x) - \frac{1}{x-1}\right]\Gamma(x) &= \zeta(x)\Gamma(x) - \Gamma(x-1)\\ &= \int_0^\infty \frac{t^{x-1}}{e^t - 1}\,dt - \int_0^\infty e^{-t}t^{x-2}\,dt = \int_0^\infty t^{x-1}\left(\frac{1}{e^t-1} - \frac{1}{te^t}\right)dt.\end{aligned}$$

Applying the dominated convergence theorem (Exercise **5.8.1**),

$$\lim_{x\to 1+}\left(\zeta(x)-\frac{1}{x-1}\right)\Gamma(x)=\int_0^\infty\left(\frac{1}{e^t-1}-\frac{1}{te^t}\right)dt. \tag{5.8.4}$$

But the integral in (5.8.4) is not easy to evaluate directly, so we abandon this approach. Instead, we use the following identity.

Theorem 5.8.3 (Sawtooth Formula). *Let $f:(1,\infty)\to\mathbf{R}$ be differentiable, decreasing, and nonnegative, and suppose that f' is continuous. Then*

$$\sum_{n=1}^\infty f(n)=\int_1^\infty f(t)\,dt+\int_1^\infty(1+\lfloor t\rfloor-t)[-f'(t)]\,dt. \tag{5.8.5}$$

Here $\lfloor t\rfloor$ is the greatest integer $\le t$, and $0\le 1+\lfloor t\rfloor-t\le 1$ is the sawtooth function (Figure 2.3 in §2.3). To get (5.8.5), break up the following integrals of nonnegative functions and integrate by parts:

$$\begin{aligned}
&\int_1^\infty f(t)\;dt+\int_1^\infty(1+\lfloor t\rfloor-t)[-f'(t)]\,dt\\
&\quad=\sum_{n=1}^\infty\left\{\int_n^{n+1}f(t)\,dt+\int_n^{n+1}(1+n-t)[-f'(t)]\,dt\right\}\\
&\quad=\sum_{n=1}^\infty\left\{\int_n^{n+1}f(t)\,dt+(1+n-t)(-f(t))|_n^{n+1}-\int_n^{n+1}f(t)\,dt\right\}\\
&\quad=\sum_{n=1}^\infty f(n).\quad\square
\end{aligned}$$

Now insert $f(t)=1/t^x$ in (5.8.5), and evaluate the integral obtaining

$$\zeta(x)=\frac{1}{x-1}+x\int_1^\infty\frac{1+\lfloor t\rfloor-t}{t^{x+1}}\,dt,\qquad x>1. \tag{5.8.6}$$

We wish to take the limit $x\to 1+$. Since the integrand is dominated by $1/t^2$ when $x>1$, the dominated convergence theorem applies. Hence,

$$\lim_{x\to 1+}\left[\zeta(x)-\frac{1}{x-1}\right]=\int_1^\infty\frac{1+\lfloor t\rfloor-t}{t^2}\,dt.$$

But

$$\int_1^\infty\frac{1+\lfloor t\rfloor-t}{t^2}\,dt=\lim_{N\nearrow\infty}\sum_{n=1}^N\int_n^{n+1}\frac{1+n-t}{t^2}\,dt$$

$$= \lim_{N \nearrow \infty} \sum_{n=1}^{N} \left(-\frac{n+1}{t} - \log t \right)\Bigg|_{n}^{n+1}$$

$$= \lim_{N \nearrow \infty} \left[1 + \frac{1}{2} + \cdots + \frac{1}{N} - \log(N+1) \right] = \gamma.$$

This completes the derivation of (5.8.3). □

The series expression for $\zeta(x)$ is valid only when $x > 1$. Below we extend the domain of ζ to $x < 1$. To this end, we seek an alternate expression for ζ. Because the expression that we will find for ζ involves Γ, first, we extend $\Gamma(x)$ to $x < 0$.

Recall (§5.1) that the gamma function is smooth and positive on $(0, \infty)$. Hence, its reciprocal $L = 1/\Gamma$ is smooth there. Since $\Gamma(x+1) = x\Gamma(x)$, $L(x) = xL(x+1)$. But $xL(x+1)$ is smooth on $x > -1$. Hence, we can use this last equation to define $L(x)$ on $x > -1$ as a smooth function vanishing at $x = 0$. Similarly, we can use $L(x) = xL(x+1) = x(x+1)L(x+2)$ to define $L(x)$ on $x > -2$ as a smooth function, vanishing at $x = 0$ and $x = -1$. Continuing in this manner, the reciprocal $L = 1/\Gamma$ of the gamma function extends to a smooth function on $\mathbf{R}$, vanishing at $x = 0, -1, -2, \ldots$. From this, it follows that Γ itself extends to a smooth function on $\mathbf{R} \setminus \{0, -1, -2, \ldots\}$. Moreover (Exercise **5.8.3**),

$$\Gamma(x) \sim \frac{(-1)^n}{n!} \cdot \frac{1}{x+n}, \qquad x \to -n, \qquad n \geq 0. \tag{5.8.7}$$

To obtain an alternate expression for ζ, start with

$$\psi(t) = \sum_{n=1}^{\infty} e^{-n^2 \pi t}, \qquad t > 0,$$

use Exercise **5.1.9**, and substitute $1/t$ for t to get, for $x > 1$,

$$\begin{aligned}
\pi^{-x/2}\Gamma(x/2)\zeta(x) &= \int_0^\infty \psi(t) t^{x/2-1}\, dt \\
&= \int_0^1 \psi(t) t^{x/2-1} dt + \int_1^\infty \psi(t) t^{x/2-1}\, dt \\
&= \int_1^\infty \left[\psi\left(\frac{1}{t}\right) t^{-x/2-1} + \psi(t) t^{x/2-1} \right] dt. \qquad (5.8.8)
\end{aligned}$$

Now by the theta functional equation (5.7.3),

$$\psi\left(\frac{1}{t}\right) = \frac{\sqrt{t}}{2} + \sqrt{t}\psi(t) - \frac{1}{2}.$$

So (5.8.8) leads to

$$\pi^{-x/2}\Gamma(x/2)\zeta(x) = \int_1^\infty t^{-x/2-1}\left(\frac{\sqrt{t}}{2} - \frac{1}{2}\right) dt + \int_1^\infty \psi(t)\left[t^{(1-x)/2} + t^{x/2}\right]\frac{dt}{t}.$$

Evaluating the first integral (recall $x > 1$), we obtain our alternate expression for ζ,

$$\pi^{-x/2}\Gamma(x/2)\zeta(x) = \frac{1}{x(x-1)} + \int_1^\infty \psi(t)\left[t^{(1-x)/2} + t^{x/2}\right]\frac{dt}{t}, \tag{5.8.9}$$

valid for $x > 1$.

Let us analyze (5.8.9). The integral on the right is a smooth function of x on $\mathbf{R}$ (Exercise **5.8.4**). Hence, the right side of (5.8.9) is a smooth function of $x \neq 0, 1$. On the other hand, $\pi^{-x/2}$ is smooth and positive, and $L(x/2) = 1/\Gamma(x/2)$ is smooth on all of $\mathbf{R}$. Thus, (5.8.9) can be used to define $\zeta(x)$ as a smooth function on $x \neq 0, 1$. Moreover, since $1/x\Gamma(x/2) = 1/2\Gamma(x/2+1)$, (5.8.9) can be used to define $\zeta(x)$ as a smooth function near $x = 0$. Now by (5.8.7), $\Gamma(x/2)(x+2n) \to 2(-1)^n/n!$ as $x \to -2n$. So multiplying (5.8.9) by $x + 2n$ and sending $x \to -2n$ yield

$$(-1)^n\pi^n(2/n!)\lim_{x\to-2n}\zeta(x) = \begin{cases} 0 & \text{if } n > 0, \\ -1 & \text{if } n = 0. \end{cases}$$

Thus, $\zeta(-2n) = 0$ for $n > 0$ and $\zeta(0) = -1/2$. Now the zeta function $\zeta(x)$ is defined for all $x \neq 1$. We summarize the results.

Theorem 5.8.4. *The zeta function can be defined, for all $x \neq 1$, as a smooth function. Moreover, $\zeta(-2n) = 0$ for $n \geq 1$, and $\zeta(0) = -1/2$.* □

Now the right side of (5.8.9) is unchanged under the substitution $x \mapsto (1-x)$. This, immediately, leads to the following.

Theorem 5.8.5 (Zeta Functional Equation). *If*

$$\xi(x) = \pi^{-x/2}\Gamma(x/2)\zeta(x),$$

then

$$\xi(x) = \xi(1-x), \qquad x \neq \dots, -4, -2, 0, 1, 3, 5, \dots. \quad \square \tag{5.8.10}$$

Since we obtained $\zeta(2n)$, $n \geq 1$, in §5.6, plugging in $x = 2n$, $n \geq 1$, into (5.8.10) leads us to the following.

Theorem 5.8.6. *For all $n \geq 1$,*

$$\zeta(1-2n) = -\frac{B_{2n}}{2n}.$$

For example, $\zeta(-1) = -1/12$. We leave the derivation of this as Exercise **5.8.5**. □ Now we know $\zeta(x)$ at all nonpositive integers and all positive even integers. Even though this result is over 200 years old, similar expressions for $\zeta(2n+1)$, $n \geq 1$, have not yet been computed. In particular, very little is known about $\zeta(3)$.

We turn to our last topic, the prime numbers. Before proceeding, the reader may wish to review the Exercises in §1.3. That there is a connection between the zeta function and the prime numbers was discovered by Euler 300 years ago.

Theorem 5.8.7 (Euler Product). *For all $x > 1$,*

$$\sum_{n=1}^{\infty} \frac{1}{n^x} = \prod_p \left(1 - \frac{1}{p^x}\right)^{-1}.$$

Here the product[17] *is over all primes.*

This follows from the fundamental theorem of arithmetic (Exercise **1.3.17**). More specifically, from (5.8.1),

$$\zeta(x)\left(1 - \frac{1}{2^x}\right) = 1 + \frac{1}{3^x} + \frac{1}{5^x} + \cdots = 1 + \sum_{2 \nmid n} \frac{1}{n^x}, \qquad x > 1, \tag{5.8.11}$$

where $2 \nmid n$ means 2 does not divide n and $n > 1$. Similarly, subtracting $1/3^x$ times (5.8.11) from (5.8.11) yields

$$\zeta(x)\left(1 - \frac{1}{2^x}\right)\left(1 - \frac{1}{3^x}\right) = 1 + \sum_{\substack{2 \nmid n \\ 3 \nmid n}} \frac{1}{n^x}, \qquad x > 1.$$

Continuing in this manner,

$$\zeta(x) \prod_{n=1}^{N} \left(1 - \frac{1}{p_n{}^x}\right) = 1 + \sum_{p_1, p_2, \ldots, p_N \nmid n} \frac{1}{n^x}, \qquad x > 1, \tag{5.8.12}$$

where $p_1, p_2, \ldots, p_N$ are the first N primes. But $p_1, p_2, \ldots, p_N \nmid n$ and $n > 1$ implies $n > N$. Hence, the series on the right side of (5.8.12) is no greater than $\sum_{n=N+1}^{\infty} 1/n^x$, which goes to zero as $N \nearrow \infty$. □

[17] This equality and its derivation are valid whether or not there are infinitely many primes.

Euler used his product to establish the infinitude of primes, as follows: Since (Exercise **5.8.9**)

$$0 < -\log(1-a) \leq 2a, \qquad 0 < a \leq 1/2, \tag{5.8.13}$$

it follows that

$$\log \zeta(x) = \sum_p -\log\left(1 - \frac{1}{p^x}\right) \leq 2\sum_p \frac{1}{p^x}, \qquad x > 1. \tag{5.8.14}$$

Now as $x \to 1+$, $\zeta(x) \to \infty$; hence, $\log \zeta(x) \to \infty$. On the other hand, $\sum_p 1/p^x \to \sum_p 1/p$, as $x \searrow 1$, by the monotone convergence theorem. We have arrived at the following.

Theorem 5.8.8. *There are infinitely many primes. In fact, there are enough of them so that*

$$\sum_p \frac{1}{p} = \infty. \quad \square$$

Our last topic is the infinitude of primes in arithmetic progressions. Let a and b be naturals. An *arithmetic progression* is a subset of $\mathbf{N}$ of the form $a\mathbf{N}+b = \{a+b, 2a+b, 3a+b, \dots\}$. Apart from 2 and 3, every prime is either in $4\mathbf{N}+1$ or $4\mathbf{N}+3$. Note that $p \in a\mathbf{N}+b$ iff a divides $p-b$, which we write as $a \mid p-b$. Here is Euler's result on primes in arithmetic progressions.

Theorem 5.8.9. *There are infinitely many primes in* $4\mathbf{N}+1$ *and in* $4\mathbf{N}+3$. *In fact, there are enough of them so that*

$$\sum_{4|p-1} \frac{1}{p} = \infty$$

and

$$\sum_{4|p-3} \frac{1}{p} = \infty.$$

We proceed by analogy with the preceding derivation. Instead of relating $\sum_p 1/p^x$ to $\log \zeta(x)$, now, we relate $\sum_{4|p-1} 1/p^x$ and $\sum_{4|p-3} 1/p^x$ to $\log L_1(x)$ and $\log L_3(x)$, where

$$L_1(x) = \sum_{4|n-1} \frac{1}{n^x}, \qquad x > 1,$$

and

$$L_3(x) = \sum_{4|n-3} \frac{1}{n^x}, \qquad x > 1.$$

By comparison with $\zeta(x)$, the series $L_1(x)$ and $L_3(x)$ are finite for $x > 1$. To make the analogy clearer, we define $\chi_1 : \mathbf{N} \to \mathbf{R}$ and $\chi_3 : \mathbf{N} \to \mathbf{R}$ by setting

$$\chi_1(n) = \begin{cases} 1, & n \in 4\mathbf{N}+1, \\ 0, & \text{otherwise,} \end{cases}$$

and

$$\chi_3(n) = \begin{cases} 1, & n \in 4\mathbf{N}+3, \\ 0, & \text{otherwise.} \end{cases}$$

Then

$$L_1(x) = \sum_{n=1}^{\infty} \frac{\chi_1(n)}{n^x}, \qquad L_3(x) = \sum_{n=1}^{\infty} \frac{\chi_3(n)}{n^x}, \qquad x > 1.$$

Proceeding further, the next step was to obtain an identity of the form

$$\sum_{n=1}^{\infty} \frac{\chi(n)}{n^x} = \prod_p \left[1 - \frac{\chi(p)}{p^x}\right]^{-1}, \qquad x > 1. \tag{5.8.15}$$

Denote the series in (5.8.15) by $L(x,\chi)$. Thus, $L_1(x) = L(x,\chi_1)$ and $L_3(x) = L(x,\chi_3)$. When $\chi = \chi_1$ or $\chi = \chi_3$, however, (5.8.15) is false, and for a very good reason, χ_1 and χ_3 are not multiplicative.

Theorem 5.8.10. *Suppose that* $\chi : \mathbf{N} \to \mathbf{R}$ *is bounded and multiplicative, i.e., suppose that* $\chi(mn) = \chi(m)\chi(n)$ *for all* $m, n \in \mathbf{N}$. *Then* (5.8.15) *holds.*

The derivation of this is completely analogous to the previous case and involves inserting factors of χ in (5.8.12). □ Having arrived at this point, Euler bypassed the failure of (5.8.15) for χ_1, χ_3 by considering, instead,

$$\chi_+ = \chi_1 + \chi_3,$$

and

$$\chi_- = \chi_1 - \chi_3.$$

Then $\chi_+(n)$ is 1 or 0 according to whether n is odd or even, $L(x,\chi_+)$ is given by (5.8.11), χ_- is given by

$$\chi_-(n) = \begin{cases} 1, & 4 \mid n-1, \\ -1, & 4 \mid n-3, \\ 0, & n \text{ even,} \end{cases}$$

and

$$L(x,\chi_-) = 1 - \frac{1}{3^x} + \frac{1}{5^x} - \frac{1}{7^x} + \dots .$$

But this is an alternating, hence convergent, series for $x > 0$ by the Leibnitz test, and $L(1,\chi_-) > 0$. Moreover (Exercise **5.8.11**), by the dominated convergence theorem,

$$\lim_{x\to 1} L(x,\chi_-) = L(1,\chi_-) > 0. \tag{5.8.16}$$

Now the key point is that χ_+ and χ_- are multiplicative (Exercise **5.8.10**), and, hence, (5.8.15) holds with $\chi = \chi_\pm$.

Proceeding, as in (5.8.14), and taking the log of (5.8.15) with $\chi = \chi_+$, we obtain

$$\log L(x, \chi_+) \le 2 \sum_p \frac{\chi_+(p)}{p^x}, \qquad x > 1.$$

Since, by (5.8.11), $\lim_{x \to 1+} L(x, \chi_+) = \infty$, sending $x \to 1+$, we conclude that

$$\lim_{x \to 1+} \sum_p \frac{\chi_+(p)}{p^x} = \infty. \tag{5.8.17}$$

Turning to χ_-, we claim it is enough to show that $\sum_p \chi_-(p) p^{-x}$ remains bounded as $x \to 1+$. Indeed, assuming this claim, we have

$$\begin{aligned} \sum_{4|p-1} \frac{1}{p} &= \sum_p \frac{\chi_1(p)}{p} \\ &= \lim_{x \searrow 1} \sum_p \frac{\chi_1(p)}{p^x} \\ &= \lim_{x \searrow 1} \frac{1}{2} \left[\sum_p \frac{\chi_+(p)}{p^x} + \sum_p \frac{\chi_-(p)}{p^x} \right] = \infty \end{aligned}$$

by the monotone convergence theorem, the claim, and (5.8.17). This is the first half of the theorem. Similarly,

$$\begin{aligned} \sum_{4|p-3} \frac{1}{p} &= \sum_p \frac{\chi_3(p)}{p} \\ &= \lim_{x \searrow 1} \sum_p \frac{\chi_3(p)}{p^x} \\ &= \lim_{x \searrow 1} \frac{1}{2} \left[\sum_p \frac{\chi_+(p)}{p^x} - \sum_p \frac{\chi_-(p)}{p^x} \right] = \infty. \end{aligned}$$

This is the second half of the theorem.

To complete the derivation, we establish the claim using

$$|-\log(1-a) - a| \le a^2, \qquad |a| \le 1/2, \tag{5.8.18}$$

which follows from the power series for log (Exercise **5.8.9**). Taking the log of (5.8.15) and using (5.8.18) with $a = \chi_-(p)/p^x$, we obtain

$$\left| \log L(x, \chi_-) - \sum_p \frac{\chi_-(p)}{p^x} \right| \le \sum_p \frac{1}{p^{2x}} \le \sum_{n=1}^{\infty} \frac{1}{n^2}, \qquad x > 1.$$

By (5.8.16), $\log L(x, \chi_-) \to \log L(1, \chi_-)$ and so remains bounded as $x \to 1+$. Since, by the last equation, $\sum_p \chi_-(p)p^{-x}$ differs from $\log L(x, \chi_-)$ by a bounded quantity, this establishes the claim. □

One hundred years after Euler's result, Dirichlet showed[18] there are infinitely many primes in any arithmetic progression $a\mathbf{N} + b$, as long as a and b have no common factor.

Exercises

5.8.1. Use the dominated convergence theorem to derive (5.8.4). (Exercise **3.5.7**.)

5.8.2. Show that

$$\gamma = \int_0^\infty \left(\frac{1}{e^t - 1} - \frac{1}{te^t} \right) dt.$$

5.8.3. Derive (5.8.7).

5.8.4. Dominate ψ by a geometric series to obtain $\psi(t) \le ce^{-\pi t}$, $t \ge 1$, where $c = 1/(1 - e^{-\pi})$. Use this to show that the integral in (5.8.9) is a smooth function of x in $\mathbf{R}$.

5.8.5. Use (5.8.10) and the values $\zeta(2n)$, $n \ge 1$, obtained in §5.6, to show that $\zeta(1 - 2n) = -B_{2n}/2n$, $n \ge 1$.

5.8.6. Let $I(x)$ denote the integral in (5.8.6). Show that $I(x)$ is finite, smooth for $x > 0$, and satisfies $I'(x) = -(x+1)I(x+1)$. Compute $I(2)$.

5.8.7. Use (5.8.9) to check that $(x - 1)\zeta(x)$ is smooth and positive on $(1 - \delta, 1+\delta)$ for δ small enough. Then differentiate $\log[(x-1)\zeta(x)]$ for $1 < x < 1+\delta$ using (5.8.6). Conclude that

$$\lim_{x \to 1} \left[\frac{\zeta'(x)}{\zeta(x)} + \frac{1}{x - 1} \right] = \gamma.$$

5.8.8. Differentiate the log of (5.8.10) to obtain $\zeta'(0) = -\frac{1}{2}\log(2\pi)$. (Use the previous Exercise, Exercise **5.5.11**, and Exercise **5.6.15**.)

5.8.9. Derive (5.8.13) and (5.8.18) using the power series for $\log(1 + a)$.

5.8.10. Show that $\chi_\pm : \mathbf{N} \to \mathbf{R}$ are multiplicative.

5.8.11. Derive (5.8.16) using the dominated convergence theorem. (Group the terms in pairs, and use the mean value theorem to show that $a^{-x} - b^{-x} \le x/a^{x+1}$, $b > a > 0$.)

---

[18] By replacing $\chi_\pm$ by the characters χ of the group $(\mathbf{Z}/a\mathbf{Z})^*$.

5.9 The Euler–Maclaurin Formula

Given a smooth function f on $\mathbf{R}$, can we find a smooth function g on $\mathbf{R}$ satisfying

$$f(x+1) - f(x) = \int_x^{x+1} g(t)\,dt, \qquad x \in \mathbf{R}? \tag{5.9.1}$$

By the fundamental theorem, the answer is yes: $g = f'$, and g is also smooth. However, this is not the only solution because $g = f' + p'$ solves (5.9.1) for any smooth periodic p, i.e., for any smooth p satisfying $p(x+1) = p(x)$ for all $x \in \mathbf{R}$.

The starting point for the Euler–Maclaurin formula is to ask the same question but with the left side in (5.9.1) modified. More precisely, given a smooth function f on $\mathbf{R}$, can we find a smooth function g on $\mathbf{R}$ satisfying

$$f(x+1) = \int_x^{x+1} g(t)\,dt, \qquad x \in \mathbf{R}? \tag{5.9.2}$$

Note that $g = 1$ works when $f = 1$. We call a g satisfying (5.9.2) an *Euler–Maclaurin derivative of f*.

It turns out the answer is yes and (5.9.2) is always solvable. To see this, let q denote a primitive of g. Then (5.9.2) becomes $f(x) = q(x) - q(x-1)$. Conversely, suppose that

$$f(x) = q(x) - q(x-1), \qquad x \in \mathbf{R}, \tag{5.9.3}$$

for some smooth q. Then it is easy to check that $g = q'$ works in (5.9.2). Thus, given f, (5.9.2) is solvable for some smooth g iff (5.9.3) is solvable for some smooth q.

In fact, it turns out that (5.9.3) is always solvable. Note, however, that q solves (5.9.3) iff $q+p$ solves (5.9.3), where p is any periodic smooth function, i.e., $p(x+1) = p(x)$, $x \in \mathbf{R}$. So the solution is not unique.

To solve (5.9.3), assume, in addition, that $f(x) = 0$ for $x < -1$, and define q by

$$q(x) = \begin{cases} f(x), & x \le 0, \\ f(x) + f(x-1), & 0 \le x \le 1, \\ f(x) + f(x-1) + f(x-2), & 1 \le x \le 2, \\ \text{and so on.} \end{cases} \tag{5.9.4}$$

Then q is well defined and smooth on $\mathbf{R}$ and (5.9.3) holds (Exercise **5.9.1**). Thus, (5.9.3) is solvable when f vanishes on $(-\infty, -1)$. Similarly, (5.9.3) is solvable when f vanishes on $(1, \infty)$ (Exercise **5.9.2**).

To obtain the general case, we write $f = f_+ + f_-$, where $f_+ = 0$ on $(-\infty, -1)$ and $f_- = 0$ on $(1, \infty)$. Then $q = q_+ + q_-$ solves (5.9.3) for f if $q_\pm$ solve (5.9.3) for $f_\pm$. Thus, to complete the solution of (5.9.3), all we need do is construct $f_\pm$.

Because we require f_+, f_- to be smooth, it is not immediately clear this can be done. To this end, we deal, first, with the special case $f = 1$, i.e., we construct $\phi_\pm$ smooth and satisfying $\phi_+ = 0$ on $(-\infty, -1)$, $\phi_- = 0$ on $(1, \infty)$, and $1 = \phi_+ + \phi_-$ on $\mathbf{R}$.

To construct $\phi_\pm$, let h denote the function in Exercise **3.5.2**. Then $h : \mathbf{R} \to \mathbf{R}$ is smooth, $h = 0$ on $\mathbf{R}^-$, and $h > 0$ on $\mathbf{R}^+$. Set

$$\phi_+(x) = \frac{h(1+x)}{h(1-x) + h(1+x)}, \qquad x \in \mathbf{R},$$

and

$$\phi_-(x) = \frac{h(1-x)}{h(1-x) + h(1+x)}, \qquad x \in \mathbf{R}.$$

Since $h(1-x) + h(1+x) > 0$ on all of $\mathbf{R}$, $\phi_\pm$ are smooth with $\phi_+ = 0$ on $(-\infty, -1)$, $\phi_- = 0$ on $(1, \infty)$, and $\phi_+ + \phi_- = 1$ on all of $\mathbf{R}$.

Now for smooth f, we may set $f_\pm = f\phi_\pm$, yielding $f = f_+ + f_-$ on all of $\mathbf{R}$. Thus, (5.9.3) is solvable for all smooth f. Hence, (5.9.2) is solvable for all smooth f.

Theorem 5.9.1. *Any smooth f on $\mathbf{R}$ has a smooth Euler–Maclaurin derivative g on $\mathbf{R}$.* □

Our main interest is to obtain a useful formula for an Euler–Maclaurin derivative g of f. To explain this, we denote $f' = Df$, $f'' = D^2 f$, $f''' = D^3 f$, and so on. Then any polynomial in D makes sense. For example, $D^3 + 2D^2 - D + 5$ is the *differential operator* that associates the smooth function f with the smooth function

$$(D^3 + 2D^2 - D + 5)f = f''' + 2f'' - f' + 5f.$$

More generally, we may consider infinite linear combinations of powers of D. For example,

$$e^{tD} f(c) = \left(1 + tD + \frac{t^2 D^2}{2!} + \frac{t^3 D^3}{3!} + \dots\right) f(c) \tag{5.9.5}$$

$$= f(c) + tf'(c) + \frac{t^2}{2!} f''(c) + \frac{t^3}{3!} f'''(c) + \dots \tag{5.9.6}$$

may sum to $f(c+t)$, since this is the Taylor series, but, for general smooth f, diverges from $f(c+t)$. When f is a polynomial of degree d, (5.9.5) does sum to $f(c+t)$. Hence, $e^{tD} f(c) = f(c+t)$. In fact, in this case, any power series in D applied to f is another polynomial of degree d, as $D^n f = 0$ for $n > d$. For example, if B_n, $n \ge 0$, are the Bernoulli numbers (§5.6), then

$$\tau(D) = 1 + B_1 D + \frac{B_2}{2!} D^2 + \frac{B_4}{4!} D^4 + \dots \tag{5.9.7}$$

may be applied to any polynomial $f(x)$ of degree d. The result $\tau(D)f(x)$, then obtained is another polynomial of, at most, the same degree.

If $\tau(D)$ is applied to $f(x) = e^{ax}$ for a real, we obtain

$$\tau(D)e^{ax} = \tau(a)e^{ax}, \tag{5.9.8}$$

where $\tau(a)$ is the Bernoulli function of §5.6,

$$\tau(a) = 1 + B_1 a + \frac{B_2}{2!}a^2 + \frac{B_4}{4!}a^4 + \dots .$$

Thus, (5.9.8) is valid only on the interval of convergence of the power series for $\tau(a)$.

Let $c(a)$ be a power series. To compute the effect of $c(D)$ on a product $e^{ax}f(x)$, where f is a polynomial, note that

$$D\left[e^{ax}x\right] = axe^{ax} + e^{ax} = e^{ax}(ax+1)$$

by the product rule. Repeating this with D^2, D^3, ...,

$$D^n\left[e^{ax}x\right] = e^{ax}(a^n x + na^{n-1}).$$

Taking linear combinations, we conclude that

$$c(D)\left(e^{ax}x\right) = e^{ax}\left[c(a)x + c'(a)\right] = \frac{\partial}{\partial a}\left[c(a)e^{ax}\right].$$

Thus, $c(D)(e^{ax}x)$ is well defined for a in the interval of convergence of $c(a)$. Similarly, one checks that $c(D)(e^{ax}x^n)$ is well defined for any $n \geq 1$ and a in the interval of convergence (Exercise **5.9.3**) and

$$c(D)\left(e^{ax}x^n\right) = \frac{\partial^n}{\partial a^n}\left[c(a)e^{ax}\right]. \tag{5.9.9}$$

We call a smooth function *elementary* if it is a product $e^{ax}f(x)$ of an exponential e^{ax} with a in the interval of convergence of $\tau(a)$ and a polynomial $f(x)$. In particular, any polynomial is elementary. Note that $\tau(D)f$ is elementary whenever f is elementary.

Theorem 5.9.2. *Let f be an elementary function. Then $\tau(D)f$ is an Euler–Maclaurin derivative,*

$$f(x+1) = \int_x^{x+1} \tau(D)f(t)\,dt, \qquad x \in \mathbf{R}. \tag{5.9.10}$$

To derive (5.9.10), start with $f(x) = e^{ax}$. If $a = 0$, (5.9.10) is clearly true. If $a \neq 0$, then by (5.9.8), (5.9.10) is equivalent to

$$e^{a(x+1)} = \int_x^{x+1} \tau(a)e^{at}\,dt = \tau(a)\cdot\frac{e^{a(x+1)} - e^{ax}}{a}$$

which is true since $\tau(a) = a/(1-e^{-a})$. Thus,

$$e^{a(x+1)} = \int_x^{x+1} \tau(a)e^{at}\,dt, \qquad a \in \mathbf{R}, x \in \mathbf{R}.$$

Now apply $\partial^n/\partial a^n$ to both sides of this last equation, differentiate under the integral sign, and use (5.9.9). You obtain (5.9.10) with $f(x) = e^{ax}x^n$. By linearity, one obtains (5.9.10) for any elementary function f. □

Given $a < b$ with $a, b \in \mathbf{Z}$, insert $x = a, a+1, a+2, \dots, b-1$ in (5.9.10), and sum the resulting equations to get the following.

Theorem 5.9.3 (Euler–Maclaurin). *For $a < b$ in $\mathbf{Z}$ and any elementary function f,*

$$\sum_{a<n\le b} f(n) = \int_a^b \tau(D)f(t)\,dt. \quad \square \tag{5.9.11}$$

The derivation of this is a triviality. The depth lies in the usefulness of the result. This arises from the fact that (5.9.11) equates a discrete sum of f on the left with a continuous sum of a related function $\tau(D)f$ on the right. Indeed, the tension between the discrete and the continuous is at the basis of many important mathematical phenomena.[19]

By inserting $a = 0$, $b = \infty$, and $f(t) = 1/(x+t)^2$, x fixed, in (5.9.11), one can derive a sharpening of Stirling's approximation (§5.5), the *Stirling series* for $\log \Gamma(x)$. Since this f is not elementary, here, one obtains a divergent series $\tau(D)f$. Instead of starting with (5.9.11), however, it will be quicker for us to derive the Stirling series from the identity (Exercise **5.6.14**)

$$\frac{d^2}{dx^2}\log \Gamma(x) = \int_0^\infty e^{-xt}\tau(t)\,dt, \qquad x > 0. \tag{5.9.12}$$

But, first, we discuss asymptotic expansions.

Let f and g be defined near $x = c$. We say that f is *big oh of g, as $x \to c$*, and we write $f(x) = O(g(x))$, as $x \to c$, if the ratio $f(x)/g(x)$ is bounded for $x \ne c$ in some interval about c. If $c = \infty$, then we require that $f(x)/g(x)$ be bounded for x sufficiently large. For example, $f(x) \sim g(x)$, as $x \to c$, implies $f(x) = O(g(x))$ and $g(x) = O(f(x))$, as $x \to c$. We write $f(x) = g(x) + O(h(x))$ to mean $f(x) - g(x) = O(h(x))$. Note that $f(x) = O(h(x))$ and $g(x) = O(h(x))$ imply $f(x) + g(x) = O(h(x))$ or, what is the same, $O(h(x)) + O(h(x)) = O(h(x))$.

We say that

$$f(x) \approx a_0 + a_1x + a_2x^2 + \dots, \tag{5.9.13}$$

is an *asymptotic expansion of f at zero* if

$$f(x) = a_0 + a_1x + \dots + a_nx^n + O(x^{n+1}), \qquad x \to 0, \tag{5.9.14}$$

---

[19] Is light composed of particles or waves?

for all $n \geq 0$. Here there is no assumption regarding the convergence of the series in (5.9.13). Although the Taylor series of a smooth function may diverge, we have the following.

Theorem 5.9.4. *If f is smooth in an interval about 0, then*

$$f(x) \approx f(0) + f'(0)x + \frac{1}{2!}f''(0)x^2 + \frac{1}{3!}f'''(0)x^3 + \frac{1}{4!}f^{(4)}(0)x^4 + \dots$$

is an asymptotic expansion at zero.

This follows from Taylor's theorem. If $a_n = f^{(n)}(0)/n!$, then from §3.5,

$$f(x) = a_0 + a_1 x + a_2 x^2 + \dots + a_n x^n + \frac{1}{(n+1)!}h_{n+1}(x)x^{n+1},$$

with h_{n+1} continuous on an interval about 0, hence bounded near 0. □

For example,

$$e^{-1/|x|} \approx 0, \qquad x \to 0,$$

since $t = 1/x$ implies $e^{-1/|x|}/x^n = e^{-|t|}t^n \to 0$ as $t \to \pm\infty$.

Actually, we will need asymptotic expansions at ∞. Let f be defined near ∞, i.e., for x sufficiently large. We say that

$$f(x) \approx a_0 + \frac{a_1}{x} + \frac{a_2}{x^2} + \dots, \tag{5.9.15}$$

is an *asymptotic expansion of f at infinity* if

$$f(x) = a_0 + \frac{a_1}{x} + \dots + \frac{a_n}{x^n} + O\left(\frac{1}{x^{n+1}}\right), \qquad x \to \infty,$$

for all $n \geq 0$. For example, $e^{-x} \approx 0$ as $x \to \infty$, since $e^{-x} = O(x^{-n})$, as $x \to \infty$, for all $n \geq 0$. Here is the Stirling series.

Theorem 5.9.5 (Stirling). *As $x \to \infty$,*

$$\log \Gamma(x) - \left[\left(x - \frac{1}{2}\right)\log x - x\right] \approx \frac{1}{2}\log(2\pi) + \frac{B_2}{2x} + \frac{B_4}{4 \cdot 3x^3} + \frac{B_6}{6 \cdot 5x^5} + \dots . \tag{5.9.16}$$

Note that, ignoring the terms with Bernoulli numbers, this result reduces to Stirling's approximation §5.5. Moreover, note that, because this is an expression for $\log \Gamma(x)$ and not $\Gamma(x)$, the terms involving the Bernoulli numbers are measures of *relative* error. Thus, the principal error term $B_2/2x = 1/12x$ equals $1/1200$ for $x = 100$ which agrees with the relative error of .08% found in Exercise **5.5.2**.

To derive (5.9.16), we will use (5.9.12) and replace $\tau(t)$ by its Bernoulli series to obtain

$$\frac{d^2}{dx^2}\log\Gamma(x) \approx \frac{1}{x} + \frac{1}{2x^2} + \frac{B_2}{x^3} + \frac{B_4}{x^5} + \dots, \qquad x \to \infty. \tag{5.9.17}$$

Then we integrate this twice to get (5.9.16).

First, we show that the portion of the integral in (5.9.12) over $(1,\infty)$ has no effect on the asymptotic expansion (5.9.17). Fix $n \geq 0$. To this end, note that

$$0 \leq \int_1^\infty e^{-xt}\tau(t)\,dt \leq \frac{1}{1-e^{-1}}\int_1^\infty e^{-xt}t\,dt = \frac{1}{1-e^{-1}}\cdot\frac{1+x}{x^2}\cdot e^{-x}$$

for $x > 0$. Thus, for all $n \geq 1$

$$\int_1^\infty e^{-xt}\tau(t)\,dt = O\left(\frac{1}{x^{n+1}}\right), \qquad x \to \infty.$$

Since τ is smooth at zero and the Bernoulli series is the Taylor series of τ, by Taylor's theorem (§3.5), there is a continuous $h_n : \mathbf{R} \to \mathbf{R}$ satisfying

$$\tau(t) = B_0 + \frac{B_1}{1!}t + \frac{B_2}{2!}t^2 + \dots + \frac{B_{n-1}}{(n-1)!}t^{n-1} + \frac{h_n(t)}{n!}t^n, \qquad t \in \mathbf{R}.$$

Then (Exercise **5.9.4**),

$$\int_0^1 e^{-xt}h_n(t)\frac{t^n}{n!}\,dt = \frac{1}{x^{n+1}}\int_0^x e^{-t}h_n(t/x)\frac{t^n}{n!}\,dt = O\left(\frac{1}{x^{n+1}}\right)$$

since $h_n(x)$ is bounded for $0 \leq x \leq 1$. Similarly (Exercise **5.9.5**),

$$\int_1^\infty e^{-xt}\frac{t^k}{k!}\,dt = O\left(\frac{1}{x^{n+1}}\right), \qquad x \to \infty, k \geq 0.$$

Now insert all this into (5.9.12), and use

$$\int_0^\infty e^{-xt}\frac{t^n}{n!}\,dt = \frac{1}{x^{n+1}}$$

to get, for fixed $n \geq 0$,

$$\begin{aligned}
\int_0^\infty e^{-xt}\tau(t)\,dt &= \int_0^1 e^{-xt}\tau(t)\,dt + \int_1^\infty e^{-xt}\tau(t)\,dt \\
&= \int_0^1 e^{-xt}\tau(t)\,dt + O\left(\frac{1}{x^{n+1}}\right) \\
&= \sum_{k=0}^{n-1} B_k \int_0^1 e^{-xt}\frac{t^k}{k!}\,dt + \int_0^1 e^{-xt}h_n(t)\frac{t^n}{n!}\,dt + O\left(\frac{1}{x^{n+1}}\right)
\end{aligned}$$

$$= \sum_{k=0}^{n-1} B_k \int_0^1 e^{-xt} \frac{t^k}{k!}\, dt + O\left(\frac{1}{x^{n+1}}\right)$$

$$= \sum_{k=0}^{n-1} \frac{B_k}{x^{k+1}} - \sum_{k=0}^{n-1} B_k \int_1^\infty e^{-xt} \frac{t^k}{k!}\, dt + O\left(\frac{1}{x^{n+1}}\right)$$

$$= \sum_{k=0}^{n-1} \frac{B_k}{x^{k+1}} + O\left(\frac{1}{x^{n+1}}\right).$$

Since $B_0 = 1$, $B_1 = 1/2$, and this is true for all $n \ge 0$, this derives (5.9.17).

To get (5.9.16), let $f(x) = (\log \Gamma(x))'' - (1/x) - (1/2x^2)$. Then by (5.9.17),

$$f(x) = \frac{B_2}{x^3} + \frac{B_4}{x^5} + \cdots + \frac{B_n}{x^{n+1}} + O\left(\frac{1}{x^{n+2}}\right). \tag{5.9.18}$$

Since the right side of this last equation is integrable over (x, ∞) for any $x > 0$, so is f. Since $-\int_x^\infty f(t)\, dt$ is a primitive of f, we obtain

$$\int_x^\infty f(t)\, dt = -[\log \Gamma(x)]' + \log x - \frac{1}{2x} - A$$

for some constant A. So integrating both sides of (5.9.18) over (x, ∞) leads to

$$-[\log \Gamma(x)]' + \log x - \frac{1}{2x} - A = \frac{B_2}{2x^2} + \frac{B_4}{4x^4} + \cdots + \frac{B_n}{nx^n} + O\left(\frac{1}{x^{n+1}}\right).$$

Similarly, integrating this last equation over (x, ∞) leads to

$$\log \Gamma(x) - x \log x + x + \frac{1}{2} \log x + Ax - B$$
$$= \frac{B_2}{2 \cdot 1x} + \frac{B_4}{4 \cdot 3x^3} + \cdots + \frac{B_n}{n \cdot (n-1)x^{n-1}} + O\left(\frac{1}{x^n}\right).$$

Noting that the right side of this last equation vanishes as $x \to \infty$, inserting $x = n$ in the left side, and comparing with Stirling's approximation in §5.5, we conclude that $A = 0$ and $B = \log(2\pi)/2$, thus obtaining (5.9.16). □

Exercises

5.9.1. Show that q, as defined by (5.9.4), is well defined, smooth on $\mathbf{R}$, and satisfies (5.9.3), when f vanishes on $(-\infty, -1)$.

5.9.2. Find a smooth q solving (5.9.3), when f vanishes on $(1, \infty)$.

5.9.3. Let c be a power series with radius of convergence R. Show that $c(D)(e^{ax}x^n)$ is well defined for $|a| < R$ and $n \geq 0$ and satisfies

$$c(D)\left(e^{ax}x^n\right) = \frac{\partial^n}{\partial a^n}[c(a)e^{ax}].$$

5.9.4. Show that $\int_0^1 e^{-xt} f(t) t^n \, dt = O\left(\frac{1}{x^{n+1}}\right)$ for any continuous bounded $f : (0,1) \to \mathbf{R}$ and $n \geq 0$.

5.9.5. For all $n \geq 0$ and $p > 0$, show that $\int_1^\infty e^{-xt} t^p \, dt \approx 0$, as $x \to \infty$.

5.9.6. Show that the Stirling series in (5.9.16) cannot converge anywhere. (If it did converge at $a \neq 0$, then the Bernoulli series would converge on all of $\mathbf{R}$.)

Chapter 6
Generalizations

The theory of the integral developed in Chapter 4 was just enough to develop all results in Chapter 5. Nevertheless, the treatment of the fundamental theorems of calculus depended strongly on the continuity of the integrand. It is natural to ask if this assumption can be relaxed and to seek the natural setting of Theorem 4.4.2 and Theorem 4.4.3.

It turns out that continuity can indeed be dropped and the theory in Chapter 4 can be pushed to provide a complete resolution. Starting one hundred years ago, Lebesgue—in his doctoral thesis *Integrale, Longueur, Aire* [11]—isolated the notion of measurability of sets and functions and used it to establish his generalizations of the fundamental theorems. Indeed, the method of exhaustion (§4.5) is a consequence of these ideas, even though it is presented earlier in this text out of necessity. The original treatment of these results was simplified and sharpened by Banach, Caratheodory, Riesz, Vitali, and others. These results, while providing a complete treatment of the fundamental theorems in one dimension, also turned out to be the spark that led to investigations of multidimensional generalizations. These investigations continue to the present day.

Throughout this chapter, there is no presumption of measurability unless explicitly stated. In particular, integrability as defined in §4.3 does not presume measurability, and we use *arbitrary* in this chapter to mean *not necessarily measurable.*

6.1 Measurable Functions and Linearity

In §4.4, we derived linearity (Theorem 4.4.5)

$$\int_a^b [f(x)+g(x)]\,dx = \int_a^b f(x)\,dx + \int_a^b g(x)\,dx \tag{6.1.1}$$

O. Hijab, *Introduction to Calculus and Classical Analysis*, Undergraduate Texts in Mathematics, DOI 10.1007/978-3-319-28400-2_6

when f and g are both integrable or both nonnegative, as long as both f and g are continuous. In this section, we establish (6.1.1) when continuity is replaced by measurability.

In §4.5, we defined measurable subsets $M \subset \mathbf{R}^2$. Now we define $M \subset \mathbf{R}$ to be *measurable* if $M \times (0,1) \subset \mathbf{R}^2$ is measurable. Since measurability in $\mathbf{R}^2$ is dilation invariant (Exercise **6.1.2**), this happens iff $M \times (0,m)$ is measurable for m positive.

Since the intersection of a sequence of measurable sets in $\mathbf{R}^2$ is measurable in $\mathbf{R}^2$ (§4.5) and

$$\left(\bigcap_{n=1}^{\infty} M_n\right) \times (0,1) = \bigcap_{n=1}^{\infty} M_n \times (0,1),$$

the same is true of measurable sets in $\mathbf{R}$. Since the union of a sequence of measurable sets in $\mathbf{R}^2$ is measurable in $\mathbf{R}^2$ (§4.5) and

$$\left(\bigcup_{n=1}^{\infty} M_n\right) \times (0,1) = \bigcup_{n=1}^{\infty} M_n \times (0,1),$$

the same is true of measurable sets in $\mathbf{R}$. Since

$$(\mathbf{R} \setminus M) \times (0,1) = (\mathbf{R} \times (0,1)) \setminus (M \times (0,1)),$$

$M \subset \mathbf{R}$ measurable implies $M^c \subset \mathbf{R}$ measurable.

We say f on (a,b) is *measurable* if $\{x : f(x) < m\} \subset (a,b)$ is measurable for all m real. Since

$$\{x : f(x) \le m\} = \bigcap_{n=1}^{\infty} \{x : f(x) < m + 1/n\},$$

and

$$\{x : f(x) < m\} = \bigcup_{n=1}^{\infty} \{x : f(x) \le m - 1/n\},$$

f is measurable iff $\{x : f(x) \le m\} \subset (a,b)$ is measurable for all m real.

Theorem 6.1.1. *If f is continuous on (a,b), then f is measurable.*

The continuity of f implies $\{x : f(x) < m\} \times (0,1)$ is open in $\mathbf{R}^2$, hence, measurable in $\mathbf{R}^2$ (§4.5); hence, $\{x : f(x) < m\}$ is measurable in $\mathbf{R}$. Since this is so for all m real, f is measurable. □

Theorem 6.1.2. *If f and g are measurable on (a,b), so are $-f$, $f+g$, and fg.*

These are Exercises **6.1.4**, **6.1.5**, and **6.1.6**. □

The next result indicates the breadth of the class of measurable functions. We say f_n, $n \ge 1$, converges to f *pointwise on* (a,b) if $f_n(x) \to f(x)$ as $n \to \infty$ for all $a < x < b$.

Theorem 6.1.3. *If f_n, $n \ge 1$, are measurable and converge pointwise on (a,b) to f, then f is measurable on (a,b).*

These results show that the class of measurable functions includes any function constructed from continuous functions using algebraic or limiting processes, in particular every function in this text. Because of this breadth, it is natural to ask whether non-measurable functions exist at all. In other words, *is every subset of* **R** *measurable?* The answer depends on the exact formulation of the axioms[1] of set theory in §1.1.

Let

$$f_{n*}(x) = \inf_{k \ge n} f_k(x), \qquad n \ge 1,$$

be the lower sequence. Then

$$\{x : f_{n*}(x) < m\} = \bigcup_{k \ge n} \{x : f_k(x) < m\}$$

hence, f_{n*} is measurable for $n \ge 1$. Now $f_{n*}(x) \nearrow f(x)$ so

$$f(x) = \sup_{n \ge 1} f_{n*}(x).$$

Then $f(x) < m$ iff $f(x) \le m - 1/k$ for some $k \ge 1$ which happens iff $f_{n*}(x) \le m - 1/k$ for all $n \ge 1$. Thus

$$\{x : f(x) < m\} = \bigcup_{k=1}^{\infty} \bigcap_{n=1}^{\infty} \{x : f_{n*}(x) \le m - 1/k\},$$

hence, f is measurable. □

Given a set $A \subset \mathbf{R}$, the *indicator function* corresponding to A is the function $\mathbf{1}_A : \mathbf{R} \to \mathbf{R}$ equal to 1 on A and 0 on A^c,

$$\mathbf{1}_A(x) = \begin{cases} 1, & x \in A, \\ 0, & x \notin A. \end{cases}$$

Then M is measurable iff $\mathbf{1}_M$ is measurable. Note $\mathbf{1}_\emptyset \equiv 0$.

A function f on (a,b) is *simple* if its range $f((a,b))$ is a finite set. Then f is simple iff

$$f(x) = a_1 \mathbf{1}_{A_1}(x) + \cdots + a_N \mathbf{1}_{A_N}(x), \qquad x \in \mathbf{R}, \qquad (6.1.2)$$

[1] Including the axiom of choice.

for some $a_1, a_2, \dots, a_N$ real and subsets $A_1, A_2, \dots, A_N$ of (a,b) (Exercise **6.1.1**). Moreover, f is measurable iff the sets $A_1, \dots, A_N$ can be chosen measurable. The sets $A_1, A_2, \dots, A_N$ need not be nonempty, and the reals $a_1, a_2, \dots, a_N$ need not be nonzero. We say (6.1.2) is in *canonical form* if $A_1, A_2, \dots, A_N$ are disjoint with union (a,b).

Theorem 6.1.4. *If f is nonnegative on (a,b), there is an increasing sequence $0 \le f_1 \le f_2 \le \cdots \le f$ of simple functions converging pointwise to f. If f is measurable, then f_n, $n \ge 1$, can be chosen measurable.*

For $n \ge 1$, let

$$A_{j,n} = \{x : (j-1)2^{-n} \le f(x) < j2^{-n}\}, \qquad j = 1, 2, \dots, n2^n,$$

and set[2]

$$f_n = \sum_{j=1}^{n2^n} (j-1)2^{-n} \mathbf{1}_{A_{j,n}}.$$

Since

$$\{x : m_1 \le f(x) < m_2\} = \{x : f(x) < m_2\} \cap \{x : f(x) < m_1\}^c,$$

$A_{j,n}$ is measurable if f is so, f_n is simple, and $f_n \le f \le f_n + 2^{-n}$; thus, $f_n \to f$ pointwise as $n \to \infty$. Now fix $n \ge 1$. We show $f_n \le f_{n+1}$ on $A_{j,n}$, for $j = 1, \dots, n2^n$. Since

$$A_{j,n} = A_{2j-1,n+1} \cup A_{2j,n+1}$$

$f(x) \in A_{j,n}$ iff $f(x) \in A_{2j-1,n+1}$ or $f(x) \in A_{2j,n+1}$. In the former case, $f_{n+1}(x) = f_n(x)$, while, in the latter case, $f_{n+1}(x) = f_n(x) + 2^{-(n+1)}$. Thus $(f_1, f_2, \dots)$ is increasing. □

Now we derive (6.1.1) for f, g nonnegative, simple, and measurable.

We begin with a single nonnegative, simple f with (6.1.2) in canonical form and $A_1, \dots, A_N$ measurable. Let G be the subgraph of f and G_j the subgraph of $a_j \mathbf{1}_{A_j}$, $j = 1, \dots, N$. Then the sets G_j are measurable in $\mathbf{R}^2$, disjoint, and $G_1 \cup \ldots \cup G_N = G$; hence, by additivity of area (Exercise **4.5.13**) and (vertical) dilation invariance,

$$\begin{aligned}
\int_a^b f(x)\,dx &= \text{area}\,(G) \\
&= \text{area}\,(G_1) + \cdots + \text{area}\,(G_N) \\
&= \int_a^b a_1 \mathbf{1}_{A_1}(x)\,dx + \cdots + \int_a^b a_N \mathbf{1}_{A_N}(x)\,dx \\
&= a_1 \int_a^b \mathbf{1}_{A_1}(x)\,dx + \cdots + a_N \int_a^b \mathbf{1}_{A_N}(x)\,dx.
\end{aligned}$$

[2] This is a *dyadic* decomposition of the range of f.

Now let f, g be nonnegative and simple with

$$f = \sum_i a_i \mathbf{1}_{A_i}, \qquad g = \sum_j b_j \mathbf{1}_{B_j},$$

both in canonical form with (A_i), (B_j) measurable. Then

$$f = \sum_{i,j} a_i \mathbf{1}_{A_i \cap B_j}, \qquad g = \sum_{i,j} b_j \mathbf{1}_{A_i \cap B_j},$$

are in canonical form, with $(A_i \cap B_j)$ measurable, and so is

$$f + g = \sum_{i,j} (a_i + b_j) \mathbf{1}_{A_i \cap B_j}.$$

By what we just derived (applied three times)

$$\begin{aligned}
\int_a^b [f(x) + g(x)]\, dx &= \sum_{i,j} (a_i + b_j) \int_a^b \mathbf{1}_{A_i \cap B_j}(x)\, dx \\
&= \sum_{i,j} a_i \int_a^b \mathbf{1}_{A_i \cap B_j}(x)\, dx + \sum_{i,j} b_j \int_a^b \mathbf{1}_{A_i \cap B_j}(x)\, dx \\
&= \int_a^b f(x)\, dx + \int_a^b g(x)\, dx.
\end{aligned}$$

This establishes (6.1.1) for f, g nonnegative, simple, and measurable.

If f, g are nonnegative and measurable, select sequences (f_n), (g_n) of nonnegative, simple, measurable functions with $f_n \nearrow f$ and $g_n \nearrow g$ pointwise as $n \to \infty$. By the monotone convergence theorem (§4.5) applied three times,

$$\begin{aligned}
\int_a^b [f(x) + g(x)]\, dx &= \lim_{n\to\infty} \int_a^b [f_n(x) + g_n(x)]\, dx \\
&= \lim_{n\to\infty} \int_a^b f_n(x)\, dx + \lim_{n\to\infty} \int_a^b g_n(x)\, dx \\
&= \int_a^b f(x)\, dx + \int_a^b g(x)\, dx.
\end{aligned}$$

This establishes (6.1.1) for f, g nonnegative and measurable.

Theorem 6.1.5. *Let f and g be measurable. If f and g are both nonnegative or both integrable on (a, b), then (6.1.1) holds.*

Above we derived the nonnegative case. Let f, g be nonnegative, measurable, and integrable, and let $A = \{x : f(x) > g(x)\} \subset (a, b)$, $f_1 = f\mathbf{1}_A$, $f_2 = f\mathbf{1}_{A^c}$, $g_1 = g\mathbf{1}_A$, $g_2 = g\mathbf{1}_{A^c}$. Then A is measurable (Exercise **6.1.3**),

$f_1 + f_2 = f$, $g_1 + g_2 = g$, $f_1 \geq g_1$, $g_2 \geq f_2$, and $(f-g)^+ = f_1 - g_1$, $(f-g)^- = g_2 - f_2$. By the nonnegative case applied four times,

$$\begin{aligned}
\int_a^b [f(x) - g(x)]\,dx &= \int_a^b [f(x) - g(x)]^+\,dx - \int_a^b [f(x) - g(x)]^-\,dx \\
&= \int_a^b [f_1(x) - g_1(x)]\,dx - \int_a^b [g_2(x) - f_2(x)]\,dx \\
&= \left(\int f_1\,dx - \int g_1\,dx\right) - \left(\int g_2\,dx - \int f_2\,dx\right) \\
&= \int_a^b f(x)\,dx - \int_a^b g(x)\,dx.
\end{aligned}$$

This establishes (6.1.1) with f, g measurable and integrable and $f \geq 0$ and $g \leq 0$.

Finally, let f, g be measurable and integrable. By what we just learned,

$$\begin{aligned}
\int_a^b [f(x) + g(x)]\,dx &= \int_a^b [f^+(x) - f^-(x) + g^+(x) - g^-(x)]\,dx \\
&= \int_a^b [f^+(x) + g^+(x)]\,dx - \int_a^b [f^-(x) + g^-(x)]\,dx \\
&= \int_a^b f^+\,dx + \int_a^b g^+\,dx - \int_a^b f^-\,dx - \int_a^b g^-\,dx \\
&= \int_a^b f(x)\,dx + \int_a^b g(x)\,dx.
\end{aligned}$$

This establishes (6.1.1) when f, g are measurable and integrable. □

Since sums of measurable functions are measurable, by induction one has linearity for finite sums

$$\int_a^b \left(\sum_{j=1}^N f_j(x)\right) dx = \sum_{j=1}^N \int_a^b f_j(x)\,dx$$

when the measurable functions $f_1, \dots, f_N$ are all integrable or all nonnegative.

Exercises

6.1.1. Show that f is simple iff (6.1.2) holds. Moreover, (6.1.2) can be taken in canonical form.

6.1.2. Let $\lambda > 0$, and suppose a bijection $f : \mathbf{R}^2 \to \mathbf{R}^2$ satisfies $\operatorname{area}(f(A)) = \lambda \operatorname{area}(A)$ for all $A \subset \mathbf{R}^2$. Show that $M \subset \mathbf{R}^2$ measurable implies $f(M)$ is measurable. In particular, this is so for any dilate or translate of M.

6.1.3. If f, g are measurable on (a, b), then $A = \{x : f(x) > g(x)\}$ is measurable. (What is the connection between A and $A_r = \{x : f(x) > r > g(x)\}$, $r \in \mathbf{Q}$?)

6.1.4. If f is measurable, so is $-f$.

6.1.5. If f, g are measurable, then $f + g$ is measurable. (What is the connection between $A = \{x : f(x) + g(x) < M\}$ and $A_r = \{x : f(x) < r\}$, $r \in \mathbf{Q}$, and $A_s = \{x : g(x) < s\}$, $s \in \mathbf{Q}$?)

6.1.6. If f, g are measurable, then fg is measurable. Start with f, g nonnegative. (What is the connection between $A = \{x : f(x)g(x) < M\}$ and $A_r = \{x : f(x) < r\}$, $r \in \mathbf{Q}$, and $A_s = \{x : g(x) < s\}$, $s \in \mathbf{Q}$?)

6.2 Limit Theorems

In this section we present the limit theorems of Chapter 5 in the broader measurable setting. The proofs are exactly as before, the only change being the use of linearity for measurable integrands Theorem 6.1.5 instead of linearity for continuous integrands Theorem 4.4.5.

There is no need to restate the monotone convergence theorem (Theorem 5.1.2) or Fatou's lemma (Exercise **5.1.2**) as they are valid for arbitrary functions.

Theorem 6.2.1 (Summation Under the Integral Sign—Positive Case). *Let f_n, $n \geq 1$, be a sequence of nonnegative functions on (a, b). If f_n, $n \geq 1$, are measurable, then*

$$\int_a^b \left[\sum_{n=1}^{\infty} f_n(x)\right] dx = \sum_{n=1}^{\infty} \int_a^b f_n(x)\, dx. \ \square$$

Next is the dominated convergence theorem.

Theorem 6.2.2 (Dominated Convergence Theorem). *Let f_n, $n \geq 1$, be a sequence of functions on (a, b). Suppose there is a function g integrable on (a, b) satisfying $|f_n(x)| \leq g(x)$ for all x in (a, b) and all $n \geq 1$. If*

$$\lim_{n \nearrow \infty} f_n(x) = f(x), \qquad a < x < b,$$

then f and f_n, $n \geq 1$, are integrable on (a, b). If g and f_n, $n \geq 1$, are measurable, then

$$\lim_{n \nearrow \infty} \int_a^b f_n(x)\, dx = \int_a^b \lim_{n \nearrow \infty} f_n(x)\, dx = \int_a^b f(x)\, dx. \ \square \tag{6.2.1}$$

Following are the consequences of the dominated convergence theorem.

Theorem 6.2.3 (Summation Under the Integral Sign — Alternating Case). *Let f_n, $n \geq 1$, be a decreasing sequence of nonnegative functions on (a,b), and suppose that f_1 is integrable on (a,b). Then f_n, $n \geq 1$, and $\sum_{n=1}^{\infty}(-1)^{n-1}f_n$ are integrable on (a,b). If f_n, $n \geq 1$, are measurable, then*

$$\int_a^b \left[\sum_{n=1}^{\infty}(-1)^{n-1}f_n(x)\right] dx = \sum_{n=1}^{\infty}(-1)^{n-1}\int_a^b f_n(x)\, dx. \ \square$$

Theorem 6.2.4 (Summation Under the Integral Sign—Absolute Case). *Let f_n, $n \geq 1$, be a sequence of functions on (a,b), and suppose that there is a function g integrable on (a,b) and satisfying $\sum_{n=1}^{\infty}|f_n(x)| \leq g(x)$ for all x in (a,b). Then f_n, $n \geq 1$, and $\sum_{n=1}^{\infty} f_n$ are integrable on (a,b). If g and f_n, $n \geq 1$, are measurable, then*

$$\int_a^b \left[\sum_{n=1}^{\infty} f_n(x)\right] dx = \sum_{n=1}^{\infty}\int_a^b f_n(x)\, dx. \ \square$$

Theorem 6.2.5 (Continuity Under the Integral Sign). *Let $f(x,t)$ be a function of two variables, defined on $(a,b) \times (c,d)$, such that $f(\cdot,t)$ is continuous for all $c < t < d$. Suppose there is a function g integrable on (c,d) satisfying $|f(x,t)| \leq g(t)$ for $a < x < b$ and $c < t < d$. Then $f(x,\cdot)$ is integrable on (c,d) for $a < x < b$ and*

$$F(x) = \int_c^d f(x,t)\, dt, \qquad a < x < b,$$

is well defined. If g and $f(x,\cdot)$, $a < x < b$, are measurable, then F is continuous. $\square$

Last is differentiation under the integral sign.

Theorem 6.2.6 (Differentiation Under the Integral Sign). *Let $f(x,t)$ be a function of two variables, defined on $(a,b) \times (c,d)$, such that*

$$\frac{\partial f}{\partial x}(x,t), \qquad a < x < b, c < t < d,$$

exists. Suppose there is a function g integrable on (c,d), such that

$$|f(x,t)| + \left|\frac{\partial f}{\partial x}(x,t)\right| \leq g(t), \qquad a < x < b, c < t < d.$$

Then $f(x,\cdot)$ and $\partial f/\partial x(x,\cdot)$ are integrable for all $a < x < b$, and

$$F(x) = \int_c^d f(x,t)\, dt, \qquad a < x < b,$$

is well defined. If g and $f(x,\cdot)$, $a < x < b$, are measurable, then F is differentiable on (a,b) and

$$F'(x) = \int_c^d \frac{\partial f}{\partial x}(x,t)\,dt, \qquad a < x < b. \ \square$$

A set $U \subset \mathbf{R}$ is *open* if for every $c \in U$, there is an open interval I containing c and contained in U. Clearly, an open interval is an open set. Conversely, by Exercise **6.2.1**, every open set $U \subset \mathbf{R}$ is a finite or countable disjoint union of open intervals. These intervals are the *component intervals* of U.

Exercises

6.2.1. Show every open set $U \subset \mathbf{R}$ is a finite or countable disjoint union of open intervals. (Given $x \in U$, look at the largest open interval $(a_x, b_x) \subset U$ containing x.)

6.2.2. F is continuous on $[a,b]$ iff for every open set $U \subset \mathbf{R}$, $F^{-1}(U)$ is an open set.

6.2.3. If F is continuous on $[a,b]$ and $U \subset (a,b)$ is an open set, then $F(U)$ is measurable.

6.2.4. If U is a countable disjoint union of open intervals $I_k = (c_k, d_k)$, $k \geq 1$, and f is arbitrary nonnegative on $\mathbf{R}$, then (4.3.5)

$$\int_{-\infty}^{\infty} \mathbf{1}_U(x) f(x)\,dx = \sum_{k=1}^{\infty} \int_{c_k}^{d_k} f(x)\,dx. \tag{6.2.2}$$

6.2.5. Let f_n, $n \geq 1$, be measurable on (a,b), and let

$$A = \{x : \lim_{n\to\infty} f_n(x) \text{ exists}\}.$$

Show A is measurable. If f equals the limit on A and zero off A, then f is measurable.

6.3 The Fundamental Theorems of Calculus

Recall (§3.7) F is a primitive of f on (a,b) if $F'(x) = f(x)$ for all x in (a,b). We previously established the following:

- The first fundamental theorem of calculus.
- Any two primitives of a continuous function differ by a constant.
- The second fundamental theorem of calculus.

In this section, we give an overview of the generalizations of these results. The proofs will occupy the rest of the chapter.

We begin with a simple example to illustrate what can happen when f is not continuous. Let $f = \mathbf{1}_{(0,1)}$, and let

$$F(x) = \int_{-\infty}^{x} f(t)\,dt = \begin{cases} 0, & x \le 0, \\ x, & 0 \le x \le 1, \\ 1, & x \ge 1. \end{cases}$$

Thus $F'(x)$ exists and equals $f(x)$ for all x real except $x = 0$ and $x = 1$.

We say f is *locally integrable on* (a,b) if f is integrable on (c,d) for all $[c,d] \subset (a,b)$. Every continuous function is locally integrable. Let f be locally integrable arbitrary on (a,b), fix c in (a,b), and let

$$F_c(x) = \int_{c}^{x} f(t)\,dt, \quad a < x < b.$$

A real x in (a,b) is a *Lebesgue point* of f if $F_c'(x)$ exists and equals $f(x)$. For any other c' in (a,b), $F_c - F_{c'}$ is a constant (Theorem 4.3.3). Thus $F_c'(x)$ exists iff $F_x'(x)$ exists, in which case they are equal.

If f is continuous, we know every point x in (a,b) is a Lebesgue point, but the above example shows that some points may fail to be Lebesgue in general. Thus the best one can expect is that the non-Lebesgue points form a negligible set in (a,b). This is made precise as follows.

Define the *length*[3] of $A \subset \mathbf{R}$ by

$$\text{length}\,(A) \equiv \text{area}\,(A \times (0,1))\,.$$

Then length$\,(I) = b - a$ for any interval $I = (a,b)$ (Theorem 4.2.1). More generally (Exercise **6.2.4**), for $U \subset \mathbf{R}$ open, length$\,(U)$ equals the sum of the lengths of its component intervals.

A set $N \subset (a,b)$ is *negligible* if length$\,(N) = 0$. Given f, g on (a,b) and $A \subset (a,b)$, we say $f(x) = g(x)$ *for almost all* $x \in A$, or $f = g$ *almost everywhere on* A, if the set $N = \{x \in A : f(x) \ne g(x)\}$ is negligible.

That this is a reasonable interpretation of "negligible" is supported by: if $f = g$ almost everywhere on (a,b), then f is measurable iff g is measurable (Exercise **6.3.4**).

For $A \subset (a,b)$ and f on (a,b), define

$$\int_A f(x)\,dx \equiv \int_a^b \mathbf{1}_A(x) f(x)\,dx$$

whenever $\mathbf{1}_A f$ is nonnegative or integrable. Clearly this depends only on the restriction of f to A. Then for arbitrary f and N negligible,

[3] This is *one-dimensional Lebesgue measure* (Exercise **6.3.5**).

$$\int_N f(x)\,dx = 0. \tag{6.3.1}$$

Indeed, by the method of exhaustion and dilation invariance,

$$\text{area}\,(N \times (0,\infty)) = \lim_{n\to\infty} \text{area}\,(N \times (0,n)) = \lim_{n\to\infty} n \cdot \text{area}\,(N \times (0,1)) = 0.$$

But for $f \geq 0$ the subgraph of $\mathbf{1}_N f$ is contained in $N \times (0,\infty)$ and $(\mathbf{1}_N f)^{\pm} = \mathbf{1}_N f^{\pm}$. Thus (6.3.1) follows for arbitrary f.

Now suppose f is defined almost everywhere on (a,b), i.e., suppose f is defined only on $(a,b) \setminus N'$ for some $N' \subset (a,b)$ negligible. If we set

$$\int_a^b f(x)\,dx \equiv \int_{(a,b)\setminus N'} f(x)\,dx, \tag{6.3.2}$$

then (6.3.1) is valid for any negligible N.

More generally, let f and g be arbitrary, defined almost everywhere on (a,b) and $f = g$ almost everywhere on (a,b). Then

$$\int_a^b f(x)\,dx = \int_a^b g(x)\,dx$$

in the sense that the left side exists iff the right side exists, in which case both sides agree. Indeed, let $\{f = g\}$ denote the set in (a,b) where the functions are both defined and agree, and let $N = (a,b) \setminus \{f = g\}$. Then N is negligible, and for $f \geq 0$ and $g \geq 0$ (here we use monotonicity and subadditivity of area),

$$\int_{\{f=g\}} f(x)\,dx \leq \int_a^b f(x)\,dx \leq \int_{\{f=g\}} f(x)\,dx + \int_N f(x)\,dx$$

hence,

$$\int_a^b f(x)\,dx = \int_{\{f=g\}} f(x)\,dx = \int_{\{f=g\}} g(x)\,dx = \int_a^b g(x)\,dx. \tag{6.3.3}$$

The general case follows by applying this to $f^{\pm}$ and $g^{\pm}$. In particular, (6.3.3) implies the integral as defined by (6.3.2) does not depend on N'.

The following establishes that almost every x in (a,b) is a Lebesgue point of f. This is the *Lebesgue differentiation theorem.*

Theorem 6.3.1 (First Fundamental Theorem of Calculus). *Let f be locally integrable arbitrary on (a,b), fix c in (a,b), and let*

$$F_c(x) = \int_c^x f(t)\,dt, \qquad a < x < b. \tag{6.3.4}$$

Then

$$F_c'(x) = f(x) \qquad \textit{for almost all } x \textit{ in } (a,b) \tag{6.3.5}$$

iff f *is measurable.*

Note existence of $F_c'(x)$ for almost all x is part of the claim. This is a direct generalization of Theorem 4.4.2 and is established in §6.6.

We now show that it is possible to have $F' = 0$ almost everywhere for nonconstant F.

To construct such an example, we use the one-dimensional version of the Cantor set (§4.1). The *Cantor set* $C \subset [0,1]$ is the set of reals of the form

$$x = \sum_{n=1}^{\infty} \frac{d_n}{3^n}, \qquad d_n = 0 \text{ or } 2, n \geq 1. \tag{6.3.6}$$

Then C is the complement of open intervals. More precisely, since $0 \leq d_n \leq 2$,

$$0 \leq \sum_{n=N+1}^{\infty} \frac{d_n}{3^n} \leq \sum_{n=N+1}^{\infty} \frac{2}{3^n} = \frac{1}{3^N},$$

so $x \notin C$ iff for some $N \geq 1$, $d_N = 1$ and d_n, $n > N$, are not all equal. Thus, $x \notin C$ iff for some $N \geq 1$

$$x = \sum_{n=1}^{N-1} \frac{d_n}{3^n} + \frac{1}{3^N} + \sum_{n=N+1}^{\infty} \frac{d_n}{3^n}$$

with d_n, $n > N$, not all equal. But this is the same as x belonging to one of the 2^{N-1} component intervals

$$\sum_{n=1}^{N-1} \frac{d_n}{3^n} + \left(\frac{1}{3^N}, \frac{2}{3^N}\right), \tag{6.3.7}$$

for some $N \geq 1$. Note x is one of the endpoints iff for some $N \geq 1$, d_n, $n > N$, are all equal. Thus C^c is the union of $(1/3, 2/3)$,$(1/9, 2/9)$,$(7/9, 8/9)$,... and

$$\text{length}\,(C^c) = \sum_{N=1}^{\infty} \frac{2^{N-1}}{3^N} = \frac{1}{3} \cdot \frac{1}{1 - \frac{2}{3}} = 1$$

by Exercise **6.2.4**, which implies length (C) = length $([0,1])-$length $(C^c) = 0$; thus, C is negligible. Another way to measure C is to note that the dilate $3C$ equals the disjoint union of C and its translate $C + 2$; hence,[4]

$$3\,\text{length}\,(C) = \text{length}\,(3C) = \text{length}\,(C \cup (C+2)) = 2\,\text{length}\,(C),$$

Thus, C is negligible.

---

[4] Interpreted appropriately, this shows $\alpha = \log_3 2$ is the *Hausdorff dimension* of C.

For x in C given by (6.3.6), define

$$F(x) = F\left(\sum_{n=1}^{\infty} \frac{d_n}{3^n}\right) = \sum_{n=1}^{\infty} \frac{d_n/2}{2^n}.$$

Then $F(c) = F(d)$ whenever (c, d) is of the form (6.3.7), so we extend the definition of F to all of $[0, 1]$ by defining F to be constant on each interval $[c, d]$. Then the *Cantor function* F is increasing and holder continuous with exponent $\alpha = \log_3 2$ (Exercise **6.3.9**) on $[0, 1]$. Since $F(0) = 0$, $F(1) = 1$, and $F'(x) = 0$ on C^c, F is an example of a nonconstant function satisfying $F'(x) = 0$ almost everywhere. See [16] for an interactive demonstration (Figure 6.1).

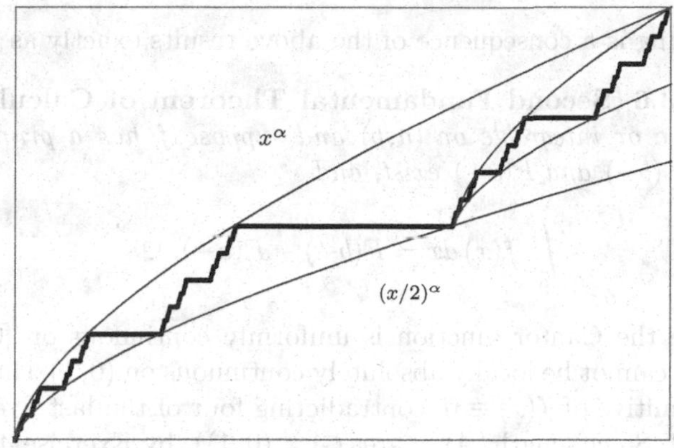

Fig. 6.1 The Cantor function

Thus functions are not determined by their derivatives without additional restrictions.

Let F be defined on $[a, b]$. We say F is *absolutely continuous on* $[a, b]$ (this is due to Vitali [14]) if for any $\epsilon > 0$, there is a $\delta > 0$ such that for any disjoint open intervals $I_k = (c_k, d_k)$, $1 \le k \le N$, in (a, b),

$$\sum_{k=1}^{N} (d_k - c_k) < \delta \quad \text{implies} \quad \sum_{k=1}^{N} |F(d_k) - F(c_k)| < \epsilon.$$

More generally, we say F is *locally absolutely continuous on* (a, b) if F is absolutely continuous on $[c, d]$ for all $[c, d] \subset (a, b)$.

Recall a primitive of f on (a, b) was defined in §3.7 to be a differentiable function F satisfying $F'(x) = f(x)$ for all x in (a, b). We now define a *primitive of* f *on* (a, b) to be a locally absolutely continuous function F on (a, b) satisfying $F'(x) = f(x)$ almost everywhere on (a, b).

Then we have the following results for primitives to be established in §6.5.

Theorem 6.3.2. *Let f be continuous on (a,b). Then $F'(x) = f(x)$ for all x in (a,b) iff F is locally absolutely continuous on (a,b) and $F'(x) = f(x)$ for almost all x in (a,b).*

Thus the definition of primitive in §3.7 and the definition above are consistent for continuous functions. In what follows, we use *primitive* as above.

Theorem 6.3.3. *A function f has a primitive F on (a,b) iff f is measurable and locally integrable on (a,b), in which case F is given by* (6.3.4).

Theorem 6.3.4. *A function F is a primitive on (a,b) of some f on (a,b) iff F is locally absolutely continuous on (a,b), in which case f is given by F'.*

Theorem 6.3.5. *Any two primitives of a locally integrable measurable f on (a,b) differ by a constant.*

The following is a consequence of the above results exactly as in §4.4.

Theorem 6.3.6 (Second Fundamental Theorem of Calculus). *Let f be nonnegative or integrable on (a,b) and suppose f has a primitive F on (a,b). Then $F(b-)$ and $F(a+)$ exist, and*

$$\int_a^b f(x)\,dx = F(b-) - F(a+). \ \square$$

Note while the Cantor function is uniformly continuous on $[0,1]$ (Exercise **6.3.9**), it cannot be locally absolutely continuous on $(0,1)$, since it would then be a primitive of $f(x) \equiv 0$, contradicting four of the last five theorems.

Let $A \subset \mathbf{R}$. Since length $(A) =$ area $(A \times (0,1))$, by Exercise **4.5.6**, given $\epsilon > 0$, there is an open $G \subset \mathbf{R}^2$ containing $A \times (0,1)$ and satisfying area $(G) \le$ length $(A) + \epsilon$. The following shows we may choose $G = U \times (0,1)$. This will be useful frequently below.

Theorem 6.3.7. *For $A \subset \mathbf{R}$ arbitrary and $\epsilon > 0$, there is an open set $U \subset \mathbf{R}$ with $A \subset U$ and*

$$\text{length}\,(A) \le \text{length}\,(U) \le \text{length}\,(A) + \epsilon. \tag{6.3.8}$$

Choose $\delta > 0$ small enough to satisfy

$$\frac{1}{1-2\delta}\,\text{length}\,(A) + \frac{\delta}{1-2\delta} \le \text{length}\,(A) + \epsilon.$$

Since length $(A) =$ area $(A \times (0,1))$, by definition of area, there is (step 2 in the proof of Theorem 4.2.1) an open paving (Q_n) of $A \times (0,1)$ satisfying

$$\sum_{n=1}^{\infty} \|Q_n\| \le \text{length}\,(A) + \delta.$$

Let $x \in A$.

Since $\{x\} \times [\delta, 1-\delta]$ is a compact rectangle, there is a finite subset S of $\mathbf{N}$ satisfying

$$\{x\} \times [\delta, 1-\delta] \subset \bigcup_{n \in S} Q_n. \tag{6.3.9}$$

By discarding the rectangles Q_n, $n \in S$ that do not intersect $\{x\} \times [\delta, 1-\delta]$, we may assume

$$(\{x\} \times [\delta, 1-\delta]) \cap Q_n \neq \emptyset, \qquad \text{for every } n \in S. \tag{6.3.10}$$

Given any finite subset S of $\mathbf{N}$, let A_S be the set of $x \in A$ satisfying (6.3.9) and (6.3.10). By the above, we conclude $A = \bigcup\{A_S : S \subset \mathbf{N}\}$. Now with $Q_n = I_n \times J_n$, $n \geq 1$, let

$$I_S = \bigcap\{I_n : n \in S\}, \qquad J_S = \bigcup\{J_n : n \in S\}, \qquad U = \bigcup\{I_S : A_S \neq \emptyset\}.$$

Then $A_S \subset I_S$. Moreover, if A_S is nonempty, then $[\delta, 1-\delta] \subset J_S$; hence, $I_S \times [\delta, 1-\delta] \subset \bigcup\{Q_n : n \in \mathbf{N}\}$. Thus $A \subset U$, U is open, and $U \times [\delta, 1-\delta] \subset \bigcup_{n=1}^{\infty} Q_n$. Hence length $(A) \leq$ length (U) and

$$(1-2\delta)\,\text{length}\,(U) = \text{area}\,(U \times [\delta, 1-\delta]) \leq \sum_{n=1}^{\infty} \|Q_n\| \leq \text{length}\,(A) + \delta.$$

Dividing by $1-2\delta$, the result follows. □

Exercises

6.3.1. A negligible set is measurable.

6.3.2. A subset of a negligible set is negligible, and a countable union of negligible sets is negligible.

6.3.3. For $A, B \subset \mathbf{R}$, the *symmetric difference* is $|A - B| = (A \cup B) \setminus (A \cap B)$. If $|A - B|$ is negligible, then A is measurable iff B is measurable.

6.3.4. Suppose $f = g$ almost everywhere on (a, b). If f is measurable on (a, b), so is g.

6.3.5. Given $A \subset \mathbf{R}$, a *paving* of A is a sequence of intervals (I_n) satisfying $A \subset \bigcup_{n=1}^{\infty} I_n$. For any interval I with endpoints a, b, let $\|I\| = b - a$. Show

$$\text{length}\,(A) = \inf\left\{\sum_{n=1}^{\infty} \|I_n\| : \text{all pavings } (I_n) \text{ of } A\right\}.$$

6.3.6. Let $0 < \alpha < 1$. If $A \subset \mathbf{R}$ satisfies $\text{length}(A \cap (a,b)) \le \alpha(b-a)$ for all $a < b$, then A is negligible (compare with Exercise **4.2.15**).

6.3.7. Let $F_0(x) = x$, $0 \le x \le 1$, and define F_n on $[0,1]$, $n \ge 1$, recursively by

$$2F_{n+1}(x) = \begin{cases} F_n(3x), & 0 \le x \le \frac{1}{3}, \\ 1, & \frac{1}{3} \le x \le \frac{2}{3}, \\ F_n(3x-2)+1, & \frac{2}{3} \le x \le 1. \end{cases}$$

Show F_n, $n \ge 0$ is increasing, piecewise linear, $F_n(0) = 0$, $F_n(1) = 1$, and $0 \le F_n'(x) \le (3/2)^n$ almost everywhere.

6.3.8. With F_n, $n \ge 0$, as in the previous exercise, let

$$e_n = \max_{0 \le x \le 1} |F_{n+1}(x) - F_n(x)|, \quad n \ge 0.$$

Show that $e_{n+1} \le e_n/2$, $n \ge 0$, and use this to show $(F_n(x))$ is Cauchy. If $F(x) \equiv \lim_{n\to\infty} F_n(x)$, $0 \le x \le 1$, show $F_n \to F$ uniformly and

$$2F(x) = \begin{cases} F(3x), & 0 \le x \le \frac{1}{3}, \\ 1, & \frac{1}{3} \le x \le \frac{2}{3}, \\ F(3x-2)+1, & \frac{2}{3} \le x \le 1. \end{cases}$$

Conclude F is the Cantor function.

6.3.9. Let $\alpha = \log_3 2$. The Cantor function is increasing and continuous and satisfies

$$0 \le F(z) - F(x) \le F(z-x), \qquad (x/2)^\alpha \le F(x) \le x^\alpha, \qquad 0 \le x \le z \le 1.$$

Conclude

$$0 \le F(z) - F(x) \le (z-x)^\alpha, \qquad 0 \le x \le z \le 1.$$

(For $(x/2)^\alpha \le F(x)$, use $\sum x_n^\alpha \ge (\sum x_n)^\alpha$. For the others, first prove by induction with F_n, $n \ge 0$, replacing F, and then take the limit.)

6.4 The Sunrise Lemma

Let G be continuous on $[a,b]$. Call an interval $(c,d) \subset (a,b)$ *balanced* if $G(c) = G(d)$. The following result, due to Riesz [12], shows that $[a,b]$ may be decomposed into balanced intervals on whose complement G is decreasing.

Recall (§6.2) every open set $U \subset \mathbf{R}$ is a finite or countable disjoint union of open intervals, the *component intervals* of U.

Theorem 6.4.1 (Sunrise Lemma). *Let G be continuous on $[a,b]$, and let*

$$U_G = \{c \in (a,b) : G(x) > G(c) \text{ for some } x > c\}.$$

Then U_G is open and $G(c) \leq G(d)$ for each component interval (c,d) of U_G (Figure 6.2).

Moreover, if $G(a) \geq G(b)$, there is an open set $W_G \subset (a,b)$ such that each component interval of W_G is balanced and $G(c) \geq G(x)$ for $a < c \notin W_G$ and $x \geq c$.

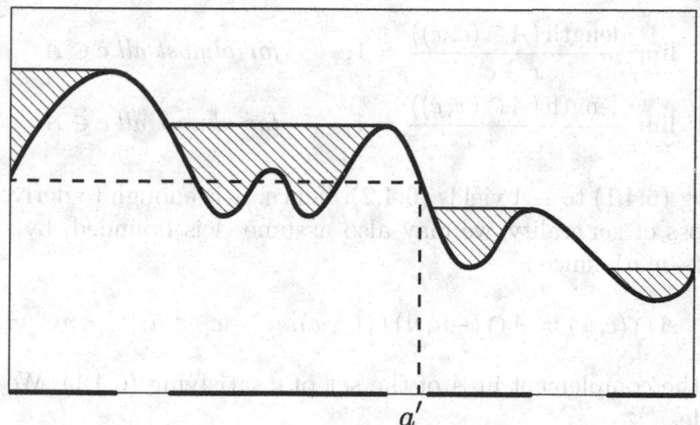

Fig. 6.2 The Sunrise Lemma

To understand the intuition behind the name and the statement, think of the graph of G as a mountainous region with the sun rising at ∞ from the right. Then U_G is the portion of the x-axis where the graph is in shadow. See [17] for an interactive demonstration.

By continuity of G, U_G is open. Let (c,d) be one of the component intervals, and suppose $G(c) > G(d)$. Let e be the largest real in $[c,d]$ satisfying $G(e) = (G(c)+G(d))/2$. Then $G(d) < G(e) < G(c)$ and $c < e < d$; hence, there exists $x > e$ satisfying $G(x) > G(e)$. If $x > d$, then $d \in U_G$, contradicting $d \notin U_G$. If $x < d$, then $G(x) > G(e) > G(d)$, so by the intermediate value property, there exists $f \in (x,d)$ satisfying $G(f) = G(e)$, contradicting the definition of e. We conclude $G(c) \leq G(d)$. Also by definition of U_G, if $a < c \notin U_G$, we have $G(x) \leq G(c)$ for $x \geq c$. This establishes the first part.

For the second part, let a' be the largest real[5] in $[a,b]$ satisfying $G(a')=G(a)$. We claim $G(a') = \max_{[a',b]} G$. If not, there is a $c > a'$ with $G(c) > G(a)$.

[5] The proof is valid if $a' = a$ or $a' = b$.

By the intermediate value property, there is a $d \in (c,b]$ with $G(d) = G(a)$, contradicting the choice of a'. Thus $G(a') = \max_{[a',b]} G$. Now apply the first part to G on $[a',b]$ yielding $U'_G \subset (a',b)$. By definition of U'_G, for each component interval (c,d) of U'_G, either $G(c) \geq G(d)$ or $c = a'$. If $c = a'$, then $G(c) = \max_{[a',b]} G$, so $G(c) \geq G(d)$. Hence $G(c) = G(d)$ for each component interval (c,d) of U'_G. Now set $W_G = (a,a') \cup U'_G$. If $a < c \notin W_G$, then $c = a'$ or $a' < c \notin U'_G$. In either case, $x \geq c$ implies $G(x) \leq G(c)$. □

Here is our first application of the sunrise lemma.

Theorem 6.4.2 (Lebesgue Density Theorem). *Let $A \subset \mathbf{R}$ be arbitrary. Then*

$$\lim_{x \to c+} \frac{\text{length}\,(A \cap (c,x))}{x - c} = 1, \qquad \textit{for almost all } c \in A \tag{6.4.1}$$

$$\lim_{x \to c-} \frac{\text{length}\,(A \cap (x,c))}{c - x} = 1, \qquad \textit{for almost all } c \in A. \tag{6.4.2}$$

Applying (6.4.1) to $-A$ yields (6.4.2); hence, it is enough to derive (6.4.1). Without loss of generality, we may also assume A is bounded, by replacing A by $A \cap (-n,n)$, since

$$A \cap (c,x) = A \cap (-n,n) \cap (c,x), \qquad |c| < n, |x| < n.$$

Let N be the complement in A of the set of c satisfying (6.4.1). We show N is negligible.

Since the ratio in (6.4.1) is ≤ 1, the limit in (6.4.1) fails to exist at c iff (§2.2) at least one right limit point is strictly less than 1. But this happens iff there is $x_j \to c+$ with

$$\frac{\text{length}\,(A \cap (c,x_j))}{x_j - c} < \alpha, \qquad j \geq 1,$$

for some $0 < \alpha < 1$. It follows that N is the union of N_α, $0 < \alpha < 1$, where

$$N_\alpha = \{c \in A : G_\alpha(x_j) > G_\alpha(c) \text{ for some sequence } x_j \to c+\}$$

and

$$G_\alpha(x) = \alpha x - \text{length}\,(A \cap (-\infty, x))\,.$$

Hence it is enough to show N_α is negligible for each $0 < \alpha < 1$.

For $a < b$, apply the first part of the Sunrise Lemma to G_α on $[a,b]$ obtaining $U_{G_\alpha} \subset (a,b)$. Then $N_\alpha \cap (a,b) \subset U_{G_\alpha}$ and $G_\alpha(d) \geq G_\alpha(c)$; hence,

$$\text{length}\,(A \cap (c,d)) \leq \alpha(d - c),$$

for each component interval (c,d) of U_G. By Exercise **6.2.4** and $N_\alpha \subset A$, summing over the component intervals (c,d) of U_G.

$$\begin{aligned}\text{length}\,(N_\alpha \cap (a,b)) &\le \sum_{(c,d)} \text{length}\,(N_\alpha \cap (c,d)) \\ &\le \sum_{(c,d)} \text{length}\,(A \cap (c,d)) \\ &\le \sum_{(c,d)} \alpha(d-c) \le \alpha(b-a).\end{aligned}$$

By Exercise **6.3.6**, this implies N_α is negligible. □

For $c \in \mathbf{R}$, define

$$D_cF(x) = \frac{F(x) - F(c)}{x - c}, \qquad x \neq c. \tag{6.4.3}$$

Note F increasing implies $D_cF(x) \ge 0$ for $x \neq c$. A function F is *Lipschitz on* $[a,b]$ *with constant* $\lambda > 0$ if

$$|F(x) - F(x')| \le \lambda |x - x'|, \qquad x, x' \in [a,b].$$

Equivalently, F is Lipschitz with constant λ if $|D_cF(x)| \le \lambda$, $a \le c < x \le b$. We say F has *constant slope* λ on (a,b) if $D_cF(x) = \lambda$, $a \le c < x \le b$.

Theorem 6.4.3 (Lipschitz Approximation). *Let F be continuous on $[a,b]$ and let $\lambda_0 = D_aF(b)$. Then for $\lambda \ge \lambda_0$, there is an open set $W_\lambda \subset (a,b)$ and F_λ continuous on $[a,b]$ such that F_λ agrees with F on W_λ^c, F_λ has constant slope λ on each component interval of W_λ, and $D_cF_\lambda(x) \le \lambda$ for $a \le c < x \le b$. If $\mu \le \lambda$ and $D_cF(x) \ge \mu$ for $a \le c < x \le b$, then $D_cF_\lambda(x) \ge \mu$ for $a \le c < x \le b$.*

If $G(x) = F(x) - \lambda x$, then $G(a) \ge G(b)$. Applying the second part of the Sunrise Lemma, $F(d) - \lambda d = F(c) - \lambda c$ for each component interval (c,d) of W_G. Given x in $[a,b]$, let $\underline{x} = \bar{x} = x$ if $x \notin W_G$, and let $\underline{x} = c$ and $\bar{x} = d$ if x is in a component interval (c,d). Define F_λ by setting

$$F_\lambda(x) - \lambda x = F(\underline{x}) - \lambda \underline{x} = F(\bar{x}) - \lambda \bar{x}, \qquad a \le x \le b,$$

and $W_\lambda \equiv W_G$. Then F_λ has constant slope λ on each component interval. Exercise **6.4.1** shows $F_\lambda(x) - \lambda x$ is continuous and decreasing on $[a,b]$; hence, $D_cF_\lambda(x) \le \lambda$ for $a \le c < x \le b$. Now assume $D_cF(x) \ge \mu$ for $a \le c < x \le b$, and let $c \le x$. If c and x are in the same component interval, then $D_cF_\lambda(x) = \lambda \ge \mu$. If not, then $\bar{c} \le \underline{x}$ and

$$D_cF_\lambda(\bar{c}) = \lambda \ge \mu, \quad D_{\bar{c}}F_\lambda(\underline{x}) = D_{\bar{c}}F(\underline{x}) \ge \mu, \quad D_{\underline{x}}F_\lambda(x) = \lambda \ge \mu,$$

which together imply $D_cF_\lambda(x) \ge \mu$. □

Let $f \ge 0$ be measurable and integrable over (a,b). For $A \subset (a,b)$, let

$$\frac{1}{\text{length}\,(A)} \int_A f(x)\,dx$$

denote *the average of f over A*. By the Lebesgue differentiation theorem,[6]

$$F(x) = \int_a^x f(t)\,dt, \qquad a < x < b,$$

implies $F'(x) = f(x)$ almost everywhere on (a, b). Thus from Exercise **6.4.3**, we obtain the following.

Theorem 6.4.4. [7] *Let $f \geq 0$ be measurable and integrable over (a, b), and let λ_0 be the average of f over (a, b). Then for all $\lambda \geq \lambda_0$, there is an open set $W_\lambda \subset (a, b)$ such that the average of f over each component interval of W_λ equals λ, and $0 \leq f \leq \lambda$ almost everywhere on W_λ^c.* □

Exercises

6.4.1. With G as in the sunrise lemma, suppose $G(a) \geq G(b)$. Define $\tilde{G}$ to be constant on $[c, d]$ for each component interval (c, d) of W_G, and $\tilde{G} = G$ elsewhere. Then $\tilde{G}$ is continuous and decreasing on $[a, b]$.

6.4.2. Given $f \geq 0$ integrable arbitrary on $\mathbf{R}$ and $\lambda > 0$, let

$$f^*(c) = \sup_{x>c} \frac{1}{x-c} \int_c^x f(t)\,dt$$

and $\{f^* > \lambda\} = \{x : f^*(x) > \lambda\}$. Show that $\{f^* > \lambda\}$ is the union over $n \geq 1$ of $U_{G_n} \subset (-n, n)$ with $G_n(x) = \int_{-n}^x f(t)\,dt - \lambda x$ on $[-n, n]$. Use this to derive the *Hardy-Littlewood maximal inequality*

$$\lambda \operatorname{length}(\{f^* > \lambda\}) \leq \int_{\{f^*>\lambda\}} f(x)\,dx.$$

6.4.3. Let F be continuous increasing on $[a, b]$ and let $\lambda_0 = D_a F(b) \geq 0$. Assume F' exists almost everywhere on (a, b). Then for all $\lambda \geq \lambda_0$, there is an open set $W_\lambda \subset (a, b)$ such that $0 \leq F'(c) \leq \lambda$ almost everywhere on W_λ^c and $D_c F(d) = \lambda$ on each component interval (c, d) of W_λ.

6.4.4. With G as in the Sunrise Lemma, assume $G(a) \geq G(b)$. If (c, d) is a component interval of W_G with $a < c$, then $G(c) > G(x)$ for $c < x \in W_G$.

---

[6] Established in §6.6

[7] This is a precursor of the *Calderón-Zygmund lemma*.

6.5 Absolute Continuity

Let F be defined on $[a,b]$. Given an open set $U \subset (a,b)$, the *variation of F over U* is

$$\mathrm{var}(F,U) = \sum_{(c,d)} |F(d) - F(c)|.$$

Here the sum is over the component intervals (c,d) of U; thus, the number of terms in the sum may be finite or infinite. Given $U \subset (a,b)$, the *total variation $v_F(U)$* is

$$v_F(U) = \sup\{\mathrm{var}(F,U') : U' \subset U \text{ open}\}.$$

When $U = (a,b)$, we write $v_F(a,b)$ instead of $v_F(U)$. We say F has *bounded variation* on $[a,b]$ if $v_F(a,b) < \infty$. Note $v_F(U) \le v_F(U')$ when $U \subset U'$.

When F is increasing on $[a,b]$, every variation in (a,b) is no greater than $F(b) - F(a)$, and in fact $v_F(a,b) = F(b) - F(a)$. By Exercise **6.5.1**, for F increasing this implies

$$v_F(U) = \mathrm{var}(F,U) = \sum_{(c,d)} F(d) - F(c)$$

whenever $U \subset (a,b)$ and the sum is over the component intervals of U.

When f is integrable over (a,b) and $F = F_c$ is given by

$$F(x) = F_c(x) = \int_c^x f(t)\,dt, \qquad a \le x \le b, \tag{6.5.1}$$

and $U' \subset U \subset (a,b)$, summing over the component intervals (c,d) of U' yields

$$\begin{aligned}\mathrm{var}(F,U') &= \sum_{(c,d)} |F(d) - F(c)| \\ &\le \sum_{(c,d)} \int_c^d |f(x)|\,dx = \int_{U'} |f(x)|\,dx \le \int_U |f(x)|\,dx.\end{aligned}$$

Taking the sup over $U' \subset U$ yields

$$v_F(U) \le \int_U |f(x)|\,dx. \tag{6.5.2}$$

Let (U_n) be a sequence of open sets in (a,b). We say F is *absolutely continuous on* $[a,b]$ if $\mathrm{length}\,(U_n) \to 0$ as $n \to \infty$ implies $v_F(U_n) \to 0$ as $n \to \infty$. Exercise **6.5.2** shows that the definitions of absolute continuity in §6.3 and here are equivalent.

Note absolute continuity on $[a,b]$ implies uniform continuity (§2.3) on $[a,b]$.

More generally, we say F is *locally absolutely continuous on* (a,b) if F is absolutely continuous on $[c,d]$ for all $[c,d] \subset (a,b)$.

If F is absolutely continuous on $[a,b]$, then (Exercise **6.5.3**) $v_F(a,b) < \infty$ and $v_F(a,b)$ equals the total variation as defined in Exercise **2.2.4**. Thus absolute continuity implies bounded variation. The converse is not true as the Cantor function is bounded variation but not absolutely continuous.

Theorem 6.5.1. *If F is continuous on $[a,b]$ and bounded variation on $[a,b]$, then there are continuous increasing functions G, H on $[a,b]$ with $F = G - H$. If F is absolutely continuous on $[a,b]$, then G and H may also be chosen absolutely continuous on $[a,b]$.*

By Exercise **6.5.3**, $v_F(a,b) < \infty$. By Exercise **2.2.6**, $G(x) = v_F(a,x)$, $H(x) = v_F(a,x) - F(x)$, $a \le x \le b$, are increasing and continuous. By Exercise **6.5.4**, $v_G(U) = v_F(U)$ for $U \subset (a,b)$ open. Thus F absolutely continuous implies G absolutely continuous. Since F_1, F_2 absolutely continuous implies[8] $F_1 + F_2$ is absolutely continuous, H is also absolutely continuous. □

The goal of this section is to derive Theorems 6.3.3, 6.3.4, 6.3.5, and 6.3.2. Here is the main result of this section.

Theorem 6.5.2. *If F is absolutely continuous on $[a,b]$, then $F'(x)$ exists for almost all x in (a,b), F' is integrable on (a,b), and*

$$F(b) - F(a) = \int_a^b F'(x)\,dx. \tag{6.5.3}$$

In fact, Exercise **6.6.4** shows

$$v_F(a,b) = \int_a^b |F'(x)|\,dx. \tag{6.5.4}$$

This says *if F is a primitive of f, then v_F is a primitive of $|f|$*. Note (6.5.4) together with Exercise **6.5.1** shows that (6.5.2) is in fact an equality.

The basic connection between integrals and absolutely continuous functions is the following.

Theorem 6.5.3. *Let f be integrable and measurable on (a,b). If (A_n) is a sequence of arbitrary subsets of (a,b) with* length $(A_n) \to 0$ *as $n \to \infty$, then*

$$\int_{A_n} f(x)\,dx \to 0, \qquad n \to \infty.$$

By decomposing $f = f^+ - f^-$, we may assume $f \ge 0$. By Theorem 6.3.7, for each $n \ge 1$, there is open $U_n \supset A_n$ with length $(U_n) \to 0$. By intersecting with (a,b), we may assume $U_n \subset (a,b)$, $n \ge 1$. It is enough to show $\int_{U_n} f(x)\,dx \to 0$. We now argue by contradiction and suppose

[8] This is most easily seen via the definition in §6.3.

$\int_{U_n} f(x)\,dx \not\to 0$. By passing to a subsequence, we can assume there is an $\epsilon > 0$ with $\int_{U_n} f(x)\,dx \geq \epsilon$, $n \geq 1$. By passing to a further subsequence, we may assume length $(U_n) \leq 2^{-n}$, $n \geq 1$. Let V_n be the union of U_j, $j \geq n+1$. Then

$$\text{length}\,(V_n) \leq \sum_{j=n+1}^{\infty} \text{length}\,(U_j) \leq \sum_{j=n+1}^{\infty} 2^{-j} = 2^{-n}, \quad n \geq 1$$

and

$$\int_{V_n} f(x)\,dx \geq \int_{U_{n+1}} f(x)\,dx \geq \epsilon, \qquad n \geq 1.$$

Let B denote the intersection of V_n, $n \geq 1$. Then length $(B) \leq$ length $(V_n) \leq 2^{-n}$, $n \geq 1$; hence, length $(B) = 0$. But $\mathbf{1}_{V_n} f$ decreases pointwise to $\mathbf{1}_B f$ as $n \to \infty$, with $0 \leq \mathbf{1}_{V_n} f \leq f$, $n \geq 1$. By the dominated convergence theorem,

$$0 = \int_B f(x)\,dx = \lim_{n\to\infty} \int_{V_n} f(x)\,dx \geq \epsilon,$$

a contradiction. □

By (6.5.2), an immediate consequence of this is the local absolute continuity of F given by (6.5.1).

We next tackle the differentiability of functions. Since $F'(c)$ is the limit of D_cF at c, the existence of $F'(c)$ (§3.1) is equivalent (§2.2) to the equality and finiteness of four quantities, the upper and lower limits of $D_cF(x)$ as $x \to c\pm$. These are the four *Dini derivatives* of F at c. Equivalently, the four Dini derivatives are equal iff all right and left limit points of D_cF at c are equal (Exercise **1.5.10** and Theorem 2.2.1).

The next two results yield estimates corresponding to two of the four Dini derivatives.

Theorem 6.5.4. *Let F be continuous on $[a,b]$, let $\lambda > 0$, and let*

$$U_\lambda = \{c \in (a,b) : D_cF(x) > \lambda \text{ for some } x > c\}.$$

Then

$$\lambda\,\text{length}\,(U_\lambda) \leq v_F(U_\lambda) \leq v_F(a,b). \tag{6.5.5}$$

To derive this, apply the first part of the sunrise lemma to $G(x) = F(x) - \lambda x$ on $[a,b]$. Then $U_\lambda = U_G$, and $G(d) \geq G(c)$ for each component interval (c,d) of U_λ, which implies $F(d) - F(c) \geq \lambda(d-c)$ for each component interval (c,d) of U_λ. Hence

$$\lambda\,\text{length}\,(U_\lambda) = \sum_{(c,d)} \lambda(d-c) \leq \sum_{(c,d)} (F(d) - F(c)) \leq v_F(U_\lambda). \ \square$$

For increasing functions, we have a complementary result.

Theorem 6.5.5. *Let F be continuous increasing on $[a,b]$, let $\lambda > 0$, and let*

$$V_\lambda = \{c \in (a,b) : D_cF(x) < \lambda \text{ for some } x < c\}.$$

Then

$$v_F(V_\lambda) \le \lambda \operatorname{length}(V_\lambda) \le \lambda \operatorname{length}([a,b]). \tag{6.5.6}$$

To derive this, apply the first part of the Sunrise Lemma to $G(x) = F(-x) + \lambda x$ on $-[a,b] = [-b,-a]$. Then $V_\lambda = -U_G$, and $G(-c) \ge G(-d)$ for each component interval $-(c,d) = (-d,-c)$ of U_G, which implies $F(d) - F(c) \le \lambda(d-c)$ for each component interval (c,d) of V_λ. Hence

$$v_F(V_\lambda) = \sum_{(c,d)} (F(d) - F(c)) \le \lambda \sum_{(c,d)} (d-c) = \lambda \operatorname{length}(V_\lambda). \ \square$$

Theorem 6.5.6. *If F is continuous and bounded variation on $\mathbf{R}$, then $F'(x)$ exists for almost all x in $\mathbf{R}$.*

Since F is the difference of two continuous increasing functions, we may assume F is continuous increasing. We first show each limit point of D_cF at c is finite, for almost all $c \in \mathbf{R}$. Let $[a,b] \subset \mathbf{R}$.

Since F is bounded variation, $v_F(a,b) < \infty$. Let $U_n = U_\lambda$ as in Theorem 6.5.4 on $[a,b]$ with $\lambda = n$. Since $n \operatorname{length}(U_n) \le v_F(a,b)$ for $n \ge 1$, $\operatorname{length}(U_n) \to 0$ as $n \to \infty$; hence,

$$\operatorname{length}\left(\bigcap_{n=1}^{\infty} U_n\right) = \lim_{n\to\infty} \operatorname{length}(U_n) = 0.$$

If a right limit point of D_cF at $c \in (a,b)$ equals ∞, then $c \in \bigcap_{n=1}^{\infty} U_n$. We conclude right limit points of D_cF at c are finite for almost all c. Similarly for left limit points.

Now all limit points of D_cF at c are equal iff all left limit points of D_cF at c are not less than all right limit points of D_cF at c and all right limit points of D_cF at c are not less than all left limit points of D_cF at c.

Let $\tilde{F}(x) = -F(-x)$. Since $D_c\tilde{F}(x) = D_{-c}F(-x)$, the second assertion follows from the first. Thus, it is enough to establish for almost all c; all left limit points of D_cF at c are not less than all right limit points of D_cF at c.

To this end, for $0 < m < n$ rational, define

$$\begin{aligned} E_m^- &= \{c \in \mathbf{R} : \text{ there is a left limit point of } D_cF \text{ less than } m\}, \\ E_n^+ &= \{c \in \mathbf{R} : \text{ there is a right limit point of } D_cF \text{ greater than } n\}, \end{aligned}$$

and

$$N = \bigcup_{m<n} E_{m,n} \equiv \bigcup_{m<n} E_m^- \cap E_n^+.$$

Let L be a left limit point of D_cF at c, and let L' be a right limit point of D_cF at c. Then $L < L'$ iff there are rationals $0 < m < n$ satisfying $L < m < n < L'$, which happens iff $c \in N$. Thus, it is enough to show $E_{m,n}$ is negligible for all $0 < m < n$.

Given a compact interval $[a, b]$, apply Theorem 6.5.5 on $[a, b]$ with $\lambda = m$ to get $V_m \subset (a, b)$ with

$$v_F(V_m) \leq m \operatorname{length}(V_m) \leq m \operatorname{length}([a, b]).$$

If (c, d) is a component interval of V_m, apply Theorem 6.5.4 on $[c, d]$ with $\lambda = n$ to get $U_{m,n} \subset (c, d)$ with

$$n \operatorname{length}(U_{m,n}) \leq v_F(U_{m,n}) \leq v_F(c, d).$$

Since $E_{m,n} \cap (c, d) \subset U_{m,n}$, it follows that

$$n \operatorname{length}(E_{m,n} \cap (c, d)) \leq v_F(c, d).$$

Since $E_{m,n} \cap (a, b) = E_{m,n} \cap V_m$, summing over all component intervals of V_m yields

$$n \operatorname{length}(E_{m,n} \cap (a, b)) \leq v_F(V_m) \leq m \operatorname{length}(V_m) \leq m(b - a)$$

hence,

$$\operatorname{length}(E_{m,n} \cap (a, b)) \leq \frac{m}{n}(b - a), \qquad a < b.$$

Exercise **6.3.6** now implies $E_{m,n}$ is negligible. □

Let F be continuous increasing on $[a, b]$, and define $F(x) = F(b)$ for $x > b$ and $F(x) = F(a)$ for $x < a$. Then F is continuous increasing on $\mathbf{R}$; hence, $F'(x) \geq 0$ exists for almost all x.

For $n \geq 1$, let

$$f_n(x) = \begin{cases} n\,(F(x + 1/n) - F(x)), & \text{if } F'(x) \text{ exists,} \\ F(x + 1) - F(x), & \text{otherwise,} \end{cases} \tag{6.5.7}$$

and let

$$f(x) = \begin{cases} F'(x), & \text{if } F'(x) \text{ exists,} \\ F(x + 1) - F(x), & \text{otherwise.} \end{cases}$$

Then $f_n(x) \to f(x)$ pointwise on $\mathbf{R}$ as $n \to \infty$, and $f = F'$ almost everywhere. Hence, we may use Fatou's lemma to conclude

$$\int_a^b f(x)\, dx \leq \liminf_{n \to \infty} \int_a^b f_n(x)\, dx.$$

But F is continuous, so applying the first fundamental theorem of calculus in Chapter 4 to F yields

$$n\int_a^{a+1/n} F(x)\,dx \to F(a) \quad \text{and} \quad n\int_b^{b+1/n} F(x)\,dx \to F(b).$$

Moreover, for $n \geq 1$,

$$\begin{aligned}\int_a^b f_n(x)\,dx &= n\int_{a+1/n}^{b+1/n} F(x)\,dx - n\int_a^b F(x)\,dx \\ &= n\int_b^{b+1/n} F(x)\,dx - n\int_a^{a+1/n} F(x)\,dx.\end{aligned}$$

Passing to the limit, we obtain

$$\int_a^b F'(x)\,dx = \int_a^b f(x)\,dx \leq \liminf_{n\to\infty}\int_a^b f_n(x)\,dx = F(b) - F(a).$$

Since every continuous bounded variation function is the difference of two continuous increasing functions, we conclude F' exists almost everywhere and is integrable, when F is continuous bounded variation. Moreover, modifying the above argument establishes (Exercise **6.5.8**)

$$\int_a^b |F'(x)|\,dx \leq v_F(a,b) \tag{6.5.8}$$

for F continuous bounded variation on $[a,b]$. In particular, applying this last result on $[c,d] \subset (a,b)$, we conclude: If F is locally absolutely continuous on (a,b), then F' exists almost everywhere; thus, F is a primitive of a measurable and locally integrable $f = F'$.

Conversely, if f is measurable and locally integrable on (a,b), the Lebesgue differentiation theorem implies F given by (6.5.1) is a primitive of f. *This establishes Theorem 6.3.3 and Theorem 6.3.4.* □

If F is Lipschitz with $|D_cF(x)| \leq \lambda$, $a \leq c < x \leq b$ and f_n, $n \geq 1$, are as in (6.5.7); then $|f_n(x)| \leq \lambda$ for all x in (a,b), $n \geq 1$. Hence, by the dominated convergence theorem,

$$\int_a^b f_n(x)\,dx \to \int_a^b f(x)\,dx = \int_a^b F'(x)\,dx.$$

Repeating the argument leading to (6.5.8) yields (6.5.3). *This establishes* (6.5.3) *for F Lipschitz.*

The end is now in sight. Given F continuous increasing on $[a,b]$, the plan is to *approximate F by F_λ Lipschitz.*

Suppose F is continuous increasing on $[a,b]$ and $\lambda \geq \lambda_0 = D_aF(b)$, and let W_λ, F_λ be as in Theorem 6.4.3. Then $F(a) = F_\lambda(a)$ and $F(b) = F_\lambda(b)$ and

$0 \le D_c F_\lambda(x) \le \lambda$ for $a \le c \le x \le b$. For each component interval (c,d) of W_λ, we have $D_c F_\lambda(d) = D_c F(d) = \lambda$; thus, $\lambda(d-c) = F(d) - F(c)$. Summing over all component intervals of W_λ,

$$\lambda \operatorname{length}(W_\lambda) = v_F(W_\lambda) \le v_F(a,b). \tag{6.5.9}$$

If $c \in U_\lambda^c$, there is a sequence $x_n \to c$ with $x_n \notin W_\lambda$, $n \ge 1$. Thus given $c \in U_\lambda^c$, there is a sequence $x_n \to c$ with $D_c F_\lambda(x_n) = D_c F(x_n)$, $n \ge 1$. If moreover $F'(c)$ and $F'_\lambda(c)$ both exist, it follows that $F'(c) = F'_\lambda(c)$.

Since F is continuous increasing, $F'(x) \ge 0$ exists for almost all x in (a,b). Since F_λ is Lipschitz, F_λ is absolutely continuous (Exercise **6.5.5**); hence, $F'_\lambda(x)$ exists for almost all x in (a,b). Let N be the negligible set on whose complement both F' and F'_λ exist, and let $\{F' \ne F'_\lambda\}$ be the set on whose complement both F' and F'_λ exist and are equal. We conclude

$$\{F' \ne F'_\lambda\} \subset W_\lambda \cup N.$$

But (6.5.3) is valid for F_λ; hence,

$$F(b) - F(a) = \int_a^b F'_\lambda(x)\, dx,$$

thus,

$$\begin{aligned}
\left| F(b) - F(a) - \int_a^b F'(x)\, dx \right| &\le \int_a^b |F'_\lambda(x) - F'(x)|\, dx \\
&\le \int_{\{F' \ne F'_\lambda\}} F'(x)\, dx + \int_{\{F' \ne F'_\lambda\}} \lambda\, dx \\
&\le \int_{W_\lambda} F'(x)\, dx + \lambda \operatorname{length}(W_\lambda). \quad (6.5.10)
\end{aligned}$$

So far, (6.5.10) is valid for any $\lambda \ge \lambda_0 = D_a F(b)$ and any F continuous increasing on $[a,b]$. But F increasing implies $v_F(a,b) < \infty$; hence, by (6.5.9), $\operatorname{length}(W_\lambda) \to 0$ as $\lambda \to \infty$. Since F' is integrable, the first term in (6.5.10) vanishes as $\lambda \to \infty$.

If F is absolutely continuous, it follows that $v_F(W_\lambda) \to 0$ as $\lambda \to \infty$. By (6.5.9) again, $\lambda \operatorname{length}(W_\lambda) \to 0$ as $\lambda \to \infty$; hence, the second term in (6.5.10) vanishes as $\lambda \to \infty$. This establishes (6.5.3) for F absolutely continuous and increasing on $[a,b]$. By Theorem 6.5.1, this establishes (6.5.3). This completes the proof of Theorem 6.5.2. □

Now let F be locally absolutely continuous on (a,b) with $F'(x) = 0$ almost everywhere on (a,b). Applying Theorem 6.5.2 on $[c,d] \subset (a,b)$ yields $F(c) = F(d)$; thus, F is constant. *This establishes Theorem 6.3.5.* □

Let f be continuous on (a,b), and suppose F is locally absolutely continuous on (a,b) with $F'(x) = f(x)$ for almost all x. Fix c in (a,b). Then F and

F_c are both primitives of f and hence differ by a constant. From Chapter 4, since f is continuous, $F_c'(x) = f(x)$ for all x in (a,b). Hence $F'(x) = f(x)$ for all x in (a,b). Conversely, suppose $F'(x) = f(x)$ for all x in (a,b). Since f is bounded on $[c,d] \subset (a,b)$, the mean value theorem implies that F is Lipschitz on $[c,d]$, hence absolutely continuous on $[c,d]$, and hence locally absolutely continuous on (a,b). *This establishes Theorem 6.3.2.* □

Exercises

6.5.1. If F is defined on $[a,b]$ and U is a countable disjoint union of open sets U_n, $n \geq 1$, in (a,b), then

$$v_F(U) = \sum_{n=1}^{\infty} v_F(U_n).$$

6.5.2. Show that the definitions of absolute continuity in §6.3 and in this section are equivalent.

6.5.3. If $a = x_0 < x_1 < \cdots < x_n = b$ is a partition of $[a,b]$, $v_F(a,b) = v_F(x_0,x_1) + \cdots + v_F(x_{n-1},x_n)$. Conclude F absolutely continuous on $[a,b]$ implies $v_F(a,b) < \infty$.

6.5.4. With $v_F(x) = v_F(a,x)$, show $v_{v_F}(U) = v_F(U)$.

6.5.5. Show F Lipschitz on $[a,b]$ implies F absolutely continuous on $[a,b]$.

6.5.6. The estimates (6.5.5) and (6.5.6) correspond to two of the four Dini derivatives. Write down and prove the estimates corresponding to the other two Dini derivatives.

6.5.7. Let C be the Cantor set, F the Cantor function, and U_λ as in (6.5.5). Let (c_k,d_k), $k \geq 1$, be the component intervals of $[0,1] \setminus C$, and let $Z = \{c_k : k \geq 1\}$. Show

$$\bigcap_{\lambda>0} U_\lambda = C \setminus Z.$$

6.5.8. Let F be continuous bounded variation on $[a,b]$. Then (6.5.8) holds.

6.6 The Lebesgue Differentiation Theorem

Now we turn to the proof of the first fundamental theorem, Theorem 6.3.1. Without loss of generality, by restricting to a subinterval $[c,d] \subset (a,b)$, we may assume f is integrable on (a,b). Then $F = F_c$ is defined on $[a,b]$.

Assume first (6.3.5) holds, and let f_n be given by (6.5.7). Then f is almost everywhere equal to the pointwise limit of the continuous functions f_n; hence, f is measurable.

Conversely, if f is measurable, $F = F_c$ is absolutely continuous on $[a, b]$; hence, F_c' exists almost everywhere on (a, b). If $g = f - F_c'$, then g is measurable integrable on (a, b), and by Theorem 6.5.2,

$$G(x) \equiv \int_c^x g(t)\,dt = 0, \qquad a \le x \le b.$$

But this implies

$$\int_U g(x)\,dx = 0$$

for every open interval U in (a, b). Let $U \subset (a, b)$ be open. Applying this to each component interval of U and summing over these intervals yields the same equality but now for any $U \subset (a, b)$ open.

Given any $A \subset (a, b)$ and $\epsilon > 0$, there is an open $U \supset A$ with length $(U) \le$ length $(A) + \epsilon$. Intersecting U with (a, b) if necessary, we may assume $U \subset (a, b)$. Hence there is a sequence of open supersets $U_n \subset (a, b)$ of A with length $(U_n) \to$ length (A).

If A is measurable, then

$$\text{length}\,(U_n \setminus A) = \text{length}\,(U_n) - \text{length}\,(A) \to 0.$$

By Theorem 6.5.3, the integral of g over $U_n \setminus A$ goes to 0 as $n \to \infty$. Since by linearity,

$$0 = \int_{U_n} g(x)\,dx = \int_{U_n \setminus A} g(x)\,dx + \int_A g(x)\,dx,$$

it follows that the integral of g over A vanishes; hence,

$$\int_A g^+(x)\,dx = \int_A g^-(x)\,dx.$$

Choosing $A = \{\pm g \ge \epsilon\} \equiv \{x : \pm g(x) \ge \epsilon\}$ yields

$$0 = \int_{\{\pm g \ge \epsilon\}} g^{\mp}(x)\,dx = \int_{\{\pm g \ge \epsilon\}} g^{\pm}(x)\,dx \ge \epsilon\,\text{length}\,(\{\pm g \ge \epsilon\}) \qquad (6.6.1)$$

hence, $\{\pm g \ge \epsilon\}$ is negligible for all $\epsilon > 0$, or $g = 0$ almost everywhere on (a, b). □

Above we derived the first fundamental theorem from the second fundamental theorem. One can instead derive the second fundamental theorem from the first fundamental theorem.

Indeed both approaches start by establishing the almost everywhere existence and integrability of F' for F continuous increasing. Then the above calculation uses the absolute continuity of G to derive the first fundamental theorem from the second fundamental theorem. Conversely, by the first fundamental theorem,

$$G(x) \equiv F(x) - \int_a^x F'(x)\,dx, \qquad a \le x \le b,$$

satisfies $G'(x) = 0$ almost everywhere, and (6.5.8) implies G is increasing. Using the absolute continuity of G (Exercise **6.6.7**), G is a constant, obtaining the second fundamental theorem.

This latter approach necessitates a direct proof of the first fundamental theorem. This standard proof, which we do not include, is based on the Lebesgue density theorem, approximation by simple functions, and the maximal inequality, Exercise **6.4.2**.

Here is a basic estimate. The proof presented is that in [13]. When F is continuous increasing, this is a consequence of the sunrise lemma via (6.5.6).

Theorem 6.6.1 (Fundamental Lemma). *Let F be continuous on $\mathbf{R}$, $\lambda > 0$ and $A \subset \mathbf{R}$ arbitrary. If $F'(c)$ exists and satisfies $|F'(c)| < \lambda$ for $c \in A$, then*

$$\text{length}\,(F(A)) \le \lambda\,\text{length}\,(A)\,.$$

Given $\epsilon > 0$, select open $U_\epsilon \supset A$ with $\text{length}\,(U_\epsilon) \le \text{length}\,(A) + \epsilon$. For each $c \in \mathbf{R}$, let

$$I_c \equiv \{x \in U_\epsilon : |D_c F(t)| < \lambda \text{ for } 0 < |t - c| \le |x - c|\}.$$

If $|F'(c)| < \lambda$, then I_c is the *largest* nonempty open interval $I_c \subset U_\epsilon$ centered at c on which $|D_c F(x)| < \lambda$ holds. Let

$$U = \bigcup\{I_c : c \in A\}.$$

Then U is open (§2.1) and $A \subset U \subset U_\epsilon$. If (a, b) is a component interval of U, then

$$(a, b) = \bigcup\{I_c : c \in A \cap (a, b)\}.$$

Let $[c, d] \subset (a, b)$. Since $\{I_x : x \in A \cap (a, b)\}$ is an open cover of $[c, d]$ and $[c.d]$ is a compact set (Theorem 2.1.5), there is a finite subcover of $[c, d]$; thus, there are $c_1 \le c_2 \le \cdots \le c_N$ in $A \cap (a, b)$ with

$$[c, d] \subset I_1 \cup I_2 \cup \ldots \cup I_N = I,$$

where $I_k = I_{c_k}$, $k = 1, \ldots, N$.

By discarding intervals, we may assume $c_1 < c_2 < \cdots < c_N$ and

$$I_k \cap I_{k+1} \cap (c_k, c_{k+1}), \qquad 1 \le k < k+1 \le N,$$

are nonempty. We may also assume $I_1, I_2, \dots, I_N$ are all bounded; otherwise, $\text{length}\,(A) = \infty$ and there is nothing to prove.

For each $k = 1, \dots, N$, define

$$F_k^{\pm}(x) = F(c_k) \pm \lambda(x - c_k), \qquad x \in \mathbf{R}.$$

The graphs of these functions are the lines in Figure 6.3. Let M_k and m_k denote the sup and inf of F_k^+ over I_k, $k = 1, \dots, N$. Since I_k is centered at c_k, M_k and m_k are also the sup and inf of F_k^- over I_k, $k = 1, \dots, N$. Moreover, $F(I_k) \subset [m_k, M_k]$, $k = 1, \dots, N$; hence, $F(I) \subset [m_*, M^*]$, where

$$M^* = \max(M_1, \dots, M_N), \qquad m_* = \min(m_1, \dots, m_N).$$

Let I^*, c^*, and $F^{*\pm}$ denote the intervals and centers and lines I_i, c_i, and $F_i^{\pm}$, where $i = \min\{k : M_k = M^*\}$, and let I_*, c_*, and $F_*^{\pm}$ denote the intervals and centers and lines I_j, c_j, and $F_j^{\pm}$, where $j = \min\{k : M_k = M_*\}$.

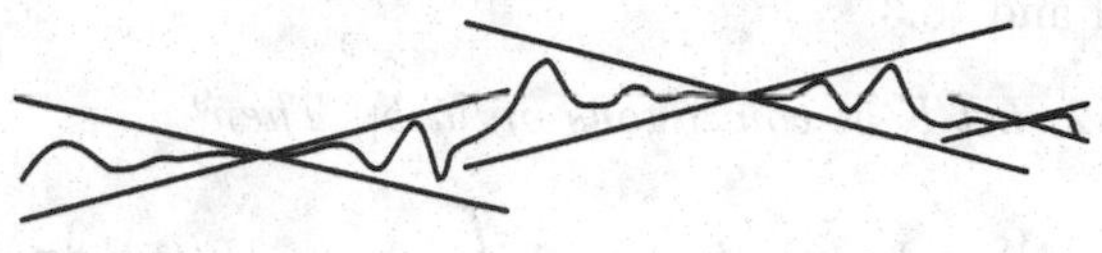

Fig. 6.3 The Fundamental Lemma [13]

Define G^*, G_* on $\mathbf{R}$ as follows: If $c^* > c_*$, $G^* = F^{*+}$, $G_* = F_*^+$; otherwise, if $c^* < c_*$, $G^* = F^{*-}$, $G_* = F_*^-$. Then

$$G_* - G^* = \lambda|c_* - c^*| + F(c_*) - F(c^*) \ge |c_* - c^*| \left(\lambda - |D_{c_*}F(c^*)|\right).$$

We claim $(m_*, M^*) \subset G_*(I)$. The realization that $F(I)$ is contained in the image $G_*(I)$ of a *single line of slope* $\pm\lambda$—as suggested by Figure 6.3—is the key idea of this proof.

Now note $|D_aF(b)| \le \lambda$ and $|D_bF(c)| \le \lambda$ together imply $|D_aF(c)| \le \lambda$. Let x be in $I_k \cap I_{k+1}$ with $c_k < x < c_{k+1}$. Then

$$|D_{c_k}F(x)| \le \lambda, \qquad |D_xF(c_{k+1})| \le \lambda$$

imply

$$|D_{c_k}F(c_{k+1})| \le \lambda, \qquad 1 \le k < k+1 \le N.$$

Applying this through all c_k between c_* and c^*, we obtain $|D_{c_*}F(c^*)| \le \lambda$. Putting this all together, $G_* \ge G^*$; hence,

$$\sup_I G_* \ge \sup_{I^*} G_* \ge \sup_{I^*} G^* = M^* \ge m_* = \inf_{I_*} G_* \ge \inf_I G_*.$$

This establishes the claim.

Since G_* is a line with absolute slope λ, we conclude

$$\text{length}\,(F([c,d])) \le \text{length}\,(F(I)) \le \text{length}\,(G_*(I)) = \lambda\,\text{length}\,(I) \le \lambda(b-a).$$

Allowing $[c,d]$ to fill (a,b), we obtain

$$\text{length}\,(F((a,b))) \le \lambda(b-a).$$

Summing over all component intervals (a,b) of U, we arrive at

$$\begin{aligned}\text{length}\,(F(A)) &\le \sum_{(a,b)} \text{length}\,(F((a,b))) \le \sum_{(a,b)} \lambda(b-a)\\ &= \lambda\,\text{length}\,(U) \le \lambda\,\text{length}\,(U_\epsilon) \le \lambda(\text{length}\,(A)+\epsilon).\end{aligned}$$

Since $\epsilon > 0$ is arbitrary, the result follows. □

Note the next result is a direct generalization of the second fundamental theorem of §4.4 and §6.3.

Theorem 6.6.2. *Let F be continuous on $[a,b]$. Then*[9]

$$\text{length}\,(F([a,b])) - \text{length}\,(F(N)) \le \int_{\{F' \text{ exists}\}} |F'(x)|\,dx \le v_F(a,b),$$

where $N = (a,b) \setminus \{F' \text{ exists}\}$.

Let $\{F' = 0\} = \{c \in (a,b) : F'(c) \text{ exists and } F'(c) = 0\}$. By the Fundamental Lemma,

$$\text{length}\,(F(\{F' = 0\})) \le \lambda(b-a)$$

for all $\lambda > 0$; hence, $F(\{F' = 0\})$ is negligible.

Given $\theta > 1$, $\mathbf{R}^+$ is the disjoint union of intervals $[\theta^n, \theta^{n+1})$, $n \in \mathbf{Z}$; hence, $\{F' \text{ exists}\}$ is the disjoint union of the measurable sets $\{F' = 0\}$ and

$$A_n = \{c \in (a,b) : F'(c) \text{ exists and } \theta^n \le |F'(c)| < \theta^{n+1}\}, \qquad n \in \mathbf{Z}.$$

Apply the fundamental lemma with $A = A_n$ and $\lambda = \theta^{n+1}$, yielding

$$\text{length}\,(F(A_n)) \le \theta^{n+1}\,\text{length}\,(A_n) \le \theta \int_{A_n} |F'(x)|\,dx, \qquad n \in \mathbf{Z}.$$

Summing over $n \in \mathbf{Z}$, we obtain

$$\text{length}\,(F(\{|F'| > 0\})) \le \sum_{n\in\mathbf{Z}} \theta \int_{A_n} |F'(x)|\,dx \le \theta \int_{\{F' \text{ exists}\}} |F'(x)|\,dx.$$

[9] N may not be negligible: For example, F' may exist nowhere!

Since $F(\{F'=0\})$ is negligible and $\theta > 1$ is arbitrary, we conclude

$$\text{length}\left(F(\{F' \text{ exists}\})\right) \le \int_{\{F' \text{ exists}\}} |F'(x)|\,dx.$$

Now $F((a,b)) = F(\{F' \text{ exists}\}) \cup F(N)$; hence,

$$\text{length}\left(F([a,b])\right) \le \text{length}\left(F(N)\right) + \int_{\{F' \text{ exists}\}} |F'(x)|\,dx.$$

By (6.5.8), the result follows. □

Let F be continuous on $[a,b]$. We say F has *Lusin's property* if $F(N)$ is negligible for every negligible $N \subset (a,b)$. By Exercise **6.6.2**, absolute continuity implies Lusin's property. Applying the above result to an absolutely continuous increasing function yields an alternate path to Theorem 6.5.2 which avoids the Lipschitz approximation argument used in §6.5.

Here is another consequence of the above result.

Theorem 6.6.3 (Banach-Zaretski). *Let F be continuous and bounded variation on $[a,b]$. Then F is absolutely continuous iff F satisfies Lusin's property.*

Indeed, if F is bounded variation, then F' exists almost everywhere, N above is negligible, and F' is integrable over (a,b). If F also satisfies Lusin's property, then $F(N)$ is negligible. Hence, for $(c,d) \subset (a,b)$,

$$|F(d) - F(c)| \le \text{length}\left(F([c,d])\right) \le \int_c^d |F'(x)|\,dx,$$

from which (6.5.2) with $f = F'$ readily follows. Thus F is absolutely continuous. □

Given F on $[a,b]$ and y real, let $\#_F(y)$ be the number of reals x in (a,b) satisfying $F(x) = y$. Then $\#_F(y)$ is zero, a natural, or ∞ for each y real. More generally, for $U \subset (a,b)$ open, let $\#_{F,U}(y)$ be the number of reals x in U satisfying $F(x) = y$. This counting function $\#_F$ is the *Banach indicatrix*. Here is another result of Banach [1].

Theorem 6.6.4. *Let F be continuous on $[a,b]$. Then*

$$\int_{-\infty}^{\infty} \#_F(y)\,dy = v_F(a,b).$$

More generally, for $U \subset (a,b)$ open,

$$\int_{-\infty}^{\infty} \#_{F,U}(y)\,dy = v_F(U).$$

This is valid whether $v_F(a,b)$ or $v_F(U)$ are finite or infinite. Also by definition of subgraph in Chapter 4, the integrals of $0 \le \#_F \le \infty$ and $0 \le \#_{F,U} \le \infty$ are well defined, without having to show $\#_F$, $\#_{F,U}$ are measurable (although they are).

Since U is the disjoint union of its component intervals (c,d), and $\#_{F,U}$ is the sum of $\#_{F,(c,d)}$ over all (c,d), the second equality follows from the first.

To establish the first equality, given $n \ge 1$, let $a = x_0 < x_1 < \cdots < x_n = b$ be a partition of (a,b) with $I_i = (x_{i-1}, x_i)$, $i = 1, \dots, n$, all of equal length, and set $J_1 = (x_0, x_1)$, $J_i = [x_{i-1}, x_i)$, $i = 2, \dots, n$. Then the intervals J_i, $i = 1, \dots, n$ are disjoint with union (a,b).

Define

$$\#_n(y) = \mathbf{1}_{F(J_1)}(y) + \mathbf{1}_{F(J_2)}(y) + \cdots + \mathbf{1}_{F(J_n)}(y).$$

Then $\#_n(y) = N$ iff y is in N of the sets $F(J_i)$, $1 \le i \le n$, which happens iff there is a partition $a < c_1 < \cdots < c_N < b$ satisfying $F(c_k) = y$, $k = 1, \dots, N$. But this implies $\#_F(y) \ge N$. Thus $\#_F(y) \ge \#_n(y)$, $n \ge 1$; hence, $\#_F(y) \ge \sup_n \#_{2^n}(y)$.

Given y real, suppose $a < c_1 < \cdots < c_N < b$ is a partition satisfying $F(c_k) = y$, $k = 1, \dots, N$. If $n \ge 1$ is selected with $(b-a)2^{-n}$ less than the mesh of $a < c_1 < \cdots < c_N < b$, exactly N subintervals J_i intersect $\{c_1, \dots, c_N\}$; then $\#_{2^n}(y) \ge N$; hence, $\sup_n \#_{2^n}(y) \ge N$. Taking the sup over all such N yields $\sup_n \#_{2^n}(y) \ge \#_F(y)$.

But $\#_{2^n}(y)$, $n \ge 1$, is increasing; Hence, $\#_{2^n}(y)$, $n \ge 1$, converges to $\sup_n \#_{2^n}(y) = \#_F(y)$. By Exercise **6.2.3**, $\#_n(y)$ is measurable for $n \ge 1$; hence, so is $\#_F(y)$.

By Exercise **6.6.1** and Exercise **6.5.3**, for $n \ge 1$,

$$\begin{aligned}\int_{-\infty}^{\infty} \#_n(y)\,dy &= \sum_{i=1}^{n} \text{length}\,(F(J_i)) \\ &= \sum_{i=1}^{n} \text{length}\,(F(I_i)) \le \sum_{i=1}^{n} v_F(I_i) = v_F(a,b).\end{aligned}$$

Sending $n \to \infty$ through powers of 2 yields, by the monotone convergence theorem,

$$\int_{-\infty}^{\infty} \#_F(y)\,dy \le v_F(a,b).$$

Conversely, let M_i and m_i denote the max and min of F over $[x_{i-1}, x_i]$, $i = 1, \dots, n$. Then given a finite disjoint union U of open intervals (c_k, d_k), $k = 1, \dots, N$, in (a,b), with $c_k \ne d_j$, $k, j = 1, \dots, N$, by Exercise **6.6.8**, we can select $n \ge 1$ such that

$$\text{var}(F,U) \le \sum_{i=1}^{n} (M_i - m_i)\,. \tag{6.6.2}$$

Thus

$$\operatorname{var}(F,U) \le \sum_{i=1}^{n}(M_i - m_i) = \sum_{i=1}^{n}\left(\sup_{J_i} F - \inf_{J_i} F\right)$$
$$= \sum_{i=1}^{n} \operatorname{length}(F(J_i)) = \int_{-\infty}^{\infty} \#_n(y)\,dy \le \int_{-\infty}^{\infty} \#_F(y)\,dy.$$

Taking the sup over all such U establishes the result. □

If $v_F(a,b) < \infty$, the above result shows $\#_F$ is integrable over $\mathbf{R}$. This can be leveraged to yield the following generalization of Exercise **4.4.25**.

Theorem 6.6.5. *Let F be absolutely continuous on $[a,b]$, and let g be nonnegative measurable on $\mathbf{R}$. Then*

$$\int_{-\infty}^{\infty} g(y)\#_F(y)\,dy = \int_a^b g(F(x))|F'(x)|\,dx.$$

To see this, start with $m < M$ and let $U = F^{-1}((m,M))$. Then

$$\int_m^M \#_F(y)\,dy = \int_{-\infty}^{\infty} \#_{F,U}(y)\,dy = v_F(U) = \int_{F^{-1}(m,M)} |F'(x)|\,dx.$$

Now let U be any open set in $\mathbf{R}$, apply this to each component interval (m,M) of U, and sum over component intervals. This yields

$$\int_U \#_F(y)\,dy = \int_{F^{-1}(U)} |F'(x)|\,dx = \int_a^b \mathbf{1}_U(F(x))|F'(x)|\,dx.$$

If $A \subset \mathbf{R}$ is measurable, select a sequence (U_n) of open supersets of A with $\operatorname{length}(U_n \setminus A) \to 0$. Then

$$\int_{U_n} \#_F(y)\,dy = \int_a^b \mathbf{1}_{U_n}(F(x))|F'(x)|\,dx \ge \int_a^b \mathbf{1}_A(F(x))|F'(x)|\,dx$$

for $n \ge 1$. Since $\#_F$ is integrable, sending $n \to \infty$, we obtain

$$\int_A \#_F(y)\,dy \ge \int_a^b \mathbf{1}_A(F(x))|F'(x)|\,dx.$$

Now replace A by $U_n \setminus A$ in this last inequality, to get

$$\int_{U_n\setminus A} \#_F(y)\,dy \ge \int_a^b \mathbf{1}_{U_n\setminus A}(F(x))|F'(x)|\,dx.$$

Since $\#_F$ is integrable, the left side goes to zero as $n \to \infty$ and hence so does the right side. Since

$$\mathbf{1}_{U_n}(F(x)) - \mathbf{1}_A(F(x)) = \mathbf{1}_{U_n \setminus A}(F(x)),$$

we conclude

$$\int_A \#_F(y)\, dy = \int_a^b \mathbf{1}_A(F(x))|F'(x)|\, dx.$$

By linearity, this implies the result for g simple measurable. By Theorem 6.1.4, the result follows. □

Exercises

6.6.1. If F is continuous on $[a,b]$ and $U \subset (a,b)$ is open, then

$$\text{length}\,(F(U)) \le v_F(U).$$

When F is continuous increasing, this is an equality.

6.6.2. If F is locally absolutely continuous on $\mathbf{R}$, then F has Lusin's property. (This result is false for the Cantor function as $F(C) = [0,1]$.)

6.6.3. Let F be continuous bounded variation on $[a,b]$. Use (6.5.8) to obtain

$$|F'(x)| \le \frac{d}{dx} v_F(a,x), \qquad \text{for almost all } x \in (a,b).$$

6.6.4. Use (6.5.2) and (6.5.3) to derive

$$\frac{d}{dx} v_F(a,x) = |F'(x)|, \qquad \text{for almost all } x \in (a,b).$$

for F absolutely continuous on $[a,b]$. Conclude (6.5.4).

6.6.5. Let F be absolutely continuous on $[a,b]$ and $A_n \subset (a,b)$ arbitrary, $n \ge 1$. Then length $(A_n) \to 0$ implies length $(F(A_n)) \to 0$.

6.6.6. Let F be absolutely continuous on $[a,b]$. Then $M \subset (a,b)$ measurable implies $F(M)$ measurable. (Since the Cantor set C is negligible, every subset of C is measurable. But for the Cantor function $F(C) = [0,1]$, so either this conclusion is false for the Cantor function, or every subset of $[0,1]$ is measurable.)

6.6.7. Let F be absolutely continuous increasing on $[a,b]$ with $F' = 0$ almost everywhere. Use (6.5.6) and Exercises **6.6.1** and **6.6.2** to directly show F is a constant.

6.6.8. With notation as in (6.6.2), if $c \in J_i$, and $d \in J_j$, $i < j$, then

$$|F(d) - F(c)| \leq \sum_{i \leq \ell \leq j} (M_\ell - m_\ell).$$

If U is a finite disjoint union of intervals (c_k, d_k), $k = 1, \dots, N$, in (a, b), with the mesh of $a = x_0 < x_1 < \cdots < x_n = b$ less than $\min\{|c_j - d_k| : j, k = 1, \dots, N\}$, then (6.6.2) holds.

Appendix A
Solutions

A.1 Solutions to Chapter 1

Solutions to Exercises 1.1

1.1.1 Suppose that $f : X \to Y$ is invertible. This means there is a $g : Y \to X$ with $g(f(x)) = x$ for all x and $f(g(y)) = y$ for all $y \in Y$. If $f(x) = f(x')$, then $x = g(f(x)) = g(f(x')) = x'$. Hence f is injective. If $y \in Y$, then $x = g(y)$ satisfies $f(x) = y$. Hence f is surjective. We conclude that f is bijective. Conversely, if f is bijective, for each $y \in Y$, let $g(y)$ denote the unique element $x \in X$ satisfying $f(x) = y$. Then by construction, $f(g(y)) = y$. Moreover, since x is the unique element of X mapping to $f(x)$, we also have $a = g(f(x))$. Thus g is the inverse of f. Hence f is invertible.

1.1.2 Suppose that $g_1 : Y \to X$ and $g_2 : Y \to X$ are inverses of f. Then $f(g_1(y)) = b = f(g_2(y))$ for all $y \in Y$. Since f is injective, this implies $g_1(y) = g_2(y)$ for all $y \in Y$, i.e., $g_1 = g_2$.

1.1.3 For the first equation in De Morgan's law, choose a in the right side. Then a lies in every set A^c, $A \in \mathcal{F}$. Hence a does not lie in any of the sets A, $A \in \mathcal{F}$. Hence a is not in $\bigcup\{A : A \in \mathcal{F}\}$, i.e., a is in the left side. If a is in the left side, then a is not in $\bigcup\{A : A \in \mathcal{F}\}$. Hence a is not in A for all $A \in \mathcal{F}$. Hence a is in A^c for all $A \in \mathcal{F}$. Thus a is in $\bigcap\{A^c : A \in \mathcal{F}\}$. This establishes the first equation. To obtain the second equation in De Morgan's law, replace $\mathcal{F}$ in the first equation by $\mathcal{F}' = \{A^c : A \in \mathcal{F}\}$.

1.1.4 $\bigcup\{x\} = \{t : t \in x\} = x$. On the other hand, $\{\bigcup x\}$ is a singleton, so it can equal x only when x is a singleton.

1.1.5 If $x \in b$, there are two cases, $x \in a$ and $x \notin a$. In the first case, $x \in a \cap b$, while in the second, $x \in a \cup b$ but not in a, so $x \in a \cup b \setminus a$.

O. Hijab, *Introduction to Calculus and Classical Analysis*, Undergraduate Texts in Mathematics, DOI 10.1007/978-3-319-28400-2

1.1.6 If $x \in A \in \mathcal{F}$, then $x \in \mathcal{F}$ by definition, so $A \subset \mathcal{F}$. If $x \in \bigcap \mathcal{F}$, then $x \in A \in \mathcal{F}$; hence $\bigcap \mathcal{F} \subset A$.

1.1.7 The elements of (a, b) are $\{a\}$ and $\{a, b\}$, so their union is $\bigcup(a, b) = \{a, b\}$, and their intersection is $\bigcap(a, b) = \{a\}$. Hence $\bigcup\bigcup(a, b) = a \cup b$, $\bigcap\bigcup(a, b) = a \cap b$, $\bigcup\bigcap(a, b) = a$, and $\bigcap\bigcap(a, b) = a$. By Exercise **1.1.5**, b is computable from a, $a \cup b$, and $a \cap b$.

1.1.8 If $a \in S(x) = x \cup \{x\}$, then either $a \in x$ or $a = x$. In the first case, $a \subset x$, since x is hierarchical. In the second, $a = x \subset x \subset S(x)$. Thus $S(x)$ is hierarchical.

1.1.9 If $a = c$ and $b = d$, then

$$(a, b) = \{\{a\}, \{a, b\}\} = \{\{c\}, \{c, d\}\} = (c, d).$$

Conversely, suppose $(a, b) = (c, d)$. By Exercise **1.1.8**,

$$a = \bigcup\bigcap(a, b) = \bigcup\bigcap(c, d) = c$$

and

$$a \cup b = \bigcup\bigcup(a, b) = \bigcup\bigcup(c, d) = c \cup d$$

and

$$a \cap b = \bigcap\bigcup(a, b) = \bigcap\bigcup(c, d) = c \cap d.$$

By Exercise **1.1.7**,

$$b = ((a \cup b) \smallsetminus a) \cup (a \cap b) = ((c \cup d) \smallsetminus c) \cup (c \cap d) = d.$$

Thus $a = c$ and $b = d$.

Solutions to Exercises 1.2

1.2.1 $a0 = a0 + (a - a) = (a0 + a) - a = (a0 + a1) - a = a(0 + 1) - a = a1 - a = a - a = 0$. This is not the only way.

1.2.2 The number 1 satisfies $1a = a1 = a$ for all a. If $1'$ also satisfied $1'b = b1' = b$ for all b, then choosing $a = 1'$ and $b = 1$ yields $1 = 11' = 1'$. Hence $1 = 1'$. Now, suppose that a has two negatives, one called $-a$ and one called b. Then $a+b = 0$, so $-a = -a+0 = -a+(a+b) = (-a+a)+b = 0+b = b$. Hence $b = -a$, so a has a unique negative. If $a \neq 0$ has two reciprocals, one called $1/a$ and one called b, $ab = 1$, so $1/a = (1/a)1 = (1/a)(ab) = [(1/a)a]b = 1b = b$.

1.2.3 Since $a + (-a) = 0$, a is the (unique) negative of $-a$ which means $-(-a) = a$. Also $a + (-1)a = 1a + (-1)a = [1 + (-1)]a = 0a = 0$, so $(-1)a$ is the negative of a or $-a = (-1)a$.

1.2.4 By the ordering property, $a > 0$, and $b > 0$ implies $ab > 0$. If $a < 0$ and $b > 0$, then $-a > 0$. Hence $(-a)b > 0$, so $ab = [-(-a)]b = (-1)(-a)b = -((-a)b) < 0$. Thus negative times positive is negative. If $a < 0$ and $b < 0$, then $-b > 0$. Hence $a(-b) < 0$. Hence $ab = a(-(-b)) = a(-1)(-b) = (-1)a(-b) = -(a(-b)) > 0$. Thus negative times negative is positive. Also $1 = 11 > 0$ whether $1 > 0$ or $1 < 0$ ($1 \neq 0$ is part of the ordering property).

1.2.5 $a < b$ implies $b-a > 0$ which implies $(b+c)-(a+c) > 0$ or $b+c > a+c$. Also $c > 0$ implies $c(b-a) > 0$ or $bc - ac > 0$ or $bc > ac$. If $a < b$ and $b < c$, then $b-a$ and $c-b$ are positive. Hence $c-a = (c-b)+(b-a)$ is positive or $c > a$. Multiplying $a < b$ by $a > 0$ and by $b > 0$ yields $aa < ab$ and $ab < bb$. Hence $aa < bb$.

1.2.6 If $0 \le a \le b$, we know, from above, that $aa \le bb$. Conversely, if $aa \le bb$, then we cannot have $a > b$ because applying **1.2.5** with the roles of a, b reversed yields $aa > bb$, a contradiction. Hence $a \le b$ iff $aa \le bb$.

1.2.7

A. By definition, $\inf A \le x$ for all $x \in A$. Hence $-\inf A \ge -x$ for all $x \in A$. Hence $-\inf A \ge y$ for all $y \in -A$. Hence $-\inf A \ge \sup(-A)$. Conversely, $\sup(-A) \ge y$ for all $y \in -A$, or $\sup(-A) \ge -x$ for all $x \in A$, or $-\sup(-A) \le x$ for all $x \in A$. Hence $-\sup(-A) \le \inf A$ which implies $\sup(-A) \ge -\inf A$. Since we already know that $\sup(-A) \le -\inf A$, we conclude that $\sup(-A) = -\inf A$. Now, replace A by $-A$ in the last equation. Since $-(-A) = A$, we obtain $\sup A = -\inf(-A)$ or $\inf(-A) = -\sup A$.

B. Now, $\sup A \ge x$ for all $x \in A$, so $(\sup A)+a \ge x+a$ for all $x \in A$. Hence $(\sup A) + a \ge y$ for $y \in A + a$, so $(\sup A) + a \ge \sup(A + a)$. In this last inequality, replace a by $-a$ and A by $A+a$ to obtain $[\sup(A+a)] - a \ge \sup(A+a-a)$, or $\sup(A+a) \ge (\sup A)+a$. Combining the two inequalities yields $\sup(A+a) = (\sup A)+a$. Replacing A and a by $-A$ and $-a$ yields $\inf(A+a) = (\inf A)+a$.

C. Now, $\sup A \ge x$ for all $x \in A$. Since $c > 0$, $c\sup A \ge cx$ for all $x \in A$. Hence $c\sup A \ge y$ for all $y \in cA$. Hence $c\sup A \ge \sup(cA)$. Now, in this last inequality, replace c by $1/c$ and A by cA. We obtain $(1/c)\sup(cA) \ge \sup A$ or $\sup(cA) \ge c\sup A$. Combining the two inequalities yields $\sup(cA) = c\sup A$. Replacing A by $-A$ in this last equation yields $\inf(cA) = c\inf A$.

Solutions to Exercises 1.3

1.3.1 Let S be the set of naturals n for which there are no naturals between n and $n+1$. From the text, we know that $1 \in S$. Assume $n \in S$. Then we claim $n+1 \in S$. Indeed, suppose that $m \in \mathbf{N}$ satisfies $n+1 < m < n+2$.

Then $m \neq 1$, so $m-1 \in \mathbf{N}$ (see §1.3) satisfies $n < m-1 < n+1$, contradicting $n \in S$. Hence $n+1 \in S$, so S is inductive. Since $S \subset \mathbf{N}$, we conclude that $S = \mathbf{N}$.

1.3.2 Fix $n \in \mathbf{N}$, and let $S = \{x \in \mathbf{R} : nx \in \mathbf{N}\}$. Then $1 \in S$ since $n1 = n$. If $x \in S$, then $nx \in \mathbf{N}$, so $n(x+1) = nx + n \in \mathbf{N}$ (since $\mathbf{N} + \mathbf{N} \subset \mathbf{N}$), so $x + 1 \in S$. Hence S is inductive. We conclude that $S \supset \mathbf{N}$ or $nm \in \mathbf{N}$ for all $m \in \mathbf{N}$.

1.3.3 Let S be the set of all naturals n such that the following holds: If $m > n$ and $m \in \mathbf{N}$, then $m - n \in \mathbf{N}$. From §1.3, we know that $1 \in S$. Assume $n \in S$. We claim $n+1 \in S$. Indeed, suppose that $m > n+1$ and $m \in \mathbf{N}$. Then $m-1 > n$. Since $n \in S$, we conclude that $(m-1)-n \in \mathbf{N}$ or $m-(n+1) \in \mathbf{N}$. Hence by definition of S, $n+1 \in S$. Thus S is inductive, so $S = \mathbf{N}$. Thus $m > n$ implies $m - n \in \mathbf{N}$. Since, for $m, n \in \mathbf{N}$, $m > n$, $m = n$, or $m < n$, we conclude that $m - n \in \mathbf{Z}$, whenever $m, n \in \mathbf{N}$. If $n \in -\mathbf{N}$ and $m \in \mathbf{N}$, then $-n \in \mathbf{N}$ and $m - n = m + (-n) \in \mathbf{N} \subset \mathbf{Z}$. If $m \in -\mathbf{N}$ and $n \in \mathbf{N}$, then $-m \in \mathbf{N}$ and $m - n = -(n + (-m)) \in -\mathbf{N} \subset \mathbf{Z}$. If n and m are both in $-\mathbf{N}$, then $-m$ and $-n$ are in $\mathbf{N}$. Hence $m - n = -((-m) - (-n)) \in \mathbf{Z}$. If either m or n equals zero, then $m - n \in \mathbf{Z}$. This shows that $\mathbf{Z}$ is closed under subtraction.

1.3.4 If n is even and odd, then $n+1$ is even. Hence $1 = (n+1) - n$ is even, say $1 = 2k$ with $k \in \mathbf{N}$. But $k \geq 1$ implies $1 = 2k \geq 2$, which contradicts $1 < 2$. If $n = 2k$ is even and $m \in \mathbf{N}$, then $nm = 2(km)$ is even. If $n = 2k - 1$ and $m = 2j - 1$ are odd, then $nm = 2(2kj - k - j + 1) - 1$ is odd.

1.3.5 For all $n \geq 1$, we establish the claim: *If* $m \in \mathbf{N}$ *and there is a bijection between* $\{1, \dots, n\}$ *and* $\{1, \dots, m\}$, *then* $n = m$. For $n = 1$, the claim is clearly true. Now, assume the claim is true for a particular n, and suppose that we have a bijection f between $\{1, \dots, n+1\}$ and $\{1, \dots, m\}$ for some $m \in \mathbf{N}$. Then by restricting f to $\{1, \dots, n\}$, we obtain a bijection g between $\{1, \dots, n\}$ and $\{1, \dots, k-1, k+1, \dots, m\}$, where $k = f(n+1)$. Now, define $h(i) = i$ if $1 \leq i \leq k-1$ and $h(i) = i-1$ if $k+1 \leq i \leq m$. Then h is a bijection between $\{1, \dots, k-1, k+1, \dots, m\}$ and $\{1, \dots, m-1\}$. Hence $h \circ g$ is a bijection between $\{1, \dots, n\}$ and $\{1, \dots, m-1\}$. By the inductive hypothesis, this forces $m - 1 = n$ or $m = n+1$. Hence the claim is true for $n+1$. Hence the claim is true, by induction, for all $n \geq 1$. Now, suppose that A is a set with n elements and with m elements. Then there are bijections $f : A \to \{1, \dots, n\}$ and $g : A \to \{1, \dots, m\}$. Hence $g \circ f^{-1}$ is a bijection from $\{1, \dots, n\}$ to $\{1, \dots, m\}$. Hence $m = n$. This shows that the number of elements of a set is well defined. For the last part, suppose that A and B have n and m elements, respectively, and are disjoint. Let $f : A \to \{1, \dots, n\}$ and $g : B \to \{1, \dots, m\}$ be bijections, and let $h(i) = i + n$, $1 \leq i \leq m$. Then $h \circ g : B \to \{n+1, \dots, n+m\}$ is a bijection. Now, define $k : A \cup B \to \{1, \dots, n+m\}$ by setting $k(x) = f(x)$ if $x \in A$ and $k(x) = h \circ g(x)$ if $x \in B$. Then k is a bijection, establishing the number of elements of $A \cup B$ is $n + m$.

1.3.6 Let $A \subset \mathbf{R}$ be finite. By induction on the number of elements, we show that $\max A$ exists. If $A = \{a\}$, then $a = \max A$. So $\max A$ exists. Now, assume that every subset with n elements has a max. If A is a set with $n+1$ elements and $a \in A$, then $B = A \setminus \{a\}$ has n elements. Hence $\max B$ exists. There are two cases: If $a \leq \max B$, then $\max B = \max A$. Hence $\max A$ exists. If $a > \max B$, then $a = \max A$. Hence $\max A$ exists. Since in either case $\max A$ exists when $\#A = n + 1$, by induction, $\max A$ exists for all finite subsets A. Since $-A$ is finite whenever A is finite, $\min A$ exists by the reflection property.

1.3.7 Let $c = \sup S$. Since $c - 1$ is not an upper bound, choose $n \in S$ with $c - 1 < n \leq c$. If $c \in S$, then $c = \max S$ and we are done. Otherwise, $c \notin S$, and $c - 1 < n < c$. Now, choose $m \in S$ with $c - 1 < n < m < c$ concluding that $m - n = (m - c) - (n - c)$ lies between 0 and 1, a contradiction. Thus $c = \max S$.

1.3.8 If we write $x = n/d$ with $n \in \mathbf{N}$ and $d \in \mathbf{N}$, then $dx = n \in N$, so S is nonempty. Conversely, if $dx = n \in \mathbf{N}$, then $x = n/d \in \mathbf{Q}$. Since $S \subset \mathbf{N}$, $d = \min S$ exists. Also if $q \in \mathbf{N}$ divides both n and d, then $x = (n/q)/(d/q)$, so $d/q \in S$. Since $d = \min S$, this implies $q = 1$; hence n and d have no common factor. Conversely, if $x = n/d$ with n and d having no common factor, then reversing the argument shows that $d = \min S$. Now $q|d$ and $q|n + kd$ imply $q|n$ and $q|d$; hence $q = 1$. Thus $x = (n + kd)/d$ with $n + kd$ and d having no common factor; hence $d = \min\{j \in \mathbf{N} : j(x + k) \in \mathbf{Z}\}$ or $d = d(x + k)$.

1.3.9 If $yq \leq x$, then $q \leq x/y$. So $\{q \in \mathbf{N} : yq \leq x\}$ is nonempty and bounded above; hence it has a max, call it q. Let $r = x - yq$. Then $r = 0$, or $r \in \mathbf{R}^+$. If $r \geq y$, then $x - y(q+1) = r - y \geq 0$, so $q+1 \in S$, contradicting the definition of q. Hence $0 \leq r < y$.

1.3.10 $1 \in A$ since $(1, a) \in f$, and $n \in A$ implies $(n, x) \in f$ for some x implies $(n + 1, g(n, x)) \in f$ implies $n + 1 \in A$. Hence A is an inductive subset of $\mathbf{N}$. We conclude that $A = \mathbf{N}$. To show that f is a mapping, we show B is empty. If B were nonempty, by Theorem 1.3.2, B would have a smallest element, call it b. We show this leads to a contradiction by using b to construct a smaller inductive set $\tilde{f}$. Either $b = 1$ or $b > 1$. If $b = 1$, let $\tilde{f} = f \setminus \{(b, z)\}$, where $z \neq a$. If $b > 1$, let $\tilde{f} = f \setminus \{(b, z)\}$, where $z \neq g(b - 1, c)$ and $(b - 1, c) \in f$. We show $\tilde{f}$ is inductive. Let (n, x) be any pair in $\tilde{f} \subset f$; then, since f is inductive, we know $(n+1, g(n, x)) \in f$. We must establish $(n+1, g(n, x))$ lies in the smaller set $\tilde{f}$. If $b = 1$, since $n + 1$ never equals 1, $(n + 1, g(n, x)) \in \tilde{f}$. If $b > 1$ and $n + 1 \neq b$, then $(n + 1, g(n, x)) \in \tilde{f}$, since f and $\tilde{f}$ differ only by $\{(b, z)\}$. If $b > 1$ and $n + 1 = b$, then $(b - 1, c) \in f$ and $(b - 1, x) \in f$; since $b - 1 \notin B$, it follows that $c = x$. Since $(b, z) \neq (n + 1, g(n, x))$, we conclude $(n + 1, g(n, x)) \in \tilde{f}$. This shows $\tilde{f}$ is inductive. Since $\tilde{f}$ is smaller than f, this is a contradiction; hence B is empty and f is a mapping satisfying $f(1) = a$, and $f(n + 1) = g(n, f(n))$, $n \geq 1$. If $h : \mathbf{N} \to \mathbf{R}$ also satisfied $h(1) = a$ and $h(n + 1) = g(n, h(n))$, $n \geq 1$, then $C = \{n \in \mathbf{N} : f(n) = h(n)\}$ is inductive, implying $f = h$.

1.3.11 By construction, we know that $a^{n+1} = a^n a$ for all $n \geq 1$. Let S be the set of $m \in \mathbf{N}$, such that $a^{n+m} = a^n a^m$ for all $n \in \mathbf{N}$. Then $1 \in S$. If $m \in S$, then $a^{n+m} = a^n a^m$, so $a^{n+(m+1)} = a^{(n+m)+1} = a^{n+m}a = a^n a^m a = a^n a^{m+1}$. Hence $m+1 \in S$. Thus S is inductive. Hence $S = \mathbf{N}$. This shows that $a^{n+m} = a^n a^m$ for all $n, m \in \mathbf{N}$. If $n = 0$, then $a^{n+m} = a^n a^m$ is clear, whereas $n < 0$ implies $a^{n+m} = a^{n+m}a^{-n}a^n = a^{n+m-n}a^n = a^m a^n$. This shows that $a^{n+m} = a^n a^m$ for $n \in \mathbf{Z}$ and $m \in \mathbf{N}$. Repeating this last argument with m, instead of n, we obtain $a^{n+m} = a^n a^m$ for all $n, m \in \mathbf{Z}$. We also establish the second part by induction on m with $n \in \mathbf{Z}$ fixed: If $m = 1$, the equation $(a^n)^m = a^{nm}$ is clear. So assume it is true for m. Then $(a^n)^{m+1} = (a^n)^m (a^n)^1 = a^{nm}a^n = a^{nm+n} = a^{n(m+1)}$. Hence it is true for $m+1$, hence by induction, for all $m \geq 1$. For $m = 0$, it is clearly true, whereas for $m < 0$, $(a^n)^m = 1/(a^n)^{-m} = 1/a^{-nm} = a^{nm}$.

1.3.12 By $1+2+\cdots+n$, we mean the function $f : \mathbf{N} \to \mathbf{R}$ satisfying $f(1) = 1$ and $f(n+1) = f(n) + (n+1)$ (Exercise **1.3.10**). Let $h(n) = n(n+1)/2$, $n \geq 1$. Then $h(1) = 1$, and

$$h(n) + (n+1) = \frac{n(n+1)}{2} + (n+1) = \frac{(n+1)(n+2)}{2} = h(n+1).$$

Thus by uniqueness, $f(n) = h(n)$ for all $n \geq 1$.

1.3.13 Since $p \geq 2$, $1 < 2^n \leq p^n$ and $n < 2^n \leq p^n$ for all $n \geq 1$. If $p^k m = p^j q$ with $k < j$, then $m = p^{j-k}q = pp^{j-k-1}q$ is divisible by p. On the other hand, if $k > j$, then q is divisible by p. Hence $p^k m = p^j q$ with m, q not divisible by p implies $k = j$. This establishes the uniqueness of the number of factors k. For existence, if n is not divisible by p, we take $k = 0$ and $m = n$. If n is divisible by p, then $n_1 = n/p$ is a natural $< p^{n-1}$. If n_1 is not divisible by p, we take $k = 1$ and $m = n_1$. If n_1 is divisible by p, then $n_2 = n_1/p$ is a natural $< p^{n-2}$. If n_2 is not divisible by p, we take $k = 2$ and $m = n_2$. If n_2 is divisible by p, we continue this procedure by dividing n_2 by p. Continuing in this manner, we obtain $n_1, n_2, \dots$ naturals with $n_j < p^{n-j}$. Since this procedure ends in n steps at most, there is some k natural or 0 for which $m = n/p^k$ is not divisible by p and n/p^{k-1} is divisible by p.

1.3.14 We want to show that $S \supset \mathbf{N}$. If this is not so, then $\mathbf{N} \setminus S$ would be nonempty and, hence, would have a least element n. Thus $k \in S$ for all naturals $k < n$. By the given property of S, we conclude that $n \in S$, contradicting $n \notin S$. Hence $\mathbf{N} \setminus S$ is empty or $S \supset \mathbf{N}$.

1.3.15 Since $m \geq p$, $m = pq + r$ with $q \in \mathbf{N}$, $r \in \mathbf{N} \cup \{0\}$, and $0 \leq r < p$ (Exercise **1.3.9**). If $r \neq 0$, multiplying by a yields $ra = ma - q(pa) \in \mathbf{N}$ since $ma \in \mathbf{N}$ and $pa \in \mathbf{N}$. Hence $r \in S_a$ is less than p contradicting $p = \min S_a$. Thus $r = 0$, or p divides m.

1.3.16 With $a = n/p$, $p \in S_a$ since $pa = n \in \mathbf{N}$, and $m \in S_a$ since $ma = nm/p \in \mathbf{N}$. By Exercise **1.3.15**, $\min S_a$ divides p. Since p is prime, $\min S_a = 1$, or $\min S_a = p$. In the first case, $1 \cdot a = a = n/p \in \mathbf{N}$, i.e., p divides n, whereas in the second case, p divides m.

1.3.17 We use induction according to Exercise **1.3.2**. For $n = 1$, the statement is true. Suppose that the statement is true for all naturals less than n. Then either n is a prime or n is composite, $n = jk$ with $j > 1$ and $k > 1$. By the inductive hypothesis, j and k are products of primes, hence so is n. Hence in either case, n is a product of primes. To show that the decomposition for n is unique except for the order, suppose that $n = p_1 \dots p_r = q_1 \dots q_s$. By the previous exercise, since p_1 divides the left side, p_1 divides the right side, hence p_1 divides one of the q_j's. Since the q_j's are prime, we conclude that p_1 equals q_j for some j. Hence $n' = n/p_1 < n$ can be expressed as the product $p_2 \dots p_r$ and the product $q_1 \dots q_{j-1} q_{j+1} \dots q_s$. By the inductive hypothesis, these p's and q's must be identical except for the order. Hence the result is true for n. By induction, the result is true for all naturals.

1.3.18 If the algorithm ends, then $r_n = 0$ for some n. Solving backward, we see that $r_{n-1} \in \mathbf{Q}$, $r_{n-2} \in \mathbf{Q}$, etc. Hence $x = r_0 \in \mathbf{Q}$. Conversely, if $x \in \mathbf{Q}$, then all the remainders r_n are rationals. Now (see Exercise **1.3.8**) write $r = N(r)/D(r)$. Then $D(r) = D(r+n)$ for $n \in \mathbf{Z}$. Moreover, since $0 \le r_n < 1$, $N(r_n) < D(r_n)$. Then

$$N(r_n) = D\left(\frac{1}{r_n}\right) = D(a_{n+1} + r_{n+1}) = D(r_{n+1}) > N(r_{n+1}).$$

As long as $r_n > 0$, the sequence $(r_0, r_1, r_2, \dots)$ continues. But this cannot continue forever, since $N(r_0) > N(r_1) > N(r_2) > \dots$ is strictly decreasing. Hence $r_n = 0$ for some n.

1.3.19 Let B be the set of $n \in \mathbf{N}$ satisfying A_n is bounded above. Since $A_1 = f^{-1}(\{1\})$ is a single natural, we have $1 \in B$. Assume $n \in B$; then A_n is bounded above, say by M. Since $A_{n+1} = A_n \cup f^{-1}(\{n+1\})$, A_{n+1} is then bounded above by the greater of M and $f^{-1}(\{n+1\})$. Hence $n+1 \in B$. Thus B is inductive; hence $B = \mathbf{N}$.

1.3.20 Since $2^{1-1} \le 1!$, the statement is true for $n = 1$. Assume it is true for n. Then $2^{(n+1)-1} = 22^{n-1} \le 2n! \le (n+1)n! = (n+1)!$.

1.3.21 Since $n!$ is the product of 1, 2, ..., n, and n^n is the product of n, n, ..., n, it is obvious that $n! \le n^n$. To prove it, use induction: $1! = 1^1$, and assuming $n! \le n^n$, we have $(n+1)! = n!(n+1) \le n^n(n+1) \le (n+1)^n(n+1) = (n+1)^{n+1}$. This establishes $n! \le n^n$ by induction. For the second part, write $n! = 1 \cdot 2 \cdot \dots \cdot n = n \cdot (n-1) \cdot \dots \cdot 1$, so $(n!)^2 = n \cdot 1 \cdot (n-1) \cdot 2 \cdot \dots 1 \cdot n$ or $(n!)^2$ equals the product of $k(n+1-k)$ over all $1 \le k \le n$. Since

$$k(n+1-k) - n = -k^2 + (n+1)k - n = -(k-n)(k-1) \ge 0,$$

the minimum of $k(n+1-k)$ over all $1 \le k \le n$ equals n. Thus $(n!)^2 \ge n \cdot n \cdot \dots \cdot n = n^n$. More explicitly, what we are doing is using

$$\left(\prod_{k=1}^{n} a(k)\right)\left(\prod_{k=1}^{n} b(k)\right) = \prod_{k=1}^{n} a(k)b(k), \qquad n \ge 1,$$

which is established by induction for $a : \mathbf{N} \to \mathbf{R}$ and $b : \mathbf{N} \to \mathbf{R}$, then inserting $a(k) = k$ and $b(k) = n+1-k$, $1 \le k \le n$.

1.3.22 The inequality $(1+a)^n \le 1 + (2^n - 1)a$ is clearly true for $n = 1$. So assume it is true for n. Then $(1+a)^{n+1} = (1+a)^n(1+a) \le (1+(2^n-1)a)(1+a) = 1 + 2^n a + (2^n - 1)a^2 \le 1 + (2^{n+1} - 1)a$ since $0 \le a^2 \le a$. Hence it is true for all n. The inequality $(1+b)^n \ge 1 + nb$ is true for $n = 1$. So suppose that it is true for n. Since $1 + b \ge 0$, then $(1+b)^{n+1} = (1+b)^n(1+b) \ge (1+nb)(1+b) = 1 + (n+1)b + nb^2 \ge 1 + (n+1)b$. Hence the inequality is true for all n.

1.3.23 Start with $n = 1$. Since $\lfloor 1/x \rfloor = a_1 = c_1 = \lfloor 1/z \rfloor$ and $\lfloor 1/x \rfloor \ge \lfloor 1/y \rfloor \ge \lfloor 1/z \rfloor$, we conclude $a_1 = b_1 = c_1$. Now assume the result is true for $n-1$, for all $0 < x < y < z < 1$. Let x' be given by $x = 1/(a_1 + x')$ and similarly y', z'. Then apply the $n-1$ case to $x' = [a_2, a_3, \dots]$, $y' = [b_2, b_3, \dots]$, $z' = [c_2, c_3, \dots]$. We conclude $a_j = b_j = c_j$, $j = 2, \dots, n$.

1.3.24 If $\#X = 1$, then $X = \{x\}$ with x nonempty, so there is $a \in x = \bigcup X$. In this case we define $f(x) = a$, establishing the case $\#X = 1$. Assume the result is true when $\#X = n$. If $\#X = n+1$, we may write $X = X' \cup \{a\}$ with a nonempty and $\#X' = n$. By the inductive hypothesis, there is $f : X' \to \bigcup X'$ with $f(x) \in x$ for $x \in X'$. Since a is nonempty, there is $b \in a$. Now extend f to all of X by defining $f(a) = b \in a$. Since $\bigcup X = \bigcup X' \cup a$, it follows that $f : X \to \bigcup X$, this establishes the case $n+1$, and hence the result follows by induction.

Solutions to Exercises 1.4

1.4.1 Since $|x| = \max(x, -x)$, $x \le |x|$. Also $-a < x < a$ is equivalent to $-x < a$ and $x < a$ and hence to $|x| = \max(x, -x) < a$. By definition of intersection, $x > -a$ and $x < a$ are equivalent to $\{x : x < a\} \cap \{x : x > -a\}$. Similarly, $|x| > a$ is equivalent to $x > a$ or $-x > a$, i.e., x lies in the union of $\{x : x > a\}$ and $\{x : x < -a\}$.

1.4.2 Clearly $|0| = 0$. If $x \ne 0$, then $x > 0$ or $-x > 0$; hence $|x| = \max(x, -x) > 0$. If $x > 0$ and $y > 0$, then $|xy| = xy = |x|\,|y|$. If $x > 0$ and $y < 0$, then xy is negative, so $|xy| = -(xy) = x(-y) = |x|\,|y|$, similarly for the other two cases.

1.4.3 If $n = 1$, the inequality is true. Assume it is true for n. Then

$$\begin{aligned}|a_1 + \cdots + a_n + a_{n+1}| &\leq |a_1 + \cdots + a_n| + |a_{n+1}| \\ &\leq |a_1| + \cdots + |a_n| + |a_{n+1}|\end{aligned}$$

by the triangle inequality and the inductive hypothesis. Hence it is true for $n+1$. By induction, it is true for all naturals n.

1.4.4 Assume first that $a > 1$. Let $S = \{x : x \geq 1 \text{ and } x^2 < a\}$. Since $1 \in S$, S is nonempty. Also $x \in S$ implies $x = x1 \leq x^2 < a$, so S is bounded above. Let $s = \sup S$. We claim that $s^2 = a$. Indeed, if $s^2 < a$, note that

$$\begin{aligned}\left(s + \frac{1}{n}\right)^2 &= s^2 + \frac{2s}{n} + \frac{1}{n^2} \\ &\leq s^2 + \frac{2s}{n} + \frac{1}{n} = s^2 + \frac{2s+1}{n} < a\end{aligned}$$

if $(2s+1)/n < a - s^2$, i.e., if $n > (2s+1)/(a-s^2)$. Since $s^2 < a$, $b = (2s+1)/(a-s^2)$ is a perfectly well defined, positive real. Since $\sup \mathbf{N} = \infty$, such a natural $n > b$ can always be found. This rules out $s^2 < a$. If $s^2 > a$, then $b = (s^2 - a)/2s$ is positive. Hence there is a natural n satisfying $1/n < b$ which implies $s^2 - 2s/n > a$. Hence

$$\left(s - \frac{1}{n}\right)^2 = s^2 - \frac{2s}{n} + \frac{1}{n^2} > a,$$

so (by Exercise **1.2.6**) $s - 1/n$ is an upper bound for S. This shows that s is not the *least* upper bound, contradicting the definition of s. Thus we are forced to conclude that $s^2 = a$. Now, if $a < 1$, then $1/a > 1$ and $1/\sqrt{1/a}$ are a positive square root of a. The square root is unique by Exercise **1.2.6**.

1.4.5 By completing the square, x solves $ax^2 + bx + c = 0$ iff x solves $(x + b/2a)^2 = (b^2 - 4ac)/4a^2$. If $b^2 - 4ac < 0$, this shows that there are no solutions. If $b^2 - 4ac = 0$, this shows that $x = -b/2a$ is the only solution. If $b^2 - 4ac > 0$, take the square root of both sides to obtain $x + b/2a = \pm(\sqrt{b^2 - 4ac})/2a$.

1.4.6 If $0 \leq a < b$, then $a^n < b^n$ is true for $n = 1$. If it is true for n, then $a^{n+1} = a^n a < b^n a < b^n b = b^{n+1}$. Hence by induction, it is true for all n. Hence $0 \leq a \leq b$ implies $a^n \leq b^n$. If $a^n \leq b^n$ and $a > b$, then by applying the previous, we obtain $a^n > b^n$, a contradiction. Hence for $a, b \geq 0$, $a \leq b$ iff $a^n \leq b^n$. For the second part, assume that $a > 1$, and let $S = \{x : x \geq 1 \text{ and } x^n \leq a\}$. Since $x \geq 1$, $x^{n-1} \geq 1$. Hence $x = x1 \leq xx^{n-1} = x^n < a$. Thus $s = \sup S$ exists. We claim that $s^n = a$. If $s^n < a$, then $b = s^{n-1}(2^n - 1)/(a - s^n)$ is a well-defined, positive real. Choose a natural $k > b$. Then $s^{n-1}(2^n - 1)/k < a - s^n$. Hence by Exercise **1.3.22**,

$$\left(s+\frac{1}{k}\right)^n = s^n\left(1+\frac{1}{sk}\right)^n$$
$$\leq s^n\left(1+\frac{2^n-1}{sk}\right) = s^n + \frac{s^{n-1}(2^n-1)}{k} < a.$$

Hence $s+1/k \in S$. Hence s is not an upper bound for S, a contradiction. If $s^n > a$, $b = ns^{n-1}/(s^n - a)$ is a well defined, positive real, so choose $k > b$. By Exercise **1.3.22**,

$$\left(s-\frac{1}{k}\right)^n = s^n\left(1-\frac{1}{sk}\right)^n$$
$$\geq s^n\left(1-\frac{n}{sk}\right)$$
$$= s^n - \frac{s^{n-1}n}{k} > a.$$

Hence by the first part of this exercise, $s - 1/k$ is an upper bound for S. This shows that s is not the least upper bound, a contradiction. We conclude that $s^n = a$. Uniqueness follows from the first part of this exercise.

1.4.7 If $t = k/(n\sqrt{2})$ is a rational p/q, then $\sqrt{2} = (kq)/(np)$ is rational, a contradiction.

1.4.8 Let (b) denote the fractional part of b. If $a = 0$, the result is clear. If $a \neq 0$, the fractional parts $\{(na) : n \geq 1\}$ of $\{na : n \geq 1\}$ are in $[0,1]$. Now, divide $[0,1]$ into finitely many subintervals, each of length, at most, ϵ. Then the fractional parts of at least two terms pa and qa, $p \neq q$, $p, q \in \mathbf{N}$, must lie in the same subinterval. Hence $|(pa) - (qa)| < \epsilon$. Since $pa - (pa) \in \mathbf{Z}$, $qa - (qa) \in \mathbf{Z}$, we obtain $|(p-q)a - m| < \epsilon$ for some integer m. Choosing $n = p - q$, we obtain $|na - m| < \epsilon$.

1.4.9 The key issue here is that $|2n^2 - m^2| = n^2|f(m/n)|$ is a nonzero integer, since all the roots of $f(x) = 0$ are irrational. Hence $|2n^2 - m^2| \geq 1$. But $2n^2 - m^2 = (n\sqrt{2} - m)(n\sqrt{2} + m)$, so $|n\sqrt{2} - m| \geq 1/(n\sqrt{2} + m)$. Dividing by n, we obtain

$$\left|\sqrt{2} - \frac{m}{n}\right| \geq \frac{1}{(\sqrt{2} + m/n)n^2}. \tag{A.1.1}$$

Now, if $|\sqrt{2} - m/n| \geq 1$, the result we are trying to show is clear. So let us assume that $|\sqrt{2} - m/n| \leq 1$. In this case, $\sqrt{2} + m/n = 2\sqrt{2} + (m/n - \sqrt{2}) \leq 2\sqrt{2} + 1$. Inserting this in the denominator of the right side of (A.1.1), the result follows.

1.4.10 By Exercise **1.4.5**, the (real) roots of $f(x) = 0$ are $\pm a$. The key issue here is that $|m^4 - 2m^2n^2 - n^4| = n^4|f(m/n)|$ is a nonzero integer, since all the roots of $f(x) = 0$ are irrational. Hence $n^4|f(m/n)| \geq 1$. But, by factoring, $f(x) = (x-a)g(x)$ with $g(x) = (x+a)(x^2 + \sqrt{2} - 1)$, so

$$\left|a - \frac{m}{n}\right| \ge \frac{1}{n^4 g(m/n)}. \tag{A.1.2}$$

Now, there are two cases: If $|a - m/n| \ge 1$, we obtain $|a - m/n| \ge 1/n^4$ since $1 \ge 1/n^4$. If $|a - m/n| \le 1$, then $0 < m/n < 3$. So from the formula for g, we obtain $0 < g(m/n) \le 51$. Inserting this in the denominator of the right side of (A.1.2), the result follows with $c = 1/51$.

1.4.11 First, we verify the absolute value properties **A**,**B**, and **C** for $x, y \in \mathbf{Z}$. **A**is clear. For $x, y \in \mathbf{Z}$, let $x = 2^k p$ and $y = 2^j q$ with p, q odd. Then $xy = 2^{j+k}pq$ with pq odd, establishing **B** for $x, y \in \mathbf{Z}$. For **C**, let $i = \min(j,k)$ and note $x + y = 0$ or $x + y = 2^i r$ with r odd. In the first case, $|x + y|_2 = 0$, whereas in the second, $|x + y|_2 = 2^{-i}$. Hence $|x + y|_2 \le 2^{-i} = \max(2^{-j}, 2^{-k}) = \max(|x|_2, |y|_2) \le |x|_2 + |y|_2$. Now, using **B** for $x, y, z \in \mathbf{Z}$, $|zx|_2 = |z|_2|x|_2$ and $|zy|_2 = |z|_2|y|_2$. Hence $|zx/zy|_2 = |zx|_2/|zy|_2 = |z|_2|x|_2/|z|_2|y|_2 = |x|_2/|y|_2 = |x/y|_2$. Hence $|\cdot|_2$ is well defined on $\mathbf{Q}$. Now, using **A**,**B**, and **C** for $x, y \in \mathbf{Z}$, one checks **A**,**B**, and **C** for $x, y \in \mathbf{Q}$.

1.4.12 Clearly $x = \sqrt{2} - 1$ satisfies $x = 1/(2 + x)$; hence by Exercise **1.3.18**, it cannot be rational.

1.4.13 Assume $x \in \mathbf{Q}$. Then $S = \{k \in \mathbf{N} : kx \in \mathbf{N}\}$ is nonempty, $d = \min S$ is well defined, and $dx \in \mathbf{N}$. Let $d' = d(x - \lfloor x \rfloor)$. Then $0 \le d' < d$, $d' \in \mathbf{N} \cup \{0\}$, and $d'x = d(x - \lfloor x \rfloor)x = dn - dx\lfloor x \rfloor \in \mathbf{N}$. If $d' > 0$, then $d' \in S$ contradicting $d = \min S$. Hence $d' = 0$ or $x \in \mathbf{N}$.

Solutions to Exercises 1.5

1.5.1 If (a_n) is increasing, $\{a_n : n \ge 1\}$ and $\{a_n : n \ge N\}$ have the same sups, similarly for decreasing. If $a_n \to L$, then $a_n^* \searrow L$. Hence $a_{n+N}^* \searrow L$ and $a_{n*} \nearrow L$. Hence $a_{(n+N)*} \nearrow L$. We conclude that $a_{n+N} \to L$.

1.5.2 If $a_n \nearrow L$, then $L = \sup\{a_n : n \ge 1\}$, so $-L = \inf\{-a_n : n \ge 1\}$, so $-a_n \searrow -L$. Similarly, if $a_n \searrow L$, then $-a_n \nearrow -L$. If $a_n \to L$, then $a_n^* \searrow L$, so $(-a_n)_* = -a_n^* \nearrow -L$, and $a_{n*} \nearrow L$, so $(-a_n)^* = -a_{n*} \searrow -L$. Hence $-a_n \to -L$.

1.5.3 First, if $A \subset \mathbf{R}^+$, let $1/A = \{1/x : x \in A\}$. Then $\inf(1/A) = 0$ implies that, for all $c > 0$, there exists $x \in A$ with $1/x < 1/c$ or $x > c$. Hence $\sup A = \infty$. Conversely, $\sup A = \infty$ implies $\inf(1/A) = 0$. If $\inf(1/A) > 0$, then $c > 0$ is a lower bound for $1/A$ iff $1/c$ is an upper bound for A. Hence $\sup A < \infty$ and $\inf(1/A) = 1/\sup A$. If $1/\infty$ is interpreted as 0, we obtain $\inf(1/A) = 1/\sup A$ in all cases. Applying this to $A = \{a_k : k \ge n\}$ yields $b_{n*} = 1/a_n^*$, $n \ge 1$. Moreover, A is bounded above iff $\inf(1/A) > 0$. Applying this to $A = \{b_{n*} : n \ge 1\}$ yields $\sup\{b_{n*} : n \ge 1\} = \infty$ since $a_n^* \searrow 0$. Hence $b_n \to \infty$. For the converse, $b_n \to \infty$ implies $\sup\{b_{n*} : n \ge 1\} = \infty$. Hence $\inf\{a_n^* : n \ge 1\} = 0$. Hence $a_n \to 0$.

1.5.4 Since $k_n \ge n$, $a_{n*} \le a_{k_n} \le a_n^*$. Since $a_n \to L$ means $a_n^* \to L$ and $a_{n*} \to L$, the ordering property implies $a_{k_n} \to L$. Now assume (a_n) is increasing. Then (a_{k_n}) is increasing. Suppose that $a_n \to L$ and $a_{k_n} \to M$. Since (a_n) is increasing and $k_n \ge n$, $a_{k_n} \ge a_n$, $n \ge 1$. Hence by ordering, $M \ge L$. On the other hand, $\{a_{k_n} : n \ge 1\} \subset \{a_n : n \ge 1\}$. Since M and L are the sups of these sets, $M \le L$. Hence $M = L$.

1.5.5 From the text, we know that all but finitely many a_n lie in $(L-\epsilon, L+\epsilon)$, for any $\epsilon > 0$. Choosing $\epsilon = L$ shows that all but finitely many terms are positive.

1.5.6 The sequence $a_n = \sqrt{n+1}-\sqrt{n} = 1/(\sqrt{n+1}+\sqrt{n})$ is decreasing. Since $a_{n^2} \le 1/n$ has limit zero, so does (a_n). Hence $a_n^* = a_n$ and $a_{n*} = 0 = a_* = a^*$.

1.5.7 We do only the case a^* finite. Since $a_n^* \to a^*$, a_n^* is finite for each $n \ge N$ beyond some $N \ge 1$. Now, $a_n^* = \sup\{a_k : k \ge n\}$, so for each $n \ge 1$, we can choose $k_n \ge n$, such that $a_n^* \ge a_{k_n} > a_n^* - 1/n$. Then the sequence (a_{k_n}) lies between (a_n^*) and $(a_n^* - 1/n)$ and hence converges to a^*. But (a_{k_n}) may not be a subsequence of (a_n) because the sequence (k_n) may not be strictly increasing. To take care of this, note that, since $k_n \ge n$, we can choose a subsequence (k_{j_n}) of (k_n) which is strictly increasing. Then $(a_p : p = k_{j_n})$ is a subsequence of (a_n) converging to a^*, similarly (or multiply by minuses) for a_*.

1.5.8 If $a_n \not\to L$, then by definition, either $a^* \ne L$ or $a_* \ne L$. For definiteness, suppose that $a_* \ne L$. Then from Exercise **1.5.7**, there is a subsequence (a_{k_n}) converging to a_*. From §1.5, if $2\epsilon = |L - a_*| > 0$, all but finitely many of the terms a_{k_n} lie in the interval $(a_* - \epsilon, a_* + \epsilon)$. Since ϵ is chosen to be half the distance between a_* and L, this implies that these same terms lie outside the interval $(L - \epsilon, L + \epsilon)$. Hence these terms form a subsequence as requested.

1.5.9 From Exercise **1.5.7**, we know that x_* and x^* are limit points. If (x_{k_n}) is a subsequence converging to a limit point L, then since $k_n \ge n$, $x_{n*} \le x_{k_n} \le x_n^*$ for all $n \ge 1$. By the ordering property for sequences, taking the limit yields $x_* \le L \le x^*$.

1.5.10 If $x_n \to L$, then $x_* = x^* = L$. Since x_* and x^* are the smallest and the largest limit points, L must be the only one. Conversely, if there is only one limit point, then $x_* = x^*$ since x_* and x^* are always limit points.

1.5.11 If $M < \infty$, then for each $n \ge 1$, the number $M - 1/n$ is not an upper bound for the displayed set. Hence there is an $x_n \in (a, b)$ with $f(x_n) > M - 1/n$. Since $f(x_n) \le M$, we see that $f(x_n) \to M$, as $n \nearrow \infty$. If $M = \infty$, for each $n \ge 1$, the number n is not an upper bound for the displayed set. Hence there is an $x_n \in (a, b)$ with $f(x_n) > n$. Then $f(x_n) \to \infty = M$.

1.5.12 Note that

$$\frac{1}{2}\left(a+\frac{2}{a}\right)-\sqrt{2}=\frac{1}{2a}\left(a-\sqrt{2}\right)^2\geq 0, \qquad a>0. \tag{A.1.3}$$

Since $e_1 = 2-\sqrt{2}$, $e_1 \geq 0$. By (A.1.3), $e_{n+1}\geq 0$ as soon as $e_n \geq 0$. Hence $e_n \geq 0$ for all $n \geq 1$ by induction. Similarly, (A.1.3) with $a = d_n$ plugged in and $d_n \geq \sqrt{2}$, $n \geq 1$, yield $e_{n+1} \leq e_n^2/2\sqrt{2}$, $n \geq 1$.

1.5.13 If $f(a) = 1/(q+a)$, then $|f(a)-f(b)| \leq f(a)f(b)|a-b|$. This implies **A**. Now, **A**implies $|x-x_n| \leq x_n|x'-x_n'| \leq x_n x_n'|x''-x_n''| \leq \dots$, where $x_n^{(k)}$ denotes x_n with k layers "peeled off." Hence

$$|x-x_n| \leq x_n x_n' x_n'' \dots x_n^{(n-1)}, \qquad n \geq 1. \tag{A.1.4}$$

Since $x_n^{(n-1)} = 1/a_n$, (A.1.4) implies **B**. For **C**, note that, since $a_n \geq 1$, $x \leq 1/[1+1/(a_2+1)] = (a_2+1)/(a_2+2)$. Let $a = (c+1)/(c+2)$. Now, if one of the a_k's is bounded by c, (A.1.4) and **C** imply $|x-x_n| \leq a$, as soon as n is large enough, since all the factors in (A.1.4) are bounded by 1. Similarly, if two of the a_k's are bounded by c, $|x-x_n| \leq a^2$, as soon as n is large enough. Continuing in this manner, we obtain **D**. If $a_n \to \infty$, we are done, by **B**. If not, then there is a c, such that $a_k \leq c$ for infinitely many n. By **D**, we conclude that the upper and lower limits of $(|x-x_n|)$ lie between 0 and a^N, for all $N \geq 1$. Since $a^N \to 0$, the upper and lower limits are 0. Hence $|x-x_n| \to 0$.

Solutions to Exercises 1.6

1.6.1 First, suppose that the decimal expansion of x is as advertised. Then the fractional part of $10^{n+m}x$ is identical to the fractional part of $10^m x$. Hence $x(10^{n+m}-10^m) \in \mathbf{Z}$, or $x \in \mathbf{Q}$. Conversely, if $x = m/n \in \mathbf{Q}$, perform long division to obtain the digits: From Exercise **1.3.9**, $10m = nd_1 + r_1$, obtaining the quotient d_1 and remainder r_1. Similarly, $10r_1 = nd_2 + r_2$, obtaining the quotient d_2 and remainder r_2. Similarly, $10r_2 = nd_3 + r_3$, obtaining d_3 and r_3, and so on. Here $d_1, d_2, \dots$ are digits (since $0 < x < 1$), and $r_1, r_2, \dots$ are zero or naturals less than n. At some point the remainders must start repeating and, therefore, the digits also.

1.6.2 We assume $d_N > e_N$, and let $x = .d_1d_2\cdots = .e_1e_2\dots$ If

$$y = .d_1d_2\dots d_{N-1}(e_N+1)00\dots,$$

then $x \geq y$. If

$$z = .e_1e_2\dots e_N 99\dots,$$

then $z \geq x$. Since $.99\cdots = 1$, $z = y$. Hence $x = y$ and $x = z$. Since $x = y$, $d_N = e_N + 1$ and $d_j = 0$ for $j > N$. Since $x = z$, $e_j = 9$ for $j > N$. Clearly, this happens iff $10^N x \in \mathbf{Z}$.

1.6.3 From Exercise **1.3.22**, $(1+b)^n \geq 1 + nb$ for $b \geq -1$. In this inequality, replace n by N and b by $-1/N(n+1)$ to obtain $[1 - 1/N(n+1)]^N \geq 1 - 1/(n+1) = n/(n+1)$. By Exercise **1.4.6**, we may take Nth roots of both sides, yielding **A**. **B** follows by multiplying **A**by $(n+1)^{1/N}$ and rearranging. If $a_n = 1/n^{1/N}$, then by **B**,

$$\begin{aligned} a_n - a_{n+1} &= \frac{(n+1)^{1/N} - n^{1/N}}{n^{1/N}(n+1)^{1/N}} \\ &\geq \frac{1}{N(n+1)^{(N-1)/N} n^{1/N}(n+1)^{1/N}} \\ &\geq \frac{1}{N(n+1)^{1+1/N}}. \end{aligned}$$

Summing over $n \geq 1$ yields **C**.

1.6.4 Since $e_1 = 2 - \sqrt{2}$ and $e_{n+1} \leq e_n^2/2\sqrt{2}$, $e_2 \leq e_1^2/2\sqrt{2} = (3 - 2\sqrt{2})/\sqrt{2}$. Similarly,

$$\begin{aligned} e_3 &\leq \frac{e_2^2}{2\sqrt{2}} \\ &\leq \frac{(3-2\sqrt{2})^2}{4\sqrt{2}} \\ &= \frac{17 - 12\sqrt{2}}{4\sqrt{2}} \\ &= \frac{1}{4(17\sqrt{2}+24)} \leq \frac{1}{100}. \end{aligned}$$

Now, assume the inductive hypothesis $e_{n+2} \leq 10^{-2^n}$. Then

$$e_{(n+1)+2} = e_{n+3} \leq \frac{e_{n+2}^2}{2\sqrt{2}} \leq e_{n+2}^2 = \left(10^{-2^n}\right)^2 = 10^{-2^{n+1}}.$$

Thus the inequality is true by induction.

1.6.5 Since $[0,2] = 2[0,1]$, given $z \in [0,1]$, we have to find $x \in C$ and $y \in C$ satisfying $x + y = 2z$. Let $z = .d_1 d_2 d_3 \dots$. Then for all $n \geq 1$, $2d_n$ is an even integer satisfying $0 \leq 2d_n \leq 18$. Thus there are digits, zero or odd (i.e., $0, 1, 3, 5, 7, 9$), d_n', d_n'', $n \geq 1$, satisfying $d_n' + d_n'' = 2d_n$. Now, set $x = .d_1'd_2'd_3' \dots$ and $y = .d_1''d_2''d_3'' \dots$.

Solutions to Exercises 1.7

1.7.1 Since B is countable, there is a bijection $f : B \to \mathbf{N}$. Then f restricted to A is a bijection between A and $f(A) \subset \mathbf{N}$. Thus it is enough to show that $C = f(A)$ is countable or finite. If C is finite, we are done, so assume C is infinite. Since $C \subset \mathbf{N}$, let $c_1 = \min C$, $c_2 = \min C \setminus \{c_1\}$, $c_3 = \min C \setminus \{c_1, c_2\}$, etc. Then $c_1 < c_2 < c_3 < \dots$. Since C is infinite, $C_n = C \setminus \{c_1, \dots, c_n\}$ is not empty, allowing us to set $c_{n+1} = \min C_n$, for all $n \geq 1$. Since (c_n) is strictly increasing, we must have $c_n \geq n$ for $n \geq 1$. If $m \in C \setminus \{c_n : n \geq 1\}$, then by construction, $m \geq c_n$ for all $n \geq 1$, which is impossible. Thus $C = \{c_n : n \geq 1\}$ and $g : \mathbf{N} \to C$ given by $g(n) = c_n$, $n \geq 1$, is a bijection.

1.7.2 With each rational r, associate the pair $f(r) = (m, n)$, where $r = m/n$ in lowest terms. Then $f : \mathbf{Q} \to \mathbf{N}^2$ is an injection. Since $\mathbf{N}^2$ is countable and an infinite subset of a countable set is countable, so is $\mathbf{Q}$.

1.7.3 If $f : \mathbf{N} \to A$ and $g : \mathbf{N} \to B$ are bijections, then $(f, g) : \mathbf{N} \times \mathbf{N} \to A \times B$ is a bijection.

1.7.4 Suppose that $[0, 1]$ is countable, and list the elements as

$$\begin{aligned} a_1 &= .d_{11}d_{12}\dots \\ a_2 &= .d_{21}d_{22}\dots \\ a_3 &= .d_{31}d_{32}\dots \\ &\dots . \end{aligned}$$

Let $a = .d_1 d_2 \dots$, where d_n is any digit chosen, such that $d_n \neq d_{nn}$, $n \geq 1$ (the "diagonal" in the above listing). Then a is not in the list, so the list is not complete, a contradiction. Hence $[0, 1]$ is not countable.

1.7.5 Note that $i + j = n + 1$ implies $i^3 + j^3 \geq (n + 1)^3/8$ since at least one of i or j is $\geq (n + 1)/2$. Sum (1.7.6) in the order of $\mathbf{N}^2$ given in §1.7:

$$\begin{aligned} \sum_{(m,n)\in\mathbf{N}^2} \frac{1}{m^3 + n^3} &= \sum_{n=1}^{\infty} \left(\sum_{i+j=n+1} \frac{1}{i^3 + j^3} \right) \\ &\leq \sum_{n=1}^{\infty} \left[\sum_{i+j=n+1} \frac{8}{(n + 1)^3} \right] \\ &= \sum_{n=1}^{\infty} \frac{8n}{(n + 1)^3} \leq 8 \sum_{n=1}^{\infty} \frac{1}{n^2} < \infty. \end{aligned}$$

1.7.6 Since $n^{-s} < 1$, the geometric series implies

$$\frac{1}{n^s - 1} = \frac{n^{-s}}{1 - n^{-s}} = \sum_{m=1}^{\infty} (n^{-s})^m.$$

Summing over $n \geq 2$,

$$\begin{aligned}\sum_{n=2}^{\infty} \frac{1}{n^s - 1} &= \sum_{n=2}^{\infty} \left(\sum_{m=1}^{\infty} n^{-sm} \right) \\ &= \sum_{m=1}^{\infty} \left(\sum_{n=2}^{\infty} n^{-ms} \right) \\ &= \sum_{m=1}^{\infty} Z(ms).\end{aligned}$$

1.7.7 Let $\tilde{a}_n = (-1)^{n+1} a_n$ and $\tilde{b}_n = (-1)^{n+1} b_n$. If $\sum \tilde{c}_n$ is the Cauchy product of $\sum \tilde{a}_n$ and $\sum \tilde{b}_n$, then

$$\tilde{c}_n = \sum_{i+j=n+1} \tilde{a}_i \tilde{b}_j = \sum_{i+j=n+1} (-1)^{i+1} (-1)^{j+1} a_i b_j = (-1)^{n+1} c_n,$$

where $\sum c_n$ is the Cauchy product of $\sum a_n$ and $\sum b_n$.

1.7.8 As in Exercise **1.5.13**, $|x_n - x_m| \leq x_n x_n' x_n'' \dots x_n^{(n-1)}$, for $m \geq n \geq 1$. Since $x_n^{(n-1)} = 1/a_n$, this yields $|x_n - x_m| \leq 1/a_n$, for $m \geq n \geq 1$. Hence if $a_n \to \infty$, $(1/a_n)$ is an error sequence for (x_n). Now, suppose that $a_n \not\to \infty$. Then there is a c with $a_n \leq c$ for infinitely many n. Hence if N_n is the number of a_k's, $k \leq n + 2$, bounded by c, $\lim_{n \nearrow \infty} N_n = \infty$. Since $x_n^{(k)} \leq (a_{k+2} + 1)/(a_{k+2} + 2)$, the first inequality above implies that

$$|x_n - x_m| \leq \left(\frac{c+1}{c+2} \right)^{N_n}, \qquad m \geq n \geq 1.$$

Set $a = (c+1)/(c+2)$. Then in this case, (a^{N_n}) is an error sequence for (x_n). For the golden mean x, note that $x = 1 + 1/x$; solving the quadratic, we obtain $x = (1 \pm \sqrt{5})/2$. Since $x > 0$, we must take the $+$.

A.2 Solutions to Chapter 2

Solutions to Exercises 2.1

2.1.1 By the theorem, select a subsequence (n_k) such that (a_{n_k}) converges to some a. Now apply the theorem to (b_{n_k}), selecting a sub-subsequence (n_{k_m}) such that $(b_{n_{k_m}})$ converges to some b. Then clearly $(a_{n_{k_m}})$ and $(b_{n_{k_m}})$ converge to a and b, respectively; hence (a_n, b_n) subconverges to (a, b).

2.1.2 For simplicity, we assume $a = 0$, $b = 1$. Let the limiting point be $L = .d_1d_2\dots$. By the construction of L, it is a limit point. Since x_* is the smallest limit point (Exercise **1.5.9**), $x_* \le L$. Now, note that if $t \in [0,1]$ satisfies $t \le x_n$ for all $n \ge 1$, then $t \le$ any limit point of (x_n). Hence $t \le x_*$. Since changing finitely many terms of a sequence does not change x_*, we conclude that $x_* \ge t$ for all t satisfying $t \le x_n$ for all but finitely n. Now, by construction, there are at most finitely many terms $x_n \le .d_1$. Hence $.d_1 \le x_*$. Similarly, there are finitely many terms $x_n \le .d_1d_2$. Hence $.d_1d_2 \le x_*$. Continuing in this manner, we conclude that $.d_1d_2\dots d_N \le x_*$. Letting $N \nearrow \infty$, we obtain $L \le x_*$. Hence $x_* = L$.

2.1.3 If $c \in \bigcup\mathcal{U}$, then there is $U \in \mathcal{U}$ with $c \in U$. Select an open interval I containing c that is contained in U. Then (§1.1) $I \subset U \subset \bigcup\mathcal{U}$, so $\bigcup\mathcal{U}$ is open.

Solutions to Exercises 2.2

2.2.1 If $\lim_{x\to c} f(x) \ne 0$, there is at least one sequence $x_n \to c$ with $x_n \ne c$, $n \ge 1$, and $f(x_n) \not\to 0$. From Exercise **1.5.8**, this means there is a subsequence (x_{k_n}) and an $N \ge 1$, such that $|f(x_{k_n})| \ge 1/N$, $n \ge 1$. But this means that, for all $n \ge 1$, the reals x_{k_n} are rationals with denominators bounded in absolute value by N. Hence $N!x_{k_n}$ are integers converging to $N!c$. But this cannot happen unless $N!x_{k_n} = N!c$ from some point on, i.e., $x_{k_n} = c$ from some point on, contradicting $x_n \ne c$ for all $n \ge 1$. Hence the result follows.

2.2.2 Let $L = \inf\{f(x) : a < x < b\}$. We have to show that $f(x_n) \to L$ whenever $x_n \to a+$. So suppose that $x_n \to a+$, and assume, first, $x_n \searrow a$, i.e., (x_n) is decreasing. Then $(f(x_n))$ is decreasing. Hence $f(x_n)$ decreases to some limit M. Since $f(x_n) \ge L$ for all $n \ge 1$, $M \ge L$. If $d > 0$, then there is an $x \in (a, b)$ with $f(x) < L + d$. Since $x_n \searrow a$, there is an $n \ge 1$ with $x_n < x$. Hence $M \le f(x_n) \le f(x) < L + d$. Since $d > 0$ is arbitrary, we conclude that $M = L$ or $f(x_n) \searrow L$. In general, if $x_n \to a+$, $x_n^* \searrow a$. Hence $f(x_n^*) \searrow L$. But $x_n \le x_n^*$; hence $L \le f(x_n) \le f(x_n^*)$. So by the ordering property, $f(x_n) \to L$. This establishes the result for the inf. For the sup, repeat the reasoning, or apply the first case to $g(x) = -f(-x)$.

2.2.3 Assume f is increasing. If $a < c < b$, then apply the previous exercise to f on (c, b), concluding that $f(c+)$ exists. Since $f(x) \geq f(c)$ for $x > c$, we obtain $f(c+) \geq f(c)$. Apply the previous exercise to f on (a, c) to conclude that $f(c-)$ exists and $f(c-) \leq f(c)$. Hence $f(c)$ is between $f(c-)$ and $f(c+)$. Now, if $a < A < B < b$ and there are N points in $[A, B]$ where f jumps by at least $\delta > 0$, then we must have $f(B) - f(A) \geq N\delta$. Hence given δ and A, B, there are, at most, finitely many such points. Choosing $A = a + 1/n$ and $B = b - 1/n$ and taking the union of all these points over all the cases $n \geq 1$, we see that there are, at most, countably many points in (a, b) at which the jump is at least δ. Now, choosing δ equal 1, 1/2, 1/3, ..., and taking the union of all the cases, we conclude that there are, at most, countably many points $c \in (a, b)$ at which $f(c+) > f(c-)$. The decreasing case is obtained by multiplying by minus.

2.2.4 This follows from

$$|f(x)| \leq |f(a)| + |f(x) - f(a)| \leq |f(a)| + v_f(a, b) \qquad a \leq x \leq b.$$

2.2.5 If f is increasing, there are no absolute values in the variation (2.2.1). But $f(d) - f(c) \leq 0$ for $d \geq c$, so the variation is $\leq F(b) - F(a)$. On the other hand, taking $I_1 = (a, b)$ shows $F(b) - F(a) \leq v_F(a, b)$. Hence $v_F(a, b) = F(b) - F(a) < \infty$. Let f, g be defined on $[a, b]$. Then $v_{-f}(a, b) = v_f(a, b)$, and by the triangle inequality, $v_{f+g}(a, b) \leq v_f(a, b) + v_g(a, b)$. Thus f, g bounded variation implies $f + g$ bounded variation.

2.2.6 Let $a \leq x < y \leq b$. If I_k, $1 \leq k \leq N$, are disjoint open intervals in (a, x), then $I_1, \dots, I_N$, (x, y) are disjoint open intervals in (a, y); hence

$$\sum_{k=1}^{N} |f(d_k) - f(c_k)| + |f(y) - f(x)| \leq v_f(a, y).$$

Taking the sup over all disjoint open intervals in (a, x) yields $v(x) + |f(y) - f(x)| \leq v(y)$. Hence v is increasing and, throwing away the absolute value, $v(x) - f(x) \leq v(y) - f(y)$, so $v - f$ is increasing. Thus $f = v - (v - f)$ is the difference of two increasing functions. Interchanging x and y yields $|v(x) - v(y)| \leq |f(x) - f(y)|$, so the continuity of v follows from that of f.

2.2.7 Look at the partition $x_i = 1/(n-i)$, $1 \leq i \leq n-1$, $x_0 = 0$, $x_n = 2$, and select x_i' irrational in (x_{i-1}, x_i), $i = 1, \dots, n$. Then $|f(x_i) - f(x_i')| = 1/(n-i)$; hence the variation corresponding to this partition is not less than the nth partial sum of the harmonic series, which diverges.

2.2.8 If $x_n \to c$ and $(f(x_n))$ subconverges to L, by replacing (x_n) by a subsequence, we may assume $(f(x_n))$ converges to L. Given $\delta > 0$, since $x_n \to c$, $c - \delta < x_n < c + \delta$ for n sufficiently large. Thus

$$\inf_{0<|x-c|<\delta} f(x) \le f(x_n) \le \sup_{0<|x-c|<\delta} f(x), \qquad n >> 1.$$

Taking the limit $n \to \infty$,

$$\inf_{0<|x-c|<\delta} f(x) \le L \le \sup_{0<|x-c|<\delta} f(x)$$

for all $\delta > 0$. Now take the supremum over δ on the left and the infimum over δ on the right to obtain $L_* \le L \le L^*$. Thus any limit point L lies in $[L_*, L^*]$. To show L^* is a limit point, for $n \ge 1$, select $\delta_n > 0$ such that

$$L^* \le \sup_{0<|x-c|<\delta_n} f(x) < L^* + \frac{1}{n}.$$

Now select x_n in $0 < |x-c| < \delta_n$ such that $\sup_{0<|x-c|<\delta_n} f(x) \le f(x_n)+1/n$. Then

$$L^* - \frac{1}{n} \le f(x_n) \le L^* + \frac{1}{n};$$

hence $f(x_n) \to L^*$ as $n \to \infty$. Thus L^* is a limit point, similarly for L_*.

Solutions to Exercises 2.3

2.3.1 Let f be a polynomial with odd degree n and highest-order coefficient a_0. Since $x^k/x^n \to 0$, as $x \to \pm\infty$, for $n > k$, it follows that $f(x)/x^n \to a_0$. Since $x^n \to \pm\infty$, as $x \to \pm\infty$, it follows that $f(\pm\infty) = \pm\infty$, at least when $a_0 > 0$. When $a_0 < 0$, the same reasoning leads to $f(\pm\infty) = \mp\infty$. Thus there are reals a, b with $f(a) > 0$ and $f(b) < 0$. Hence by the intermediate value property, there is a c with $f(c) = 0$.

2.3.2 By definition of μ_c, $|f(x) - f(c)| \le \mu_c(\delta)$ when $|x - c| < \delta$. Since $|x - c| < 2|x - c|$ for $x \ne c$, choosing $\delta = 2|x - a|$ yields $|f(x) - f(c)| \le \mu_c(2|x - c|)$, for $x \ne c$. If $\mu_c(0+) \ne 0$, setting $\epsilon = \mu_c(0+)/2$, $\mu_c(1/n) \ge 2\epsilon$. By the definition of the sup in the definition of μ_c, this implies that, for each $n \ge 1$, there is an $x_n \in (a, b)$ with $|x_n - c| \le 1/n$ and $|f(x_n) - f(c)| \ge \epsilon$. Since the sequence (x_n) converges to c and $f(x_n) \not\to f(c)$, it follows that f is not continuous at c. This shows **A** implies **B**.

2.3.3 Let $A = f((a, b))$. To show that A is an interval, it is enough to show that $(\inf A, \sup A) \subset A$. By definition of inf and sup, there are sequences $m_n \to \inf A$ and $M_n \to \sup A$ with $m_n \in A$, $M_n \in A$. Hence there are reals c_n, d_n with $f(c_n) = m_n$, $f(d_n) = M_n$. Since $f([c_n, d_n])$ is a compact interval, it follows that $[m_n, M_n] \subset A$ for all $n \ge 1$. Hence $(\inf A, \sup A) \subset A$. For the second part, if $f((a, b))$ is not an open interval, then there is a $c \in (a, b)$ with $f(c)$ a max or a min. But this cannot happen: since f is strictly monotone, we can always find an x and y to the right and to the left of c such that $f(x)$ and $f(y)$ are larger and smaller than $f(c)$. Thus $f((a,b))$ is an open interval.

2.3.4 Let $a = \sup A$. Since $a \geq x$ for $x \in A$ and f is increasing, $f(a) \geq f(x)$ for $x \in A$, or $f(a)$ is an upper bound for $f(A)$. Since $a = \sup A$, there is a sequence $(x_n) \subset A$ with $x_n \to a$ (Theorem 1.5.4). By continuity, $f(x_n) \to f(a)$. Now, let M be any upper bound for $f(A)$. Then $M \geq f(x_n)$, $n \geq 1$. Hence $M \geq f(a)$. Thus $f(a)$ is the least upper bound for $f(A)$, similarly for inf.

2.3.5 Let $y_n = f(x_n)$. From Exercise **2.3.4**, $f(x_n^*) = \sup\{f(x_k) : k \geq n\} = y_n^*$, $n \geq 1$. Since $x_n^* \to x^*$ and f is continuous, $y_n^* = f(x_n^*) \to f(x^*)$. Thus $y^* = f(x^*)$, similarly for lower stars.

2.3.6 Remember x^r is defined as $(x^m)^{1/n}$ when $r = m/n$. Since

$$\left[(x^{1/n})^m\right]^n = (x^{1/n})^{mn} = \left[\left(x^{1/n}\right)^n\right]^m = x^m,$$

$x^r = (x^{1/n})^m$ also. With $r = m/n \in \mathbf{Q}$ and p in $\mathbf{Z}$, $(x^r)^p = \left[(x^{1/n})^m\right]^p = (x^{1/n})^{mp} = x^{mp/n} = x^{rp}$. Now, let $r = m/n$ and $s = p/q$ with m,n,p,q integers and $nq \neq 0$. Then $[(x^r)^s]^{nq} = (x^r)^{snq} = x^{rsnq} = x^{mp}$. By the uniqueness of roots, $(x^r)^s = (x^{mp})^{1/nq} = x^{rs}$. Similarly, $(x^r x^s)^{nq} = x^{rnq}x^{snq} = x^{rnq+snq} = x^{(r+s)nq} = (x^{r+s})^{nq}$. By the uniqueness of roots, $x^r x^s = x^{r+s}$.

2.3.7 We are given $a^b = \sup\{a^r : 0 < r < b, r \in \mathbf{Q}\}$, and we need to show that $a^b = c$, where $c = \inf\{a^s : s > b, s \in \mathbf{Q}\}$. If r, s are rationals with $r < b < s$, then $a^r < a^s$. Taking the sup over all $r < b$ yields $a^b \leq a^s$. Taking the inf over all $s > b$ implies $a^b \leq c$. On the other hand, choose $r < b < s$ rational with $s - r < 1/n$. Then $c \leq a^s < a^r a^{1/n} \leq a^b a^{1/n}$. Taking the limit as $n \nearrow \infty$, we obtain $c \leq a^b$.

2.3.8 In this solution, r, s, and t denote rationals. Given b, let (r_n) be a sequence of rationals with $r_n \to b-$. If $t < bc$, then $t < r_n c$ for n large. Pick one such r_n and call it r. Then $s = t/r < c$ and $t = rs$. Thus $t < bc$ iff t is of the form rs with $r < b$ and $s < c$. By Exercise **2.3.6**, $(a^b)^c = \sup\{(a^b)^s : 0 < s < c\} = \sup\{(\sup\{a^r : 0 < r < b\})^s : 0 < s < c\} = \sup\{a^{rs} : 0 < r < b, 0 < s < c\} = \sup\{a^t : 0 < t < bc\} = a^{bc}$.

2.3.9 Since $f(x + x') = f(x)f(x')$, by induction, we obtain $f(nx) = f(x)^n$. Hence $f(n) = f(1)^n = a^n$ for n natural. Also $a = f(1) = f(1+0) = f(1)f(0) = af(0)$. Hence $f(0) = 1$. Since $1 = f(n-n) = f(n)f(-n) = a^n f(-n)$, we obtain $f(-n) = a^{-n}$ for n natural. Hence $f(n) = a^n$ for $n \in \mathbf{Z}$. Now, $(f(m/n))^n = f(n(m/n)) = f(m) = a^m$, so by the uniqueness of roots, $f(m/n) = a^{m/n}$. Hence $f(r) = a^r$ for $r \in \mathbf{Q}$. If x is real, choose rationals $r_n \to x$. Then $a^{r_n} = f(r_n) \to f(x)$. Since we know that a^x is continuous, $a^{r_n} \to a^x$. Hence $f(x) = a^x$.

2.3.10 Given $\epsilon > 0$, we seek $\delta > 0$, such that $|x-1| < \delta$ implies $|(1/x)-1| < \epsilon$. By the triangle inequality, $|x| = |(x-1)+1| \geq 1 - |x-1|$. So $\delta < 1$ and $|x-1| < \delta$ imply $|x| > 1 - \delta$ and

$$\left|\frac{1}{x} - 1\right| = \frac{|x-1|}{|x|} < \frac{\delta}{1-\delta}.$$

Solving $\delta/(1-\delta) = \epsilon$, we have found a δ, $\delta = \epsilon/(1+\epsilon)$, satisfying $0 < \delta < 1$ and the ϵ-δ criterion.

2.3.11 Let A_n be the set of all real roots of all polynomials with degree d and rational coefficients $a_0, a_1, \dots, a_d$, *with denominators bounded by* n and satisfying

$$|a_0| + |a_1| + \cdots + |a_d| + d \leq n.$$

Since each polynomial has finitely many roots and there are finitely many polynomials involved here, for each n, the set A_n is finite. But the set of algebraic numbers is $\bigcup_{n=1}^{\infty} A_n$; hence it is countable.

2.3.12 Let b be a root of f, $b \neq a$, and let $f(x) = (x-b)g(x)$. If b is rational, then the coefficients of g are necessarily rational (this follows from the construction of g in the text) and $0 = f(a) = (a-b)g(a)$. Hence $g(a) = 0$. But the degree of g is less than the degree of f. This contradiction shows that b is irrational.

2.3.13 Write $f(x) = (x-a)g(x)$. By the previous exercise, $f(m/n)$ is never zero. Since $n^d f(m/n)$ is an integer,

$$n^d|f(m/n)| = n^d|m/n - a|\,|g(m/n)| \geq 1,$$

or

$$\left|a - \frac{m}{n}\right| \geq \frac{1}{n^d|g(m/n)|}. \tag{A.2.1}$$

Since g is continuous at a, choose $\delta > 0$ such that $\mu_a(\delta) < 1$, i.e., such that $|x-a| < \delta$ implies $|g(x)-g(a)| \leq \mu_a(\delta) < 1$. Then $|x-a| < \delta$ implies $|g(x)| < |g(a)| + 1$. Now, we have two cases: either $|a - m/n| \geq \delta$ or $|a - m/n| < \delta$. In the first case, we obtain the required inequality with $c = \delta$. In the second case, we obtain $|g(m/n)| < |g(a)| + 1$. Inserting this in the denominator of the right side of (A.2.1) yields

$$\left|a - \frac{m}{n}\right| \geq \frac{1}{n^d(|g(a)|+1)},$$

which is the required inequality with $c = 1/[|g(a)| + 1]$. Now, let

$$c = \min\left(\delta, \frac{1}{|g(a)|+1}\right).$$

Then in either case, the required inequality holds with this choice of c.

2.3.14 Let a be the displayed real. Given a natural d, we show that a is not algebraic of order d. This means, given any $\epsilon > 0$, we have to find M/N such that

$$\left| a - \frac{M}{N} \right| < \frac{\epsilon}{N^d}. \tag{A.2.2}$$

Thus the goal is as follows: given any $d \geq 1$ and $\epsilon > 0$, find M/N satisfying (A.2.2).

The simplest choice is to take M/N equal to the nth partial sum s_n in the series for a. In this case, $N = 10^{n!}$. The question is, which n? To figure this out, note that $k! \geq n!k$ when $k \geq n+1$, and $10^{n!} \geq 2$, so

$$\begin{aligned} |a - s_n| &= \sum_{k=n+1}^{\infty} \frac{1}{10^{k!}} \leq \sum_{k=n+1}^{\infty} \frac{1}{10^{n!k}} \\ &= \frac{1}{(10^{n!})^{n+1}} \sum_{k=0}^{\infty} \frac{1}{(10^{n!})^k} \leq \frac{1}{(10^{n!})^{n+1}} \sum_{k=0}^{\infty} \frac{1}{2^k} = \frac{1}{(10^{n!})^n} \cdot \frac{2}{10^{n!}}. \end{aligned}$$

Thus if we select $n \geq 1$ satisfying $n \geq d$ and $2/10^{n!} < \epsilon$, we obtain

$$\left| a - \frac{M}{N} \right| < \frac{\epsilon}{N^n} \leq \frac{\epsilon}{N^d};$$

hence $s_n = M/N$ satisfies (A.2.2).

2.3.15 From §1.6, we know that $\sum n^{-r}$ converges when $r = 1 + 1/N$. Given $s > 1$ real, we can always choose N with $1 + 1/N < s$. The result follows by comparison.

2.3.16 To show that $b^{\log_a c} = c^{\log_a b}$, apply $\log_a$ obtaining $\log_a b \log_a c$ from either side. For the second part, $\sum 1/5^{\log_3 n} = \sum 1/n^{\log_3 5}$ which converges since $\log_3 5 > 1$.

2.3.17 Such an example cannot be continuous by the results of §2.3. Let $f(x) = x + 1/2$, $0 \leq x < 1/2$, and $f(x) = x - 1/2$, $1/2 \leq x < 1$, $f(1) = 1$. Then f is a bijection, hence invertible.

2.3.18 First, assume f is increasing. Then by Exercise **2.2.3**, $f(c-) = f(c) = f(c+)$ for all except, at most, countably many points $c \in (a, b)$, where there are at worst jumps. Hence f is continuous for all but at most countably many points at which there are at worst jumps. If f is of bounded variation, then $f = g - h$ with g and h bounded increasing. But then f is continuous wherever both g and h are continuous. Thus the set of discontinuities of f is, at most, countable with the discontinuities at worst jumps.

2.3.19 Let $M = \sup f((a, b))$. Then $M > -\infty$. By Theorem 1.5.4, there is a sequence (x_n) in (a, b) satisfying $f(x_n) \to M$. Now, from Theorem 2.1.2, (x_n) subconverges to some x in (a, b) or to a or to b. If (x_n) subconverges to

a or b, then $(f(x_n))$ subconverges to $f(a+)$ or $f(b-)$, respectively. But this cannot happen since $M > f(a+)$ and $M > f(b-)$. Hence (x_n) subconverges to some $a < x < b$. Since f is continuous, $(f(x_n))$ must subconverge to $f(x)$. Hence $f(x) = M$. This shows that M is finite and M is a max.

2.3.20 Fix y and set $h(x) = xy - f(x)$. Then by superlinearity

$$h(\infty) = \lim_{x\to\infty}(xy - f(x)) = \lim_{x\to\infty} x\left(y - \frac{f(x)}{x}\right) = -\infty;$$

similarly $h(-\infty) = -\infty$. Thus Exercise **2.3.19** applies and the sup is attained in the definition of $g(y)$. Now, for $x > 0$ fixed and $y_n \to \infty$, $g(y_n) \geq xy_n - f(x)$. Hence

$$\frac{g(y_n)}{y_n} \geq x - \frac{f(x)}{y_n}.$$

It follows that the lower limit of $(g(y_n)/y_n)$ is $\geq x$. Since x is arbitrary, it follows that the lower limit is ∞. Hence $g(y_n)/y_n \to \infty$. Since (y_n) was any sequence converging to ∞, we conclude that $\lim_{y\to\infty} g(y)/|y| = \infty$. Similarly, $\lim_{y\to-\infty} g(y)/|y| = \infty$. Thus g is superlinear.

2.3.21 Suppose that $y_n \to y$. We want to show that $g(y_n) \to g(y)$. Let $L^* \geq L_*$ be the upper and lower limits of $(g(y_n))$. For all z, $g(y_n) \geq zy_n - f(z)$. Hence $L_* \geq zy - f(z)$. Since z is arbitrary, taking the sup over all z, we obtain $L_* \geq g(y)$. For the reverse inequality, let (y'_n) be a subsequence of (y_n) satisfying $g(y'_n) \to L^*$. Pick, for each $n \geq 1$, x'_n with $g(y'_n) = x'_n y'_n - f(x'_n)$. From §2.1, (x'_n) subconverges to some x, possibly infinite. If $x = \pm\infty$, then superlinearity (see previous solution) implies the subconvergence of $(g(y'_n))$ to $-\infty$. But $L^* \geq L_* \geq g(y) > -\infty$, so this cannot happen. Thus (x'_n) subconverges to a finite x. Hence by continuity, $g(y'_n) = x'_n y'_n - f(x'_n)$ subconverges to $xy - f(x)$ which is $\leq g(y)$. Since by construction, $g(y'_n) \to L^*$, this shows that $L^* \leq g(y)$. Hence $g(y) \leq L_* \leq L^* \leq g(y)$ or $g(y_n) \to g(y)$.

2.3.22 Note that $0 \leq f(x) \leq 1$ and $f(x) = 1$ iff $x \in \mathbf{Z}$. We are supposed to take the limit in m first, then n. If $x \in \mathbf{Q}$, then there is an $N \in \mathbf{N}$, such that $n!x \in \mathbf{Z}$ for $n \geq N$. Hence $f(n!x) = 1$ for $n \geq N$. For such an x, $\lim_{m\to\infty}[f(n!x)]^m = 1$, for every $n \geq N$. Hence the double limit is 1 for $x \in \mathbf{Q}$. If $x \notin \mathbf{Q}$, then $n!x \notin \mathbf{Q}$, so $f(n!x) < 1$ and so $[f(n!x)]^m \to 0$, as $m \nearrow \infty$, for every $n \geq 1$. Hence the double limit is 0 for $x \notin \mathbf{Q}$.

2.3.23 In the definition of $\mu_c(\delta)$, we are to maximize $|(1/x) - (1/c)|$ over all $x \in (0,1)$ satisfying $|x - c| < \delta$ or $c - \delta < x < c + \delta$. In the first case, if $\delta \geq c$, then $c - \delta \leq 0$. Hence all points x near and to the right of 0 satisfy $|x - c| < \delta$. Since $\lim_{x\to 0+}(1/x) = \infty$, in this case, $\mu_c(\delta) = \infty$. In the second case, if $0 < \delta < c$, then x varies between $c - \delta > 0$ and $c + \delta$. Hence

$$\left|\frac{1}{x} - \frac{1}{c}\right| = \frac{|x - c|}{xc}$$

is largest when the numerator is largest ($|x-c|=\delta$) and the denominator is smallest ($x=c-\delta$). Thus

$$\mu_c(\delta)=\begin{cases}\delta/(c^2-c\delta), & 0<\delta<c,\\ \infty, & \delta\ge c.\end{cases}$$

Now, $\mu_I(\delta)$ equals the sup of $\mu_c(\delta)$ for all $c\in(0,1)$. But, for δ fixed and $c\to 0+$, $\delta\ge c$ eventually. Hence $\mu_I(\delta)=\infty$ for all $\delta>0$. Hence $\mu_I(0+)=\infty$, or f is not uniformly continuous on $(0,1)$.

2.3.24 Follow the proof of the uniform continuity theorem. If $\mu(0+)>0$, set $\epsilon=\mu(0+)/2$. Then since μ is increasing, $\mu(1/n)\ge 2\epsilon$ for all $n\ge 1$. Hence for each $n\ge 1$, by the definition of the sup in the definition of $\mu(1/n)$, there is a $c_n\in\mathbf{R}$ with $\mu_{c_n}(1/n)>\epsilon$. Now, by the definition of the sup in $\mu_{c_n}(1/n)$, for each $n\ge 1$, there is an $x_n\in\mathbf{R}$ with $|x_n-c_n|<1/n$ and $|f(x_n)-f(c_n)|>\epsilon$. Then by compactness (§2.1), (x_n) subconverges to some real x or to $x=\pm\infty$. It follows that (c_n) subconverges to the same x. Hence $\epsilon<|f(x_n)-f(c_n)|$ subconverges to $|f(x)-f(x)|=0$, a contradiction.

2.3.25 If $\sqrt{2}^{\sqrt{2}}$ is rational, we are done. Otherwise $a=\sqrt{2}^{\sqrt{2}}$ is irrational. In this case, let $b=\sqrt{2}$. Then $a^b=(\sqrt{2}^{\sqrt{2}})^{\sqrt{2}}=\sqrt{2}^2=2$ is rational. In fact $\sqrt{2}^{\sqrt{2}}$ is transcendental, a result due to Gelfond.

A.3 Solutions to Chapter 3

Solutions to Exercises 3.1

3.1.1 Since $a>0$, $f(0)=0$. If $a=1$, we already know that f is not differentiable at 0. If $a<1$, then $g(x)=(f(x)-f(0))/(x-0)=|x|^a/x$ satisfies $|g(x)|\to\infty$ as $x\to 0$, so f is not differentiable at 0. If $a>1$, then $g(x)\to 0$ as $x\to 0$. Hence $f'(0)=0$.

3.1.2 Since $\sqrt{2}$ is irrational, $f(\sqrt{2})=0$. Hence $(f(x)-f(\sqrt{2}))/(x-\sqrt{2})=f(x)/(x-\sqrt{2})$. Now, if x is irrational, this expression vanishes, whereas if x is rational with denominator d,

$$q(x)\equiv\frac{f(x)-f(\sqrt{2})}{x-\sqrt{2}}=\frac{1}{d^3(x-\sqrt{2})}.$$

By Exercise **1.4.9** it seems that the limit will be zero, as $x\to\sqrt{2}$. To prove this, suppose that the limit of $q(x)$ is not zero, as $x\to\sqrt{2}$. Then there exists at least one sequence $x_n\to\sqrt{2}$ with $q(x_n)\not\to 0$. It follows that there is a $\delta>0$ and a sequence $x_n\to\sqrt{2}$ with $|q(x_n)|\ge\delta$. But this implies all the reals x_n are rational. If d_n is the denominator of x_n, $n \geq 1$, we obtain, from Exercise **1.4.9**,

$$|q(x_n)| \leq \frac{d_n^2}{d_n^3 c} = \frac{1}{d_n c}.$$

Since $d_n \to \infty$ (Exercise **2.2.1**), we conclude that $q(x_n) \to 0$, contradicting our assumption. Hence our assumption must be false, i.e., $\lim_{x\to\sqrt{2}} q(x) = 0$ or $f'(\sqrt{2}) = 0$.

3.1.3 Since f is superlinear (Exercise **2.3.20**), $g(y)$ is finite, and the max is attained at some critical point x. Differentiating $xy - ax^2/2$ with respect to x yields $0 = y - ax$, or $x = y/a$ for the critical point, which, as previously said, must be the global max. Hence

$$g(y) = (y/a)y - a(y/a)^2/2 = y^2/2a.$$

Since $f'(x) = ax$ and $g'(y) = y/a$, it is clear they are inverses.

3.1.4 Suppose that $g'(\mathbf{R})$ is bounded above. Then $g'(x) \leq c$ for all x. Hence $g(x) - g(0) = g'(z)(x - 0) \leq cx$ for $x > 0$, which implies $g(x)/x \leq c$ for $x > 0$, which contradicts superlinearity. Hence $g'(\mathbf{R})$ is not bounded above. Similarly, $g'(\mathbf{R})$ is not bounded below.

3.1.5 To show that $f'(c)$ exists, let $x_n \to c$ with $x_n \neq c$ for all $n \geq 1$. Then for each $n \geq 1$, there is a y_n strictly between c and x_n, such that $f(x_n) - f(c) = f'(y_n)(x_n - c)$. Since $x_n \to c$, $y_n \to c$, and $y_n \neq c$ for all $n \geq 1$. Since $\lim_{x\to c} f'(x) = L$, it follows that $f'(y_n) \to L$. Hence $(f(x_n) - f(c))/(x_n - c) \to L$. Since (x_n) was arbitrary, we conclude that $f'(c) = L$.

3.1.6 To show that $f'(c) = f'(c+)$, let $x_n \to c+$. Then for each $n \geq 1$, there is a y_n between c and x_n, such that $f(x_n) - f(c) = f'(y_n)(x_n - c)$. Since $x_n \to c+$, $y_n \to c+$. It follows that $f'(y_n) \to f'(c+)$. Hence $(f(x_n) - f(c))/(x_n - c) \to f'(c+)$, i.e., $f'(c) = f'(c+)$, similarly for $f'(c-)$.

3.1.7 If $a = x_0 < x_1 < x_2 < \cdots < x_n = b$ is a partition, the mean value theorem says $f(x_k) - f(x_{k-1}) = f'(z_k)(x_k - x_{k-1})$ for some z_k between x_{k-1} and x_k, $1 \leq k \leq n$. Since $|f'(x)| \leq I$, we obtain $|f(x_k) - f(x_{k-1})| \leq I(x_k - x_{k-1})$. Summing over $1 \leq k \leq n$, we see that the variation corresponding to this partition is $\leq I(b - a)$. Since the partition was arbitrary, the result follows.

3.1.8 Let $c \in \mathbf{Q}$. We have to show that, for some $n \geq 1$, $f(c) \geq f(x)$ for all x in $(c - 1/n, c + 1/n)$. If this were not the case, for each $n \geq 1$, we can find a real x_n satisfying $|x_n - c| < 1/n$ and $f(x_n) > f(c)$. But, by Exercise **2.2.1**, we know that $f(x_n) \to 0$ since $x_n \to c$ and $x_n \neq c$, contradicting $f(c) > 0$. Hence c must be a local maximum.

3.1.9 If f is even, then $f(-x) = f(x)$. Differentiating yields $-f'(-x) = f'(x)$, or f' is odd, similarly if f is odd.

3.1.10 Let $g(x) = (f(x) - f(r))/(x - r)$, $x \neq r$, and $g(r) = f'(r)$. Then g is continuous and $f(r) = 0$ iff $f(x) = (x - r)g(x)$.

3.1.11 As in the previous exercise, set

$$g(x) = \frac{f(x)}{\prod_{j=1}^{d}(x - r_j)}.$$

Then g is continuous away from $r_1, \dots, r_d$. If $f(r_j) = 0$, then

$$\lim_{x \to r_j} g(x) = \frac{f'(r_j)}{\prod_{j \neq i}(r_i - r_j)}.$$

Since g has removable singularities at r_j, g can be extended to be continuous there. With this extension, we have $f(x) = (x - r_1)\dots(x - r_d)g(x)$. Conversely, if $f(x) = (x - r_1)\dots(x - r_d)g(x)$, then $f(r_j) = 0$.

3.1.12 If $a < b$ and $f(a) = f(b) = 0$, then by the mean value theorem, there is a $c \in (a, b)$ satisfying $f'(c) = 0$. Thus between any two roots of f, there is a root of f'.

Solutions to Exercises 3.2

3.2.1 Let $f(x) = e^x$. Since $f(x) - f(0) = f'(c)x$ for some $0 < c < x$, we have $e^x - 1 = e^c x$. Since $c > 0$, $e^c > 1$. Hence $e^x - 1 \geq x$ for $x \geq 0$.

3.2.2 Let $f(x) = 1 + x^\alpha$, $g(x) = (1 + x)^\alpha$. Then $f'(c)/g'(c) = (c/(1 + c))^{\alpha - 1} \geq 1$; hence $(f(x) - f(0))/(g(x) - g(0)) \geq 1$. For the second part, $(a + b)^\alpha = a^\alpha(1 + (b/a))^\alpha \leq a^\alpha(1 + (b/a)^\alpha) = a^\alpha + b^\alpha$.

3.2.3 If $f(x) = \log(1 + x)$, then by l'Hôpital's rule $\lim_{x \to 0} \log(1 + x)/x = \lim_{x \to 0} 1/(1 + x) = 1$. This proves the first limit. If $a \neq 0$, set $x_n = a/n$. Then $x_n \to 0$ with $x_n \neq 0$ for all $n \geq 1$. Hence

$$\lim_{n \nearrow \infty} n \log(1 + a/n) = a \lim_{n \nearrow \infty} \frac{\log(1 + x_n)}{x_n} = a.$$

By taking exponentials, we obtain $\lim_{n \nearrow \infty}(1 + a/n)^n = e^a$ when $a \neq 0$. If $a = 0$, this is immediate, so the second limit is true for all a. Now, if $a_n \to a$, then for some $N \geq 1$, $a - \epsilon \leq a_n \leq a + \epsilon$ for all $n \geq N$. Hence

$$\left(1 + \frac{a - \epsilon}{n}\right)^n \leq \left(1 + \frac{a_n}{n}\right)^n \leq \left(1 + \frac{a + \epsilon}{n}\right)^n$$

for $n \ge N$. Thus the upper and lower limits of the sequence in the middle lie between $e^{a-\epsilon}$ and $e^{a+\epsilon}$. Since $\epsilon > 0$ is arbitrary, the upper and lower limits must be both equal e^a. Hence we obtain the third limit.

3.2.4 Let $b = x/(n+1)$, $v = (n+1)/n$, and $e_n = (1+x/n)^n$. Then $|b| < 1$, so by (3.2.2), $(1+b)^{nv} \ge (1+vb)^n$. Hence $e_{n+1} \ge e_n$.

3.2.5 If $f(x) = (1+x^2)^{-1/2}$, then $f(x) - f(0) = f'(c)(x-0)$ with $0 < c < x$. Since $f'(c) = -c(1+c^2)^{-3/2} > -c > -x$ for $x > 0$,

$$\frac{1}{\sqrt{1+x^2}} - 1 = f'(c)x \ge -x^2,$$

which implies the result.

3.2.6 With $f(t) = -t^{-x}$ and $f'(t) = xt^{-x-1}$, the left side of the displayed inequality equals $f(2j) - f(2j-1)$, which equals $f'(c)$ with $2j-1 < c < 2j$. Hence the left side is $xc^{-x-1} \le x(2j-1)^{-x-1}$.

3.2.7 Let $g(x) = f(x)^2$. Then $g'(x) \le 2g(x)$, so $e^{-2x}g(x)$ has nonpositive derivative and $e^{-2x}g(x) \le g(0) = 0$. Thus f is identically zero.

3.2.8 First, suppose that $L = 0$. If there is no such x, then f' is never zero. Hence $f' > 0$ on (a,b) or $f' < 0$ on (a,b), contradicting $f'(c) < 0 < f'(d)$. Hence there is an x satisfying $f'(x) = 0$. In general, let $g(x) = f(x) - Lx$. Then $f'(c) < L < f'(d)$ implies $g'(c) < 0 < g'(d)$, so the general case follows from the case $L = 0$.

3.2.9 From Exercise **3.1.4**, $g'(\mathbf{R})$ is not bounded above nor below. But g' satisfies the intermediate value property. Hence the range of g' is an interval. Hence $g'(\mathbf{R}) = \mathbf{R}$.

3.2.10 Note first $f_d(1) = 1$ and

$$f_d'(t) = \left(\frac{d-1}{d}\right)^2 t^{-1/d} - \frac{1}{d^2} t^{-(d+1)/d} \le \left(\frac{d-1}{d}\right)^2, \qquad t \ge 1.$$

By the mean value theorem,

$$f_d(t) - 1 = f_d(t) - f_d(1) \le \left(\frac{d-1}{d}\right)^2 (t-1), \qquad t \ge 1.$$

Solutions to Exercises 3.3

3.3.1 Since $f'(x) = (1/2) - (1/x^2)$, the only positive critical point (Figure A.1) is $x = \sqrt{2}$. Moreover, $f(\infty) = f(0+) = \infty$, so $\sqrt{2}$ is a global minimum over $(0,\infty)$, and $f(\sqrt{2}) = \sqrt{2}$. Also $f''(x) = 2/x^3 > 0$, so f is convex.

3.3.2 If $a < x < y < b$, then

$$f((1-t)x + ty) = (1-t)f(x) + tf(y), \qquad 0 < t < 1. \tag{A.3.1}$$

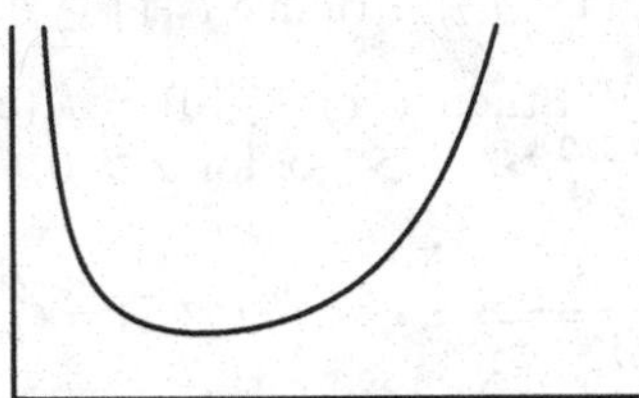

Fig. A.1 The graph of $(x + 2/x)/2$

Differentiate with respect to t to get

$$f'((1-t)x + ty) = \frac{f(y) - f(x)}{y - x}, \qquad 0 < t < 1. \tag{A.3.2}$$

Thus f' is constant on (x, y) and hence on (a, b). Conversely, suppose f' is a constant m. Then by the MVT, $f(y) - f(x) = m(y - x)$, hence (A.3.2) holds, and thus

$$\frac{d}{dt} f((1-t)x + ty) = f(y) - f(x), \qquad 0 < t < 1,$$

which implies

$$f((1-t)x + ty) = f(x) + t(f(y) - f(x)), \qquad 0 < t < 1.$$

But this is (A.3.1).

3.3.3 If $a < x < y < b$ and $0 \le t \le 1$, then $f((1-t)x + ty) \le (1-t)f(x) + tf(y)$, so

$$g(f((1-t)x + ty)) \le g((1-t)f(x) + tf(y)) \le (1-t)g(f(x)) + tg(f(y)).$$

3.3.4 If $a < b < c$ and $t = (b - a)/(c - a)$, then $b = (1-t)a + tc$. Hence by convexity,

$$f(b) \le (1-t)f(a) + tf(c).$$

Subtracting $f(a)$ from both sides, and then dividing by $b - a$, yields $s[a, b] \le s[a, c]$. Instead, if we subtract both sides from $f(c)$ and then divide by $c - b$, we obtain $s[a, c] \le s[b, c]$.

3.3.5 Exercise **3.3.4** says $x \mapsto s[c, x]$ is an increasing function of x. Hence

$$f'_+(c) = \lim_{t \to c+} s[c, t] = \inf\{s[c, t] : t > c\} \le s[c, x], \quad x > c$$

exists. Similarly

$$f'_-(d) = \lim_{t\to d-} s[t,d] = \sup\{s[t,d] : t < d\} \geq s[x,d], \quad x < d$$

exists. Since $s[c,x] \leq s[x,d]$ by Exercise **3.3.4**, (3.3.12) follows. Also since $s[y,c] \leq s[c,x]$ for $y < c < x$, inserting $c = d$ in the last two inequalities, we conclude that $f'_-(c) \leq f'_+(c)$. Moreover, since $t < x < s < y$ implies $s[t,x] \leq s[s,y]$, let $t \to x-$ and $s \to y-$ to get $f_-(x) \leq f_-(y)$; hence f_- is increasing, similarly for f_+.

3.3.6 The inequality (3.3.12) implies $f'_+(c) \leq s[c,x] \leq f'_-(d)$. Multiplying this inequality by $(x-c)$ and letting $x \to c+$ yield $f(c+) = f(c)$. Similarly multiplying $f'_+(c) \leq s[x,d] \leq f'_-(d)$ by $(x-d)$ and letting $x \to d-$ yield $f(d-) = f(d)$. Since c and d are any reals in (a,b), we conclude f is continuous on (a,b).

3.3.7 Multiply $f'_+(c) \leq s[c,x]$ by $(x-c)$ for $x > c$ and rearrange to get

$$f(x) \geq f(c) + f'_+(c)(x-c), \qquad x \geq c.$$

Since $f'_+(c) \geq f'_-(c)$, this implies

$$f(x) \geq f(c) + f'_-(c)(x-c), \qquad x \geq c.$$

Similarly, multiply $f'_-(c) \geq s[y,c]$ by $(y-c)$ for $y < c$ and rearrange to get

$$f(y) \geq f(c) + f'_-(c)(y-c), \qquad y \leq c.$$

Since $f'_+(c) \geq f'_-(c)$ and $y - c \leq 0$, this implies

$$f(y) \geq f(c) + f'_+(c)(y-c), \qquad y \leq c.$$

Thus

$$f(x) \geq f(c) + f'_\pm(c)(x-c), \qquad a < x < b.$$

If f is differentiable at c, the second inequality follows since $f'_+(c) = f'(c) = f'_-(c)$.

3.3.8 If p is a subdifferential of f at c, rearranging the inequality

$$f(x) \geq f(c) + p(x-c), \qquad a < x < b,$$

yields

$$\frac{f(x)-f(c)}{x-c} \geq p \geq \frac{f(y)-f(c)}{y-c}$$

for $y < c < x$. Letting $x \to c+$ and $y \to c-$, we conclude

$$f'_-(c) \leq p \leq f'_+(c).$$

Conversely, assume f is convex; then $f'_\pm(c)$ exist and are subdifferentials of f at c. If $x \ge c$ and $f'_-(c) \le p \le f'_+(c)$, we have

$$f(x) \ge f(c) + f'_+(c)(x-c) \ge f(c) + p(x-c), \qquad c \le x < b.$$

Similarly, if $x \le c$, we have

$$f(x) \ge f(c) + f'_-(c)(x-c) \ge f(c) + p(x-c), \qquad a < x \le c.$$

Hence p is a subdifferential of f at c.

3.3.9 If c is a maximum of f, then $f(c) \ge f(x)$ for $a < x < b$. Let p be a subdifferential of f at c: $f(x) \ge f(c) + p(x-c)$ for $a < x < b$. Combining these inequalities yields $f(x) \ge f(x) + p(x-c)$ or $0 \ge p(x-c)$ for $a < x < b$. Hence $p = 0$; hence $f(x) \ge f(c)$ from the subdifferential inequality. Thus $f(x) = f(c)$ for $a < x < b$.

3.3.10 We are given that $f(c) - g(c) \ge f(x) - g(x)$ for all $a < x < b$. Let p be a subdifferential of f at c: $f(x) \ge f(c) + p(x-c)$, $a < x < b$. Combining these inequalities yields $g(x) \ge g(c) + p(x-c)$ or p is a subdifferential of g at c. Hence $p = g'(c)$ by Exercise **3.3.8**. Hence f has a unique subdifferential at c. Hence by Exercise **3.3.8** again, f is differentiable at c and $f'(c) = g'(c)$.

3.3.11 Since f_j, $j = 1, \dots, n$, is convex, we have

$$f_j((1-t)x+ty) < (1-t)f_j(x) + tf_j(y) \le (1-t)f(x) + tf(y), \qquad 0 < t < 1,$$

for each $j = 1, \dots, n$. Maximizing the left side over $j = 1, \dots, n$, the result follows.

3.3.12 The sup in (3.3.10) is attained by Exercise **2.3.20**; since by Exercise **2.3.21**, g is continuous, the same reasoning applies to (3.3.11); hence the sup is attained there as well. Fix $a < b$ and $0 \le t \le 1$. Then

$$x[(1-t)a+tb] - f(x) = (1-t)[xa - f(x)] + t[xb - f(x)] \le (1-t)g(a) + tg(b).$$

Since this is true for every x, taking the sup of the left side over x yields the convexity of g. Now, fix y, and suppose that $g(y) = xy - f(x)$, i.e., x attains the max in the definition (3.3.10) of $g(y)$. Since $g(z) \ge xz - f(x)$ for all z, we get

$$g(z) \ge xz - f(x) = xz - (xy - g(y)) = g(y) + x(z-y)$$

for all z. This shows x is a subdifferential of g at y.

3.3.13 Since $x \mapsto -x$ is a bijection and f is even,

$$g(-y) = \sup_{-\infty<x<\infty} (x(-y) - f(x))$$

$$= \sup_{-\infty<x<\infty} ((-x)(-y) - f(-x))$$
$$= \sup_{-\infty<x<\infty} (xy - f(x)) = g(y).$$

Thus g is even. If $y \geq 0$ and $x \geq 0$, then $xy - f(x) \geq (-x)y - f(x)$, so when $y \geq 0$, the sup can be restricted over $x \geq 0$.

3.3.14 Since $p > 1$, f is superlinear, so the max exists. By Exercise **3.3.13**, we need to consider only $y \geq 0$ and consequently only $x \geq 0$ when maximizing $xy - f(x)$. If $y = 0$, we obtain $g(0) = 0$, whereas if $y > 0$, we need to consider only $x > 0$ since $f(x) \geq 0$. To find the critical points, solve $0 = (xy - f(x))' = y - f'(x)$ for $x > 0$ obtaining $x^{p-1} = y$ or $x = y^{1/(p-1)}$. Plugging this into $xy - f(x)$ yields the required g, using $(p-1)(q-1) = 1$ and $p(q-1) = q$. Finally, f' and g' are odd, and, for $x \geq 0$, $f'(x) = x^{p-1}$ and $g'(y) = y^{q-1}$ are inverses since $(p-1)(q-1) = 1$.

3.3.15 Again, it is enough to restrict to $y \geq 0$ and $x \geq 0$. Also $(xy - f(x))' = 0$ iff $y = f'(x)$, i.e., $y = e^x$. Thus $x > 0$ is a critical point if $y > 1$ and $x = \log y$, which gives $xy - f(x) = y \log y - y + 1$. If $0 \leq y \leq 1$, the function $x \mapsto xy - f(x)$ has no critical points in $(0, \infty)$. Hence it is maximized at $x = 0$, i.e., $g(y) = 0$ when $0 \leq y \leq 1$. If $y > 1$, we obtain the critical point $x = \log y$, the corresponding critical value $y \log y - y + 1$, and the endpoint values 0 and $-\infty$. To see which of these three values is largest, note that $(y \log y - y + 1)' = \log y > 0$ for $y > 1$, and thus $y \log y - y + 1 \geq 0$ for $y \geq 1$. Hence $g(y) = y \log y - y + 1$ for $y \geq 1$ and $g(y) = 0$ for $0 \leq y \leq 1$.

3.3.16 Since f is convex, it is continuous (Exercise **3.3.6**). Thus by Exercise **2.3.20**, g is well defined and superlinear. By Exercise **3.3.12**, g is convex. It remains to derive the formula for $f(x)$. To see this, note, by the formula for $g(y)$, that $f(x) + g(y) \geq xy$ for all x and all y, which implies $f(x) \geq \max_y [xy - g(y)]$. To obtain equality, we need to show the following: For each x, there is a y satisfying $f(x) + g(y) = xy$. To this end, fix x; by Exercise **3.3.8**, f has a subdifferential p at x. Hence $f(t) \geq f(x) + p(t - x)$ for all t which yields $xp \geq f(x) + (pt - f(t))$. Taking the sup over all t, we obtain $xp \geq f(x) + g(p)$. Since we already know $f(x) + g(p) \geq xp$ by the definition (3.3.10) of g, we conclude $f(x) + g(p) = xp$. Hence $f(x) = \max_y (xy - g(y))$, i.e., (3.3.11) holds. Note that when f is the Legendre transform of g, then f is necessarily convex; hence if f is not convex, the result cannot possibly be true.

3.3.17 The only if part was carried out in Exercise **3.3.12**. Now fix y and suppose x is a subdifferential of g at y. Then $g(z) \geq g(y) + x(z - y)$ for all z. This implies $xy \geq g(y) + (xz - g(z))$ for all z. Maximizing over z and appealing to Exercise **3.3.16**, we obtain $xy \geq g(y) + f(x)$. Since we already know by the definition (3.3.10) of g that $xy \leq g(y) + f(x)$, we conclude x achieves the maximum in the definition of g. For a counterexample for non-convex f, let $f(x) = 1 - x^2$ for $|x| \leq 1$, $f(x) = x^2 - 1$ for $|x| \geq 1$. Then f is superlinear and continuous, so its Legendre transform g is well defined and convex. In fact $g(y) = |y|$, $|y| \leq 2$, $g(y) = 1 + y^2/4$, $|y| \geq 2$. The set of subdifferentials of g at $y = 0$ is $[-1, 1]$, while x attains the max in (3.3.10) for $g(0)$ iff $x = \pm 1$. It may help to graph f and g.

3.3.18 Since f is convex, f' is increasing. By Exercise **2.2.3**, this implies f' can only have jump discontinuities. By Exercise **3.2.8**, f' satisfies the intermediate value property; hence it cannot have jump discontinuities. Hence f' is continuous.

3.3.19 Fix y and suppose the maximum in the definition (3.3.10) of $g(y)$ is attained at x_1 and x_2. By strict convexity of f, if $x = (x_1 + x_2)/2$, we have

$$g(y) = \frac{1}{2}g(y) + \frac{1}{2}g(y) = \frac{1}{2}(x_1 y - f(x_1)) + \frac{1}{2}(x_2 y - f(x_2)) < xy - f(x),$$

contradicting the definition of $g(y)$. Thus there can only be one real x at which the sup is attained; hence by Exercise **3.3.17**, there is a unique subdifferential of g at y. By Exercise **3.3.8**, this shows $g'_+(y) = g'_-(y)$; hence g is differentiable at y. Since g is convex by Exercise **3.3.12**, we conclude g' is continuous by Exercise **3.3.18**.

3.3.20 We already know g is superlinear, differentiable, and convex, and $xy = f(x) + g(y)$ iff x attains the maximum in the definition (3.3.10) of $g(y)$ iff x is a subdifferential of g at y iff $x = g'(y)$. Similarly, since f is the Legendre transform of g, we know $xy = f(x) + g(y)$ iff y attains the maximum in (3.3.11) iff y is a subdifferential of f at x iff $y = f'(x)$. Thus f' is the inverse of g'. By the inverse function theorem for continuous functions §2.3, it follows that g' is strictly increasing; hence g is strictly convex.

3.3.21 Here f' does not exist at 0. However, the previous exercise suggests that g' is trying to be the inverse of f' which suggests that $f'(0)$ should be defined to be (Figure A.2) the line segment $[-1, 1]$ on the vertical axis. Of course, with such a definition, f' is no longer a function, but something more general ($f'(0)$ is the set of subdifferentials at 0; see Exercise **3.3.8**).

3.3.22 Since $(e^x)'' > 0$, e^x is convex. Hence $e^{t \log a + (1-t) \log b} \leq te^{\log a} + (1 - t)e^{\log b}$. But this simplifies to $a^t b^{1-t} \leq ta + (1-t)b$.

3.3.23 By Exercise **3.3.20**, we know g is superlinear, differentiable, and strictly convex, with $g'(f'(x)) = x$ for all x. If g' is differentiable, differentiating yields $g''[f'(x)]f''(x) = 1$, so $f''(x)$ never vanishes. Since convexity implies $f''(x) \geq 0$, we obtain $f''(x) > 0$ for all x. Conversely, if $f''(x) > 0$, by the inverse function theorem for derivatives §3.2, g is twice differentiable with $g''(x) = (g')'(x) = 1/(f')'[g'(x)] = 1/f''[g'(x)]$. Hence g is twice differentiable and

$$g''(x) = \frac{1}{f''[g'(x)]}.$$

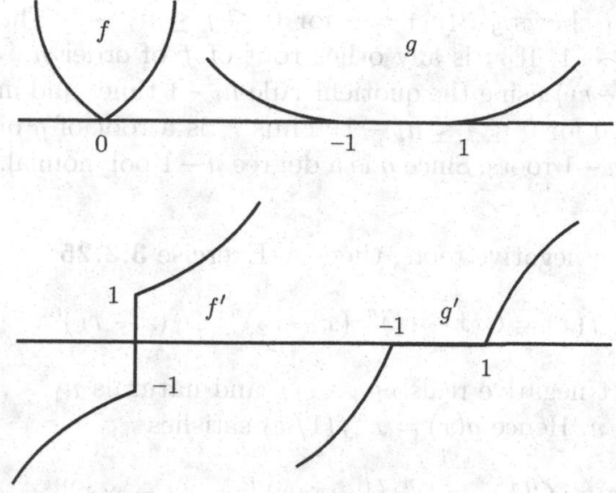

Fig. A.2 The graphs of f, g, f', g' (Exercise **3.3.21**)

Since f is smooth, whenever g is n times differentiable, g' is $n-1$ times differentiable; hence by the right side of this last equation, g'' is $n-1$ times differentiable; hence g is $n+1$ times differentiable. By induction, it follows that g is smooth. For the counterexample, let $f(x) = x^4/4$. Although $f''(0) = 0$, since $f''(x) > 0$ for $x \neq 0$, it follows that f is strictly convex on $(-\infty, 0)$ and on $(0, \infty)$. From this it is easy to conclude (draw a picture) that f is strictly convex on $\mathbf{R}$. Also f is superlinear and smooth, but $g(y) = (3/4)|y|^{4/3}$ (Exercise **3.3.14**) is not smooth at 0.

3.3.24 Since $(f')^{(j)}(r_i) = f^{(j+1)}(r_i) = 0$ for $0 \leq j \leq n_i - 2$, it follows that r_i is a root of f' of order $n_i - 1$. Also by Exercise **3.1.12**, there are $k-1$ other roots $s_1, \dots, s_{k-1}$. Since

$$(n_1 - 1) + (n_2 - 1) + \cdots + (n_k - 1) + k - 1 = n - 1,$$

the result follows. Note if these roots of f are in (a, b), then so are these roots of f'.

3.3.25 If $f(x) = (x - r_1)^{n_1} g(x)$, differentiating j times, $0 \leq j \leq n_1 - 1$, shows r_1 is a root of f of order n_1. Since the advertised f has the form $f(x) = (x - r_i)^{n_i} g_i(x)$ for each $1 \leq i \leq k$, each r_i is a root of order n_i; hence f has n roots. Conversely, we have to show that a degree n polynomial having n roots must be of the advertised form. This we do by induction. If $n = 1$, then $f(x) = ax + b$ and $f(r) = 0$ imply $ar + b = 0$; hence $b = -ar$; hence $f(x) = a(x - r)$. Assume the result is true for $n - 1$, and let f be a degree n polynomial having n roots. If r_1 is a root of f of order n_1, define $g(x) = f(x)/(x - r_1)$. Differentiating $f(x) = (x - r_1)g(x)$ $j+1$ times yields

$$f^{(j+1)}(x) = jg^{(j)}(x) + (x - r_1)g^{(j+1)}(x).$$

Inserting $x = r_1$ shows $g^{(j)}(r_1) = 0$ for $0 \le j \le n_1 - 2$. Thus r_1 is a root of g of order $n_1 - 1$. If r_i is any other root of f of order n_i, differentiating $g(x) = f(x)/(x - r_1)$ using the quotient rule $n_i - 1$ times and inserting $x = r_i$ show $g^{(j)}(r_i) = 0$ for $0 \le j \le n_i - 1$. Thus r_i is a root of g of order n_i. We conclude g has $n - 1$ roots. Since g is a degree $n - 1$ polynomial, by induction, the result follows.

3.3.26 If f has n negative roots, then by Exercise **3.3.25**

$$f(x) = C(x - r_1)^{n_1}(x - r_2)^{n_2} \dots (x - r_k)^{n_k}$$

for some distinct negative reals $r_1, \dots, r_k$ and naturals $n_1, \dots, n_k$ satisfying $n_1 + \dots + n_k = n$. Hence $g(x) = x^n f(1/x)$ satisfies

$$\begin{aligned} g(x) &= C(1 - r_1 x)^{n_1}(1 - r_2 x)^{n_2} \dots (1 - r_k x)^{n_k} \\ &= C' \left(x - \frac{1}{r_1}\right)^{n_1} \left(x - \frac{1}{r_2}\right)^{n_2} \dots \left(x - \frac{1}{r_k}\right)^{n_k} \end{aligned}$$

which shows g has n negative roots.

3.3.27 Since the a_j's are positive, f has n negative roots by Exercise **3.3.25**, establishing **A**. **B** follows from Exercise **3.3.24**, and **C** follows from Exercise **3.3.26**. Since the jth derivative of x^k is nonzero iff $j \le k$, the only terms *in* f that do not vanish upon $n - k - 1$ differentiations are

$$x^n + \binom{n}{1} p_1 x^{n-1} + \dots + \binom{n}{k+1} p_{k+1} x^{n-k-1}.$$

This implies

$$g(x) = \frac{n!}{(k+1)!}\left(x^{k+1} + \binom{k+1}{1} p_1 x^k + \dots + \binom{k+1}{k} p_k x + p_{k+1}\right)$$

which implies

$$h(x) = \frac{n!}{(k+1)!}\left(1 + \binom{k+1}{1} p_1 x + \dots + \binom{k+1}{k} p_k x^k + p_{k+1} x^{k+1}\right).$$

Differentiating these terms $k - 1$ times yields p. By Exercise **3.3.24**, p has two roots. This establishes **D**. Since a quadratic with roots has nonnegative discriminant (Exercise **1.4.5**), the result follows.

3.3.28 Since $p_k^2 \ge p_{k-1}p_{k+1}$, $1 \le k \le n - 1$, we have $p_1^2 \ge p_2$ or $p_1 \ge p_2^{1/2}$. Assume $p_{k-1}^{1/(k-1)} \ge p_k^{1/k}$. Then

$$p_k^2 \ge p_{k-1}p_{k+1} \ge p_k^{(k-1)/k} p_{k+1}$$

which implies

$$p_k^{(k+1)/k} \geq p_{k+1}.$$

Taking the $(k+1)$-st root, we obtain $p_k^{1/k} \geq p_{k+1}^{1/(k+1)}$. If we have $p_1 = p_2^{1/2} = \dots = p_n^{1/n} = m$, then from the previous exercise, $f(x)$ equals

$$x^n + \binom{n}{1} m x^{n-1} + \dots + \binom{n}{n-1} m^{n-1} x + m^n = (x+m)^n$$

by the binomial theorem. Hence all the a_j's equal m.

3.3.29 Since p, p', p'', ... are polynomials in x with integer coefficients and f'/f, f''/f, ... are sums of products of p, p', p'', ..., plugging $x = a$ yields $f(t,a) = 1$; hence the result follows.

3.3.30 Referring to the previous exercise, with $p(x) = x(a-x)$ and $f(t,x) = \exp(tp(x))$, we know $f^{(k)}(t,0)$ and $f^{(k)}(t,a)$ are polynomials in t with integer coefficients for $k \geq 0$. Setting $g(t,x) = f(t,bx)$, it follows that $g^{(k)}(t,0)$ and $g^{(k)}(t,p)$ are polynomials in t with integer coefficients for $k \geq 0$. But these coefficients are $g_n^{(k)}(0)$ and $g_n^{(k)}(p)$.

Solutions to Exercises 3.4

3.4.1 Since $n! \geq 2^{n-1}$ (Exercise **1.3.20**),

$$\sum_{k \geq n+1} \frac{1}{k!} \leq \sum_{k \geq n+1} 2^{1-k} = 2^{1-n}.$$

Now choose $n = 15$. Then $2^{n-1} > 10^4$, so adding the terms of the series up to $n = 15$ yields accuracy to four decimals. Adding these terms yields $e \sim 2.718281829$ where $\sim$ means that the error is $< 10^{-4}$.

3.4.2 If $n \geq 100$, then

$$n! \geq 101 \cdot 102 \cdot \dots \cdot n \geq 100 \cdot 100 \cdot \dots \cdot 100 = 100^{n-100}.$$

Hence $(n!)^{1/n} \geq 100^{(n-100)/n}$, which clearly approaches 100. Thus the lower limit of $((n!)^{1/n})$ is ≥ 100. Since 100 may be replaced by any N, the result follows.

3.4.3 $a_n = f^{(n)}(0)/n! = b_n$.

3.4.4 Since $\sinh(\pm\infty) = \pm\infty$, $\sinh(\mathbf{R}) = \mathbf{R}$. Since $\sinh' = \cosh > 0$, sinh is bijective, hence invertible. Note that $\cosh^2 - \sinh^2 = 1$, so

$$\cosh^2(\operatorname{arcsinh} x) = 1 + x^2.$$

The derivative of arcsinh : $\mathbf{R} \to \mathbf{R}$ is (by the IFT)

$$\operatorname{arcsinh}'(x) = \frac{1}{\sinh'(\operatorname{arcsinh} x)} = \frac{1}{\cosh(\operatorname{arcsinh} x)} = \frac{1}{\sqrt{1+x^2}}.$$

Since $1/\sqrt{1+x^2}$ is smooth, so is arcsinh. Now, cosh is superlinear since $\cosh x \geq e^{|x|}/2$ and strictly convex since $\cosh'' = \cosh > 0$. Hence the max in

$$g(y) = \max_{-\infty<x<\infty} (xy - \cosh(x))$$

is attained at $x = \operatorname{arcsinh} y$. We obtain $g(y) = y \operatorname{arcsinh} y - \sqrt{1+y^2}$.

3.4.5 With $a_n = (-1)^n/4^n(n!)^2$, use the ratio test,

$$\frac{|a_n|}{|a_{n+1}|} = 4(n+1)^2 \to \infty.$$

Hence the radius ρ equals ∞.

3.4.6 Here neither the ratio test nor the root test works. If $|x| \geq 1$, by the nth term test, the series diverges, whereas if $|x| < 1$, the series converges absolutely by comparison with the geometric series. Hence $R = 1$.

3.4.7 Inserting $-x$ for x in the series yields $f(-x) = a_0 - a_1x + a_2x^2 - \dots$. Hence

$$f^e(x) = \frac{f(x) + f(-x)}{2} = a_0 + a_2x^2 + \dots .$$

But f is even iff $f = f^e$, so the result follows. The odd case is similar.

3.4.8 Establish the first identity by induction. If $k = 1$, we have

$$\left(x\frac{d}{dx}\right)\left(\frac{1}{1-x}\right) = x\left(\frac{1}{1-x}\right)' = \frac{x}{(1-x)^2} = -\frac{1}{1-x} + \frac{1}{(1-x)^2}.$$

Now assume the identity is true for k; differentiate it to get

$$\begin{aligned}\left(x\frac{d}{dx}\right)^{k+1}\left(\frac{1}{1-x}\right) &= x\frac{d}{dx}\sum_{j=0}^{k}\frac{a_j}{(1-x)^{j+1}} \\ &= \sum_{j=0}^{k}\frac{(j+1)xa_j}{(1-x)^{j+2}} \\ &= \sum_{j=0}^{k}\frac{-(j+1)a_j}{(1-x)^{j+1}} + \sum_{j=0}^{k}\frac{(j+1)a_j}{(1-x)^{j+2}}.\end{aligned}$$

This establishes the inductive step. For the second assertion, note

$$\sum_{n=1}^{\infty} \frac{n^k}{2^n} = \left(x\frac{d}{dx}\right)^k \left(\frac{1}{1-x}\right)\Bigg|_{x=1/2}$$

by differentiating the geometric series under the summation sign. The result follows by plugging $x = 1/2$ into the first assertion.

Solutions to Exercises 3.5

3.5.1 By Taylor's Theorem, $f(c+t) = f(c) + f'(c)t + f''(\eta)t^2/2$ with η between c and $c+t$. Since $f(c+t) \geq 0$ and $f''(\eta) \leq 1/2$, we obtain

$$0 \leq f(c) + f'(c)t + t^2/4, \qquad -\infty < t < \infty.$$

Hence the quadratic $Q(t) = f(c) + f'(c)t + t^2/4$ has at most one solution. But this implies (Exercise **1.4.5**) $f'(c)^2 - 4(1/4)f(c) \leq 0$, which gives the result.

3.5.2 For $n = 1$, **A**is true since we can choose $R_1(x) = 1$. Also **B** is true for $n = 1$ since $h(0) = 0$. Now, assume that **A**and **B** are true for n. Then

$$\lim_{x\to 0+} \frac{h^{(n)}(x) - h^{(n)}(0)}{x} = \lim_{x\to 0+} \frac{R_n(x)}{x} e^{-1/x} = \lim_{t\to\infty} tR_n(1/t)e^{-t} = 0$$

since $t^d e^{-t} \to 0$, as $t \to \infty$, and R_n is rational. Since $\lim_{x\to 0-}[h^{(n)}(x) - h^{(n)}(0)]/x = 0$, this establishes **B** for $n+1$. Now, establish **A**for $n+1$ using the product rule and the fact that the derivative of a rational function is rational. Thus **A**and **B** hold by induction for all $n \geq 1$.

3.5.3 Apply the binomial theorem with $v = -1/2$ to obtain

$$\frac{1}{\sqrt{1+x}} = 1 - \frac{1}{2}x + \frac{1}{2}\cdot\frac{3}{4}x^2 - \frac{1}{2}\cdot\frac{3}{4}\cdot\frac{5}{6}x^3 + \dots .$$

Now, replace x by $-x^2$.

3.5.4 If $f(x) = \log(1+x)$, then $f(0) = 0$, $f'(x) = 1/(1+x)$, $f''(x) = -1/(1+x)^2$, and $f^{(n)}(x) = (-1)^{n-1}(n-1)!/(1+x)^n$ for $n \geq 1$. Hence $f^{(n)}(0)/n! = (-1)^{n-1}/n$ or

$$\log(1+x) = x - \frac{x^2}{2} + \frac{x^3}{3} - \frac{x^4}{4} + \dots .$$

3.5.5 If $1/(1+x) = \sum_{n=0}^{\infty} a_n x^n$ and $\log(1+x) = \sum_{n=1}^{\infty} b_n x^n$, then $\log(1+x)/(1+x) = \sum_{n=1}^{\infty} c_n x^n$, where

$$c_n = a_0 b_n + a_1 b_{n-1} + \cdots + a_{n-1}b_1 + a_n b_0.$$

Since $a_n = (-1)^n$, $b_n = (-1)^{n-1}/n$, we obtain

$$c_n = \sum_{i=0}^{n} a_{n-i} b_i = \sum_{i=0}^{n} (-1)^{n-i}(-1)^{i-1}/i$$
$$= -(-1)^n \left(1 + \frac{1}{2} + \frac{1}{3} + \cdots + \frac{1}{n}\right).$$

3.5.6 First, $f(x) = f(0) + f'(0)x + h(x)x^2/2$ with h continuous and $h(0) = f''(0) = q$. So

$$f\left(\frac{x}{\sqrt{n}}\right) = 1 + \frac{h(x/\sqrt{n})x^2}{2n}.$$

Now, apply Exercise **3.2.3** with $a_n = h(x/\sqrt{n})x^2/2 \to qx^2/2$ to obtain the result.

3.5.7 Define $h(t)$ by $e^t = 1 + t + t^2 h(t)/2$, $t \neq 0$. Then $e^t - 1 = t(1 + th(t)/2)$, and, by the exponential series, $\lim_{t\to 0} h(t) = 1$. Now,

$$\frac{1}{e^t - 1} - \frac{1}{t} = \frac{1}{t[1 + th(t)/2]} - \frac{1}{t}$$
$$= \frac{1}{t}\left[\frac{1}{1 + th(t)/2} - 1\right] = \frac{1}{t} \cdot \frac{th(t)/2}{1 + th(t)/2} = \frac{-h(t)/2}{1 + th(t)/2}.$$

This shows that the limit is $-1/2$.

3.5.8 This follows from the Lagrange form of the remainder in Taylor's theorem: for $|x - c| < d$,

$$\frac{|f^{(n+1)}(\eta)|}{(n+1)!}|x - c|^{n+1} \le C \cdot \frac{|x - c|^{n+1}}{d^{n+1}} \to 0.$$

3.5.9 With $a_j = f^{(j)}(c)/j!$, by (3.5.7), we have

$$\sum_{j=1}^{\infty} |a_j| d^j = g(|c| + d).$$

Thus the $(n+1)$-st remainder is no larger than the tail $\sum_{j=n+1}^{\infty} a_j (x - c)^j$ which is no larger in absolute value than

$$|x - c|^{n+1} \sum_{j=n+1}^{\infty} |a_j| d^{j-n-1} \le \frac{|x - c|^{n+1}}{d^{n+1}} \sum_{j=n+1}^{\infty} |a_j| d^j \le \frac{|x - c|^{n+1}}{d^{n+1}} g(|c| + d).$$

3.5.10 For $n = 0$, this is immediate. If $n \geq 1$,

$$\begin{aligned}\binom{-1/2}{n} &= \frac{(-\frac{1}{2})(-\frac{1}{2}-1)\dots(-\frac{1}{2}-n+1)}{1\cdot 2\cdot\dots\cdot n}\\ &= \frac{(-1)^n 1\cdot 3\cdot 5\cdot\dots\cdot(2n-1)}{2^n n!}\\ &= \frac{(-1)^n 1\cdot 2\cdot 3\cdot 4\cdot\dots\cdot(2n-1)\cdot(2n)}{2^n n!\cdot 2\cdot 4\cdot 6\cdot\dots\cdot(2n)}\\ &= \frac{(-1)^n (2n)!}{2^n n!\cdot 2^n\cdot 1\cdot 2\cdot 3\cdot\dots\cdot n}\\ &= \frac{(-1)^n(2n)!}{4^n(n!)^2}.\end{aligned}$$

Solutions to Exercises 3.6

3.6.1 Since $\sin^2 + \cos^2 = 1$, $\tan^2 + 1 = 1/\cos^2$, or $\cos^2 = 1/(1+\tan^2)$. Hence $\cos^2(\theta/2) = 1/(1+t^2)$, which gives

$$\cos\theta = 2\cos(\theta/2)^2 - 1 = \frac{2}{1+t^2} - 1 = \frac{1-t^2}{1+t^2}.$$

Also

$$\sin\theta = 2\sin(\theta/2)\cos(\theta/2) = 2t\cos^2(\theta/2) = \frac{2t}{1+t^2}.$$

Also

$$\tan\theta = \sin\theta/\cos\theta = \frac{2t}{1-t^2}.$$

3.6.2 A straightforward calculation using $\cos^2\theta + \sin^2\theta = 1$.

3.6.3 Use the addition formulae

$$\begin{aligned}R(\cos\phi, \sin\phi) &= (\cos\theta\cos\phi - \sin\theta\sin\phi, \sin\theta\cos\phi + \cos\theta\sin\phi)\\ &= (\cos(\phi+\theta), \sin(\phi+\theta)).\end{aligned}$$

3.6.4 By the previous exercises, we may translate and rotate the right triangle without affecting the validity of the claim. Thus we may assume the right angle vertex is the origin. Since every point in the plane is of the form $(r\cos\theta, r\sin\theta)$, we may rotate the triangle so that its sides are on the axes. Then the result is immediate.

3.6.5 f is differentiable at all nonzero reals, hence continuous there. Since $|f(x)| \le |x|$, f is also continuous at $x = 0$. Compute the variation of f corresponding to the partition $x_k = 2/(k\pi)$, $k = 1, \dots, n$. Since $f(x_k) = 0$ for k even and $f(x_k) = \pm 2/k\pi$ for k odd, the variation is larger than $(2/\pi)(1/2 + 1/3 + \cdots + 1/n)$. Hence f is not of bounded variation near 0.

3.6.6 If $x \ne 0$, then $f'(x) = 2x\sin(1/x) - \cos(1/x)$. If $x = 0$, then $f'(0) = \lim_{x\to 0} f(x)/x = \lim_{x\to 0} x\sin(1/x) = 0$. Hence $|f'(x)| \le 1 + 2|x|$ for all x. By Exercise **3.1.7**, f is of bounded variation on any bounded interval.

3.6.7 If f were injective on $(0,\epsilon)$, then by the IFT for continuous functions applied on $[0,\epsilon]$, f would be monotone. By the IFT for differentiable functions, this implies $f' \ge 0$ on $(0,\epsilon)$ or $f' \le 0$ on $(0,\epsilon)$. But $f'(x) = 1 + 4x\sin(1/x) - 2\cos(1/x)$ for $x \ne 0$. Since $\cos(1/x)$ oscillates between ± 1 arbitrarily close to 0, f' takes on positive and negative values arbitrarily close to 0.

3.6.8 It is enough to show that $x^n/n! \ge x^{n+1}/(n+1)!$ for $0 \le x \le 3$ and $n \ge 3$. But, simplifying, we see that the inequality holds iff $x \le n+1$, which is true, since $x \le 3 \le n+1$.

3.6.9 $\sin(\pi/3) = \sin(\pi - \pi/3)$. Hence

$$\sin(\pi/3) = \sin(2\pi/3) = 2\sin(\pi/3)\cos(\pi/3),$$

or $\cos(\pi/3) = 1/2$, which implies $\sin(\pi/3) = \sqrt{3}/2$ and $\tan(\pi/3) = \sqrt{3}$. We obtain $\sin(\pi/2 - x) = \cos x$ and $\cos(\pi/2 - x) = \sin x$. Hence $\sin(\pi/6) = 1/2$, $\cos(\pi/6) = \sqrt{3}/2$, and $\tan(\pi/6) = 1/\sqrt{3}$. Also $0 = \cos(\pi/2) = 2\cos(\pi/4)^2 - 1$, so $\cos(\pi/4) = 1/\sqrt{2}$, $\sin(\pi/4) = 1/\sqrt{2}$, and $\tan(\pi/4) = 1$.

3.6.10 Let $\theta = \pi/9$. Then $\sin(8\theta) = \sin(\theta)$ since $8\theta + \theta = 9\theta = \pi$. Hence

$$\begin{aligned}\sin\theta &= \sin(8\theta)\\ &= 2\sin(4\theta)\cos(4\theta)\\ &= 4\sin(2\theta)\cos(2\theta)\cos(4\theta)\\ &= 8\sin(\theta)\cos(\theta)\cos(2\theta)\cos(4\theta).\end{aligned}$$

Now, divide both sides by $8\sin\theta$.

3.6.11 Let $x = \cos(2\theta) = \cos(4\pi/5) < 0$. Then

$$\begin{aligned}0 &= \sin(\theta + 4\theta)\\ &= \sin\theta\cos(4\theta) + \cos\theta\sin(4\theta)\\ &= \sin\theta(2x^2 - 1) + 2\cos\theta\sin(2\theta)x\\ &= \sin\theta(2x^2 - 1) + 4x\cos^2\theta\sin\theta\\ &= \sin\theta(2x^2 - 1 + 2x(1 + x)) = \sin\theta(4x^2 + 2x - 1).\end{aligned}$$

Solving the quadratic yields $x = (-1 - \sqrt{5})/4$ hence $\cos(\pi/5) = (1 + \sqrt{5})/4$.

3.6.12 Let s_n denote the sum on the left. Since $2\cos a \sin b = \sin(a+b) - \sin(a-b)$ from (3.6.3), with $a = x$ and $b = x/2$,

$$\begin{aligned}\sin(x/2)s_1 &= \sin(x/2) + 2\cos x \sin(x/2)\\ &= \sin(x/2) + \sin(3x/2) - \sin(x/2) = \sin(3x/2).\end{aligned}$$

Thus the result is true when $n = 1$. Assuming the result is true for n and repeating the same reasoning with $a = (n+1)x$ and $b = x/2$,

$$\begin{aligned}\sin(x/2)s_{n+1} &= \sin(x/2)\,(s_n + 2\cos((n+1)x))\\ &= \sin((n+1/2)x) + 2\sin(x/2)\cos((n+1)x)\\ &= \sin((n+1/2)x) + \sin((n+3/2)x) - \sin((n+1/2)x)\\ &= \sin((n+3/2)x).\end{aligned}$$

This derives the result for $n+1$. Hence the result is true for all $n \geq 1$.

3.6.13 Divide $2\cos(2x) = 2\cos^2 x - 2\sin^2 x$ by $\sin(2x) = 2\sin x \cos x$.

3.6.14 The first identity is established using the double-angle formula. When $n = 2$, the identity (3.6.5) says $(x^4 - 1) = (x^2-1)(x^2+1)$ and is true. Now, assume the validity of the identity (3.6.5) for n. To obtain (3.6.5) with $2n$ replacing n, replace x by x^2 in (3.6.5) and use the first identity. Then

$$\begin{aligned}\frac{x^{4n}-1}{x^4-1} &= \prod_{k=1}^{n-1}\left[x^4 - 2x^2\cos(k\pi/n) + 1\right]\\ &= \prod_{k=1}^{n-1}\left[x^2 - 2x\cos(k\pi/2n) + 1\right] \cdot \prod_{k=1}^{n-1}\left[x^2 - 2x\cos(\pi - k\pi/2n) + 1\right]\\ &= \prod_{k=1}^{n-1}\left[x^2 - 2x\cos(k\pi/2n) + 1\right] \cdot \prod_{k=1}^{n-1}\left[x^2 - 2x\cos((2n-k)\pi/2n) + 1\right]\\ &= \prod_{k=1}^{n-1}\left[x^2 - 2x\cos(k\pi/2n) + 1\right] \cdot \prod_{k=n+1}^{2n-1}\left[x^2 - 2x\cos(k\pi/2n) + 1\right]\\ &= \prod_{\substack{k\neq n\\ 1\leq k\leq 2n-1}}\left[x^2 - 2x\cos(k\pi/2n) + 1\right]\\ &= \frac{1}{(x^2+1)} \cdot \prod_{k=1}^{2n-1}\left[x^2 - 2x\cos(k\pi/2n) + 1\right].\end{aligned}$$

Multiplying by $(x^4 - 1) = (x^2-1)(x^2+1)$, we obtain the result.

3.6.15 Let $a_n = 1/n$ and $c_n = \cos(nx)$, $n \geq 1$. Then for $x \notin 2\pi\mathbf{Z}$, by Exercise **3.6.12**, the sequence $b_n = c_1 + \cdots + c_n$, $n \geq 1$, is bounded. Hence by the Dirichlet test, $\sum \cos(nx)/n$ converges.

3.6.16 Fix $a \in \mathbf{R}$. We claim $f(a) = f(0)$. Since f is continuous at a, given $\epsilon > 0$, we can select $\delta > 0$ such that $|x-a| < \delta$ implies $|f(x)-f(a)| < \epsilon$. Since $p^* = 0$, we can select a period p less than δ. Then by periodicity $f(0) = f(np)$ for all integers n. Select n such that $np \leq a < (n+1)p$. Then $|np-a| < \delta$, so $|f(np) - f(a)| < \epsilon$. Thus $|f(0) - f(a)| < \epsilon$. Since ϵ is arbitrary, we conclude $f(0) = f(a)$. Thus f is constant.

Solutions to Exercises 3.7

3.7.1 With $dv = e^x dx$ and $u = \cos x$, $v = e^x$ and $du = -\sin x\, dx$. So

$$I = \int e^x \cos x\, dx = e^x \cos x + \int e^x \sin x\, dx.$$

Repeat with $dv = e^x dx$ and $u = \sin x$. We get $v = e^x$ and $du = \cos x\, dx$. Hence

$$\int e^x \sin x\, dx = e^x \sin x - \int e^x \cos x\, dx.$$

Now, insert the second equation into the first to yield

$$I = e^x \cos x + [e^x \sin x - I].$$

Solving for I yields

$$\int e^x \cos x\, dx = \frac{1}{2}\left(e^x \cos x + e^x \sin x\right).$$

3.7.2 Let $u = \arcsin x$. Then $x = \sin u$, so $dx = \cos u\, du$. So

$$\int e^{\arcsin x} dx = \int e^u \cos u\, du = \frac{1}{2} e^u \left(\sin u + \cos u\right)$$

by Exercise **3.7.1**. Since $\cos u = \sqrt{1-x^2}$, we obtain

$$\int e^{\arcsin x} dx = \frac{1}{2} e^{\arcsin x} \left(x + \sqrt{1-x^2}\right).$$

3.7.3

$$\begin{aligned}
\int \frac{x+1}{\sqrt{1-x^2}} dx &= \int \frac{x}{\sqrt{1-x^2}} dx + \int \frac{dx}{\sqrt{1-x^2}} \\
&= -\frac{1}{2} \int \frac{d(1-x^2)}{\sqrt{1-x^2}} + \arcsin x \\
&= -(1-x^2)^{1/2} + \arcsin x = \arcsin x - \sqrt{1-x^2}.
\end{aligned}$$

3.7.4 If $u = \arctan x$, then

$$\int \frac{\arctan x}{1+x^2} dx = \int u\, du = \frac{1}{2}u^2 = \frac{1}{2}(\arctan x)^2.$$

3.7.5 If $u = (\log x)^2$ and $dv = x^2 dx$, then $v = x^3/3$ and $du = 2(\log x)dx/x$. Hence

$$\int x^2(\log x)^2 dx = \frac{1}{3}x^3(\log x)^2 - \frac{2}{3}\int x^2 \log x\, dx.$$

If $u = \log x$ and $dv = x^2 dx$, then $v = x^3/3$ and $du = dx/x$. Hence

$$\int x^2 \log x\, dx = \frac{1}{3}x^3 \log x - \frac{1}{3}\int x^2 dx = \frac{1}{3}x^3 \log x - \frac{1}{9}x^3.$$

Now, insert the second integral into the first equation, and rearrange to obtain

$$\int x^2(\log x)^2 dx = \frac{x^3}{27}\left(9\log^2 x - 6\log x + 2\right).$$

3.7.6 Take $u = \sqrt{1-e^{-2x}}$. Then $u^2 = 1 - e^{-2x}$, so $2udu = 2e^{-2x}dx = 2(1-u^2)dx$. Hence

$$\int \sqrt{1-e^{-2x}} dx = \int \frac{u^2 du}{1-u^2} = \int \frac{du}{1-u^2} - u = \frac{1}{2}\log\left(\frac{1+u}{1-u}\right) - u$$

which simplifies to

$$\int \sqrt{1-e^{-2x}} dx = \log\left(1+\sqrt{1-e^{-2x}}\right) + x - \sqrt{1-e^{-2x}}.$$

3.7.7 Since $|\sin| \le 1$, $F'(0) = \lim_{x\to 0} F(x)/x = \lim_{x\to 0} x\sin(1/x) = 0$. Moreover,

$$F'(x) = 2x\sin(1/x) - \cos(1/x), \qquad x \neq 0.$$

So F' is not continuous at zero.

3.7.8 If f is of bounded variation, then its discontinuities are, at worst, jumps (Exercise **2.3.18**). But if $f = F'$, then f is a derivative and Exercise **3.1.6** says f cannot have any jumps. Thus f must be continuous.

3.7.9 Let $\theta = \arcsin\sqrt{x}$. Then $x = \sin^2\theta$ and the left side equals 2θ. Now, $2x - 1 = 2\sin^2\theta - 1 = -\cos(2\theta)$, so the right side equals

$$\pi/2 - \arcsin(\cos(2\theta)) = \arccos[\cos(2\theta)] = 2\theta.$$

3.7.10 Since $(\arcsin x)' = 1/\sqrt{1-x^2}$ and the derivative of the exhibited series is the Taylor series (Exercise **3.5.3**) of $1/\sqrt{1-x^2}$, the result follows from the theorem in this section.

3.7.11 Integration by parts, with $u = f(x)$ and $dv = \sin x\,dx$, $v = -\cos x$, and $du = f'(x)dx$. So

$$\int f(x)\sin x dx = -f(x)\cos x + \int f'(x)\cos x\,dx.$$

Repeating with $u = f'(x)$ and $dv = \cos x\,dx$,

$$\int f(x)\sin x dx = -f(x)\cos x + f'(x)\sin x + \int f''(x)\sin x\,dx.$$

But this is a recursive formula. Repeating this procedure with derivatives of f replacing f, we obtain the result.

3.7.12 Divide the first equation in (3.6.2) by the second equation in (3.6.2). You obtain the tangent formula. Set $a = \arctan(1/5)$ and $b = \arctan(1/239)$. Then $\tan a = 1/5$, so $\tan(2a) = 5/12$, so $\tan(4a) = 120/119$. Also $\tan b = 1/239$, so

$$\begin{aligned}\tan(4a-b) &= \frac{\tan(4a)-\tan b}{1+\tan(4a)\tan b} = \frac{(120/119)-(1/239)}{1+(120/119)(1/239)} \\ &= \frac{120\cdot 239 - 119}{119\cdot 239 + 120} = 1.\end{aligned}$$

Hence $4a - b = \pi/4$.

3.7.13 Since

$$\arctan x = x - \frac{x^3}{3} + \frac{x^5}{5} - \frac{x^7}{7} + \dots$$

is alternating with decreasing terms (as long as $0 < x < 1$), plugging in $x = 1/5$ and adding the first two terms yield $\arctan(1/5) = .1973$ with an error less than the third term which is less than 1×10^{-4}. Now, plugging $x = 1/239$ into the first term yields .00418 with an error less than the second term which is less than 10^{-6}. Since 16 times the first error plus 4 times the second error is less than 10^{-2}, $\pi = 16\arctan(1/5) - 4\arctan(1/239) = 3.14$ with an error less than 10^{-2}.

3.7.14 If $\theta = \arcsin(\sin 100)$, then $\sin(\theta) = \sin 100$ and $|\theta| \le \pi/2$. But $32\pi = 32 \times 3.14 = 100.48$, and $31.5\pi = 98.9$, with an error less than $32 \times 10^{-2} = .32$. Hence we are sure that $31.5\pi < 100 < 32\pi$ or $-\pi/2 < 100 - 32\pi < 0$, i.e., $\theta = 100 - 32\pi$.

3.7.15 Let $u = 1 - x^2$. Then $du = -2xdx$. Hence

$$\int \frac{-4x}{1-x^2}dx = \int \frac{2du}{u} = 2\log u = 2\log(1-x^2).$$

3.7.16 Completing the square, $x^2 - \sqrt{2}x + 1 = (x - 1/\sqrt{2})^2 + 1/2$. So with $u = \sqrt{2}x - 1$ and $v = (\sqrt{2}x - 1)^2 + 1 = 2x^2 - 2\sqrt{2}x + 2 = 2(x^2 - \sqrt{2}x + 1)$,

$$\begin{aligned}
\int \frac{4\sqrt{2} - 4x}{x^2 - \sqrt{2}x + 1}dx &= \int \frac{8\sqrt{2} - 8x}{(\sqrt{2}x - 1)^2 + 1}dx \\
&= \int \frac{4\sqrt{2}}{(\sqrt{2}x - 1)^2 + 1}dx - 2\int \frac{2\sqrt{2}(\sqrt{2}x - 1)}{(\sqrt{2}x - 1)^2 + 1}dx \\
&= \int \frac{4du}{u^2 + 1} - 2\int \frac{dv}{v} \\
&= 4\arctan u - 2\log v + 2\log 2 \\
&= 4\arctan(\sqrt{2}x - 1) - 2\log(x^2 - \sqrt{2}x + 1).
\end{aligned}$$

3.7.17 Let $s_n(x)$ denote the nth partial sum in (3.7.3). Then if $0 < x < 1$, by the Leibnitz test,

$$s_{2n}(x) \le \log(1+x) \le s_{2n-1}(x), \qquad n \ge 1.$$

In this last inequality, the number of terms in the partial sums is finite. Letting $x \nearrow 1$, we obtain

$$s_{2n}(1) \le \log 2 \le s_{2n-1}(1), \qquad n \ge 1.$$

Now, let $n \nearrow \infty$.

A.4 Solutions to Chapter 4

Solutions to Exercises 4.1

4.1.1 The first subrectangle thrown out has area $(1/3) \times 1 = 1/3$, the next two each has area $1/9$, the next four each has area $1/27$, and so on. So the areas of the removed rectangles sum to $(1/3)(1 + (2/3) + (2/3)^2 + \dots) = 1$. Hence the area of what is left, C', is zero. At the nth stage, the width of each of the remaining rectangles in C'_n is 3^{-n}. Since $C' \subset C'_n$, no rectangle in C' can have width greater than 3^{-n}. Since $n \ge 1$ is arbitrary, no open rectangle can lie in C'.

4.1.2 Here the widths of the removed rectangles sum to $\alpha/3+2\alpha/3^2+4\alpha/3^3+\cdots=\alpha$, so the area of what is left, C^α, is $1-\alpha>0$. At the nth stage, the width of each of the remaining rectangles in C^α_n is $3^{-n}\alpha$. Since $C^\alpha\subset C^\alpha_n$, no rectangle in C^α can have width greater than $3^{-n}\alpha$. Since $n\ge 1$ is arbitrary, no open rectangle can lie in C^α.

4.1.3 Here all expansions are ternary. Since $[0,2]\times[0,2]=2C_0$, given $(x,y)\in C_0$, we have to find $(x',y')\in C$ and $(x'',y'')\in C$ satisfying $x'+x''=2x$ and $y'+y''=2y$. Let $x=.d_1d_2d_3\dots$ and $y=.e_1e_2e_3\dots$. Then for all $n\ge 1$, $2d_n$ and $2e_n$ are 0, 2, or 4. Thus there are digits d'_n, d''_n, e'_n, $e''_n, n\ge 1$, equalling 0 or 2 and satisfying $d'_n+d''_n=2d_n$ and $e'_n+e''_n=2e_n$. Now, set $x'=.d'_1d'_2d'_3\dots$, $y'=.e'_1e'_2e'_3\dots$, $x''=.d''_1d''_2d''_3\dots$, $y''=.e''_1e''_2e''_3\dots$.

Solutions to Exercises 4.2

4.2.1 If Q is a rectangle, then so is $-Q$ and $\|Q\|=\|-Q\|$. If (Q_n) is a paving of A, then $(-Q_n)$ is a paving of $-A$, so

$$\operatorname{area}(-A)\le\sum_{n=1}^{\infty}\|-Q_n\|=\sum_{n=1}^{\infty}\|Q_n\|.$$

Since area (A) is the inf of the sums on the right, we obtain area $(-A)\le$ area (A). Applying this to $-A$, instead of A, yields area $(A)\le$ area $(-A)$ which yields reflection invariance, when combined with the previous inequality. For monotonicity, if $A\subset B$ and (Q_n) is a paving of B, then (Q_n) is a paving of A. So area $(A)\le\sum_{n=1}^\infty\|Q_n\|$. Since the inf of the sums on the right over all pavings of B equals area (B), area $(A)\le$ area (B).

4.2.2 Let L be a line segment. If L is vertical, we already know that area $(L)=0$. Otherwise, by translation and dilation invariance, we may assume that $L=\{(x,y):0\le x\le 1, y=mx\}$. If $Q_i=[(i-1)/n,i/n]\times[m(i-1)/n,mi/n]$, $i=1,\dots,n$, then $(Q_1,\dots,Q_n)$ is a paving of L and $\sum_{i=1}^n\|Q_i\|=m/n$. Since $n\ge 1$ is arbitrary, we conclude that area $(L)=0$. Since any line L is a countable union of line segments, by subadditivity, area $(L)=0$. Or just use rotation invariance to rotate L into the y-axis.

4.2.3 Assume first the base is horizontal. Write $P=A\cup B\cup C$, where A and B are triangles with horizontal bases and C is a rectangle, all intersecting only along their edges. Since the sum of the naive areas of A, B, and C is the naive area of P, subadditivity yields

$$\begin{aligned}\operatorname{area}(P)&\le\operatorname{area}(A)+\operatorname{area}(B)+\operatorname{area}(C)\\&=\|A\|+\|B\|+\|C\|=\|P\|.\end{aligned}$$

To obtain the reverse inequality, draw two triangles B and C with horizontal bases, such that $P \cup B \cup C$ is a rectangle and P, B, and C intersect only along their edges. Then the sum of the naive areas of P, B, and C equals the naive area of $P \cup B \cup C$, so by subadditivity of area,

$$\begin{aligned}\|P\| + \|B\| + \|C\| &= \|P \cup B \cup C\| \\ &= \text{area}\,(P \cup B \cup C) \\ &\leq \text{area}\,(P) + \text{area}\,(B) + \text{area}\,(C) \\ &\leq \text{area}\,(P) + \|B\| + \|C\|.\end{aligned}$$

Canceling $\|B\|$ and $\|C\|$, we obtain the reverse inequality $\|P\| \leq \text{area}\,(P)$. The case with a vertical base is analogous.

4.2.4 If T is the trapezoid, T can be broken up into the union of a rectangle and two triangles. As before, by subadditivity, this yields area $(T) \leq \|T\|$. Also two triangles can be added to T to obtain a rectangle. As before, this yields $\|T\| \leq \text{area}\,(T)$.

4.2.5 By extending the sides of the rectangles, $A \cup B$ can be decomposed into the union of finitely many rectangles intersecting only along their edges (seven rectangles in the general case). Moreover, since $A \cap B$ is one of these rectangles and is counted twice, using subadditivity, we obtain area $(A \cup B) \leq$ area (A) + area (B) − area $(A \cap B)$. To obtain the reverse inequality, add two rectangles C and D, to fill in the corners, obtaining a rectangle $A \cup B \cup C \cup D$, and proceed as before.

4.2.6 If Q is a rectangle, then $H(Q)$ is a (possibly degenerate) rectangle and $\|H(Q)\| = |k| \cdot \|Q\|$. If (Q_n) is a paving of A, then $(H(Q_n))$ is a paving of $H(A)$. So

$$\text{area}\,[H(A)] \leq \sum_{n=1}^{\infty} \|H(Q_n)\| = \sum_{n=1}^{\infty} |k| \cdot \|Q_n\| = |k| \sum_{n=1}^{\infty} \|Q_n\|.$$

Taking the inf over all pavings of A yields area $[H(A)] \leq |k| \cdot \text{area}\,(A)$. This establishes the result for H when $k = 0$. When $k \neq 0$, H is a bijection. In this case, replace in the last inequality A by $H^{-1}(A)$ and k by $1/k$ to obtain $|k| \cdot \text{area}\,(A) \leq \text{area}\,(H(A))$. Thus area $[H(A)] = |k| \cdot \text{area}\,(A)$ for $k \neq 0$, hence for all k real. The case V is similar.

4.2.7 Let $T(x,y) = (ax + by, cx + dy)$, $T'(x,y) = (a'x + b'y, c'x + d'y)$. Then

$$\begin{aligned}T \circ T'(x,y) &= (a(a'x + b'y) + b(c'x + d'y), c(a'x + b'y) + d(c'x + d'y)) \\ &= (a''x + b''y, c''x + d''y)\end{aligned}$$

where

$$\begin{pmatrix} a'' & b'' \\ c'' & d'' \end{pmatrix} = \begin{pmatrix} aa' + bc' & ab' + bd' \\ ca' + dc' & cb' + dd'. \end{pmatrix}$$

This shows $T'' = T \circ T'$ is linear. Now

$$\det(T'') = a''d'' - b''c'' = (aa' + bc')(cb' + dd') - (ab' + bd')(ca' + dc')$$

simplifies to $(ad - bc)(a'd' - b'c')$.

4.2.8 If $\text{area}\,(T(B)) = |\det(T)|\,\text{area}\,(B)$ for all B and $\text{area}\,(T'(A)) = |\det(T')|\,\text{area}\,(A)$ for all A, replace B by $T'(A)$ to obtain

$$\begin{aligned}\text{area}\,(T \circ T'(A)) &= \text{area}\,(T(T'(A))) = |\det(T)|\,\text{area}\,(T'(A)) \\ &= |\det(T)||\det(T')|\,\text{area}\,(A) = |\det(T \circ T')|\,\text{area}\,(A).\end{aligned}$$

4.2.9 Using matrix notation,

$$\begin{pmatrix} a & b \\ 0 & d \end{pmatrix} = \begin{pmatrix} 1 & 0 \\ 0 & d \end{pmatrix} \circ \begin{pmatrix} 1 & b \\ 0 & 1 \end{pmatrix} \circ \begin{pmatrix} a & 0 \\ 0 & 1 \end{pmatrix} \quad \text{and} \quad \begin{pmatrix} a & 0 \\ c & d \end{pmatrix} = \begin{pmatrix} a & 0 \\ 0 & 1 \end{pmatrix} \circ \begin{pmatrix} 1 & 0 \\ c & 1 \end{pmatrix} \circ \begin{pmatrix} 1 & 0 \\ 0 & d \end{pmatrix}.$$

Thus $U = V \circ S \circ H$ and $L = H \circ S \circ V$. Now

$$\begin{pmatrix} a & 0 \\ c & f \end{pmatrix} \circ \begin{pmatrix} 1 & e \\ 0 & 1 \end{pmatrix} = \begin{pmatrix} a & ae \\ c & f + ec \end{pmatrix} = \begin{pmatrix} a & b \\ c & d \end{pmatrix},$$

if $e = b/a$ and $f = d - ec$. Hence every T may be expressed as $L \circ U$ when $a \neq 0$. When $a = 0$,

$$\begin{pmatrix} 0 & b \\ c & d \end{pmatrix} = \begin{pmatrix} 0 & 1 \\ 1 & 0 \end{pmatrix} \circ \begin{pmatrix} c & d \\ 0 & b \end{pmatrix},$$

so in case we have $T = F \circ U$.

4.2.10 If (a, b) is in the unit disk, then $a^2 + b^2 < 1$. Hence $|a| < 1$. Hence

$$\sqrt{\left(\sqrt{2} - a\right)^2 + (0 - b)^2} \geq |\sqrt{2} - a| \geq \sqrt{2} - |a| \geq \sqrt{2} - 1.$$

Hence $d(D, \{(\sqrt{2}, 0)\}) > 0$. For the second part, let (a_n) denote a sequence of rationals converging to $\sqrt{2}$. Then $(a_n, 0) \in \mathbf{Q} \times \mathbf{Q}$, and

$$d((a_n, 0), (\sqrt{2}, 0)) \to 0, \qquad n \nearrow \infty.$$

4.2.11 For $1 > h > 0$, let $D_h^+ = \{(x, y) \in D^+ : y > h\}$. Then $D^+ \setminus D_h^+$ is contained in a rectangle with area h, and D_h^+ and $D^- = \{(x, y) \in D : y < 0\}$ are well separated. By reflection invariance, $\text{area}\,(D^+) = \text{area}\,(D^-)$. So by subadditivity,

$$\begin{aligned}\text{area}\,(D) &\geq \text{area}\,(D_h^+ \cup D^-) = \text{area}\,(D_h^+) + \text{area}\,(D^-) \\ &\geq \text{area}\,(D^+) - \text{area}\,(D^+ \setminus D_h^+) + \text{area}\,(D^+) \\ &\geq 2 \cdot \text{area}\,(D^+) - h.\end{aligned}$$

Since $h > 0$ is arbitrary, we obtain $\text{area}\,(D) \geq 2 \cdot \text{area}\,(D^+)$. The reverse inequality follows by subadditivity.

4.2.12 The derivation is similar to the derivation of area $(C) = 0$ presented at the end of §4.2.

4.2.13 If T is the triangle joining $(0,0)$, (a,b), $(a,-b)$, where

$$(a,b) = (\cos(\pi/n), \sin(\pi/n)),$$

then area $(T) = \sin(\pi/n)\cos(\pi/n)$. Since D_n is the union of n triangles T_1, ..., T_n, each having the same area as T (rotation invariance), subadditivity yields area $(D_n) \leq n\sin(\pi/n)\cos(\pi/n) = n\sin(2\pi/n)/2$. The reverse inequality is obtained by shrinking each triangle T_i toward its center and using well-separated additivity.

4.2.14 Let α denote the inf on the right side. Since (T_n) is a cover, we obtain area $(A) \leq \sum_{n=1}^\infty$ area $(T_n) = \sum_{n=1}^\infty \|T_n\|$. Hence area $(A) \leq \alpha$. On the other hand, if (Q_n) is any paving of A, write each $Q_n = T_n \cup T_n'$ as the union of two triangles to obtain a triangular paving $(T_n) \cup (T_n')$. Hence

$$\alpha \leq \sum_{n=1}^\infty \|T_n\| + \|T_n'\| = \sum_{n=1}^\infty \|Q_n\|.$$

Taking the inf over all pavings (Q_n), we obtain $\alpha \leq$ area (A).

4.2.15 Assume not. Then for *every* rectangle Q, area $(Q \cap A) \leq \alpha \cdot$ area (Q). If (Q_n) is any paving of A, then $(Q_n \cap A)$ is a cover of A. So

$$\text{area}\,(A) \leq \sum_{n=1}^\infty \text{area}\,(Q_n \cap A) \leq \alpha \sum_{n=1}^\infty \text{area}\,(Q_n).$$

Taking the inf of the right side over all (Q_n), area $(A) \leq \alpha \cdot$ area (A). Since $\alpha < 1$, this yields area $(A) = 0$.

Solutions to Exercises 4.3

4.3.1 Apply dilation invariance to $g(x) = f(x)/x$. Then

$$\begin{aligned}\int_0^\infty f(kx)x^{-1}\,dx &= k\int_0^\infty f(kx)(kx)^{-1}\,dx = k\int_0^\infty g(kx)\,dx \\ &= k \cdot \frac{1}{k}\int_0^\infty g(x)\,dx = \int_0^\infty f(x)x^{-1}\,dx.\end{aligned}$$

4.3.2 Let G be the subgraph of f over (a,b), and let $H(x,y) = (-x,y)$. Then $H(G)$ equals $\{(-x,y) : a < x < b, 0 < y < f(x)\}$. But this equals $\{(x,y) : -b < x < -a, 0 < y < f(-x)\}$, which is the subgraph of $f(-x)$ over $(-b,-a)$. Thus by Exercise **4.2.6**,

$$\int_a^b f(x)\,dx = \text{area}\,(G) = \text{area}\,(H(G)) = \int_{-b}^{-a} f(-x)\,dx.$$

4.3.3 Since f is uniformly continuous on $[a,b]$, there is a $\delta > 0$, such that $\mu(\delta) < \epsilon/(b-a)$. Here μ is the uniform modulus of continuity of f over $[a,b]$. In §2.3, we showed that, with this choice of δ, for any partition $a = x_0 < x_1 < \cdots < x_n = b$ of mesh $< \delta$ and choice of intermediate points $x_1^\#, \dots, x_n^\#$, the piecewise constant function $g(x) = f(x_i^\#)$, $x_{i-1} < x < x_i$, $1 \le i \le n$, satisfies $|f(x) - g(x)| < \epsilon/(b-a)$ over (a,b). Now, by additivity ,

$$\int_a^b [g(x) \pm \epsilon/(b-a)]\,dx = \sum_{i=1}^n f(x_i^\#)(x_i - x_{i-1}) \pm \epsilon,$$

and

$$g(x) - \epsilon/(b-a) < f(x) < g(x) + \epsilon/(b-a), \quad a < x < b.$$

Hence by monotonicity,

$$\sum_{i=1}^n f(x_i^\#)(x_i - x_{i-1}) - \epsilon \le \int_a^b f(x)\,dx \le \sum_{i=1}^n f(x_i^\#)(x_i - x_{i-1}) + \epsilon,$$

which is to say that

$$\left| I - \sum_{i=1}^n f(x_i^\#)(x_i - x_{i-1}) \right| \le \epsilon.$$

4.3.4 Since $f(x) \le x = g(x)$ on $(0,1)$ and the subgraph of g is a triangle, by monotonicity, $\int_0^1 f(x)\,dx \le \int_0^1 x\,dx = 1/2$. On the other hand, the subgraph of g equals the union of the subgraphs of f and $g - f$. So by subadditivity,

$$1/2 = \int_0^1 g(x)\,dx \le \int_0^1 f(x)\,dx + \int_0^1 [g(x) - f(x)]\,dx.$$

But $g(x) - f(x) > 0$ iff $x \in \mathbf{Q}$. So the subgraph of $g - f$ is a countable (§1.7) union of line segments. By subadditivity, again, the area $\int_0^1 [g(x) - f(x)]\,dx$ of the subgraph of $g - f$ equals zero. Hence $\int_0^1 f(x)\,dx = 1/2$.

4.3.5 Suppose that g is a constant $c \geq 0$, and let G be the subgraph of f. Then the subgraph of $f + c$ is the union of the rectangle $Q = (a, b) \times (0, c]$ and the vertical translate $G + (0, c)$. Thus by subadditivity and translation invariance,

$$\begin{aligned} \int_a^b [f(x) + g(x)]\, dx &= \text{area}\,[Q \cup (G + (0, c))] \\ &\leq \text{area}\,(G) + \text{area}\,(Q) \\ &= \int_a^b f(x)\, dx + \int_a^b g(x)\, dx. \end{aligned}$$

Let αQ denote the centered dilate of Q, $0 < \alpha < 1$. Then αQ and $G + (0, c)$ are well separated. So

$$\begin{aligned} \int_a^b [f(x) + g(x)]\, dx &\geq \text{area}\,[\alpha Q \cup (G + (0, c))] \\ &= \text{area}\,(G) + \alpha^2\, \text{area}\,(Q) \\ &= \int_a^b f(x)\, dx + \alpha^2 \int_a^b g(x)\, dx. \end{aligned}$$

Let $\alpha \to 1$ to get the reverse inequality. Thus the result is true when g is constant. If g is piecewise constant over a partition $a = x_0 < x_1 < \cdots < x_n = b$, then apply the constant case to the intervals (x_{i-1}, x_i), $i = 1, \ldots, n$, and sum.

4.3.6 By subadditivity, $\int_0^\infty f(x)\, dx \leq \sum_{n=1}^\infty \int_{n-1}^n f(x)\, dx = \sum_{n=1}^\infty c_n$. For the reverse inequality,

$$\int_0^\infty f(x)\, dx \geq \int_0^N f(x)\, dx = \sum_{n=1}^N \int_{n-1}^n f(x)\, dx = \sum_{n=1}^N c_n.$$

Now, let $N \nearrow \infty$. This establishes the nonnegative case. Now, apply the nonnegative case to $|f|$. Then f is integrable iff $\sum |c_n| < \infty$. Now, apply the nonnegative case to f^+ and f^-, and subtract to get the result for the integrable case.

4.3.7 To be Riemann integrable, the Riemann sum must be close to a specific real I for any partition with small enough mesh and any choice of intermediate points $x_1^\#, \ldots, x_n^\#$. But for any partition $a = x_0 < x_1 < \cdots < x_n = b$, with any mesh size, we can choose intermediate points that are irrational, leading to a Riemann sum of 1. We can also choose intermediate points that are rational, leading to a Riemann sum of 0. Since no real I can be close to 0 and 1, simultaneously, f is not Riemann integrable.

4.3.8 Apply the integral test to $f(x) = g(x\delta)$ to get

$$\int_1^\infty g(x\delta)\,dx \le \sum_{n=1}^\infty g(n\delta) \le \int_1^\infty g(x\delta)\,dx + g(\delta).$$

By dilation invariance,

$$\int_\delta^\infty g(x)\,dx \le \delta \sum_{n=1}^\infty g(n\delta) \le \int_\delta^\infty g(x)\,dx + \delta g(\delta).$$

Since g is bounded, $\delta g(\delta) \to 0$, as $\delta \to 0+$. Now, let $\delta \to 0+$, and use continuity at the endpoints.

4.3.9 If f is even and nonnegative or integrable, by Exercise **4.3.2**,

$$\begin{aligned}\int_{-b}^b f(x)\,dx &= \int_{-b}^0 f(x)\,dx + \int_0^b f(x)\,dx \\ &= \int_0^b f(-x)\,dx + \int_0^b f(x)\,dx \\ &= 2\int_0^b f(x)\,dx.\end{aligned}$$

If f is odd and integrable,

$$\begin{aligned}\int_{-b}^b f(x)\,dx &= \int_{-b}^0 f(x)\,dx + \int_0^b f(x)\,dx \\ &= \int_0^b f(-x)\,dx + \int_0^b f(x)\,dx = 0.\end{aligned}$$

4.3.10 By the previous exercise, $\int_{-\infty}^\infty e^{-a|x|}\,dx = 2\int_0^\infty e^{-ax}\,dx$. By the integral test, $\int_1^\infty e^{-ax}\,dx \le \sum_{n=1}^\infty e^{-an} = 1/(1-e^{-a}) < \infty$. Hence

$$\int_0^\infty e^{-ax}\,dx = \int_0^1 e^{-ax}\,dx + \int_1^\infty e^{-ax}\,dx \le 1 + \frac{1}{1-e^{-a}}.$$

4.3.11 Since f is superlinear, for any $M > 0$, there is a $b > 0$, such that $f(x)/x > M$ for $x > b$. Hence $\int_b^\infty e^{sx}e^{-f(x)}\,dx \le \int_b^\infty e^{(s-M)x}\,dx = \int_b^\infty e^{-(M-s)|x|}\,dx$. Similarly, there is an $a < 0$ such that $f(x)/(-x) > M$ for $x < a$. Hence $\int_{-\infty}^a e^{sx}e^{-f(x)}\,dx \le \int_{-\infty}^a e^{(s+M)x}\,dx = \int_{-\infty}^a e^{-(M+s)|x|}\,dx$. Since f is continuous, f is bounded on (a,b), hence integrable over (a,b). Thus with $g(x) = e^{sx}e^{-f(x)}$,

$$\int_{-\infty}^{\infty} e^{sx}e^{-f(x)}\,dx = \int_{-\infty}^{a} g(x)\,dx + \int_{a}^{b} g(x)\,dx + \int_{b}^{\infty} g(x)\,dx$$
$$\leq \int_{-\infty}^{a} e^{-(M+s)|x|}\,dx + \int_{a}^{b} e^{sx}e^{-f(x)}\,dx + \int_{b}^{\infty} e^{-(M-s)|x|}\,dx,$$

which is finite, as soon as M is chosen $> |s|$.

4.3.12 Suppose that there was such a δ. First, choose $f(x) = 1$, for all $x \in \mathbf{R}$, in (4.3.13) to get $\int_{-\infty}^{\infty} \delta(x)\,dx = 1$. Thus δ is integrable over $\mathbf{R}$. Now, let f equal 1 at all points except at zero, where we set $f(0) = 0$. Then (4.3.13) fails because the integral is still 1, but the right side vanishes. However, this is of no use, since we are assuming (4.3.13) for f continuous only. Because of this, we let $f_n(x) = 1$ for $|x| \geq 1/n$ and $f_n(x) = n|x|$ for $|x| \leq 1/n$. Then f_n is continuous and nonnegative. Hence by monotonicity,

$$0 \leq \int_{|x|\geq 1/n} \delta(x)\,dx \leq \int_{-\infty}^{\infty} \delta(x)f_n(x)\,dx = f_n(0) = 0,$$

for all $n \geq 1$. This shows that $\int_{|x|\geq 1/n} \delta(x)\,dx = 0$ for all $n \geq 1$. But, by continuity at the endpoints,

$$\begin{aligned}
1 = \int_{-\infty}^{\infty} \delta(x)\,dx &= \int_{-\infty}^{0} \delta(x)\,dx + \int_{0}^{\infty} \delta(x)\,dx \\
&= \lim_{n\nearrow\infty} \left(\int_{-\infty}^{-1/n} \delta(x)\,dx + \int_{1/n}^{\infty} \delta(x)\,dx \right) \\
&= \lim_{n\nearrow\infty} (0+0) = 0,
\end{aligned}$$

a contradiction.

4.3.13 By Exercise **3.3.7** with $x = c \pm \delta$,

$$f(c\pm\delta) - f(c) \geq \pm f'_{\pm}(c)\delta.$$

Since f'_+ is increasing,

$$f'_+(c)\delta \geq \int_{c-\delta}^{c} f'_+(x)\,dx.$$

Also since f'_- is increasing,

$$f'_-(c)\delta \leq \int_{c}^{c+\delta} f'_-(x)\,dx.$$

Combining these inequalities yields (4.3.14). Now note that $f'_{\pm}$ are both increasing and therefore bounded on $[a,b]$ between $f'_{\pm}(a)$ and $f'_{\pm}(b)$, hence integrable on (a,b). Select $n \ge 1$ and let $\delta = (b-a)/n$ and $a = x_0 < x_1 < \dots < x_n = b$ be the partition of $[a,b]$ given by $x_i = a + i\delta$, $i = 0, \dots, n$. Then applying (4.3.14) at $c = x_i$ yields

$$f(x_{i+1}) - f(x_i) \ge \int_{x_{i-1}}^{x_i} f'_+(x)\,dx.$$

Summing over $1 \le i \le n-1$, we obtain

$$f(b) - f(a+\delta) \ge \int_a^{b-\delta} f'_+(x)\,dx.$$

Let $\delta \to 0$. Since f is continuous (Exercise **3.3.6**) and integrable, continuity at the endpoints implies

$$f(b) - f(a) \ge \int_a^b f'_+(x)\,dx.$$

Similarly,

$$f(x_i) - f(x_{i-1}) \le f'_-(x_i)\delta \le \int_{x_i}^{x_{i+1}} f'_-(x)\,dx.$$

Summing over $1 \le i \le n-1$, we obtain

$$f(b-\delta) - f(a) \le \int_{a+\delta}^{b} f'_-(x)\,dx.$$

Now let $\delta \to 0$. Since $f'_-(t) \le f'_+(t)$, the result follows.

4.3.14 From §4.3, we know $\lim_{n\to\infty} F(n\pi)$ exists; it follows that any subsequence $(F(N_n\pi))$ is convergent. Given $b_n \to \infty$, let $N_n = \lfloor b_n/\pi \rfloor$. Then $|b_n - N_n\pi| \le \pi$ and

$$|F(b_n) - F(N_n\pi)| = \left| \int_{N_n\pi}^{b_n} \frac{\sin x}{x}\,dx \right| \le \frac{|b_n - N_n\pi|}{N_n\pi} \le \frac{1}{N_n}.$$

It follows that $F(\infty)$ exists. Since $F : (0,\infty) \to \mathbf{R}$ is continuous, it follows F is bounded as in Exercise **2.3.24**.

Solutions to Exercises 4.4

4.4.1 $F(x) = e^{-sx}/(-s)$ is a primitive of $f(x) = e^{-sx}$, and e^{-sx} is positive. So

$$\int_0^\infty e^{-sx}\,dx = \frac{1}{-s} e^{-sx} \Big|_0^\infty = \frac{1}{s}, \qquad s > 0.$$

4.4.2 x^r/r is a primitive of x^{r-1} for $r \neq 0$, and $\log x$ is a primitive, when $r = 0$. Thus $\int_0^1 dx/x = \log x|_0^1 = 0 - (-\infty) = \infty$ and $\int_1^\infty dx/x = \log x|_1^\infty = \infty - 0 = \infty$. Hence all three integrals are equal to ∞, when $r = 0$. Now,

$$\int_0^1 x^{r-1}\, dx = \frac{1}{r} - \frac{1}{r} \lim_{x\to 0+} x^r = \begin{cases} \frac{1}{r}, & r > 0, \\ \infty, & r < 0. \end{cases}$$

Also

$$\int_1^\infty x^{r-1}\, dx = \frac{1}{r} \lim_{x\to\infty} x^r - \frac{1}{r} = \begin{cases} -\frac{1}{r}, & r < 0, \\ \infty, & r > 0. \end{cases}$$

Since $\int_0^\infty = \int_0^1 + \int_1^\infty$, $\int_0^\infty x^{r-1}\, dx = \infty$ in all cases.

4.4.3 Pick c in (a, b). Since any primitive F differs from F_c by a constant, it is enough to verify the result for F_c. But f bounded and (a, b) bounded imply f integrable. So $F_c(a+)$ and $F_c(b-)$ exist and are finite by continuity at the endpoints.

4.4.4 By the fundamental theorem, for $T > 1$,

$$f(T) - f(1+) = f(T-) - f(1+) = \int_1^T f'(x)\, dx.$$

The result follows by sending $T \to \infty$ and using continuity at the endpoints.

4.4.5 By the integral test, $\sum_{n=1}^\infty f'(n) < \infty$ iff $\int_1^\infty f'(x)\, dx < \infty$ which by the previous exercise happens iff $f(\infty) < \infty$. Similarly $\sum_{n=1}^\infty f'(n)/f(n) < \infty$ iff $\log f(\infty) < \infty$.

4.4.6 Take $u = 1/x$ and $dv = \sin x\, dx$. Then $du = -dx/x^2$ and $v = -\cos x$. So

$$\int_1^b \frac{\sin x}{x}\, dx = \frac{-\cos x}{x}\bigg|_1^b + \int_1^b \frac{\cos x}{x^2}\, dx.$$

But, by (4.3.1), $\cos x/x^2$ is integrable over $(1, \infty)$. So by continuity at the endpoints,

$$\lim_{b\to\infty} \int_1^b \frac{\sin x}{x}\, dx = \cos 1 + \int_1^\infty \frac{\cos x}{x^2}\, dx.$$

Since $F(b) - \int_1^b \sin x(dx/x)$ does not depend on b, $F(\infty)$ exists and is finite.

4.4.7 The function $g(t) = e^{-t}$ is strictly monotone with $g((0, \infty)) = (0, 1)$. Now, apply substitution.

4.4.8 Let $u = x^n$ and $dv = e^{-sx}\, dx$. Then $du = nx^{n-1}\, dx$, and $v = e^{-sx}/(-s)$. Hence

$$\int_0^\infty e^{-sx}x^n\,dx = \left.\frac{e^{-sx}x^n}{-s}\right|_0^\infty + \frac{n}{s}\int_0^\infty e^{-sx}x^{n-1}\,dx.$$

If we call the integral on the left I_n, this says that $I_n = (n/s)I_{n-1}$. Iterating this down to $n=0$ yields $I_n = n!/s^n$ since $I_0 = 1/s$ from Exercise **4.4.1**.

4.4.9 Call the integrals I_s and I_c. Let $u = \sin(sx)$ and $dv = e^{-nx}\,dx$. Then $du = s\cos(sx)\,dx$ and $v = e^{-nx}/(-n)$. So

$$I_s = \left.\frac{e^{-nx}\sin(sx)}{-n}\right|_0^\infty + \frac{s}{n}\int_0^\infty e^{-nx}\cos(sx)\,dx = \frac{s}{n}I_c.$$

Now, let $u = \cos(sx)$ and $dv = e^{-nx}\,dx$. Then $du = -s\sin(sx)\,dx$ and $v = e^{-nx}/(-n)$. So

$$I_c = \left.\frac{e^{-nx}\cos(sx)}{-n}\right|_0^\infty - \frac{s}{n}\int_0^\infty e^{-nx}\sin(sx)\,dx = \frac{1}{n} - \frac{s}{n}I_s.$$

Thus $nI_s = sI_c$, and $nI_c = 1 - sI_s$. Solving, we obtain $I_s = s/(n^2+s^2)$ and $I_c = n/(n^2+s^2)$.

4.4.10 Let $u = t^{x-1}$ and $dv = e^{-t^2/2}t\,dt$. Then $du = (x-1)t^{x-2}\,dt$, and $v = -e^{-t^2/2}$. So

$$\begin{aligned}\int_0^\infty e^{-t^2/2}t^x\,dx &= -e^{-t^2/2}t^{x-1}\Big|_0^\infty + (x-1)\int_0^\infty e^{-t^2/2}t^{x-2}\,dt\\ &= (x-1)\int_0^\infty e^{-t^2/2}t^{x-2}\,dt.\end{aligned}$$

If $I_n = \int_0^\infty e^{-t^2/2}t^n\,dt$, then

$$\begin{aligned}I_{2n+1} &= 2n\cdot I_{2n-1} = 2n\cdot(2n-2)I_{2n-3}\\ &= \cdots = 2n\cdot(2n-2)\ldots 4\cdot 2\cdot I_1 = 2^n n! I_1.\end{aligned}$$

But, substituting $u = t^2/2$, $du = t\,dt$ yields

$$I_1 = \int_0^\infty e^{-t^2/2}t\,dt = \int_0^\infty e^{-u}\,du = 1.$$

4.4.11 Let $u = (1-t)^n$ and $dv = t^{x-1}\,dt$. Then $du = -n(1-t)^{n-1}\,dt$, and $v = t^x/x$. So

$$\begin{aligned}\int_0^1 (1-t)^n t^{x-1}\,dt &= \left.\frac{(1-t)^n t^x}{x}\right|_0^1 + \frac{n}{x}\int_0^1 (1-t)^{n-1}t^{x+1}\,dt\\ &= \frac{n}{x}\int_0^1 (1-t)^{n-1}t^{x+1}\,dt.\end{aligned}$$

Thus integrating by parts increases the x by 1 and decreases the n by 1. Iterating this n times,

$$\int_0^1 (1-t)^n t^{x-1}\,dt = \frac{n\cdot(n-1)\cdot\dots\cdot 1}{x\cdot(x+1)\cdot\dots\cdot(x+n-1)}\cdot\int_0^1 t^{x+n-1}\,dt.$$

But $\int_0^1 t^{x+n-1}\,dt = 1/(x+n)$. So

$$\int_0^1 (1-t)^n t^{x-1}\,dt = \frac{n!}{x\cdot(x+1)\cdot\dots\cdot(x+n)}.$$

4.4.12 Let $t = -\log x$, $x = e^{-t}$. Then from Exercise **4.4.7** and Exercise **4.4.8**,

$$\int_0^1 (-\log x)^n\,dx = \int_0^\infty e^{-t}t^n\,dt = n!.$$

4.4.13 Let $I_n = \int_{-1}^1 (x^2-1)^n\,dx$ and $u = (x^2-1)^n$, $dv = dx$. Then

$$\begin{aligned} I_n &= x(x^2-1)^n\big|_{-1}^1 - 2n\int_{-1}^1 (x^2-1)^{n-1}x^2\,dx \\ &= -2nI_n - 2nI_{n-1}. \end{aligned}$$

Solving for I_n, we obtain

$$I_n = -\frac{2n}{2n+1}I_{n-1} = \dots = (-1)^n\frac{2n\cdot(2n-2)\cdot\dots\cdot 2}{(2n+1)\cdot(2n-1)\cdot\dots\cdot 3}\cdot 2$$

since $I_0 = 2$.

4.4.14 Let $f(x) = (x^2-1)^n$. Then $P_n(x) = f^{(n)}(x)/2^n n!$. Note that $f(\pm 1) = 0$, $f'(\pm 1) = 0$, ..., and $f^{(n-1)}(\pm 1) = 0$, since all these derivatives have at least one factor (x^2-1) by the product rule. Hence integrating by parts,

$$\int_{-1}^1 f^{(n)}(x)f^{(n)}(x)\,dx = -\int_{-1}^1 f^{(n-1)}(x)f^{(n+1)}(x)\,dx$$

increases one index and decreases the other. Iterating, we get

$$\int_{-1}^1 P_n(x)^2\,dx = \frac{1}{(2^n n!)^2}\int_{-1}^1 \left[f^{(n)}(x)\right]^2\,dx = \frac{(-1)^n}{(2^n n!)^2}\int_{-1}^1 f(x)f^{(2n)}(x)\,dx.$$

But f is a polynomial of degree $2n$ with highest-order coefficient 1. Hence $f^{(2n)}(x) = (2n)!$. So

$$\int_{-1}^{1} P_n(x)^2\,dx = \frac{(-1)^n(2n)!}{(2^n n!)^2}\int_{-1}^{1}(x^2-1)^n\,dx.$$

Now, inserting the result of the previous exercise and simplifying leads to $2/(2n+1)$.

4.4.15 Apply the fundamental theorem to the result of Exercise **3.7.11**.

4.4.16 With $\pi = a/b$ and

$$g_n(x) = \frac{(bx)^n(a-bx)^n}{n!},$$

it is clear that $I_n > 0$ since $\sin x$ and $g_n(x)$ are positive on $(0,\pi)$. Moreover $I_n \in \mathbf{Z}$ follows from Exercise **3.3.30** and Exercise **4.4.15**. Finally, $I_n \le 2M^n/n!$ where $M = \max bx(a-bx)$ over $[0,\pi]$, which goes to 0 as $n \to \infty$. But there are no integers between 0 and 1; hence π is irrational.

4.4.17 By the integral test, $\zeta(s)$ differs from $\int_1^\infty x^{-s}\,dx$ by, at most, 1. But $\int_1^\infty x^{-s}\,dx$ converges for $s > 1$ by Exercise **4.4.2**.

4.4.18 Here $f(x) = 1/x$ and $\int_1^{n+1} f(x)\,dx = \log(n+1)$. Since $1/(n+1) \to 0$, by the integral test,

$$\begin{aligned}
\gamma &= \lim_{n\nearrow\infty}\left(1+\frac{1}{2}+\frac{1}{3}+\cdots+\frac{1}{n}-\log n\right)\\
&= \lim_{n\nearrow\infty}\left[1+\frac{1}{2}+\frac{1}{3}+\cdots+\frac{1}{n}+\frac{1}{n+1}-\log(n+1)\right]\\
&= \lim_{n\nearrow\infty}\left[f(1)+f(2)+\cdots+f(n)-\int_1^{n+1} f(x)\,dx\right]
\end{aligned}$$

exists and satisfies $0 < \gamma < 1$.

4.4.19 Call the integrals I_n^c and I_n^s. Then $I_0^s = 0$, and $I_n^c = 0$, $n \ge 0$, since the integrand is odd. Also $I_n^s = 2\int_0^\pi x\sin(nx)\,dx$ since the integrand is even. Now, for $n \ge 1$,

$$\begin{aligned}
\int_0^\pi x\sin(nx)\,dx &= -\left.\frac{x\cos(nx)}{n}\right|_0^\pi + \frac{1}{n}\int_0^\pi \cos(nx)\,dx\\
&= -\frac{\pi\cos(n\pi)}{n} + \frac{1}{n}\cdot\left.\frac{\sin(nx)}{n}\right|_0^\pi\\
&= \frac{(-1)^{n-1}\pi}{n}.
\end{aligned}$$

Thus $I_n^s = 2\pi(-1)^{n-1}/n$, $n \ge 1$.

4.4.20 By oddness, $\int_{-\pi}^{\pi} \sin(nx)\cos(mx)\,dx = 0$ for all $m, n \geq 0$. For $m \neq n$, using (3.6.3),

$$\begin{aligned}\int_{-\pi}^{\pi} \sin(nx)\sin(mx)\,dx &= \frac{1}{2}\int_{-\pi}^{\pi} [\cos((n-m)x) - \cos((n+m)x)]\,dx \\ &= \frac{1}{2}\left(\frac{\sin((n-m)x)}{n-m} - \frac{\sin((n+m)x)}{n+m}\right)\Bigg|_{-\pi}^{\pi} \\ &= 0.\end{aligned}$$

For $m = n$,

$$\begin{aligned}\int_{-\pi}^{\pi} \sin(nx)\sin(mx)\,dx &= \frac{1}{2}\int_{-\pi}^{\pi} [1 - \cos(2nx)]\,dx \\ &= \frac{1}{2}[x - \sin(2nx)/2n]\Bigg|_{-\pi}^{\pi} = \pi.\end{aligned}$$

Similarly, for $\int_{-\pi}^{\pi} \cos(nx)\cos(mx)\,dx$. Hence

$$\int_{-\pi}^{\pi} \cos(nx)\cos(mx)\,dx = \int_{-\pi}^{\pi} \sin(nx)\sin(mx)\,dx = \begin{cases} 0, & n \neq m, \\ \pi, & n = m. \end{cases}$$

4.4.21 By additivity, $q(t) = at^2 + 2bt + c$ where $a = \int_a^b g(t)^2\,dt$, $b = \int_a^b f(t)g(t)\,dt$, and $c = \int_a^b f(t)^2\,dt$. Since q is nonnegative, q has at most one root. Hence $b^2 - ac \leq 0$, which is Cauchy–Schwarz.

4.4.22 By substituting $t = n(1-s)$, $dt = nds$, and equation (2.3.2),

$$\begin{aligned}\int_0^n \frac{1-(1-t)^n}{t}\,dt &= \int_0^1 \frac{1-s^n}{1-s}\,ds \\ &= \int_0^1 \left[1 + s + \cdots + s^{n-1}\right]ds \\ &= 1 + \frac{1}{2} + \cdots + \frac{1}{n}.\end{aligned}$$

4.4.23 Continuity of F was established in §4.3. Now, f is continuous on the subinterval (x_{i-1}, x_i). Hence $F(x) = F(x_{i-1}) + F_{x_{i-1}}(x)$ is differentiable by the first fundamental theorem.

4.4.24 For any f, let $I(f) = \int_a^b f(x)\,dx$. If $g : [a,b] \to \mathbf{R}$ is nonnegative and continuous, we can (§2.3) find a piecewise constant $g_\epsilon \geq 0$, such that $g_\epsilon(x) \leq g(x) + \epsilon \leq g_\epsilon(x) + 2\epsilon$ on $a \leq x \leq b$. By monotonicity,

$$I(g_\epsilon) \leq I(g + \epsilon) \leq I(g_\epsilon + 2\epsilon).$$

But, by Exercise **4.3.5**, $I(g+\epsilon) = I(g) + \epsilon(b-a)$ and $I(g_\epsilon + 2\epsilon) = I(g_\epsilon) + 2\epsilon(b-a)$. Hence

$$I(g_\epsilon) \le I(g) + \epsilon(b-a) \le I(g_\epsilon) + 2\epsilon(b-a),$$

or

$$|I(g) - I(g_\epsilon)| \le \epsilon(b-a).$$

Similarly, since $f(x) + g_\epsilon(x) \le f(x) + g(x) + \epsilon \le f(x) + g_\epsilon(x) + 2\epsilon$,

$$|I(f+g) - I(f) - I(g_\epsilon)| \le \epsilon(b-a),$$

where we have used Exercise **4.3.5**, again. Thus $|I(f+g) - I(f) - I(g)| \le 2\epsilon(b-a)$. Since ϵ is arbitrary, we conclude that $I(f+g) = I(f) + I(g)$.

4.4.25 Let $m_i = g(t_i)$, $i = 0, 1, \dots, n+1$. For each $i = 1, \dots, n+1$, define $\#_i : (m, M) \to \{0, 1\}$ by setting $\#_i(x) = 1$ if x is between m_{i-1} and m_i and $\#_i(x) = 0$, otherwise. Since the m_i's may not be increasing, for a given x, more than one $\#_i(x)$, $i = 1, \dots, n+1$, may equal one. In fact, for any x not equal to the m_i's,

$$\#(x) = \#_1(x) + \cdots + \#_{n+1}(x).$$

Since G is strictly monotone on (t_{i-1}, t_i),

$$\int_{t_{i-1}}^{t_i} f(g(t))|g'(t)|\,dt = \int_{m_{i-1}}^{m_i} f(x)\,dx = \int_m^M f(x)\#_i(x)\,dx.$$

Now, add these equations over $1 \le i \le n+1$ to get

$$\int_a^b f(g(t))|g'(t)|\,dt = \sum_{i=1}^{n+1} \int_m^M f(x)\#_i(x)\,dx$$
$$= \int_m^M \sum_{i=1}^{n+1} f(x)\#_i(x)\,dx = \int_m^M f(x)\#(x)\,dx.$$

Here the last equality follows from the fact that $\#(x)$ and $\sum_{i=1}^{n+1} \#_i(x)$ differ only on finitely many points in (m, M).

4.4.26 Let $v_f(a,b)$ be the total variation of F over $[a,b]$ and assume first f' is continuous on an open interval containing $[a,b]$. If $a = x_0 < x_1 < \cdots < x_n = b$ is a partition, then

$$|f(x_i) - f(x_{i-1})| = \left| \int_{x_{i-1}}^{x_i} f'(x)\,dx \right| \le \int_{x_{i-1}}^{x_i} |f'(x)|\,dx$$

by the fundamental theorem. Summing this over $1 \le i \le n$ yields the first part. Also since this was any partition, taking the sup over all partitions shows

$$v_f(a,b) \le \int_a^b |f'(x)|\,dx.$$

Call this last integral I. To show that $v_f(a,b)$ equals I, given ϵ, we will exhibit a partition whose variation is within ϵ of I. Now, since $|f'|$ is continuous over $[a,b]$, $|f'|$ is Riemann integrable. Hence (Exercise **4.3.3**), given $\epsilon > 0$, there is a partition $a = x_0 < x_1 < \cdots < x_n = b$ whose corresponding Riemann sum $\sum_{i=1}^n |f'(x_i^\#)|(x_i - x_{i-1})$ is within ϵ of I, for any choice of intermediate points $x_i^\#$, $i = 1,\dots,n$. But, by the mean value theorem,

$$\sum_{i=1}^n |f(x_i) - f(x_{i-1})| = \sum_{i=1}^n |f'(x_i^\#)|(x_i - x_{i-1})$$

for some intermediate points $x_i^\#$, $i = 1,\dots,n$. Thus the variation of this partition is within ϵ of I and we conclude $v_f(c,d) = \int_c^d |f'(x)|\,dx$ for all $[c,d] \subset (a,b)$. Now $v_f(a,b) = \int_a^b |f'(x)|\,dx$ follows.

4.4.27 Let p denote the integral. Using the equation $y = \sqrt{1-x^2}$ for the upper-half unit circle, we have $y' = -x/\sqrt{1-x^2}$; hence $\sqrt{1+y'^2} = 1/\sqrt{1-x^2}$, hence the formula for p. Since $x^2 < x$ on $(0,1)$, it follows that $\sqrt{1-x^2} > \sqrt{1-x}$ on $(0,1)$. Thus

$$p = 2\int_0^1 \frac{dx}{\sqrt{1-x^2}} < 2\int_0^1 \frac{dx}{\sqrt{1-x}} = -4\sqrt{1-x}\,\Big|_0^1 = 4.$$

Here we used the fact that the integral is even on $[-1,1]$ and the primitive of $2/\sqrt{1-x}$ is $-4\sqrt{1-x}$.

4.4.28 If $y \ge 0$, then $\sin\theta \ge 0$, so $0 \le \theta \le \pi$ and the length of the counterclockwise arc joining $(1,0)$ to (x,y) is

$$L = \int_{\cos\theta}^1 \frac{dx}{\sqrt{1-x^2}}.$$

Now the substitution $x = \cos t$ transforms the integral to $\int_0^\theta dt = \theta$. When $y < 0$, $\sin\theta < 0$, hence $\pi < \theta < 2\pi$. Now the lower-half unit circle is $y = -\sqrt{1-x^2}$; hence $\sqrt{1+y'^2} = 1/\sqrt{1-x^2}$, so

$$L = \pi + L' = \pi + \int_{-1}^{\cos\theta} \frac{dx}{\sqrt{1-x^2}}.$$

Now the substitution $x = \cos t$ transforms the integral to $\int_\pi^\theta dt = \theta - \pi$; hence the result follows.

4.4.29 $\theta'(x) = -1/\sqrt{1-x^2}$ follows from the FTC. θ is continuous by continuity at the endpoints, and θ is strictly decreasing since $\theta' < 0$. Since c is the inverse of θ, $c' = -\sqrt{1-c^2}$ follows from the IFT. Since $s = \sqrt{1-c^2}$, this implies $c' = -s$. Also $s' = (1/2)(1-c^2)^{-1/2}(-2cc') = -cc'/s = c$.

4.4.30 We use the integral formula for the remainder in Taylor's theorem to get, for $|x-c| \le d$,

$$R_{n+1}(x,c) \le (n+1)\frac{|x-c|^{n+1}}{d^{n+1}} \le \int_0^1 g(c+s(x-c))(1-s)^n\,ds.$$

But this last integral is no larger than $\int_0^1 g(c+s(x-c))dx$; hence the remainder goes to zero.

4.4.31 Integration by parts applied on $(0,b)$ yields that $(fG)(0+)$ exists and

$$\int_0^b f(x)g(x)\,dx = f(b)G(b) - (fG)(0+) - \int_0^b f'(x)G(x)\,dx. \qquad \text{(A.4.1)}$$

Now suppose $|G(x)| \le M$. Since $f' \le 0$, by continuity at the endpoints,

$$\begin{aligned}\int_1^\infty |f'(x)G(x)|\,dx &\le M \lim_{b\to\infty}\int_1^b (-f'(x))\,dx \\ &= M \lim_{b\to\infty}(f(1) - f(b)) = Mf(1).\end{aligned}$$

Thus $f'G$ is integrable over $(0,\infty)$. Sending $b \to \infty$ in (A.4.1), the result follows since $f(\infty) = 0$.

Solutions to Exercises 4.5

4.5.1 Let $Q = (a,b) \times (c,d)$. If $(x,y) \in Q$ and $(x',y') \notin Q$, then the distance from (x,y) to (x',y') is no smaller than the distance from (x,y) to the boundary of Q, which, in turn, is no smaller than the minimum of $|x-a|$ and $|x-b|$.

4.5.2 Let $Q_n = (-1/n,1/n) \times (-1/n,1/n)$, $n \ge 1$. Then Q_n is an open set for each $n \ge 1$, and $\bigcap_{n=1}^\infty Q_n$ is a single point $\{(0,0)\}$, which is not open.

4.5.3 If Q is compact, then Q^c is a union of four open rectangles. So Q^c is open. So Q is closed. If C_n, $n \ge 1$, is closed, then C_n^c is open. So

$$\left(\bigcap_{n=1}^\infty C_n\right)^c = \bigcup_{n=1}^\infty C_n^c$$

is open. Hence $\bigcap_{n=1}^\infty C_n$ is closed. Let $Q_n = [0,1] \times [1/n, 1]$, $n \ge 1$. Then Q_n is closed, but

$$\bigcup_{n=1}^\infty Q_n = [0,1] \times (0,1]$$

is not.

4.5.4 It is enough to show that C^c is open. But C^c is the union of the four sets (draw a picture) $(-\infty, a) \times \mathbf{R}$, $(b, \infty) \times \mathbf{R}$, $(a,b) \times (-\infty, 0)$, and $\{(x,y) : a < x < b, y > f(x)\}$. The first three sets are clearly open, whereas the fourth is shown to be open using the continuity of f, exactly as in the text. Thus C is closed. Since C contains the subgraph of f, area $(C) \ge \int_a^b f(x)\,dx$. On the other hand, C is contained in the union of the subgraph of $f + \epsilon/(1+x^2)$ with L_a and L_b. Thus

$$\begin{aligned}
\text{area}\,(C) &\le \int_a^b [f(x) + \epsilon/(1+x^2)]\,dx + \text{area}\,(L_a) + \text{area}\,(L_b) \\
&= \int_a^b f(x)\,dx + \epsilon \int_a^b \frac{dx}{1+x^2} \le \int_a^b f(x)\,dx + \epsilon \int_{-\infty}^\infty \frac{dx}{1+x^2} \\
&= \int_a^b f(x)\,dx + \epsilon\pi.
\end{aligned}$$

Since $\epsilon > 0$ is arbitrary, the result follows.

4.5.5 Distance is always nonnegative, so ($\iff$ means iff),

$$\begin{aligned}
C \text{ is closed } &\iff C^c \text{ is open} \\
&\iff d((x,y), (C^c)^c) > 0 \text{ iff } (x,y) \in C^c \\
&\iff d((x,y), C) > 0 \text{ iff } (x,y) \in C^c \\
&\iff d((x,y), C) = 0 \text{ iff } (x,y) \in C.
\end{aligned}$$

This is the first part. For the second, let $(x,y) \in G_n$. If $\alpha = 1/n - d((x,y),C) > 0$ and $|x - x'| < \epsilon$, $|y - y'| < \epsilon$, then $d((x',y'),C) \le d((x,y),C) + 2\epsilon = 1/n + 2\epsilon - \alpha$, by the triangle inequality. Thus for $\epsilon < \alpha/2$, $(x',y') \in G_n$. This shows that $Q_\epsilon \subset G_n$, where Q_ϵ is the open rectangle centered at (x,y) with sides of length 2ϵ. Thus G_n is open, and $\bigcap_{n=1}^\infty G_n = \{(x,y) : d((x,y),C) = 0\}$, which equals C.

4.5.6 Given $\epsilon > 0$, it is enough to find an open superset G of A satisfying area $(G) \le$ area $(A) + 2\epsilon$. If area $(A) = \infty$, $G = \mathbf{R}^2$ will do. If area $(A) < \infty$, choose a paving (Q_n), such that $\sum_{n=1}^\infty$ area $(Q_n) <$ area $(A) + \epsilon$. For each $n \ge 1$, let Q_n' be an open rectangle containing Q_n and satisfying area $(Q_n') \le$ area $(Q_n) + \epsilon 2^{-n}$. (For each $n \ge 1$, such a rectangle Q_n' can be obtained by dilating Q_n° slightly.) Then $G = \bigcup_{n=1}^\infty Q_n'$ is open, G contains A, and

$$\text{area}(G) \le \sum_{n=1}^{\infty} \text{area}(Q'_n) \le \sum_{n=1}^{\infty} (\text{area}(Q_n) + \epsilon 2^{-n}) < \text{area}(A) + 2\epsilon.$$

For the second part, let $\alpha = \inf\{\text{area}(G) : A \subset G, G \text{ open}\}$. Choosing G as above, $\alpha \le \text{area}(G) \le \text{area}(A) + \epsilon$ for all ϵ. Hence $\alpha \le \text{area}(A)$. Conversely, monotonicity implies that $\text{area}(A) \le \text{area}(G)$ for any superset G. Hence $\text{area}(A) \le \alpha$.

4.5.7 For each $\epsilon > 0$, by Exercise **4.5.6**, choose G_ϵ open such that $A \subset G_\epsilon$ and $\text{area}(G_\epsilon) \le \text{area}(A) + \epsilon$. Let $I = \bigcap_{n=1}^{\infty} G_{1/n}$. Then I is interopen, and $\text{area}(I) \le \inf_{n\ge 1} \text{area}(G_{1/n}) \le \inf_{n\ge1}(\text{area}(A) + 1/n) = \text{area}(A)$. But $I \supset A$. So $\text{area}(I) \ge \text{area}(A)$.

4.5.8 If M is measurable, select an interopen superset $I \supset M$ satisfying $\text{area}(I) = \text{area}(M)$. With $A = I$ in the definition of measurability, we obtain $\text{area}(I - M) = 0$. Conversely, suppose such I exists and let A be arbitrary. Then

$$A \cap M^c \subset (A \cap I^c) \cup (I \cap M^c);$$

hence

$$\text{area}(A \cap M^c) \le \text{area}(A \cap I^c) + \text{area}(I \cap M^c) = \text{area}(A \cap I^c).$$

Since I is measurable

$$\begin{aligned}\text{area}(A) &\ge \text{area}(A \cap I) + \text{area}(A \cap I^c)\\ &\ge \text{area}(A \cap M) + \text{area}(A \cap M^c).\end{aligned}$$

Hence M is measurable.

4.5.9 We already know that the intersection of a sequence of measurable sets is measurable. By De Morgan's law (§1.1), M_n measurable implies the complement M_n^c is measurable. So

$$\left(\bigcup_{n=1}^{\infty} M_n\right)^c = \bigcap_{n=1}^{\infty} M_n^c$$

is measurable. So the complement $\bigcup_{n=1}^{\infty} M_n$ is measurable.

4.5.10 Let P_k, $k = 0, \dots, n$, denote the vertices of D'_n. It is enough to show that the closest approach to O of the line joining P_k and P_{k+1} is at the midpoint $M = (P_k + P_{k+1})/2$, where the distance to O equals 1. Let $\theta_k = k\pi/n$. Then the distance squared from the midpoint to O is given by

$$\frac{[\cos(2\theta_k) + \cos(2\theta_{k+1})]^2 + [\sin(2\theta_k) + \sin(2\theta_{k+1})]^2}{4\cos(\theta_1)^2}$$

$$= \frac{2 + 2[\cos(2\theta_k)\cos(2\theta_{k+1}) + \sin(2\theta_k)\sin(2\theta_{k+1})]}{4\cos^2(\theta_1)}$$
$$= \frac{2 + 2\cos(2\theta_1)}{4\cos^2(\theta_1)} = 1.$$

Thus the distance to the midpoint is 1. To show that this is the minimum, check that the line segments OM and P_kP_{k+1} are perpendicular.

4.5.11 Here $a_n = n\sin(\pi/n)\cos(\pi/n)$, and $a'_n = n\tan(\pi/n)$. So $a_na'_n = n^2\sin^2(\pi/n)$. But $a_{2n} = 2n\sin(\pi/2n)\cos(\pi/2n) = n\sin(\pi/n)$. So $a_{2n} = \sqrt{a_na'_n}$. Also

$$\begin{aligned}
\frac{1}{a_{2n}} + \frac{1}{a'_n} &= \frac{1}{n\sin(\pi/n)} + \frac{1}{n\tan(\pi/n)} \\
&= \frac{\cos(\pi/n) + 1}{n\sin(\pi/n)} = \frac{2\cos^2(\pi/2n)}{2n\sin(\pi/2n)\cos(\pi/2n)} \\
&= \frac{2}{2n\tan(\pi/2n)} = \frac{2}{a'_{2n}}.
\end{aligned}$$

4.5.12 In the definition of measurable, replace M and A by $A \cup B$ and A, respectively. Then $A \cap M$ is replaced by A, and $A \cap M^c$ is replaced by B.

4.5.13 From the previous exercise and induction,

$$\text{area}\left(\bigcup_{n=1}^{\infty} A_n\right) \geq \text{area}\left(\bigcup_{n=1}^{N} A_n\right) = \sum_{n=1}^{N} \text{area}\,(A_n).$$

Let $N \nearrow \infty$ to get

$$\text{area}\left(\bigcup_{n=1}^{\infty} A_n\right) \geq \sum_{n=1}^{\infty} \text{area}\,(A_n).$$

Since the reverse inequality follows from subadditivity, we are done.

4.5.14 Note that A and $B \setminus A$ are disjoint and their union is $A \cup B$. But $B \setminus A$ and $A \cap B$ are disjoint and their union is B. So

$$\begin{aligned}
\text{area}\,(A \cup B) &= \text{area}\,(A) + \text{area}\,(B \setminus A) \\
&= \text{area}\,(A) + \text{area}\,(B) - \text{area}\,(A \cap B).
\end{aligned}$$

The general formula, the *inclusion–exclusion principle*, is that the area of a union equals the sum of the areas of the sets minus the sum of the areas of their double intersections plus the sum of the areas of their triple intersections minus the sum of the areas of their quadruple intersections, etc.

4.5.15 By Exercise **4.5.6**, given $\epsilon > 0$, there is an open superset G of M satisfying $\text{area}\,(G) \le \text{area}\,(M) + \epsilon$. If M is measurable and $\text{area}\,(M) < \infty$, replace A in (4.5.4) by G to get $\text{area}\,(G) = \text{area}\,(M) + \text{area}\,(G \setminus M)$. Hence $\text{area}\,(G \setminus M) \le \epsilon$. If $\text{area}\,(M) = \infty$, write $M = \bigcup_{n=1}^\infty M_n$ with $\text{area}\,(M_n) < \infty$ for all $n \ge 1$. For each $n \ge 1$, choose an open superset G_n of M_n satisfying $\text{area}\,(G_n \setminus M_n) \le \epsilon 2^{-n}$. Then $G = \bigcup_{n=1}^\infty G_n$ is an open superset of M, and $\text{area}\,(G \setminus M) \le \sum_{n=1}^\infty \text{area}\,(G_n \setminus M_n) \le \epsilon$. This completes the first part. Conversely, suppose, for all $\epsilon > 0$, there is an open superset G of M satisfying $\text{area}\,(G \setminus M) \le \epsilon$, and let A be arbitrary. Since G is measurable,

$$\begin{aligned}
&\text{area}\,(A \cap M) + \text{area}\,(A \cap M^c) \\
&\qquad \le \text{area}\,(A \cap G) + \text{area}\,(A \cap G^c) + \text{area}\,(A \cap (G \setminus M)) \\
&\qquad \le \text{area}\,(A) + \epsilon.
\end{aligned}$$

Thus $\text{area}\,(A \cap M) + \text{area}\,(A \cap M^c) \le \text{area}\,(A)$. Since the reverse inequality follows by subadditivity, M is measurable.

4.5.16 When A is a rectangle, the result is obvious (draw a picture). In fact, for a rectangle Q and $Q' = Q + (a,b)$, $\text{area}\,(Q \cap Q') \ge (1-\epsilon)^2\, \text{area}\,(Q)$. To deal with general A, let Q be as in Exercise **4.2.15** with α to be determined below, and let $A' = A + (a,b)$. Then by subadditivity and translation invariance,

$$\begin{aligned}
\text{area}\,(Q \cap Q') &\le \text{area}\,((Q \cap A) \cap (Q' \cap A')) \\
&\quad + \text{area}\,(Q \setminus (Q \cap A)) + \text{area}\,(Q' \setminus (Q' \cap A')) \\
&= \text{area}\,((Q \cap A) \cap (Q' \cap A')) + 2 \cdot \text{area}\,(Q \setminus (Q \cap A)) \\
&\le \text{area}\,(A \cap A') + 2 \cdot \text{area}\,(Q \setminus (Q \cap A)).
\end{aligned}$$

But, from Exercise **4.2.15** and the measurability of A, $\text{area}\,[Q \setminus (Q \cap A)] < (1-\alpha)\,\text{area}\,(Q)$. Hence

$$\begin{aligned}
\text{area}\,(A \cap A') &\ge \text{area}\,(Q \cap Q') - 2(1-\alpha)\,\text{area}\,(Q) \\
&\ge (1-\epsilon)^2\,\text{area}\,(Q) - 2(1-\alpha)\,\text{area}\,(Q).
\end{aligned}$$

Thus the result follows as soon as one chooses $2(1-\alpha) < (1-\epsilon)^2$.

4.5.17 Since $\text{area}\,(A \cap N) = 0$,

$$\text{area}\,(A) \ge \text{area}\,(A \cap N^c) = \text{area}\,(A \cap N) + \text{area}\,(A \cap N^c);$$

hence N is measurable.

4.5.18 If $\text{area}\,[A \cap (A + (a,b))] > 0$, then $A \cap (A + (a,b))$ is nonempty. If $(x,y) \in A \cap (A + (a,b))$, then $(x,y) = (x',y') + (a,b)$ with $(x',y') \in A$. Hence $(a,b) \in A - A$. Since Exercise **4.5.16** says that $\text{area}\,[A \cap (A + (a,b))] > 0$ for all $(a,b) \in Q_\epsilon$, the result follows.

A.5 Solutions to Chapter 5

Solutions to Exercises 5.1

5.1.1 Let $f_n(x) = 1/n$ for all $x \in \mathbf{R}$. Then $\int_{-\infty}^{\infty} f_n(x)\,dx = \infty$ for all $n \geq 1$, and $f_n(x) \searrow f(x) = 0$.

5.1.2 Let $I_n = \int_a^b f_n(x)\,dx$ and $I = \int_a^b f(x)\,dx$. We have to show that $I_* \geq I$. The lower sequence is

$$g_n(x) = \inf\{f_k(x) : k \geq n\}, \qquad n \geq 1.$$

Then $(g_n(x))$ is nonnegative and increasing to $f(x)$, $a < x < b$. So the monotone convergence theorem applies. So $J_n = \int_a^b g_n(x)\,dx \to \int_a^b f(x)\,dx = I$. Since $f_n(x) \geq g_n(x)$, $a < x < b$, $I_n \geq J_n$. Hence $I_* \geq J_* = I$.

5.1.3 Given x fixed, $|x - n| \geq 1$ for n large enough. Hence $f_0(x - n) = 0$. Hence $f_n(x) = 0$ for n large enough. Thus $f(x) = \lim_{n \nearrow \infty} f_n(x) = 0$. But, by translation invariance,

$$\int_{-\infty}^{\infty} f_n(x)\,dx = \int_{-\infty}^{\infty} h(x)\,dx = \int_{-1}^{1} [1 - x^2]\,dx = \frac{4}{3} > 0.$$

Since $\int_{-\infty}^{\infty} f(x)\,dx = 0$, here, the inequality in Fatou's lemma is strict.

5.1.4 By Exercise **3.2.4**, $(1 - t/n)^n \nearrow e^{-t}$ as $n \nearrow \infty$. To take care of the upper limit of integration that changes with n, let

$$f_n(t) = \begin{cases} \left(1 - \frac{t}{n}\right)^n t^{x-1}, & 0 < t < n, \\ 0, & t \geq n. \end{cases}$$

Then by the monotone convergence theorem,

$$\begin{aligned} \Gamma(x) &= \int_0^{\infty} e^{-t} t^{x-1}\,dt \\ &= \int_0^{\infty} \lim_{n \nearrow \infty} f_n(t)\,dt \\ &= \lim_{n \nearrow \infty} \int_0^{\infty} f_n(t)\,dt \\ &= \lim_{n \nearrow \infty} \int_0^{n} \left(1 - \frac{t}{n}\right)^n t^{x-1}\,dt. \end{aligned}$$

5.1.5 By Exercise **4.4.11**,

$$\int_0^n \left(1-\frac{t}{n}\right)^n t^{x-1}\,dt = n^x \int_0^1 (1-s)^n s^{x-1}\,ds = \frac{n^x n!}{x\cdot(x+1)\cdot\dots\cdot(x+n)}.$$

For the second limit, replace x by $x+1$ and note $n/(x+1+n)\to 1$ as $n\to\infty$.

5.1.6 Convexity of f_n means that $f_n((1-t)x+ty)\le(1-t)f_n(x)+tf_n(y)$ for all $a<x<y<b$ and $0\le t\le 1$. Letting $n\nearrow\infty$, we obtain $f((1-t)x+ty)\le(1-t)f(x)+tf(y)$ for all $a<x<y<b$ and $0\le t\le 1$, which says that f is convex. Now, let

$$f_n(x)=\log\left(\frac{n^x n!}{x\cdot(x+1)\cdot\dots\cdot(x+n)}\right).$$

Then $\log\Gamma(x)=\lim_{n\nearrow\infty} f_n(x)$ by (5.1.3) and

$$\frac{d^2}{dx^2}f_n(x)=\frac{d^2}{dx^2}\left(x\log n+\log(n!)-\sum_{k=0}^n \log(x+k)\right)=\sum_{k=0}^n\frac{1}{(x+k)^2},$$

which is positive. Thus f_n is convex. So $\log\Gamma$ is convex.

5.1.7 Since $\log\Gamma(x)$ is convex,

$$\log\Gamma((1-t)x+ty)\le(1-t)\log\Gamma(x)+t\log\Gamma(y)$$

for $0<x<y<\infty, 0\le t\le 1$. Since e^x is convex and increasing,

$$\begin{aligned}\Gamma((1-t)x+ty)&=\exp(\log\Gamma((1-t)x+ty))\\&\le\exp((1-t)\log\Gamma(x)+t\log\Gamma(y))\\&\le(1-t)\exp(\log\Gamma(x))+t\exp(\log\Gamma(y))\\&=(1-t)\Gamma(x)+t\Gamma(y)\end{aligned}$$

for $0<x<y<\infty$ and $0\le t\le 1$.

5.1.8 Use summation under the integral sign, with $f_n(t)=t^{x-1}e^{-nt}$, $n\ge 1$. Then substituting $s=nt$, $ds=ndt$,

$$\begin{aligned}\int_0^\infty \frac{t^{x-1}}{e^t-1}\,dt&=\int_0^\infty\sum_{n=1}^\infty t^{x-1}e^{-nt}\,dt\\&=\sum_{n=1}^\infty\int_0^\infty t^{x-1}e^{-nt}\,dt\\&=\sum_{n=1}^\infty n^{-x}\int_0^\infty s^{x-1}e^{-s}\,ds\\&=\zeta(x)\Gamma(x).\end{aligned}$$

5.1.9 Use summation under the integral sign, with $f_n(t) = t^{x-1}e^{-n^2\pi t}$, $n \ge 1$. Then substituting $s = n^2\pi t$, $ds = n^2\pi dt$,

$$\begin{aligned}\int_0^\infty \psi(t)t^{x/2-1}\,dt &= \int_0^\infty \sum_{n=1}^\infty e^{-n^2\pi t}t^{x/2-1}\,dt \\ &= \sum_{n=1}^\infty \int_0^\infty e^{-n^2\pi t}t^{x/2-1}\,dt \\ &= \sum_{n=1}^\infty \pi^{-x/2}n^{-x}\int_0^\infty e^{-s}s^{x/2-1}\,ds \\ &= \pi^{-x/2}\zeta(x)\Gamma(x/2).\end{aligned}$$

5.1.10 By Exercise **4.4.7**,

$$\int_0^1 t^{x-1}(-\log t)^{n-1}\,dt = \int_0^\infty e^{-xs}s^{n-1}\,ds = \frac{\Gamma(n)}{x^n}.$$

Here we used the substitutions $t = e^{-s}$, then $xs = r$.

5.1.11 Recall that $\log t < 0$ on $(0,1)$ and $0 < \log t < t$ on $(1,\infty)$. From the previous exercise, with $f(t) = e^{-t}t^{x-1}|\log t|^{n-1}$,

$$\begin{aligned}\int_0^\infty e^{-t}t^{x-1}|\log t|^{n-1}\,dt &= \int_0^1 f(t)\,dt + \int_1^\infty f(t)\,dt \\ &\le \int_0^1 t^{x-1}|\log t|^{n-1}\,dt + \int_1^\infty e^{-t}t^{x-1}t^{n-1}\,dt \\ &= \frac{\Gamma(n)}{x^n} + \Gamma(x+n-1).\end{aligned}$$

5.1.12 If $x_n \searrow 1$, then $k^{-x_n} \nearrow k^{-1}$ for $k \ge 1$. So by the monotone convergence theorem for series, $\zeta(x_n) = \sum_{k=1}^\infty k^{-x_n} \to \sum_{k=1}^\infty k^{-1} = \zeta(1) = \infty$. If $x_n \to 1+$, then $x_n^* \searrow 1$ (§1.5) and $\zeta(1) \ge \zeta(x_n) \ge \zeta(x_n^*)$. So $\zeta(x_n) \to \zeta(1) = \infty$. Thus $\zeta(1+) = \infty$. Similarly, $\psi(0+) = \infty$.

5.1.13 Since $\tau(t) = \sum_{n=0}^\infty te^{-nt}$, use summation under the integral sign:

$$\int_0^\infty e^{-xt}\tau(t)\,dt = \sum_{n=0}^\infty \int_0^\infty e^{-xt}te^{-nt}\,dt = \sum_{n=0}^\infty \frac{1}{(x+n)^2}.$$

Here we used the substitution $s = (x+n)t$, and $\Gamma(2) = 1$.

5.1.14 The problem, here, is that the limits of integration depend on n. So the monotone convergence theorem is not directly applicable. To remedy this, let $f_n(x) = f(x)$ if $a_n < x < b_n$, and let $f_n(x) = 0$ if $a < x \le a_n$ or $b_n \le x < b$. Then $f_n(x) \nearrow f(x)$ (draw a picture). Hence by the monotone convergence theorem,

$$\int_{a_n}^{b_n} f(x)\,dx = \int_a^b f_n(x)\,dx \to \int_a^b f(x)\,dx.$$

5.1.15 Differentiate the log of both sides using the fundamental theorem.

Solutions to Exercises 5.2

5.2.1 Dividing yields

$$t^6 - 4t^5 + 5t^4 - 4t^2 + 4 + \frac{4}{1+t^2}.$$

Integrate over $(0,1)$.

5.2.2 First, for $x > 0$,

$$e^{-sx}\left(1 + \frac{x^2}{3!} + \frac{x^4}{5!} + \dots\right) \le e^{-sx}\sum_{n=0}^{\infty}\frac{x^{2n}}{(2n)!} \le e^{-sx}e^{x}.$$

So with $g(x) = e^{-(s-1)x}$, we may use summation under the integral sign to get

$$\begin{aligned}
\int_0^\infty e^{-sx}\frac{\sin x}{x}\,dx &= \int_0^\infty \sum_{n=0}^{\infty}(-1)^n\frac{x^{2n}}{(2n+1)!}e^{-sx}\,dx \\
&= \sum_{n=0}^{\infty}\frac{(-1)^n}{(2n+1)!}\int_0^\infty e^{-sx}x^{2n}\,dx \\
&= \sum_{n=0}^{\infty}\frac{(-1)^n}{(2n+1)!}\cdot\frac{\Gamma(2n+1)}{s^{2n+1}} \\
&= \sum_{n=0}^{\infty}\frac{(-1)^n(1/s)^{2n+1}}{(2n+1)} \\
&= \arctan\left(\frac{1}{s}\right).
\end{aligned}$$

Here we used (3.6.4).

5.2.3 Since $|f_n(x)| \le g(x)$ for all $n \ge 1$, taking the limit yields $|f(x)| \le g(x)$. Since g is integrable, so is f.

5.2.4 $J_0(x)$ is a power series; hence it may be differentiated term by term. The calculation is made simpler by noting that $x[xJ_0'(x)]' = x^2J_0''(x) + xJ_0'(x)$. Then

$$xJ_0'(x) = \sum_{n=1}^{\infty}(-1)^n \frac{2n\,x^{2n}}{4^n(n!)^2},$$

and

$$\begin{aligned} x^2J_0''(x) + xJ_0'(x) = x(xJ_0'(x))' &= \sum_{n=1}^{\infty}(-1)^n \frac{4n^2x^{2n}}{4^n(n!)^2} \\ = \sum_{n=0}^{\infty}(-1)^{n+1}\frac{4(n+1)^2x^{2n+2}}{4^{n+1}((n+1)!)^2} &= -x^2\sum_{n=0}^{\infty}(-1)^n\frac{x^{2n}}{4^n(n!)^2} = -x^2J_0(x). \end{aligned}$$

5.2.5 With $u = \sin^{n-1} x$ and $dv = \sin x\, dx$,

$$I_n = \int \sin^n x\, dx = -\cos x \sin^{n-1} x + (n-1)\int \sin^{n-2} x \cos^2 x\, dx.$$

Inserting $\cos^2 x = 1 - \sin^2 x$,

$$I_n = -\cos x \sin^{n-1} x + (n-1)(I_{n-2} - I_n).$$

Solving for I_n,

$$I_n = -\frac{1}{n}\cos x \sin^{n-1} x + \frac{n-1}{n} I_{n-2}.$$

5.2.6 Using the double-angle formula,

$$\begin{aligned} \sin x &= 2\cos(x/2)\sin(x/2) \\ &= 4\cos(x/2)\cos(x/4)\sin(x/4) \\ &= \cdots = 2^n\cos(x/2)\cos(x/4)\dots\cos(x/2^n)\sin(x/2^n). \end{aligned}$$

Now, let $n \nearrow \infty$, and use $2^n\sin(x/2^n) \to x$.

5.2.7 The integral on the left equals $n!$. So the left side is $\sum_{n=0}^{\infty}(-1)^n$, which has no sum. The series on the right equals e^{-x}. So the right side equals $\int_0^\infty e^{-2x}\, dx = 1/2$.

5.2.8 Now, for $x > 0$,

$$\frac{\sin(sx)}{e^x - 1} = \sum_{n=1}^{\infty} e^{-nx}\sin(sx),$$

and

$$\sum_{n=1}^{\infty} e^{-nx}|\sin(sx)| \le \sum_{n=1}^{\infty} e^{-nx}|s|x = |s|x/(e^x - 1) = g(x),$$

which is integrable ($\int_0^\infty g(x)\,dx = |s|\Gamma(2)\zeta(2)$ by Exercise **5.1.8**). Hence we may use summation under the integral sign to obtain

$$\begin{aligned}\int_0^\infty \frac{\sin(sx)}{e^x - 1}\,dx &= \sum_{n=1}^{\infty}\int_0^\infty e^{-nx}\sin(sx)\,dx\\ &= \sum_{n=1}^{\infty}\frac{s}{n^2+s^2}.\end{aligned}$$

Here we used Exercise **4.4.9**.

5.2.9 Writing $\sinh(sx) = (e^{sx} - e^{-sx})/2$ and breaking the integral into two pieces leads to infinities. So we proceed, as in the previous exercise. For $x > 0$, use the mean value theorem to check

$$\left|\frac{\sinh x}{x}\right| \le \cosh x \le e^x.$$

So

$$\frac{\sinh(sx)}{e^x - 1} = \sum_{n=1}^{\infty} e^{-nx}\sinh(sx),$$

and

$$\sum_{n=1}^{\infty} e^{-nx}|\sinh(sx)| \le \sum_{n=1}^{\infty} |s|xe^{-nx}e^{|s|x} = g(x),$$

which is integrable when $|s| < 1$ ($\int_0^\infty g(x)\,dx = |s|\sum_{n=1}^\infty \Gamma(2)/(n-|s|)^2$). Using summation under the integral sign, we obtain

$$\begin{aligned}\int_0^\infty \frac{\sinh(sx)}{e^x - 1}\,dx &= \sum_{n=1}^{\infty}\int_0^\infty e^{-nx}\sinh(sx)\,dx\\ &= \sum_{n=1}^{\infty}\frac{1}{2}\int_0^\infty e^{-nx}(e^{sx} - e^{-sx})\,dx\\ &= \frac{1}{2}\sum_{n=1}^{\infty}\left(\frac{1}{n-s} - \frac{1}{n+s}\right)\\ &= \sum_{n=1}^{\infty}\frac{s}{n^2+s^2}.\end{aligned}$$

5.2.10 Clearly $s_n < \pi$ for it is enough to show the nth tail is $< 1/4(n+1)^2 16^{n+1}$. Now

$$\frac{1}{8k+1} - \frac{1}{8k+p} < \frac{p-1}{64k^2}.$$

Applying this with $p = 4, 5, 6$, the nth tail is less than

$$\frac{2(4-1)+(5-1)+(6-1)}{64(n+1)^2 16^{n+1}} \sum_{k=0}^{\infty} \frac{1}{16^k} = \frac{1}{4(n+1)^2 16^{n+1}}.$$

5.2.11 This follows immediately from Exercise **1.3.23**. Here is python code:

```
from fractions import Fraction

f = Fraction

def partialsum(n):
    p = 0
    for k in range(0,n+1):
        s  = f(4,8*k+1) - f(2,8*k+4)
        s -= f(1,8*k+5) + f(1,8*k+6)
        s *= f(1,16**k)
        p += s
    return p

def tail(n):
    return f(1,4*16**(n+1)*(n+1)**2)

def cf(r):
    n = r.numerator
    d = r.denominator
    if n%d == 0:
        return [int(n/d)]
    else:
        return [int(n//d)] + cf(f(d,n%d))

def CF(r):
    n = r.numerator
    d = r.denominator
    if n%d == 0:
        return int(n/d)
    elif n < d:
        return cf(f(d,n%d))
    else:
        return str(int(n//d)) + ' + ' + str(cf(f(d,n%d)))
```

```
for n in range(0,6):
    e = tail(n)
    s = partialsum(n)
    S = s + e
    print n,"\n",s,"\n",S,"\n",CF(s),"\n",CF(S),"\n\n"
```

5.2.12 We have to show that $x_n \to x$ implies $J_\nu(x_n) \to J_\nu(x)$. But $g(t) = 1$ is integrable over $(0, \pi)$ and dominates the integrands below. So we can apply the dominated convergence theorem,

$$J_\nu(x_n) = \frac{1}{\pi}\int_0^\pi \cos(\nu t - x_n \sin t)\, dt \to \frac{1}{\pi}\int_0^\pi \cos(\nu t - x \sin t)\, dt = J_\nu(x).$$

5.2.13 It is enough to show that ψ is continuous on (a, ∞) for all $a > 0$, for then ψ is continuous on $(0, \infty)$. We have to show that $x_n \to x > a$ implies $\psi(x_n) \to \psi(x)$. But $x_n > a$, $n \geq 1$, implies $e^{-k^2\pi x_n} \leq e^{-k\pi a}$, $k \geq 1, n \geq 1$, and $\sum g_k = \sum e^{-k\pi a} < \infty$. So the dominated convergence theorem for series applies, and

$$\psi(x_n) = \sum_{k=1}^{\infty} e^{-k^2\pi x_n} \to \sum_{k=1}^{\infty} e^{-k^2\pi x} = \psi(x).$$

5.2.14 Set $\tilde{f}_n(x) = f_n(x)$ if $a_n < x < b_n$, and $\tilde{f}_n(x) = 0$ if $a < x < a \leq a_n$ or $b_n \leq x < b$. Then $|\tilde{f}_n(x)| \leq g(x)$ on (a, b), and $\tilde{f}_n(x) \to f(x)$ for any x in (a, b). Hence by the dominated convergence theorem,

$$\int_{a_n}^{b_n} f_n(x)\, dx = \int_a^b \tilde{f}_n(x)\, dx \to \int_a^b f(x)\, dx.$$

5.2.15 Use the Taylor series for cos:

$$J_0(x) = \frac{1}{\pi}\int_0^\pi \cos(x \sin t)\, dt = \sum_{n=0}^{\infty} (-1)^n \frac{x^{2n}}{(2n)!\pi} \int_0^\pi \sin^{2n} t\, dt.$$

But

$$\begin{aligned}\frac{1}{\pi}\int_0^\pi \sin^{2n} t\, dt &= \frac{2}{\pi}\int_0^{\pi/2} \sin^{2n} t\, dt = \frac{2}{\pi} I_{2n} \\ &= \frac{(2n-1)\cdot(2n-3)\cdot\dots\cdot 1}{2n\cdot(2n-2)\cdot\dots\cdot 2} = \frac{(2n)!}{2^{2n}(n!)^2}.\end{aligned}$$

Inserting this in the previous expression, one obtains the series for $J_0(x)$. Here for x fixed, we used summation under the integral sign with $g(t) = e^x$. Since $\int_0^\pi g(t)\,dt = e^x\pi$, this applies.

5.2.16 By Exercise **4.4.22**,

$$\begin{aligned}\lim_{n\nearrow\infty} & \left[\int_0^1 \frac{1-(1-t/n)^n}{t}\,dt - \int_1^n \frac{(1-t/n)^n}{t}\,dt\right] \\ &= \lim_{n\nearrow\infty}\left[\int_0^n \frac{1-(1-t/n)^n}{t}\,dt - \int_1^n \frac{1}{t}\,dt\right] \\ &= \lim_{n\nearrow\infty}\left(1+\frac{1}{2}+\cdots+\frac{1}{n}-\log n\right) \\ &= \gamma.\end{aligned}$$

For the second part, since $(1-t/n)^n \to e^{-t}$, we obtain the stated formula by switching the limits and the integrals. To justify the switching, by the mean value theorem with $f(t) = (1-t/n)^n$,

$$0 \le \frac{1-(1-t/n)^n}{t} = \frac{f(0)-f(t)}{t} = -f'(c) = (1-c/n)^{n-1} \le 1.$$

So we may choose $g(t) = 1$ for the first integral. Since $(1-t/n)^n \le e^{-t}$, we may choose $g(t) = e^{-t}/t$ for the second integral.

5.2.17 Use Euler's continued fraction formula with $a_0 = x$, $a_1 = -x^2/3$, $a_2 = -x^2/5$, $a_3 = -x^2/7, \dots$.

Solutions to Exercises 5.3

5.3.1 By convexity of e^x,

$$a^{1-t}b^t = e^{(1-t)\log a + t\log b} \le (1-t)e^{\log a} + te^{\log b} = (1-t)a + tb.$$

Thus $0 < b < b' < a' < a$. The rest follows as in §5.3.

5.3.2 If $a_n \to a$ and $b_n \to b$ with $a > b > 0$, then there is a $c > 0$ with $a_n > c$ and $b_n > c$ for all $n \ge 1$. Hence

$$f_n(\theta) = \frac{1}{\sqrt{a_n^2\cos^2\theta + b_n^2\sin^2\theta}} \le \frac{1}{\sqrt{c^2\cos^2\theta + c^2\sin^2\theta}} = \frac{1}{c}.$$

Hence we may apply the dominated convergence theorem with $g(\theta) = 2/c\pi$.

5.3.3 Note the map $\theta \mapsto x'$ is smooth on $(0, \pi/2)$ with range in $(-1, 1)$. Since $\arccos : (-1,1) \to (0,\pi)$ is smooth, the map $G(\theta) = \theta' = \arccos(x')$ is well defined and smooth on $(0,\pi/2)$. From the proof, we know $G'(\theta)=2b'\lambda>0$, so G is strictly increasing. Since $G(0+)=G(0)=0$ and $G(\pi/2-)=G(\pi/2)=\pi$, the result follows.

5.3.4 $b\,dt=b^2\sec^2\theta\,d\theta=b^2(1+\tan^2\theta)\,d\theta=(b^2+t^2)\,d\theta$. Moreover $a^2\cos^2\theta+b^2\sin^2\theta=\cos^2\theta(a^2+t^2)$ and $b^2\sec^2\theta=b^2+b^2\tan^2\theta=b^2+t^2$. Thus

$$a^2\cos^2\theta+b^2\sin^2\theta=b^2\cdot\frac{a^2+t^2}{b^2+t^2}$$

and the result follows.

5.3.5 Since the arithmetic and geometric means of $a=1+x$ and $b=1-x$ are $(1,\sqrt{1-x^2})$, $M(1+x,1-x)=M(1,\sqrt{1-x^2})$. So the result follows from

$$\cos^2\theta+(1-x^2)\sin^2\theta=1-x^2\sin^2\theta.$$

5.3.6 By the binomial theorem,

$$\frac{1}{\sqrt{1-x^2\sin^2\theta}}=\sum_{n=0}^{\infty}(-1)^n\binom{-1/2}{n}x^{2n}\sin^{2n}\theta.$$

By Exercise **3.5.10**, $\binom{-1/2}{n}=(-1)^n4^{-n}\binom{2n}{n}$. So this series is positive. Hence we may apply summation under the integral sign. From Exercise **5.2.15**, $I_{2n}=(2/\pi)\int_0^{\pi/2}\sin^{2n}\theta\,d\theta=4^{-n}\binom{2n}{n}$. Integrating the series term by term, we get the result.

5.3.7 With $t=x/s$, $dt=-xds/s^2$, and $f(t)=1/\sqrt{(1+t^2)(x^2+t^2)}$, $f(t)\,dt=-f(s)\,ds$. So

$$\begin{aligned}\frac{1}{M(1,x)}&=\frac{2}{\pi}\int_0^\infty\frac{dt}{\sqrt{(1+t^2)(x^2+t^2)}}\\&=\frac{2}{\pi}\int_0^{\sqrt{x}}f(t)\,dt+\frac{2}{\pi}\int_{\sqrt{x}}^\infty f(t)\,dt\\&=\frac{2}{\pi}\int_0^{\sqrt{x}}f(t)\,dt+\frac{2}{\pi}\int_0^{\sqrt{x}}f(s)\,ds\\&=\frac{4}{\pi}\int_0^{\sqrt{x}}\frac{dt}{\sqrt{(1+t^2)(x^2+t^2)}}\\&=\frac{4}{\pi}\int_0^{1/\sqrt{x}}\frac{dr}{\sqrt{(1+(xr)^2)(1+r^2)}}.\end{aligned}$$

For the last integral, we used $t=xr$, $dt=xdr$.

5.3.8 The AGM iteration yields $(1+x,1-x)\mapsto(1,x')\mapsto((1+x')/2,\sqrt{x'})$.

5.3.9 Now, $x < M(1,x) < 1$, and $x' < M(1,x') < 1$. So

$$\begin{aligned}\left|\frac{1}{M(1,x)} - \frac{1}{Q(x)}\right| &= \frac{1-M(1,x')}{M(1,x)} \\ &\le \frac{1-x'}{x} = \frac{1-x'^2}{x(1+x')} = \frac{x}{1+x'} \le x.\end{aligned}$$

5.3.10 We already know that

$$Q\left(\frac{1-x'}{1+x'}\right) = \frac{1}{2}Q(x).$$

Substitute $x = 2\sqrt{y}/(1+y)$. Then $x' = (1-y)/(1+y)$. So solving for y yields $y = (1-x')/(1+x')$.

5.3.11 From the integral formula, $M(1,x)$ is strictly increasing and continuous. So $M(1,x')$ is strictly decreasing and continuous. So $Q(x)$ is strictly increasing and continuous. Moreover, $x \to 0$ implies $x' \to 1$ implies $M(1,x) \to 0$ and $M(1,x') \to 1$, which implies $Q(x) \to 0$. Thus $Q(0+) = 0$. If $x \to 1-$, then $x' \to 0+$. Hence $M(1,x) \to 1$ and $M(1,x') \to 0+$. So $Q(x) \to \infty$. Thus $Q(1-) = \infty$; hence $M(1,\cdot) : (0,1) \to (0,1)$ and $Q : (0,1) \to (0,\infty)$ are strictly increasing bijections.

5.3.12 $M(a,b) = 1$ is equivalent to $M(1,b/a) = 1/a$ which is uniquely solvable for b/a, hence for b by the previous exercise.

5.3.13 Let $x = b/a = f(a)/a$. Then the stated asymptotic equality is equivalent to

$$\lim_{a\to\infty}\left[\log(x/4) + \frac{\pi a}{2}\right] = 0.$$

Since $0 < b < 1$, $a \to \infty$ implies $x \to 0$ and $M(1,x) = M(1,b/a) = 1/a$ by homogeneity, this follows from (5.3.7).

5.3.14 By multiplying out the d factors in the product, the only terms with x^{d-1} are $a_j x^{d-1}$, $1 \le j \le d$; hence $dp_1 = a_1 + \cdots + a_d$; hence p_1 is the arithmetic mean. If $x = 0$ is inserted, the identity reduces to $a_1 a_2 \dots a_d = p_d$. If $a_1 = a_2 = \cdots = a_d = 1$, the identity reduces to the binomial theorem; hence $p_k(1,1,\dots,1) = 1$, $1 \le k \le d$. The arithmetic and geometric mean inequality is then an immediate consequence of Exercise **3.3.28**.

5.3.15 Since $a_1 \ge a_2 \ge \cdots \ge a_d > 0$, replacing $a_2,\dots,a_{d-1}$ by a_1 in p_1 increases p_1. Similarly, replacing $a_2,\dots,a_{d-1}$ by a_d in p_d decreases p_d. Thus

$$\frac{a_1'}{a_d'} \le \frac{a_1 + a_1 + \cdots + a_1 + a_d}{d(a_1 a_d \dots a_d a_d)^{1/d}} = f_d\left(\frac{a_1}{a_d}\right),$$

where f_d is as in Exercise **3.2.10**. The result follows.

5.3.16 Note by Exercise **3.3.28**, $(a'_1, \dots, a'_d) = (p_1, \dots, p_d^{1/d})$ implies $a'_1 \ge a'_2 \ge \cdots \ge a'_d > 0$. Now a'_1 is the arithmetic mean and a_1 is the largest; hence $a_1 \ge a'_1$. Also a'_d is the geometric mean and a_d is the smallest; hence $a_d \le a'_d$. Hence $(a_1^{(n)})$ is decreasing and $(a_d^{(n)})$ is increasing and thus both sequences converge to limits $a_{1*} \ge a_d^*$. If we set $I_n = [a_d^{(n)}, a_1^{(n)}]$, we conclude the intervals I_n are nested $I_1 \supset I_2 \supset \cdots \supset [a_d^*, a_{1*}]$, and, for all $n \ge 0$, the reals $a_1^{(n)}, \dots, a_d^{(n)}$ all lie in I_n. By applying the inequality in Exercise **5.3.15** repeatedly, we conclude

$$0 \le \frac{a_1^{(n)}}{a_d^{(n)}} - 1 \le \left(\frac{d-1}{d}\right)^{2n} \left(\frac{a_1}{a_d} - 1\right), \qquad n \ge 0.$$

Letting $n \to \infty$, we conclude $a_d^* = a_{1*}$. Denoting this common value by m, we conclude $a_j^{(n)} \to m$ as $n \to \infty$, for all $1 \le j \le d$. The last identity follows from the fact that the limit of a sequence is unchanged if the first term of the sequence is discarded.

Solutions to Exercises 5.4

5.4.1 If $x = \sqrt{2t}$, $dx = dt/\sqrt{2t}$, and $t = x^2/2$. So

$$\sqrt{\frac{\pi}{2}} = \int_0^\infty e^{-x^2/2}\, dx = \int_0^\infty e^{-t} \frac{dt}{\sqrt{2t}} = \frac{1}{\sqrt{2}} \Gamma(1/2).$$

Hence $(1/2)! = \Gamma(3/2) = (1/2)\Gamma(1/2) = \sqrt{\pi}/2$.

5.4.2 Since $(x-s)^2 = x^2 - 2xs + s^2$,

$$e^{-s^2/2} L(s) = \int_{-\infty}^\infty e^{-(x-s)^2/2}\, dx = \int_{-\infty}^\infty e^{-x^2/2}\, dx = \sqrt{2\pi}$$

by translation invariance.

5.4.3 By differentiation under the integral sign,

$$L^{(n)}(s) = \int_{-\infty}^\infty e^{sx} x^n e^{-x^2/2}\, dx.$$

So

$$L^{(2n)}(0) = \int_{-\infty}^\infty x^{2n} e^{-x^2/2}\, dx.$$

To justify this, note that for $|s| < b$ and $f(s,x) = e^{sx - x^2/2}$,

$$\begin{aligned}\sum_{k=0}^{n}\left|\frac{\partial^k}{\partial s^k}f(s,x)\right| &= e^{sx-x^2/2}\sum_{k=0}^{n}|x|^k \\ &\le n!e^{sx-x^2/2}\sum_{k=0}^{n}\frac{|x|^k}{k!} \\ &\le n!e^{(b+1)|x|-x^2/2} = g(x)\end{aligned}$$

and g is even and integrable ($\int_{-\infty}^{\infty} g(x)\,dx \le 2n!L(b+1)$). Since the integrand is odd for n odd, $L^{(n)}(0) = 0$ for n odd. On the other hand, the exponential series yields

$$L(s) = \sqrt{2\pi}e^{s^2/2} = \sqrt{2\pi}\sum_{n=0}^{\infty}\frac{s^{2n}}{2^n n!} = \sum_{n=0}^{\infty}L^{(2n)}(0)\frac{s^{2n}}{(2n)!}.$$

Solving for $L^{(2n)}(0)$, we obtain the result.

5.4.4 With $f(s,x) = e^{-x^2/2}\cos(sx)$,

$$\begin{aligned}|f(s,x)| + \left|\frac{\partial}{\partial s}f(s,x)\right| &= e^{-x^2/2}(|\cos(sx)| + |x|\,|\sin(sx)|) \\ &\le e^{-x^2/2}(1+|x|) = g(x),\end{aligned}$$

which is integrable since $\int_{-\infty}^{\infty} g(x)\,dx = \sqrt{2\pi}+2$. Thus with $u = \sin(sx)$ and $dv = -xe^{-x^2/2}dx$, $v = e^{-x^2/2}$, $du = s\cos(sx)\,dx$. So

$$\begin{aligned}F'(s) &= -\int_{-\infty}^{\infty} e^{-x^2/2}x\sin(sx)\,dx \\ &= uv|_{-\infty}^{\infty} - s\int_{-\infty}^{\infty} e^{-x^2/2}\cos(sx)\,dx \\ &= -sF(s).\end{aligned}$$

Integrating $F'(s)/F(s) = -s$ over $(0,s)$ yields $\log F(s) = -s^2/2 + \log F(0)$ or $F(s) = F(0)e^{-s^2/2}$.

5.4.5 With $f(a,x) = e^{-x-a/x}/\sqrt{x}$ and $a \ge \epsilon > 0$,

$$|f(a,x)| + \left|\frac{\partial}{\partial a}f(a,x)\right| \le \begin{cases} e^{-\epsilon/x}\left(\dfrac{1}{\sqrt{x}} + \dfrac{1}{x\sqrt{x}}\right), & 0 < x < 1, \\ e^{-x}, & x > 1.\end{cases}$$

But the expression on the right is integrable over $(0,\infty)$. Hence we may differentiate under the integral sign on (ϵ,∞), hence on $(0,\infty)$. Thus with $x = a/t$,

$$H'(a) = -\int_0^\infty e^{-x-a/x} \frac{dx}{x\sqrt{x}}$$
$$= -\int_0^\infty e^{-a/t-t} \frac{dt}{\sqrt{at}} = -\frac{1}{\sqrt{a}} H(a).$$

Integrating $H'(a)/H(a) = -1/\sqrt{a}$, we get $\log H(a) = -2\sqrt{a} + \log H(0)$ or $H(a) = H(0)e^{-2\sqrt{a}}$.

5.4.6 Let $x = y\sqrt{q}$. Then $dx = \sqrt{q}\, dy$. Hence

$$\int_{-\infty}^{\infty} e^{-x^2/2q}\, dx = \sqrt{q} \int_{-\infty}^{\infty} e^{-y^2/2}\, dy = \sqrt{2\pi q}.$$

5.4.7 Inserting $g(x) = e^{-x^2\pi}$ and $\delta = \sqrt{t}$ in Exercise **4.3.8** and $\pi = 1/2q$ in the previous exercise yields

$$\lim_{t\to 0+} \sqrt{t}\,\psi(t) = \int_0^\infty e^{-x^2\pi}\, dx = \frac{1}{2}\sqrt{2\pi q} = \frac{1}{2}.$$

5.4.8 It is enough to show that ζ is smooth on (a, ∞) for any $a > 1$. Use differentiation N times under the summation sign to get

$$\zeta^{(k)}(s) = \sum_{n=1}^{\infty} \frac{(-1)^k \log^k n}{n^s}, \qquad s > a, N \geq k \geq 0.$$

To justify this, let $f_n(s) = n^{-s}$, $n \geq 1$, $s > 1$. Since $\log n/n^\epsilon \to 0$ as $n \nearrow \infty$, for any $\epsilon > 0$, the sequence $(\log n/n^\epsilon)$ is bounded, which means that there is a constant $C_\epsilon > 0$, such that $|\log n| \leq C_\epsilon n^\epsilon$ for all $n \geq 1$. Hence

$$\sum_{k=0}^{N} \left| f_n^{(k)}(s) \right| \leq \sum_{k=0}^{N} \frac{|\log^k n|}{n^s} \leq (N+1) \frac{C_\epsilon^{N+1} n^{N\epsilon}}{n^a} = \frac{C}{n^{a-N\epsilon}}.$$

Then if we choose ϵ small enough, so that $a - N\epsilon > 1$, the dominating series $\sum g_n = C \sum n^{N\epsilon - a}$ converges.

5.4.9 Again, we show that ψ is smooth on (a, ∞) for all $a > 0$. Use differentiation N times under the summation sign to get

$$\psi^{(k)}(t) = \sum_{n=1}^{\infty} (-1)^k \pi^k n^{2k} e^{-n^2\pi t}, \qquad t > a, N \geq k \geq 0.$$

To justify this, let $f_n(t) = e^{-n^2\pi t}$, $n \geq 1$, $t > a$. Since $x^{N+1}e^{-x} \to 0$ as $x \to \infty$, the function $x^{N+1}e^{-x}$ is bounded. Thus there is a constant $C_N > 0$, such that $x^N e^{-x} \leq C_N/x$ for $x > 0$. Inserting $x = n^2\pi t$, $t > a$,

$$\sum_{k=0}^{N}\left|f_n^{(k)}(t)\right| \le \sum_{k=0}^{N} n^{2k}\pi^k e^{-n^2\pi t} \le \frac{(N+1)C_N}{n^2\pi\cdot a^{N+1}}.$$

Then the dominating series $\sum g_n = \sum (N+1)C_N/\pi a^{N+1}n^2$ converges.

5.4.10 Differentiating under the integral sign leads only to bounded functions of t. So J_ν is smooth. Computing, we get

$$J_\nu'(x) = \frac{1}{\pi}\int_0^\pi \sin t\,\sin(\nu t - x\sin t)\,dt$$

and

$$J_\nu''(x) = -\frac{1}{\pi}\int_0^\pi \sin^2 t\,\cos(\nu t - x\sin t)\,dt.$$

Now, integrate by parts with $u = -x\cos t - \nu$, $dv = (\nu - x\cos t)\cos(\nu t - x\sin t)\,dt$, $du = x\sin t\,dt$, and $v = \sin(\nu t - x\sin t)$:

$$\begin{aligned}
x^2J_\nu''(x) + (x^2-\nu^2)J_\nu(x) &= \frac{1}{\pi}\int_0^\pi (x^2\cos^2 t - \nu^2)\cos(\nu t - x\sin t)\,dt\\
&= \frac{1}{\pi}\int_0^\pi u\,dv = \frac{1}{\pi}uv\Big|_0^\pi - \frac{1}{\pi}\int_0^\pi v\,du\\
&= -\frac{1}{\pi}\int_0^\pi x\sin t\,\sin(\nu t - x\sin t)\,dt\\
&= -xJ_\nu'(x).
\end{aligned}$$

Here ν must be an integer to make the uv term vanish at π.

5.4.11 Differentiating under the integral sign,

$$F^{(n)}(s) = \int_{-\infty}^{\infty} x^n e^{sx}e^{-f(x)}\,dx.$$

Since $|x|^n \le n!e^{|x|}$, with $h(s,x) = e^{sx}e^{-f(x)}$ and $|s| < b$,

$$\sum_{n=0}^{N}\left|\frac{\partial^n h}{\partial s^n}\right| \le \sum_{n=0}^{N}|x|^n e^{s|x|}e^{-f(x)} \le (N+1)!e^{(b+1)|x|-f(x)} = g(x)$$

and g is integrable by Exercise **4.3.11**. This shows that F is smooth. Differentiating twice,

$$[\log F(s)]'' = \frac{F''(s)F(s) - F'(s)^2}{F(s)^2}.$$

Now, use the Cauchy–Schwarz inequality (Exercise **4.4.21**) with the functions $e^{(sx-f(x))/2}$ and $xe^{(sx-f(x))/2}$ to get $F''(s)F(s) \ge F'(s)^2$. Hence $[\log F(s)]'' \ge 0$, or $\log F(s)$ is convex.

5.4.12 With $u = e^{-sx}$ and $dv = \sin x(dx/x)$, $du = -se^{-sx}\,dx$, and $v = F(t) = \int_0^t \sin r(dr/r)$. So integration by parts yields the first equation. Now, change variables $y = sx$, $sdx = dy$ in the integral on the right yielding

$$\int_0^b e^{-sx}\frac{\sin x}{x}\,dx = -e^{-sb}F(b) + \int_0^{b/s} e^{-y}F(y/s)\,dy.$$

Let $b \to \infty$. Since F is bounded,

$$\int_0^\infty e^{-sx}\frac{\sin x}{x}\,dx = \int_0^\infty e^{-y}F(y/s)\,dy.$$

Now, let $s \to 0+$, and use the dominated convergence theorem. Since $F(y/s) \to F(\infty)$, as $s \to 0+$, for all $y > 0$,

$$\lim_{s\to 0+}\int_0^\infty e^{-sx}\frac{\sin x}{x}\,dx = \int_0^\infty e^{-y}F(\infty)\,dy = F(\infty) = \lim_{b\to\infty}\int_0^b \frac{\sin x}{x}\,dx.$$

But, from the text, the left side is

$$\lim_{x\to 0+}\arctan\left(\frac{1}{x}\right) = \arctan(\infty) = \frac{\pi}{2}.$$

Solutions to Exercises 5.5

5.5.1 Without loss of generality, assume that $a = \max(a, b, c)$. Then $(b/a)^n \le 1$ and $(c/a)^n \le 1$. So

$$\lim_{n\nearrow\infty}(a^n + b^n + c^n)^{1/n} = a\lim_{n\nearrow\infty}(1 + (b/a)^n + (c/a)^n)^{1/n} = a.$$

For the second part, replace a, b, and c in the first part by e^a, e^b, and e^c. Then take the log. For the third part, given $\epsilon > 0$, for all but finitely many $n \ge 1$, we have $\log(a_n) \le (A+\epsilon)n$ or $a_n \le e^{n(A+\epsilon)}$. Similarly, $b_n \le e^{n(B+\epsilon)}$, $c_n \le e^{n(C+\epsilon)}$ for all but finitely many $n \ge 1$. Hence the upper limit of $\log(a_n + b_n + c_n)/n$ is $\le \max(A, B, C) + \epsilon$. Similarly, the lower limit of $\log(a_n + b_n + c_n)/n \ge \max(A, B, C) - \epsilon$. Since ϵ is arbitrary, the result follows.

5.5.2 The relative error is about .083%; here is python code:

```
from math import exp, log, atan2, factorial

def stirling(n):
    p  = atan2(1,1)*4
```

```
    s  = (n + .5)*log(n) - n
    s  = exp(s)
    s *= (2 * p)**(.5)
    return s

f = float(factorial(100))
s = stirling(100)
e = 100*(f-s)/f

print "100! = ",f,"\n"
print "stirling = ",s,"\n"
print "percentage error = ",e,"\n"
```

5.5.3 In the asymptotic for $\binom{n}{k}$, replace n by $2n$ and k by n. Since $t = n/2n = 1/2$ and $H(1/2) = 0$, we get $1/\sqrt{\pi n}$.

5.5.4 Straight computation.

5.5.5 $H'(t,p) = \log(t/p) - \log[(1-t)/(1-p)]$ equals zero when $t = p$. Since $H''(t,p) = 1/t + 1/(1-t)$, H is convex. So $t = p$ is a global minimum.

5.5.6 Straight computation.

5.5.7 Since $(q^n)^{x^2} = e^{nx^2 \log q}$, the limit is

$$\sup\{x^2 \log q : a < x < b\} = a^2 \log q,$$

by the theorem. Here $\log q < 0$.

5.5.8 Since $\Gamma(s+1) = s\Gamma(s)$,

$$\begin{aligned} f(s+1) &= 3^{3s+3} \frac{\Gamma(s+1)\Gamma(s+1+1/3)\Gamma(s+1+2/3)}{\Gamma(3s+3)} \\ &= \frac{3^3 s(s+1/3)(s+2/3)}{(3s+2)(3s+1)3s} \cdot f(s) = f(s). \end{aligned}$$

Inserting the asymptotic for $\Gamma(s+n)$ yields $2\pi\sqrt{3}$ for the limit. The general case is similar.

5.5.9 Take the log and divide by n to get

$$\frac{1}{n}\sum_{k=0}^{n-1} \log \Gamma(s+k/n) = \frac{n-1}{2n}\log(2\pi) + \log \Gamma(ns) - (ns - 1/2) \cdot \frac{\log n}{n}.$$

Using Stirling's approximation for $\log \Gamma(ns)$, the result follows.

5.5.10 Here

$$L_n(ny) = \int_{-\infty}^{\infty} e^{n(xy - f(x))}\, dx,$$

and $g(y) = \max\{xy - f(x) : x \in \mathbf{R}\}$ exists by Exercise **2.3.20** and the sup is attained at some c. Fix $y \in \mathbf{R}$ and select $M > 0$ such that $-(M \pm y) < g(y)$ and $M \pm y > 0$. Since $f(x)/|x| \to \infty$ as $|x| \to \infty$, we can choose b such that $b > 1$, $b > c$, and $f(x) \geq Mx$ for $x \geq b$. Similarly, we can choose a such that $a < -1$, $a < c$, and $f(x) \geq M(-x)$ for $x \leq a$. Write $L_n(ns) = I_n^- + I_n^0 + I_n^+ = \int_{-\infty}^a + \int_a^b + \int_b^\infty$. The second theorem in §5.5 applies to I_n^0; hence

$$\lim_{n\to\infty} \frac{1}{n} \log(I_n^0) = \max\{xy - f(x) : a < x < b\} = g(y)$$

since the max over $\mathbf{R}$ is attained within (a, b). Now

$$I_n^- \leq \int_{-\infty}^{a} e^{n(M+y)x}\, dx = \frac{e^{n(M+y)a}}{n(M+y)},$$

so

$$\lim_{n\to\infty} \frac{1}{n} \log(I_n^-) \leq (M+y)a < -(M+y) < g(y).$$

Similarly,

$$I_n^+ \leq \int_{b}^{\infty} e^{-n(M-y)x}\, dx = \frac{e^{-n(M-y)b}}{n(M-y)},$$

so

$$\lim_{n\to\infty} \frac{1}{n} \log(I_n^+) \leq -(M-y)b < -(M-y) < g(y).$$

By Exercise **5.5.1**, we conclude that

$$\lim_{n\nearrow\infty} \frac{1}{n} \log L_n(ny) = \lim_{n\nearrow\infty} \frac{1}{n} \log\left(I_n^- + I_n^0 + I_n^+\right) = g(y).$$

5.5.11 The log of the duplication formula is

$$2s\log 2 + \log\Gamma(s) + \log\Gamma(s+1/2) - \log\Gamma(2s) = \log(2\sqrt{\pi}).$$

Differentiating,

$$2\log 2 + \frac{\Gamma'(s)}{\Gamma(s)} + \frac{\Gamma'(s+1/2)}{\Gamma(s+1/2)} - 2\frac{\Gamma'(2s)}{\Gamma(2s)} = 0.$$

Inserting $s = 1/2$, we obtain the result.

5.5.12 Insert $s = 1/4$ in the duplication formula to get

$$\sqrt{2}\frac{\Gamma(1/4)\Gamma(3/4)}{\Gamma(1/2)} = 2\sqrt{\pi}.$$

Now recall that $\Gamma(1/2) = \sqrt{\pi}$. To obtain the formula for $1/M(1, 1/\sqrt{2})$, replace $\Gamma(3/4)$ in the formula in the text by $\pi\sqrt{2}/\Gamma(1/4)$.

Solutions to Exercises 5.6

5.6.1 Using (5.6.9), replacing x by $x/2\pi$, and setting $A_{2n} = \zeta(2n)/(2\pi)^{2n}$, we have

$$\frac{x}{1-e^{-x}} = 1 + \frac{1}{2} + 2A_2x^2 + 2A_4x^4 + 2A_6x^6 + \dots;$$

thus

$$x = \left(x - \frac{x^2}{2} + \frac{x^3}{6} - \frac{x^4}{24} + \dots\right)\left(1 + \frac{x}{2} + 2A_2x^2 - 2A_4x^4 + 2A_6x^6 + \dots\right).$$

Multiplying out yields $A_2 = 1/24$, $A_4 = 1/1440$, $A_6 = 1/60480$, $A_8 =$ ·2419200; hence $\zeta(2) = \pi^2/6$, $\zeta(4) = \pi^4/90$, $\zeta(6) = \pi^6/945$, and $\zeta(8) = \pi^8/9450$.

5.6.2 Let $b_k = B_k/k!$, and suppose that $|b_k| \le 2^k$ for $k \le n-2$. Then (5.6.8) reads

$$\sum_{k=0}^{n-1} \frac{b_k(-1)^{n-1-k}}{(n-k)!} = 0$$

which implies ($n! \ge 2^{n-1}$)

$$\begin{aligned}|b_{n-1}| &\le \sum_{k=0}^{n-2} \frac{|b_k|}{(n-k)!} \le \sum_{k=0}^{n-2} \frac{2^k}{(n-k)!} \\ &\le \sum_{k=0}^{n-2} \frac{2^k}{2^{n-k-1}} \le 2^{n-1}.\end{aligned}$$

Thus $|b_n| \le 2^n$ for all $n \ge 1$ by induction. Hence the radius of convergence by the root test is at least $1/2$. Also from the formula for $\zeta(2n) > 0$, the Bernoulli numbers are alternating.

5.6.3 The left inequality follows from (5.6.11) since $b_1 = \sum_{n=1}^{\infty} a_n$. For the right inequality, use $1 + a_n \le e^{a_n}$. So

$$\prod_{n=1}^{\infty}(1 + a_n) \le \prod_{n=1}^{\infty} e^{a_n} = \exp\left(\sum_{n=1}^{\infty} a_n\right).$$

5.6.4 From (5.1.3),

$$\begin{aligned}\Gamma(x) &= \lim_{n\nearrow\infty} \frac{n^x n!}{x(1+x)(2+x)\dots(n+x)} \\ &= \frac{1}{x}\lim_{n\nearrow\infty} \frac{e^{x(\log n-1-1/2-\dots-1/n)}e^x e^{x/2}\dots e^{x/n}}{(1+x)(1+x/2)\dots(1+x/n)} \\ &= \frac{e^{-\gamma x}}{x}\prod_{n=1}^{\infty}\left(\frac{e^{x/n}}{1+\frac{x}{n}}\right).\end{aligned}$$

For the second product, use (5.1.4) instead.

5.6.5 For $0 < x < 1$ use (5.1.3) with x and $1-x$ replacing x. Then $\Gamma(x)\Gamma(1-x)$ equals

$$\begin{aligned}&\lim_{n\nearrow\infty} \frac{n^x n! n^{1-x} n!}{x(1+x)(2+x)\dots(n+x)(1-x)(2-x)\dots(n+1-x)} \\ &= \frac{1}{x}\lim_{n\nearrow\infty}\frac{n(n!)^2}{(1-x^2)(4-x^2)\dots(n^2-x^2)(n+1-x)} \\ &= \frac{1}{x}\lim_{n\nearrow\infty}\frac{1}{(1-x^2)(1-x^2/4)\dots(1-x^2/n^2)(1+(1-x)/n)} \\ &= \left[x\prod_{n=1}^{\infty}\left(1-\frac{x^2}{n^2}\right)\right]^{-1} = \frac{\pi}{\sin(\pi x)}.\end{aligned}$$

5.6.6 The series $B(x)$ is the alternating version of the Bernoulli series (5.6.7), and the Taylor series for $\sin(x/2)$ is the alternating version of the Taylor series for $\sinh(x/2)$. But the Bernoulli series times $\sinh(x/2)$ equals $(x/2)\cosh(x/2)$. Hence (Exercise **1.7.7**),

$$B(x)\sin(x/2) = (x/2)\cos(x/2).$$

Dividing by $\sin(x/2)$, we obtain the series for $(x/2)\cot(x/2)$.

5.6.7 If $\beta > 1$, then $B(x)$ would converge at 2π. But $(x/2)\cot(x/2) = B(x)$ is infinite at 2π.

5.6.8 Taking the log of (5.6.14),

$$\log[\sin(\pi x)] - \log(\pi x) = \sum_{n=1}^{\infty}\log\left(1-\frac{x^2}{n^2}\right).$$

Differentiating under the summation sign,

$$\pi\cot(\pi x) - \frac{1}{x} = \sum_{n=1}^{\infty}\frac{2x}{x^2-n^2}.$$

To justify this, let $f_n(x) = \log(1 - x^2/n^2)$ and let $|x| < b < 1$. Then $\log(1 - t) = t + t^2/2 + t^3/3 + \cdots \le t + t^2 + t^3 + \cdots = t/(1-t)$. So

$$|f_n(x)| + |f_n'(x)| \le \frac{x^2}{n^2 - x^2} + \frac{2|x|}{n^2 - x^2} \le \frac{b^2 + 2b}{n^2 - b^2} = g_n,$$

which is summable. Since this is true for all $b < 1$, the equality is valid for $|x| < 1$.

5.6.9 By Exercise **3.6.13**, $\cot x - 2\cot(2x) = \tan x$. Then the series for $\tan x$ follows from the series for $\cot x$ in Exercise **5.6.6** applied to $\cot x$ and $2\cot(2x)$.

5.6.10 By Exercise **5.6.4**,

$$\log \Gamma(x) = -\gamma x - \log x + \sum_{n=1}^{\infty} \left[\frac{x}{n} - \log\left(1 + \frac{x}{n}\right)\right].$$

Differentiating, we obtain the result. To justify this, let $f_n(x) = x/n - \log(1 + x/n)$, $f_n'(x) = 1/n - 1/(x+n)$. Then $t - \log(1+t) = t^2/2 - t^3/3 + \cdots \le t^2/2$ for $t > 0$. Hence $f_n(x) \le x^2/n^2$. So

$$|f_n(x)| + |f_n'(x)| \le \frac{b^2 + b}{n^2} = g_n,$$

which is summable, when $0 < x < b$. Since b is arbitrary, the result is valid for $x > 0$. For the second identity, use the infinite product for $x!$ instead.

5.6.11 From the previous exercise,

$$\frac{d}{dx}\log(x!) = -\gamma + \sum_{n=1}^{\infty}\left(\frac{1}{n} - \frac{1}{x+n}\right) = -\gamma + \sum_{n=1}^{\infty} \frac{x}{n(n+x)}.$$

Differentiating r times, we obtain the identity. To justify this differentiation under the summation sign, let $h(x) = x/(n(n+x))$ and consider $-1 + \epsilon \le x \le 1/\epsilon$ for ϵ small. Then

$$|h(x)| + |h'(x)| + \cdots + |h^{(r)}(x)| \le \frac{|x|}{n(n+x)} + \frac{1}{(n+x)^2} + \cdots + \frac{r!}{(n+x)^{r+1}}.$$

When $n = 1$, this is no larger than $(r+1)!/\epsilon^{r+1}$. When $n \ge 2$, this is no larger than $(r+1)!/\epsilon(n-1)^2$. Since this is the general term of a convergent series, the identity is justified on $[-1+\epsilon, 1/\epsilon]$. Since $\epsilon > 0$ is arbitrarily small, the identity follows. The inequality follows by separating out the first term.

5.6.12 Inserting $x = 0$ in the previous exercise yields

$$\frac{1}{r!}\frac{d^r}{dx^r}\log(x!)\bigg|_{x=0} = (-1)^r \zeta(r)/r.$$

This yields the coefficients of the Taylor series. To show the series converges to the function, note the integral form of the remainder in Taylor's theorem §4.4 and use the inequality in the previous exercise. You get (here $-1 < x \le 1$)

$$\left|\frac{x^{r+1}}{(r+1)!}h_{r+1}(x)\right| \le \int_0^1 \left[\frac{1}{(sx+1)^{r+1}} + \zeta(2)\right](1-s)^r\,ds.$$

But for $1 \ge x > -1$ and $0 < s < 1$, $1 - s < 1 + sx$, so $(1-s)/(1+sx) < 1$; hence $(1-s)^r/(1+sx)^r \to 0$ as $r \to \infty$. Thus the integral goes to zero as $r \to \infty$ by the dominated convergence theorem. For the second series, plug $x = 1$ into the first.

5.6.13 Inserting $x = 1$ in the series for $\Gamma'(x)/\Gamma(x)$ yields a telescoping series. So we get $\Gamma'(1) = -\gamma$. Inserting $x = 2$ yields

$$-\gamma - \frac{1}{2} + 1 - \frac{1}{3} + \frac{1}{2} - \frac{1}{4} + \frac{1}{3} - \frac{1}{5} + \frac{1}{4} - \frac{1}{6} + \cdots = 1 - \gamma.$$

Since Γ is strictly convex, this forces the min to lie in $(1,2)$.

5.6.14 Differentiate the series in Exercise **5.6.10**, and compare with the series in Exercise **5.1.13**. Here on $0 < x < b$, we may take $g_n = (b+1)/n^2$, $n \ge 1$.

5.6.15 Substituting $(1-x)/2 \mapsto x$, we see that the stated equality is equivalent to

$$\lim_{x\to 0+}\left[\frac{\Gamma'(x)}{\Gamma(x)} + \frac{1}{x}\right] = -\gamma.$$

Move the $1/x$ to the left in Exercise **5.6.10**, and then take the limit $x \to 0+$. Under this limit, the series collapses to zero by the dominated convergence theorem for series (here, $g_n = b/n^2$ for $0 < x < b$).

Solutions to Exercises 5.7

5.7.1 θ_0 is strictly increasing since it is the sum of strictly increasing monomials. Since θ_0 is continuous and $\theta_0(0+) = 1$, $\theta_0(1-) = \infty$, $\theta_0((0,1)) = (1,\infty)$. Similarly, for θ_+.

5.7.2 Multiply (5.7.21) by $s\theta_0(s)^2 = \theta_0^2(1/s)$. You get (5.7.22).

5.7.3 The AGM of $\theta_0^2(q)$ and $\theta_-^2(q)$ equals 1: $M(\theta_0^2(q), \theta_-^2(q)) = 1$ for $0 < q < 1$. Since θ_0 is strictly increasing, θ_0^2 is strictly increasing. This forces θ_-^2 to be strictly decreasing. Moreover, $\theta_-(0+) = 1$. Hence $\theta_-^2(0+) = 1$, and $q \to 1-$ implies $M(\infty, \theta_-^2(1-)) = 1$ or $\theta_-^2(1-) = 0$. Thus θ_-^2 maps $(0,1)$ onto $(0,1)$. Since θ_- is continuous and $\theta_-(0+) = 1$, we also have θ_- strictly decreasing and $\theta_-((0,1)) = (0,1))$.

5.7.4 Since $\sigma(6) = \sigma(7) = 0$ and $\sigma(2n) = \sigma(n)$, $\sigma(12) = \sigma(14) = 0$. Also $\sigma(13) = 8$ since

$$13 = (\pm 2)^2 + (\pm 3)^2 = (\pm 3)^2 + (\pm 2)^2.$$

To show that $\sigma(4n-1) = 0$, we show that $4n - 1 = i^2 + j^2$ cannot happen. Note that $4n - 1$ is odd when exactly one of i or j is odd and the other is even. Say $i = 2k$ and $j = 2\ell + 1$. Then

$$4n - 1 = 4k^2 + 4\ell^2 + 4\ell + 1 = 4(k^2 + \ell^2 + \ell) + 1,$$

an impossibility. Hence $\sigma(4n-1) = 0$, so $\sigma(11) = \sigma(15) = 0$.

5.7.5 Let $m = M(a,b)$, $a' = a/m$, $b' = b/m$. Then $b/a = b'/a'$ and $M(a',b') = 1$, so $(a',b') = (\theta_0^2(q), \theta_-^2(q))$. Hence

$$\begin{aligned} a - b = m(a' - b') = m\left(\theta_0^2(q) - \theta_-^2(q)\right) &= 2M(a,b) \sum_{n \text{ odd}} \sigma(n) q^n \\ &= 8M(a,b)q \times \left(1 + 2q^4 + q^8 + \dots\right). \end{aligned}$$

Replacing q by q^{2^n} yields the result.

5.7.6 Let $f_n(t,x) = e^{-n^2\pi t}\cos(nx)$, $n \ge 1$. Then $\partial f_n/\partial t = -n^2\pi f_n$, and $\partial^2 f_n/\partial x^2 = -n^2 f_n$, $n \ge 1$. Thus to obtain the heat equation, we need only to justify differentiation under the summation sign. But, for $t \ge 2a > 0$,

$$|f_n| + \left|\frac{\partial f_n}{\partial t}\right| + + \left|\frac{\partial f_n}{\partial x}\right| + \left|\frac{\partial^2 f_n}{\partial x^2}\right| \le 4n^2 \pi e^{-2an^2} = g_n,$$

which is summable since xe^{-ax} is bounded for $x > 0$; hence $xe^{-2ax} \le e^{-ax}$.

Solutions to Exercises 5.8

5.8.1 Assume that $1 < x < 2$. The integrand $f(x,t)$ is positive, thus increasing; hence $\le f(2,t)$. By Exercise **3.5.7**, $f(2,t)$ is asymptotically equal to $t/2$, as $t \to 0+$ which is bounded. Also the integrand is asymptotically equal to te^{-t}, as $t \to \infty$ which is integrable. Since $f(2,t)$ is continuous, $g(t) = f(2,t)$ is integrable. Hence we may switch the limit with the integral.

5.8.2 The limit of the left side of (5.8.4) is γ, by (5.8.3), and $\Gamma(1) = 1$.

5.8.3 Since $\Gamma(x+n+1) = (x+n)\Gamma(x+n) = (x+n)(x+n-1)\dots x\Gamma(x)$, for $x \in (-n-1, -n) \cup (-n, -n+1)$,

$$(x+n)\Gamma(x) = \frac{\Gamma(x+n+1)}{x(x+1)\dots(x+n-1)}.$$

Letting $x \to -n$, we get $(x+n)\Gamma(x) \to (-1)^n/n!$.

5.8.4 For $t \geq 1$,

$$\psi(t) \leq \sum_{n=1}^{\infty} e^{-n\pi t} = \frac{e^{-\pi t}}{1 - e^{-\pi t}} \leq ce^{-\pi t}$$

with $c = 1/(1 - e^{-\pi})$. Now, the integrand $f(x,t)$ in (5.8.9) is a smooth function of x with $\partial^n f/\partial x^n$ continuous in (x,t) for all $n \geq 0$. Moreover, for $b > 1$ and $0 < x < b$ and each $n \geq 0$,

$$\begin{aligned}\left|\frac{\partial^n f}{\partial x^n}\right| &\leq \psi(t) 2^{-n} |\log t|^n \left[t^{(1-x)/2} + t^{x/2}\right] \\ &\leq ce^{-\pi t} 2^{-n+1} |\log t|^n t^{b/2} = g_n(t),\end{aligned}$$

which is integrable over $(1, \infty)$. Hence we may repeatedly apply differentiation under the integral sign to conclude that the integral in (5.8.9) is smooth.

5.8.5 Inserting $x = 2n$ in (5.8.10), we get

$$\begin{aligned}\zeta(1-2n) &= \frac{\pi^{-n}\Gamma(n)\zeta(2n)}{\pi^{n-1/2}\Gamma(-n+1/2)} \\ &= \pi^{-2n+1/2}(n-1)! \\ &\quad \times \frac{(-n+1/2)(-n+3/2)\ldots(-n+n-1/2)}{\Gamma(-n+n+1/2)} \\ &\quad \times \frac{(-1)^{n-1}B_{2n}2^{2n-1}\pi^{2n}}{(2n)!} = -\frac{B_{2n}}{2n}.\end{aligned}$$

5.8.6 Here $f(x,t) = (1 + [t] - t)/t^{x+1}$, and $I(x) = \int_1^\infty f(x,t)\,dt$, $x > 1$. Since, for $b > x > a > 1$,

$$|f(x,t)| + \left|\frac{\partial f}{\partial x}(x,t)\right| \leq \frac{b+2}{t^{a+1}} = g(t), \quad t > 1,$$

and g is integrable, we may differentiate under the integral sign, obtaining $I'(x) = (x+1)I(x+1)$ for $a < x < b$. Since $a < b$ are arbitrary, this is valid for $x > 0$. Inserting $x = 2$ in (5.8.6) yields $\pi^2/6 = \zeta(2) = 1 + 2I(2)$ or $I(2) = \pi^2/12 - 1/2$.

5.8.7 The right side of (5.8.9) is smooth, except at $x = 0, 1$. Hence $(x-1)\zeta(x)$ is smooth, except (possibly) at $x = 0$. By (5.8.6),

$$\log((x-1)\zeta(x)) = \log\Bigl(1 + x(x-1)I(x)\Bigr).$$

So

$$\frac{\zeta'(x)}{\zeta(x)} + \frac{1}{x-1} = \frac{d}{dx}\log((x-1)\zeta(x)) = \frac{(2x-1)I(x) + x(x-1)I'(x)}{1 + x(x-1)I(x)}.$$

Taking the limit $x \to 1$, we approach $I(1) = \gamma$.

5.8.8 (5.8.10) says that $\pi^{-x/2}\Gamma(x/2)\zeta(x) = \pi^{-(1-x)/2}\Gamma((1-x)/2)\zeta(1-x)$. Differentiating the log of (5.8.10) yields

$$-\frac{1}{2}\log\pi + \frac{1}{2}\frac{\Gamma'(x/2)}{\Gamma(x/2)} + \frac{\zeta'(x)}{\zeta(x)} = \frac{1}{2}\log\pi - \frac{1}{2}\frac{\Gamma'((1-x)/2)}{\Gamma((1-x)/2)} - \frac{\zeta'(1-x)}{\zeta(1-x)}.$$

Now, add $1/(x-1)$ to both sides, and take the limit $x \to 1$. By the previous exercise, the left side becomes $-\log\pi/2 + \Gamma'(1/2)/2\Gamma(1/2) + \gamma$. By Exercise **5.6.15**, the right side becomes $\log\pi/2 + \gamma/2 - \zeta'(0)/\zeta(0)$. But $\zeta(0) = -1/2$, and by Exercise **5.5.11** and Exercise **5.6.13**,

$$\begin{aligned}\frac{1}{2}\frac{\Gamma'(1/2)}{\Gamma(1/2)} &= \frac{1}{2}\frac{\Gamma'(1)}{\Gamma(1)} - \log 2\\ &= -\frac{\gamma}{2} - \log 2.\end{aligned}$$

So

$$-\frac{\log\pi}{2} + \left(-\frac{\gamma}{2} - \log 2\right) + \gamma = \frac{\log\pi}{2} + \gamma/2 + 2\zeta'(0).$$

Hence $\zeta'(0) = -\log(2\pi)/2$.

5.8.9 For $0 < a \le 1/2$,

$$-\log(1-a) = a + \frac{a^2}{2} + \frac{a^3}{3} + \cdots \le a + a^2 + a^3 + \cdots = \frac{a}{1-a} \le 2a.$$

On the other hand, by the triangle inequality, for $|a| \le 1/2$,

$$\begin{aligned}|-\log(1-a) - a| &= \left|\frac{a^2}{2} + \frac{a^3}{3} + \dots\right|\\ &\le \frac{|a|^2}{2} + \frac{|a|^3}{3} + \dots\\ &\le \frac{1}{2}\left(|a|^2 + |a|^3 + \dots\right) = \frac{1}{2}\cdot\frac{a^2}{1-|a|} \le a^2.\end{aligned}$$

5.8.10 m and n are both odd iff mn is odd. So $\chi_+(m)$ and $\chi_+(n)$ are both equal 1 iff $\chi_+(mn) = 1$. Since χ_+ equals to 0 or 1, this shows that $\chi_+(mn) = \chi_+(m)\chi_+(n)$. For χ_-, m or n is even iff mn is even. So

$$\chi_-(mn) = \chi_-(m)\chi_-(n) \tag{A.5.1}$$

when either n or m is even. If n and m are both odd and $m = 4i+3$, $n = 4j+3$, then $mn = (4i+3)(4j+3) = 4(4ij+3i+3j+2)+1$ which derives (A.5.1) when $\chi_-(m) = \chi_-(n) = -1$. The other *three* cases are similar.

5.8.11 With $f_n(x) = 1/(4n-3)^x - 1/(4n-1)^x$,

$$L(s,\chi_-) = \sum_{n=1}^{\infty} f_n(x), \qquad x > 0.$$

Then by the mean value theorem, $f_n(x) \le x/(4n-3)^{x+1}$. Hence $|f_n(x)| \le b/(4n-3)^{a+1} = g_n$ for $0 < a < x < b$. Since $4n-3 \ge n$, $\sum g_n \le b\zeta(a+1)$. So the dominated convergence theorem applies.

Solutions to Exercises 5.9

5.9.1 On $n-1 \le x \le n$, $q(x) = \sum_{k=0}^{n} f(x-k)$, so $q(n) = \sum_{k=0}^{n} f(n-k)$. On $n \le x \le n+1$, $q(x) = \sum_{k=0}^{n+1} f(x-k) = \sum_{k=0}^{n} f(x-k) + f(x-n-1)$, so $q(n) = \sum_{k=0}^{n} f(n-k) + f(-1)$. Since $f(-1) = 0$, q is well defined at all integers. If $n-1 \le x \le n$, $q(x) = \sum_{k=0}^{n} f(x-n)$, and $q(x-1) = \sum_{k=0}^{n} f(x-1-k) = \sum_{k=1}^{n+1} f(x-k)$. So $q(x) - q(x-1) = f(x) - f(x-n-1) = f(x)$ since $f = 0$ on $[-2,-1]$. Thus q solves (5.9.3). To show that q is smooth on $\mathbf{R}$, assume that q is smooth on $(-\infty, n)$. Then $q(x-1) + f(x)$ is smooth on $(-\infty, n+1)$. Hence so is $q(x)$ by (5.9.3). Thus q is smooth on $\mathbf{R}$.

5.9.2 If $f(x) = 0$ for $x > 1$, the formula reads

$$q(x) = \begin{cases} -f(x+1), & x \ge -1, \\ -f(x+1) - f(x+2), & -1 \ge x \ge -2, \\ -f(x+1) - f(x+2) - f(x+3), & -2 \ge x \ge -3, \\ \text{and so on.} \end{cases}$$

5.9.3 We show, by induction on $n \ge 0$, that

$$c(D)\left[e^{ax}x^n\right] = \frac{\partial^n}{\partial a^n}\left[c(a)e^{ax}\right] \tag{A.5.2}$$

for all $|a| < R$ and convergent series $c(a)$ on $(-R, R)$. Clearly, this is so for $n = 0$. Assume that (A.5.2) is true for $n-1$ and check (by induction over k) that $D^k(e^{ax}x^n) = xD^k(e^{ax}x^{n-1}) + kD^{k-1}(e^{ax}x^{n-1})$, $k \ge 0$. Taking linear combinations, we get $c(D)(e^{ax}x^n) = xc(D)(e^{ax}x^{n-1}) + c'(D)(e^{ax}x^{n-1})$. By the inductive hypothesis applied to c and c', we obtain

$$
\begin{aligned}
c(D)(e^{ax}x^n) &= \frac{\partial^{n-1}}{\partial a^{n-1}}\left[xc(a)e^{ax} + c'(a)e^{ax}\right] \\
&= \frac{\partial^{n-1}}{\partial a^{n-1}} \cdot \frac{\partial}{\partial a}[c(a)e^{ax}] \\
&= \frac{\partial^{n}}{\partial a^{n}}[c(a)e^{ax}].
\end{aligned}
$$

Thus (A.5.2) is true for all $n \ge 0$.

5.9.4 Change variable $xt = s$, $xdt = ds$. Then with $|f(t)| \le C$, $0 < t < 1$,

$$
\begin{aligned}
\left| x^{n+1} \int_0^1 e^{-xt} f(t) t^n \, dt \right| &= \left| \int_0^x e^{-s} f(s/x) s^n \, ds \right| \\
&\le C \int_0^\infty e^{-s} s^n \, ds = C\Gamma(n+1).
\end{aligned}
$$

This shows that the integral is $O(x^{-n-1})$.

5.9.5 Note that $\int_1^\infty e^{-xt}\,dt = e^{-x}/x$ and, by differentiation under the integral sign,

$$
\int_1^\infty e^{-xt} t^p \, dt = (-1)^p \frac{d^p}{dx^p} \int_1^\infty e^{-xt} \, dt = (-1)^p \frac{d^p}{dx^p} \frac{e^{-x}}{x} = R(x) e^{-x}
$$

for some rational function R. But $e^{-x} = O(x^{-n})$ for all $n \ge 1$ implies $R(x)e^{-x} = O(x^{-n})$ for all $n \ge 1$. Thus the integral is ≈ 0.

5.9.6 If the Stirling series converged at some point a, then $B_n/n(n-1)a^{n-1}$, $n \ge 1$, would be bounded by the nth term test. Then the Bernoulli series would be dominated by

$$
\sum \left| \frac{B_n}{n!} x^n \right| \le \sum \frac{C}{|a|(n-2)!} (|a|x)^n
$$

which converges for all x. But we know (§5.6) that the radius of convergence of the Bernoulli series is 2π.

A.6 Solutions to Chapter 6

Solutions to Exercises 6.1

6.1.1 If (6.1.2) holds, then the range of f is finite. Conversely, if the range of f is $\{c_1, \dots, c_N\}$, let $A_j = \{x : f(x) = c_j\}$, $j = 1, \dots, N$.

6.1.2 Let $M \subset \mathbf{R}^2$ be measurable. Since $f(A \cap B) = f(A) \cap f(B)$ and $f(A^c) = f(A)^c$,

$$\begin{aligned}
\text{area}\,(A) &= \lambda\,\text{area}\left(f^{-1}(A)\right) \\
&= \lambda\,\text{area}\left(f^{-1}(A) \cap M\right) + \lambda\,\text{area}\left(f^{-1}(A) \cap M^c\right) \\
&= \text{area}\left(f(f^{-1}(A) \cap M)\right) + \text{area}\left(f(f^{-1}(A) \cap M^c)\right) \\
&= \text{area}\,(A \cap f(M)) + \text{area}\,(A \cap f(M^c)) \\
&= \text{area}\,(A \cap f(M)) + \text{area}\,(A \cap f(M)^c)\,.
\end{aligned}$$

Thus $f(M)$ is measurable.

6.1.3 This follows from the countability of $\mathbf{Q}$ and

$$\begin{aligned}
\{x : f(x) > g(x)\} &= \bigcup_{r \in \mathbf{Q}} \{x : f(x) > r > g(x)\} \\
&= \bigcup_{r \in \mathbf{Q}} \{x : f(x) > r\} \cap \{x : r > g(x)\}.
\end{aligned}$$

6.1.4 If f is measurable, then

$$\{x : -f(x) < M\} = \{x : f(x) > -M\} = \{x : f(x) \leq -M\}^c$$

is measurable.

6.1.5 This follows from the countability of $\mathbf{Q}$ and

$$\{x : f(x) + g(x) < M\} = \bigcup_{\substack{r,s \in \mathbf{Q} \\ r+s<M}} \{x : f(x) < r\} \cap \{x : g(x) < s\}.$$

6.1.6 For $f \geq 0$, $g \geq 0$, this follows from the countability of $\mathbf{Q}$ and

$$\{x : f(x)g(x) < M\} = \bigcup_{\substack{r,s \in \mathbf{Q} \\ rs<M}} \{x : f(x) < r\} \cap \{x : g(x) < s\}.$$

For general f, g, write $f = f^+ - f^-$, $g = g^+ - g^-$.

Solutions to Exercises 6.2

6.2.1 Given $x \in U$, let $a_x = \inf\{t \in U : (t,x) \subset U\}$ and $b_x = \sup\{t \in U : (x,t) \subset U\}$. Then $-\infty \leq a_x < x < b_x \leq \infty$ and $I_x = (a_x, b_x) \subset U$. If $a_x \in U$, then there is an open interval $J_x \subset U$ containing a_x. But then $I_x \cup J_x \subset U$, contradicting the definition of a_x. Thus $a_x \notin U$. Similarly $b_x \notin U$. Moreover the same argument shows I_x, $I_{x'}$ are either disjoint or identical, for all x, x' in U. Since $\{I_x : x \in U\}$ contain distinct rationals, $\{I_x : x \in U\}$ is finite or countable.

6.2.2 Suppose F is continuous, U is open, and $F(c) \in U$. Given $\epsilon > 0$, there is a $\delta > 0$ such that $|x - c| < \delta$ implies $|F(x) - F(c)| < \epsilon$. Choose ϵ small enough so that the interval $(F(c) - \epsilon, F(c) + \epsilon)$ is in U. Then the interval $(c - \delta, c + \delta)$ is in $F^{-1}(U)$. Thus $F^{-1}(U)$ is open. Conversely, the interval $U = (F(c) - \epsilon, F(c) + \epsilon)$ is an open set, so its inverse image $F^{-1}(U)$ is an open set containing c. Thus there is an open interval I with $c \in I \subset F^{-1}(U)$. Selecting $\delta > 0$ such that $(c - \delta, c + \delta) \subset I$ establishes the continuity of F.

6.2.3 If I is an open interval, then $F(I)$ is an interval, hence measurable. Now

$$F\left(\bigcup_{k=1}^{\infty} I_k\right) = \bigcup_{k=1}^{\infty} F(I_k);$$

hence U open implies $F(U)$ measurable.

6.2.4 Let $U_n = I_1 \cup \ldots \cup I_n$. Apply (4.3.5) to $f\mathbf{1}_{U_n}$ and the partition

$$-\infty \leq c_1 < d_1 \leq c_2 < d_2 \leq \cdots \leq c_n < d_n \leq \infty$$

to get

$$\int_{U_n} f(x)\,dx = \sum_{k=1}^{n} \int_{c_k}^{d_k} f(x)\,dx.$$

Now send $n \to \infty$ and use the monotone convergence theorem.

6.2.5 The limit of $(f_n(x))$ exists iff $(f_n(x))$ is Cauchy iff

$$\inf_{n\geq 1} \sup_{j\geq n} |f_j(x) - f_n(x)| = 0.$$

Thus $(f_n(x))$ is not cauchy iff x is in

$$A^c = \bigcup_{m=1}^{\infty} \bigcap_{n=1}^{\infty} \bigcup_{j=n}^{\infty} \{x : |f_j(x) - f_n(x)| > 1/m\}.$$

But the countable unions and intersections and complements of measurable sets are measurable. Since A is measurable, $f_n\mathbf{1}_A$, $n \geq 1$, is measurable and converges pointwise to f. Thus f is measurable.

Solutions to Exercises 6.3

6.3.1 length (M) = area $(M \times (0,1))$, so M negligible in $\mathbf{R}$ implies $M \times (0,1)$ is negligible in $\mathbf{R}^2$; hence $M \times (0,1)$ is measurable in $\mathbf{R}^2$ (Exercise **4.5.17**); hence M is measurable in $\mathbf{R}$.

6.3.2 $A \subset B$ implies $A \times (0,1) \subset B \times (0,1)$ and

$$\bigcup_{n=1}^{\infty} (A_n \times (0,1)) = \left(\bigcup_{n=1}^{\infty} A_n \right) \times (0,1).$$

Now use monotonicity and subadditivity of area.

6.3.3 If $|A - B|$ is negligible, then so is $A - B$ and $B - A$. But B is the union of $B - A$ and $A - (A - B)$. Thus A measurable implies B measurable.

6.3.4 The symmetric difference of $\{x : f(x) < m\}$ and $\{x : g(x) < m\}$ is contained in $\{x : f(x) \neq g(x)\}$.

6.3.5 Let I denote the inf. If (I_n) is a paving of A in $\mathbf{R}$, then $(I_n \times (0,1))$ is a paving of $A \times (0,1)$ in $\mathbf{R}^2$; hence

$$\begin{aligned} \text{length}\,(A) &= \text{area}\,(A \times (0,1)) \\ &\leq \sum_{n=1}^{\infty} \|I_n \times (0,1)\| \\ &= \sum_{n=1}^{\infty} \|I_n\|, \end{aligned}$$

and thus length $(A) \leq I$. On the other hand, given $\epsilon > 0$, there is an open $U \subset \mathbf{R}$ containing A with

$$\text{length}\,(U) \leq \text{length}\,(A) + \epsilon.$$

Write U as the countable disjoint union of intervals I_n, $n \geq 1$, to get by Exercise **6.2.3**

$$I \leq \sum_{n=1}^{\infty} \|I_n\| = \text{length}\,(U) \leq \text{length}\,(A) + \epsilon.$$

The result follows.

6.3.6 Let $\bigcup_{k=1}^{\infty} I_k$ be a paving of N by intervals. Then length $(A \cap I_k) \leq \alpha$ length (I_k), so

$$\text{length}\,(A) \leq \sum_{k=1}^{\infty} \text{length}\,(A \cap I_k) \leq \alpha \sum_{k=1}^{\infty} \text{length}\,(I_k)\,.$$

Taking the inf over all pavings, length $(A) \leq \alpha$ length (A). Hence length $(A) = 0$.

6.3.7 By induction, $F_n([0,1]) \subset [0,1]$, $n \geq 0$. Assume F_n is increasing. Since $0 \leq x \leq 1/3 \leq y \leq 2/3 \leq z \leq 1$ implies $F_{n+1}(x) \leq F_{n+1}(y) \leq F_{n+1}(z)$, it is enough to show F_{n+1} is increasing on each subinterval $[0,1/3]$, $[1/3,2/3]$, $[2/3,1]$. But this is clear from the definition. Thus by induction F_n, $n \geq 0$, is increasing. Piecewise linearity is also clear by induction, as is $0 \leq F_n'(x) \leq (3/2)^n$.

6.3.8 $e_{n+1} \leq e_n/2$ is clear by induction, so for $m \geq 1$, $e_{n+m} \leq 2^{-m}e_n$; hence

$$|F_{n+m}(x) - F_n(x)| \leq \sum_{k=n}^{n+m-1} |F_{k+1}(x) - F_k(x)| \leq \sum_{k=n}^{\infty} 2^{-k}e_0 = 2^{-n+1}e_0.$$

Hence $(F_n(x))$ is Cauchy. Let $m \to \infty$ in the last inequality to get

$$|F(x) - F_n(x)| \leq 2^{-n+1}e_0.$$

Thus $F_n \to F$ uniformly on $[0,1]$, and F satisfies the recursive identity. If $x = (d+y)/3$ with d equal to 0 or 2 and $0 \leq y \leq 1$, then by the identity, $F(x) = (d/2)/2 + F(y)/2$. Thus

$$F\left(\sum_{k=1}^{N} \frac{d_k}{3^k} + \frac{y}{3^N}\right) = \sum_{k=1}^{N} \frac{d_k/2}{2^k} + \frac{F(y)}{2^N}.$$

Let $N \to \infty$ to conclude F is the Cantor function.

6.3.9 Since F_n is increasing and $F_n \to F$, F is increasing. Since $F([0,1]) = [0,1]$, F cannot have any jump discontinuities; hence F is continuous.

We establish

$$F_n(z) - F_n(x) \leq F_n(z-x), \qquad 0 \leq x \leq z \leq 1, n \geq 0, \tag{A.6.1}$$

by induction. For $n = 0$, (A.6.1) is clear. Assume (A.6.1) is valid for n. To verify (A.6.1) for $n+1$, there are six cases. (1) $x \leq z \leq 1/3$, (2) $2/3 \leq x \leq z$, (3) $1/3 \leq x \leq z \leq 2/3$, (4) $x \leq 1/3 \leq 2/3 \leq z$, (5) $1/3 \leq x \leq 2/3 \leq z$, (6) $x \leq 1/3 \leq z \leq 2/3$. Cases (1) and (2) are straightforward, there is nothing to prove in case (3), case (4) depends on whether $3z - 2 \geq 3x$ or not, case (5) is reduced to case (2) by replacing x by $2/3$, and case (6) is reduced to case (1) by replacing z by $1/3$. This establishes (A.6.1). Sending $n \to \infty$ yields $F(z) - F(x) \leq F(z-x)$.

We establish

$$F_n(x) \leq x^\alpha, \qquad 0 \leq x \leq 1, n \geq 0, \tag{A.6.2}$$

by induction. For this let $f(x) = (x - 2/3)^\alpha + (1/3)^\alpha$ and $g(x) = x^\alpha$ for $2/3 \leq x \leq 1$. Then $f'(x) \geq g'(x)$ and $f(1) = g(1)$. By the mean value theorem, $f(x) \leq g(x)$; hence (Figure 6.1)

$$\left(x - \frac{2}{3}\right)^\alpha + \left(\frac{1}{3}\right)^\alpha \le x^\alpha, \qquad \frac{2}{3} \le x \le 1. \tag{A.6.3}$$

For $n = 0$ (A.6.2) is clear. Assume (A.6.2) is valid for n. Then there are three cases in verifying the validity of (A.6.2) for $n+1$, the last case using (A.6.3). This establishes (A.6.2). Sending $n \to \infty$ yields $F(x) \le x^\alpha$.

The inequality $F(x) \ge (x/2)^\alpha$ is established directly, because $F_n(x) \ge (x/2)^\alpha$ is false for all $n \ge 0$. For this we use the inequality

$$\sum x_n^\alpha \ge \left(\sum x_n\right)^\alpha, \qquad x_n \ge 0. \tag{A.6.4}$$

Let $\Delta = \{\sum x_n = 1, x_n \ge 0\}$ denote the simplex of all nonnegative sequences (x_n) summing to 1 and let $h(x_1, x_2, \dots) = \sum x_n^\alpha$. Since $(x^\alpha)'' = \alpha(\alpha - 1)x^{\alpha-2} \le 0$, x^α is concave, so h is concave. Since $h = 1$ at the vertices of Δ, we have $h \ge 1$ on Δ, yielding (A.6.4). By (A.6.4),

$$F(x) = \sum_{n=1}^\infty \frac{d_n/2}{2^n} = 2^{-\alpha} \sum_{n=1}^\infty \left(\frac{d_n}{3^n}\right)^\alpha \ge 2^{-\alpha} \left(\sum_{n=1}^\infty \frac{d_n}{3^n}\right)^\alpha = (x/2)^\alpha.$$

Solutions to Exercises 6.4

6.4.1 Let $c \notin W_G$. If $x_n \to c+$, then $c \le \underline{x}_n \le x_n$, so $\tilde{G}(x_n) = G(\underline{x}_n) \to G(c) = \tilde{G}(c)$. Similarly if $x_n \to c-$. Thus $\tilde{G}$ is continuous on W_G^c. Since $\tilde{G}$ is constant on each component interval, $\tilde{G}$ is continuous on W_G. Thus $\tilde{G}$ is continuous on $[a, b]$.

Suppose $c < x$ with $\tilde{G}(c) \ne \tilde{G}(x)$. Then $\bar{c} \le \underline{x}$.

If $a < c$, then $a < \bar{c} \notin W_G$, so $\tilde{G}(c) = G(\bar{c}) \ge G(\underline{x}) = \tilde{G}(x)$. If $c = a$ is the left endpoint of a component interval, let d be the corresponding right endpoint. Then $d \le \underline{x}$ and $a < d \notin W_G$, so $\tilde{G}(c) = G(c) = G(d) \ge G(\underline{x}) = \tilde{G}(x)$. If $c = a$ is not the left endpoint of a component interval, there exists $c < c_n < x$, $c_n \notin W_G$, $c_n \to c$. Then $c_n \le \underline{x}$ and $a < c_n \notin W_G$ so $\tilde{G}(c) = G(c) = \lim_{n\to\infty} G(c_n) \ge G(\underline{x}) = \tilde{G}(x)$. Thus $\tilde{G}$ is decreasing.

6.4.2 $c \in \{f^* > \lambda\}$ iff there exists $n \ge 1$ with $|c| < n$ and there exists $x > c$ with $|x| < n$ satisfying $G_n(x) > G_n(c)$ which happens iff $c \in U_{G_n}$. By the sunrise lemma, if (c, d) is one of the component intervals of $U_{G_n} \subset (-n, n)$, $G_n(d) \ge G_n(c)$ or

$$\lambda(d - c) \le \int_c^d f(x)\, dx.$$

Summing over all component intervals,

$$\lambda\,\text{length}\,(U_{G_n}) \le \int_{U_{G_n}} f(x)\,dx.$$

But $U_{G_n} \subset \mathbf{R}$, $n \ge 1$, is increasing. Now use the method of exhaustion and the monotone convergence theorem.

6.4.3 Apply the second part of the sunrise lemma to $G(x) = F(x) - \lambda x$ on $[a,b]$ with $\lambda \ge \lambda_0$, and set $W_\lambda = W_G$. Then we have $G(x) \le G(c)$ for $x \ge c$ and $a < c \notin W_\lambda$, which implies $D_cF(x) \le \lambda$ for $x \ge c$ and $a < c \notin W_\lambda$. Moreover $D_cF(d) = \lambda$ for each component interval (c,d) of W_λ. Since F is increasing, we have $0 \le D_cF(x) \le \lambda$ for $x \ge c$ and $a < c \notin W_\lambda$. But this implies $0 \le F'(c) \le \lambda$ at every Lebesgue point of F; hence the result follows.

6.4.4 Let (c,d) be a component interval of W_G with $a < c$. Then $G(c) = G(d)$ and $G(c) \ge G(x)$ for $x \ge c$. If $c < e \in W_G$ and $G(c) = G(e)$, then there is $x > e > c$ with $G(x) > G(e) = G(c)$ implying $c \in W_G$. But this contradicts $c \notin W_G$.

Solutions to Exercises 6.5

6.5.1 If $U' \subset U$, then each component interval of U' lies in U_n for exactly one $n \ge 1$. Hence

$$\text{var}(F, U') = \sum_{n=1}^{\infty} \text{var}(F, U' \cap U_n).$$

Since $\text{var}(F, U' \cap U_n) \le v_F(U_n)$ and $U' \subset U$ is arbitrary, we have

$$v_F(U) \le \sum_{n=1}^{\infty} v_F(U_n).$$

Conversely, given $\epsilon > 0$, for each $n \ge 1$, select $U'_n \subset U_n$ satisfying $\text{var}(F, U'_n) \ge v_F(U_n) - \epsilon 2^{-n}$. Let U' be the union of U'_n, $n \ge 1$. Then $U' \cap U_n = U'_n$, $n \ge 1$, so

$$v_F(U) \ge \text{var}(F, U') = \sum_{n=1}^{\infty} \text{var}(F, U'_n) \ge \sum_{n=1}^{\infty} v_F(U_n) - \epsilon.$$

Since $\epsilon > 0$ is arbitrary, the result follows.

6.5.2 Suppose F is absolutely continuous in the sense of §6.3. Then given $\epsilon > 0$, there exists $\delta > 0$ such that $\text{var}(F, U) < \epsilon$ for any finite disjoint union U of intervals satisfying length $(U) < \delta$. Suppose $U_n \subset (a,b)$ satisfies length $(U_n) \to 0$ and select $N \ge 1$ such that length $(U_n) < \delta$ for $n \ge N$. For each $n \ge 1$, select $U'_n \subset U_n$ satisfying $v_F(U_n) < \text{var}(F, U'_n) + \epsilon$. For each

$n \geq 1$, order the component intervals of U'_n, and let U_n^k denote the union of the first k of these intervals. (If U'_n has only finitely many component intervals, then U_n^k doesn't depend on k for k large.) Now U_n^k is a finite disjoint union of intervals and length $(U_n^k) < \delta$; hence $\text{var}(F, U_n^k) < \epsilon$ for $k \geq 1$, $n \geq 1$. Letting $k \to \infty$ yields $\text{var}(F, U'_n) \leq \epsilon$, $n \geq 1$; hence

$$v_F(U_n) < \text{var}(F, U'_n) + \epsilon \leq 2\epsilon, \qquad n \geq N.$$

This establishes $v_F(U_n) \to 0$; thus F is absolutely continuous in the sense of §6.5. Conversely, suppose F is not absolutely continuous in the sense of §6.3. Then there exists $\epsilon > 0$ and finite disjoint unions U_n satisfying length $(U_n) < 1/n$ and $v_F(U_n) \geq \text{var}(F, U_n) \geq \epsilon$. Hence F is not absolutely continuous in the sense of §6.5.

6.5.3 First we note $\text{var}(F, (c,d)) = |F(d) - F(c)|$. Let $a = x_0 < x_1 < \cdots < x_n = b$ be a partition of $[a, b]$. If $U = (c, d)$, by the triangle inequality

$$\text{var}(F, U) \leq \sum_{k=1}^{n} \text{var}(F, U \cap (x_{k-1}, x_k)).$$

Applying this inequality to each component interval of an open $U \subset (a, b)$ and summing over these intervals yields the same inequality but now for open $U \subset (a, b)$. Thus

$$\text{var}(F, U) \leq \sum_{k=1}^{n} v_F(x_{k-1}, x_k).$$

Taking the sup over all $U \subset (a, b)$ yields

$$v_F(a, b) \leq \sum_{k=1}^{n} v_F(x_{k-1}, x_k).$$

On the other hand, $\bigcup_{k=1}^{n}(x_{k-1}, x_k) \subset (a, b)$. Since (x_{k-1}, x_k), $k \geq 1$, are disjoint, by Exercise **6.5.1**,

$$v_F(a, b) \geq v_F\left(\bigcup_{k=1}^{n}(x_{k-1}, x_k)\right) = \sum_{k=1}^{n} v_F(x_{k-1}, x_k).$$

Thus $v_F(a, b) = v_F(x_0, x_1) + \cdots + v_F(x_{n-1}, x_n)$. Since F is absolutely continuous, there is an $n \geq 1$ such that $v_F(U) < 1$ when length $(U) \leq 1/n$. Select the partition to have mesh $< 1/n$. Then $v_F(x_{k-1}, x_k) < 1$, $k = 1, \ldots, n$; hence $v_F(a, b) < n$.

6.5.4 If U is open, since v_F is increasing, by Exercise **6.5.1** applied twice, Exercise **6.5.3**, and summing over the component intervals of U,

$$v_{v_F}(U) = \sum_{(c,d)} v_F(a, d) - v_F(a, c) = \sum_{(c,d)} v_F(c, d) = v_F(U).$$

6.5.5 Since F is Lipschitz, $\text{var}(F,(c,d)) = |F(d)-F(c)| \le M(d-c)$. Applying this to the component intervals of an open U and summing over these intervals yield $\text{var}(F,U) \le M \,\text{length}\,(U)$. Thus $v_F(U) \le M\,\text{length}\,(U)$. This implies absolute continuity.

6.5.6 If we call the set in Theorem 6.5.4 U_λ^+ and the set in Theorem 6.5.5 V_λ^-, the other two sets are

$$\begin{aligned} U_\lambda^- &= \{c \in (a,b) : D_cF(x) > \lambda \text{ for some } x < c\} \\ V_\lambda^+ &= \{c \in (a,b) : D_cF(x) < \lambda \text{ for some } x > c\}, \end{aligned}$$

and the estimates they satisfy are the same as those of U_λ^+ and V_λ^-, respectively. The proof follows by applying the results for U_λ^+ and V_λ^- to $\tilde{F}(x) = -F(-x)$.

6.5.7 Let $0 \le c < x \le 1$. Then $c \notin C \setminus Z$ iff there is $\delta > 0$ with $(c, c+\delta) \subset C^c$ which happens iff the graph of F on $(c,1)$ is below the line passing through $(c, F(c))$ with slope $\lambda_0 \equiv (1-F(c))/\delta$. Hence $0 \le D_cF(x) \le \lambda_0$ and $c \notin W_{\lambda_0}$. Conversely, suppose $c \in C \setminus Z$ and let $c = \sum_{n=1}^\infty d_n/3^n$. Then for all $n \ge 1$, there is an $N \ge n$ with $d_N = 0$. If we let $x_n = c + 2/3^N$, we obtain $x_n \in C$, $x_n > c$, with $2D_cF(x_n) = (3/2)^N \ge (3/2)^n$. Hence $c \in W_\lambda$ for all $\lambda > 0$.

6.5.8 We assume F is bounded variation continuous on $[a,b]$ and F has been extended to all of $\mathbf{R}$ by $F(x) = F(b)$ for $x > b$ and $F(x) = F(a)$ for $x < a$. Then $v_F(a,x) = v_F(a,b)$ for $x > b$. For $n \ge 1$, with f_n as before,

$$\begin{aligned} \int_a^b |f_n(x)|\,dx &\le n \int_a^b v_F(x, x+1/n)\,dx \\ &= n\left(\int_a^b v_F(a, x+1/n)\,dx - \int_a^b v_F(a,x)\,dx\right) \\ &= n\left(\int_{a+1/n}^{b+1/n} v_F(a,x)\,dx - \int_a^b v_F(a,x)\,dx\right) \\ &= n\left(\int_b^{b+1/n} v_F(a,x)\,dx - \int_a^{a+1/n} v_F(a,x)\,dx\right) \\ &= v_F(a,b) - n\int_a^{a+1/n} v_F(a,x)\,dx \le v_F(a,b). \end{aligned}$$

Now apply Fatou's lemma as before to get

$$\int_a^b f(x)\,dx \le \liminf_{n\to\infty} \int_a^b f_n(x)\,dx \le v_F(a,b).$$

Solutions to Exercises 6.6

6.6.1 Given an interval $[a,b]$, there are c and d in $[a,b]$ with $\max_{[a,b]} F = F(d)$ and $\min_{[a,b]} F = F(c)$. Then $F((a,b))$ is an interval and

$$\text{length}\,(F((a,b))) = \max_{[a,b]} F - \min_{[a,b]} F = F(d) - F(c) \le v_F(a,b).$$

Applying this to each component interval (c,d) of an open set U and summing over these intervals yield

$$\text{length}\,(F(U)) \le \sum_{(c,d)} \text{length}\,(F((c,d))) \le \sum_{(c,d)} v_F(c,d) = v_F(U). \qquad \text{(A.6.5)}$$

6.6.2 Given $a < b$, F is absolutely continuous on $[a,b]$ and $N \cap (a,b)$ is negligible. Select open supersets $U_n \subset (a,b)$ of $N \cap (a,b)$ satisfying length $(U_n) \to 0$. Then by (A.6.5),

$$\text{length}\,(F(N \cap (a,b))) \le \text{length}\,(F(U_n)) \le v_F(U_n) \to 0.$$

Thus $F(N \cap (a,b))$ is negligible for all $a < b$; hence $F(N)$ is negligible.

6.6.3 In (6.5.8) replace $[a,b]$ by $[x, x+\delta]$ and divide by $\delta > 0$. If x is a Lebesgue point of $|F'(x)|$ and $v_F(a,\cdot)$ is differentiable at x, the result follows.

6.6.4 By (6.5.2) with $f = F'$ and (6.5.3), we have

$$v_F(a,x) - v_F(a,c) = v_F(c,x) \le \int_c^x |F'(t)|\,dt, \qquad a < c < x < b.$$

On the other hand, for $x > c$, $|F(x) - F(c)| = \text{var}(F,(c,x)) \le v_F(c,x) = v_F(a,x) - v_F(a,c)$. Hence

$$|F(x) - F(c)| \le v_F(a,x) - v_F(a,c) \le \int_c^x |F'(t)|\,dt, \qquad a < c < x < b.$$

Since v_F is absolutely continuous, v_F is differentiable almost everywhere. Divide by $x - c$ and let $x \to c+$. The Lebesgue differentiation theorem yields

$$\frac{d}{dc} v_F(a,c) = |F'(c)|, \qquad \text{almost everywhere in } (a,b).$$

Then (6.5.4) follows from (6.5.2) with $f = v_F$.

6.6.5 Suppose not. Then there is $\epsilon > 0$ and (A_n) with length $(A_n) \to 0$ and length $(F(A_n)) \ge \epsilon > 0$. For each $n \ge 1$, choose $U_n \supset A_n$ with length $(U_n) \to 0$. Then by (A.6.5),

$$v_F(U_n) \ge \text{length}\,(F(U_n)) \ge \text{length}\,(F(A_n)) \ge \epsilon > 0, \qquad n \ge 1.$$

But this contradicts absolute continuity.

6.6.6 There is a sequence of open sets $M \subset U_n \subset (a,b)$ with length $(U_n) \to$ length (M). If M is measurable, then length $(U_n \setminus M) \to 0$. For $n \ge 1$, select open $U_n \setminus M \subset V_n \subset (a,b)$ satisfying length $(V_n) \to 0$. By absolute continuity,

$$\begin{aligned}\text{length}\,(F(U_n) \setminus F(M)) &\le \text{length}\,(F(U_n \setminus M))\\ &\le \text{length}\,(F(V_n)) \le v_F(V_n) \to 0.\end{aligned}$$

Let I denote the intersection of $F(U_n)$, $n \ge 1$. Then length $(I \setminus F(M)) = 0$ and $I \supset M$. Since U_n is open and F is continuous, by Exercise **6.2.2**, $F(U_n)$ is measurable; hence I is measurable. By Exercise **6.3.3**, the result follows.

6.6.7 While this is an immediate consequence of Theorem 6.6.2, the goal here is to derive the result directly from the sunrise lemma, avoiding the fundamental lemma. Let $A = \{F' = 0\}$. Then $A \subset V_\lambda$; hence by (6.5.6) and Exercise **6.6.1**,

$$\text{length}\,(F(A)) \le \text{length}\,(F(V_\lambda)) \le v_F(V_\lambda) \le \lambda(b-a)$$

for all $\lambda > 0$; hence $F(A)$ is negligible. Also by Exercise **6.6.2**, $N = (a,b) \setminus A$ is negligible, so is $F(N)$. Hence $F([a,b])$ is negligible or F is constant.

6.6.8 Since $d \in J_j$, $|F(d) - F(x_{j-1})| \le M_j - m_j$. Since $c \in J_i$, $|F(x_i) - F(c)| \le M_i - m_i$. Now include the intermediate endpoints using the triangle inequality to get

$$|F(d) - F(c)| \le (M_i - m_i) + \sum_{i<\ell<j} |F(x_\ell) - F(x_{\ell-1})| + (M_j - m_j).$$

(If there are no intervals J_ℓ between J_i and J_j, the sum is zero.) Hence

$$|F(d) - F(c)| \le \sum_{i \le \ell \le j} (M_\ell - m_\ell).$$

For each interval (c_k, d_k), let $c_k \in J_i$, $i = i(k)$, $d_k \in J_j$, $j = j(k)$. Then

$$\sum_{k=1}^{N} |F(d_k) - F(c_k)| \le \sum_{k=1}^{N} \sum_{i(k) \le \ell \le j(k)} (M_\ell - m_\ell).$$

If the mesh of $a = x_0 < x_1 < \cdots < x_n = b$ is $< \delta$, then each J_i contains at most one endpoint; hence the intervals $[i(k), j(k)] \cap \mathbf{N}$, $[i(k'), j(k')] \cap \mathbf{N}$ do not overlap for $k \ne k'$. (6.6.2) follows.

References

[1] Stefan Banach, *Sur les lignes rectifiables et les surfaces dont l'aire est finie*, Fund. Math. **7** (1925), 225–236.
[2] Richard Bellman, *Analytic Number Theory*, Benjamin, Reading, MA, 1980.
[3] Garrett Birkhoff, *A Source Book in Classical Analysis*, Harvard University Press, Cambridge, MA, 1973.
[4] J. M. Borwein and P. B. Borwein, *Pi and the AGM*, Wiley-Interscience, New York, NY, 1987.
[5] D. Bailey, P. Borwein, and S. Plouffe, *On the rapid computation of various polylogarithmic constants*, Math. Comput. **66** (1997), 903–913.
[6] C. Carathéodory, *Über das lineare mass von punktmengen, eine verallgemeinerung des längenbegriffs*, Nachr. Gesell. Wiss. Göttingen (1914), 404–426.
[7] John H. Conway, *On Numbers and Games*, A. K. Peters, Natick, MA, 2001.
[8] H.-D. Ebbinghaus et al, *Numbers*, Springer-Verlag, New York, NY, 1990.
[9] Leonard Euler, *An Introduction to the Analysis of the Infinite*, Springer-Verlag, New York, NY, 1990.
[10] Anthony W. Knapp, *Group representations and harmonic analysis from Euler to Langlands, part I*, Notices of the AMS **43** (1996), 410–415.
[11] Henri Lebesgue, *Integrale, Longueur, Aire*, Doctoral Thesis, University of Paris, 1902.
[12] Frédéric Riesz, *Sur un théorème de maximum de Mm. Hardy et Littlewood*, Journal of the London Mathematical Society **7** (1932), 10–13.
[13] Dale E. Varberg, *On absolutely continuous functions*, Amer. Math. Monthly **72** (1965), 831–841.
[14] G. Vitali, *Sulle funzioni integrali*, Atti dell' Accademia delle Scienze di Torino **40** (1905), 1021–1034.
[15] E. T. Whittaker and G. N. Watson, *Modern Analysis*, Cambridge University Press, London, UK, 1927.
[16] `http://demonstrations.wolfram.com/CantorFunction/`.
[17] `http://demonstrations.wolfram.com/RieszsRisingSunLemma/`.

O. Hijab, *Introduction to Calculus and Classical Analysis*, Undergraduate Texts in Mathematics, DOI 10.1007/978-3-319-28400-2

Index

O. Hijab, *Introduction to Calculus and Classical Analysis*, Undergraduate Texts in Mathematics, DOI 10.1007/978-3-319-28400-2

Printed in the United States
By Bookmasters